*An Introduction to
Optical Waveguides*

An Introduction to Optical Waveguides

M. J. ADAMS

Department of Electronics
The University, Southampton

A Wiley–Interscience Publication

JOHN WILEY & SONS
Chichester · New York · Brisbane · Toronto

Copyright © 1981 by John Wiley & Sons Ltd.

All rights Reserved.

No part of this book may be reproduced by any means, nor transmitted, nor translated into machine language without the written permission of the publisher.

British Library Cataloguing in Publication Data:

Adams, M. J.
   An introduction to optical waveguides.
   1. Optical wave guide theory
   I. Title
   535.8'9   QC448   80–42059

ISBN 0 471 27969 2

Typeset by Macmillan India Ltd, Bangalore
and printed in the United States of America
by Vail-Ballou Press Inc., Binghamton, N.Y.

Contents

Preface ix

Bibliography xi

Historical Introduction xiii

1. Two-dimensional Conducting-wall Waveguides 1
   1.1 Transverse Electromagnetic Waves 1
       1.1.1 Plane-polarized TEM wave in an unbounded dielectric ... 1
       1.1.2 Parallel strip transmission line 2
   1.2 Ray–Mode Connections 3
       1.2.1 Ray treatment of two parallel plane mirrors 5
       1.2.2 Mode treatment of two parallel mirrors 7
       1.2.3 Normalized variables 7

2. Three-layer Dielectric Slab Waveguides 10
   2.1 Reflection and Refraction at a Plane Dielectric Interface ... 10
       2.1.1 Snell's laws 11
       2.1.2 Fresnel's laws 11
       2.1.3 Normal incidence 13
       2.1.4 Brewster angle 14
       2.1.5 Critical angle 15
       2.1.6 Phase shift on total internal reflection 15
       2.1.7 The Goos–Haenchen shift 16
       2.1.8 Summary and numerical example 18
   2.2 Ray Treatment of the Dielectric Slab Waveguide: Bound Rays ... 20
       2.2.1 Eigenvalue equation 21
       2.2.2 Effective width 22
       2.2.3 Group velocity 24
       2.2.4 Losses 25
   2.3 Electromagnetic Mode Treatment of the Dielectric Slab Waveguide: Guided Modes ... 28
       2.3.1 Transverse electric (TE) guided modes 29
       2.3.2 Transverse magnetic (TM) guided modes 30
       2.3.3 Mode numbers and cut-offs 31
       2.3.4 Normalization in terms of power flow 32
       2.3.5 Weakly-guiding symmetric slab waveguide 35
       2.3.6 Waveguiding properties of heterostructure injection lasers ... 42
   2.4 Effects of Loss and Gain 49
       2.4.1 Mode stability 49
       2.4.2 Mode cut-offs 52

  2.4.3 Pure gain-guidance 53
  2.4.4 Application to stripe-geometry lasers 55
 2.5 Metal-clad and Hollow Waveguides 59
  2.5.1 Waveguide classification 59
  2.5.2 Asymmetric metal-clad dielectric waveguide 61
  2.5.3 Symmetric metal-clad guide 68
  2.5.4 Hollow dielectric guide 71
 2.6 Multilayer Waveguides 75
  2.6.1 Four-layers asymmetric slab theory 75
  2.6.2 Applications of the four-layer slab 77
  2.6.3 Five-layer symmetric slab theory 82
  2.6.4 The 'W'-guide 86

3. Two-dimensional Parabolic-index Media 90
 3.1 Ray Treatment of Parabolic-index Media 91
  3.1.1 Local plane wave derivation of ray equations 92
  3.1.2 The ray trajectory and transit time 93
  3.1.3 Phase shift at a caustic 96
  3.1.4 The eigenvalue equation 96
 3.2 Electromagnetic Mode Treatment of Parabolic-index Media 97
  3.2.1 The vector wave equations for general graded-index media 98
  3.2.2 TE solutions for parabolic-index media 99
  3.2.3 TM solutions for parabolic-index media 102
  3.2.4 Group delay and radiation confinement factor 102
 3.3 Variation of Loss or Gain 104
  3.3.1 General beam modes 105
  3.3.2 Mode stability 106
  3.3.3 Phase front curvature 108
  3.3.4 Equilibrium solutions 109

4. Other Graded-index Two-dimensional Guides 113
 4.1 The Exponential Profile 113
  4.1.1 Ray treatment of the strongly asymmetric profile 114
  4.1.2 Mode treatment of the strongly asymmetric profile 116
  4.1.3 The effects of a dielectric cover 117
  4.1.4 The symmetric exponential profile 119
 4.2 The Linear Profile 122
  4.2.1 Ray treatment of the strongly asymmetric profile 122
  4.2.2 Mode treatment of the strongly asymmetric profile 123
  4.2.3 The effects of a dielectric cover 125
  4.2.4 The symmetric linear profile 126
 4.3 The Epstein-layer model 128
  4.3.1 Mode treatment of the general profile 130
  4.3.2 Modal results for the symmetric profile 132
  4.3.3 Ray paths for the symmetric profile 134

5. Approximate Methods for Two-dimensional Graded-index Guides 136
 5.1 Perturbation Theory 136
  5.1.1 Perturbation solution of the ray equations 136

	5.1.2 First-order perturbation theory for the scalar wave equation	139
	5.1.3 Polynomial index profiles	141
	5.1.4 The cladded-parabolic profile	143
5.2	The WKB Approximation	145
	5.2.1 Approximate solution near a caustic	146
	5.2.2 Solutions in the presence of two caustics	147
	5.2.3 Solutions in the presence of an index discontinuity	148
	5.2.4 'Buried' modes near an index discontinuity	149
	5.2.5 WKB tunnelling coefficient for leaky waves	152
5.3	Variational Methods	155
	5.3.1 The variational principle for eigenvalues	156
	5.3.2 A general relationship between phase and group velocities	157
	5.3.3 An alternative stationary expression	160
	5.3.4 Solution by the Rayleigh–Ritz method	161
	5.3.5 Results for specific index profiles	163
5.4	Other Numerical Methods	166
	5.4.1 Series solution	166
	5.4.2 Multilayer 'staircase' approximation	169
	5.4.3 Evanescent field theory	172
	5.4.4 'Exact' solution for the cladded-parabolic profile	174

6. Guides of Rectangular Cross-section 178
   6.1 Rectangular Conducting-wall Waveguides 178
       6.1.1 TE modes 180
       6.1.2 TM modes 181
       6.1.3 Modal characteristics 181
   6.2 Rectangular Dielectric Waveguides 183
       6.2.1 Approximate modal analysis 184
       6.2.2 The effective index method 188
       6.2.3 Slab-coupled guides 190
       6.2.4 Hollow guides 198
   6.3 Three-dimensional Graded-index Waveguides 201
       6.3.1 Diffused channel waveguides 202
       6.3.2 Strip-loaded diffused guides 206
       6.3.3 Stripe-geometry heterostructure lasers 208

7. Circular Waveguides and Step-index Fibres 213
   7.1 Propagation in Hollow Circular Pipes 213
       7.1.1 Geometric optics derivation of the eigenvalue equation 214
       7.1.2 Maxwell's equations in circular cylindrical coordinates ... 217
       7.1.3 TE and TM modes 219
   7.2 Step-index Optical Fibres 223
       7.2.1 Circular dielectric rod 223
       7.2.2 Weakly-guiding optical fibres 228
       7.2.3 Properties of LP modes 233
       7.2.4 Relevant properties of practical fibres 239
       7.2.5 Effects of small departures from circularity 250
   7.3 Step-index Guides of More Complicated Structure 260
       7.3.1 Hollow and metallic waveguides 260
       7.3.2 Multilayer guides 267

8. Graded-index Optical Fibres ... 278
8.1 Parabolic-index Media in Three Dimensions ... 279
8.1.1 Ray equations for general graded-index media ... 279
8.1.2 Geometric optics results for parabolic index-profiles ... 285
8.1.3 Vector field theory for general graded-index media ... 290
8.1.4 Modal results for parabolic index-profiles ... 293
8.2 The WKB Approximation in Graded-index Fibre Analysis ... 297
8.2.1 Formal theory ... 297
8.2.2 Alpha-profiles and their optimization ... 302
8.2.3 Wavelength-dependence of multimode fibre bandwidths ... 305
8.2.4 More complicated index profiles ... 315
8.2.5 Leaky modes on fibres of arbitrary index profile ... 320
8.2.6 Impulse response ... 328
8.2.7 Leaky mode attenuation ... 330
8.3 Variational Methods ... 339
8.3.1 The conventional approach ... 339
8.3.2 An alternative approach ... 344
8.3.3 Results for alpha-profiles ... 348
8.4 Perturbation Theory ... 350
8.4.1 Fourth-order polynomial profiles ... 351
8.4.2 Small departures from desired profiles ... 353
8.4.3 Cladded α-profiles ... 357
8.5 Summary of Methods for Graded-index Fibre Analysis ... 360
8.5.1 Multimode fibres ... 360
8.5.2 Monomode fibres ... 366

References ... 370

Index ... 397

Preface

The reasons for adding another book on optical waveguide theory to the number of excellent texts already available (see Bibliography) are many. In the first place, it has been my experience when asked to recommend an introductory treatment for someone new to the field that it is difficult to recommend any one book; each is more than adequate in its treatment of certain selected areas but none covers all aspects (including up-to-date results) from a sufficiently general point of view. Similarly, when called upon to advise on background reading to supplement lecture courses in this field, one again faces the same dilemma and usually ends by recommending individual parts of several texts. The main differences between these texts stem from the varying interests of workers in the field of planar or rectangular guides, as opposed to those in the branch of the subject dealing with circularly-symmetric structures. Whereas the former are usually motivated by interests in integrated optics or semiconductor lasers, the latter are almost always concerned with optical fibres. This divergence of interests has led to a somewhat artificial separation of the subject into two halves with the consequence that new results or techniques derived first in one half may take some time to become well known amongst workers in the other area. Hence there is certainly a need to encourage cross-fertilization of ideas between the two halves by emphasizing the overall unity of the subject. In this respect there is an important contribution to be made in unifying the notation employed so as to make as clear as possible where appropriate the similarities of approaches between planar, rectangular, and circular waveguides.

In a recent book review Felix Kapron has noted that 'usually, books written by theoreticians in the field of fiber optics cover some elementary waveguide optics in perhaps a novel fashion, and then present a unified view of the authors' papers to date.' Whilst I have tried to avoid this fairly obvious trap, it is inevitable in a book of this size that certain areas of the subject have had to be excluded. For example, no discussion is given of the effects of bends, of mode coupling, or of far-field patterns of waveguides. Thus the book presents an introduction only to the theory of *perfect* optical waveguides. The layout of the book is straightforward: Chapters 1–5 deal with two-dimensional waveguides, i.e. planar guides with one direction of propagation and one of confinement. The sequence of topics covered is to progress from conducting-wall guides through uniform-core dielectric guides to graded-index structures. Chapter 6 repeats this sequence for waveguides of rectangular cross-section, whilst Chapters 7 and 8 deal respectively with uniform-core and graded-index optical fibres. The references to the literature, although rather comprehensive, are not intended to be exhaustive.

The list of acknowledgements to people who have helped me learn about optical waveguides would be too long to even attempt. To all of these I offer thanks, whilst accepting full responsibility for my errors of understanding or interpretation. Especial thanks are due to my colleagues in the optical fibre research group at Southampton for their tolerance during the long period of writing this book. I am overwhelmingly grateful to my wife for her encouragement and forbearance during this time, and further indebted to her for typing (and retyping) the manuscript with unfailing accuracy and patience.

Southampton   M. J. ADAMS
August 1980

Bibliography

A. Books on Optical Waveguides

1. Kapany, N. S. (1967). *Fibre Optics*, Academic Press, N.Y.
2. Kapany, N. S. and J. J. Burke (1972). *Optical Waveguides*, Academic Press, N.Y.
3. Marcuse, D. (1972). *Light Transmission Optics*, Van Nostrand Reinhold, N.Y.
4. Marcuse, D. (1974). *Theory of Dielectric Optical Waveguides*, Academic Press, N.Y.
5. Arnaud, J. A. (1976). *Beam and Fiber Optics*, Academic Press, N.Y.
6. Unger, H.-G. (1977). *Planar Optical Waveguides and Fibres*, O.U.P., Oxford.
7. Sodha, M. S. and A. K. Ghatak (1977). *Inhomogeneous Optical Waveguides*, Plenum Press, London.

B. Books on Optical Fibre Communications

1. Barnoski, M. K. (Ed.) (1976). *Fundamentals of Optical Fibre Communications*, Academic Press, N.Y.
2. Ostrowsky, D. B. (Ed.) (1978). *Fibre and Integrated Optics*, Plenum Press, N.Y.
3. Elion, G. R. and H. A. Elion (1978). *Fiber Optics in Communications Systems*, Marcel Dekker AG, Basel.
4. Midwinter, J. E. (1979). *Optical Fibres for Transmission*, Wiley, New York.
5. Sandbank, C. P. (Ed.) (1979). *Optical Fibre Communications Systems*, Wiley, Chichester.
6. Miller, S. E. and A. G. Chynoweth (Eds.) (1979). *Optical Fibre Telecommunications*, Academic Press, N.Y.
7. Bendow, B. and S. S. Mitra (Eds.) (1979). *Fibre Optics*, Plenum Press, N.Y.
8. Howes, M. J. and D. V. Morgan (Eds.) (1980). *Optical Fibre Communications*, Wiley, Chichester.
9. CSELT staff (1980). *Optical Fibre Communication*, Larrotto and Bella, Torino.
10. Cozannet, A., J. Fleuret, H. Maitre, and M. Rousseau (1981). *Optique et Telecommunications*, Eyralles, Paris.

C. Books on Integrated Optics and Semiconductor Lasers

1. Tamir, T. (Ed.) (1975). *Integrated Optics: Topics in Applied Physics, Vol. 7* Springer-Verlag, Berlin.
2. Kressel, H. J. and J. K. Butler (1977). *Semiconductor Lasers and Heterojunction LED's*, Academic Press, N.Y.
3. Casey, H. C. and M. B. Panish (1978). *Heterostructure lasers: Part A—Fundamental Principles; Part B—Materials and Operating Characteristics*, Academic Press, N.Y.
4. Thompson, G. H. B. (1980). *Physics of Semiconductor Laser Diodes*, Wiley, Chichester.
5. Kressel, H. (Ed.) (1980). *Semiconductor Devices for Optical Communications*, Springer-Verlag, Berlin.

D. Collections of Reprints

1. Marcuse, D. (Ed.) (1973). *Integrated Optics*, IEEE Press, N.Y.
2. Clarricoats, P. J. B. (Ed.) (1975). *Optical Fibre Waveguides*, Peter Peregrinus Ltd, on behalf of IEE, London.
3. Gloge, D. (Ed.) (1976). *Optical Fibre Technology*, IEEE Press, N.Y.
4. Butler, J. K. (Ed.) (1979). *Semiconductor Injection Lasers*, IEEE Press, N.Y.
5. Clarricoats, P. J. B. (Ed.) (1980). *Progress in Optical Communication*, Peter Peregrinus Ltd, on behalf of IEE, London.
6. Kao, K. C. (1980). *Optical Fiber Technology*, IEEE Press, N. Y.

E. General Waveguide Texts

1. Brekhovskikh, L. M. (1960). *Waves In Layered Media*, Academic Press, N.Y.
2. Shevchenko, V. V. (1971). *Continuous Transitions in Open Waveguides*, Golem Press, Colorado.
3. Lewin, L. (1975). *Theory of Waveguides: Techniques for the Solution of Waveguide Problems*, Newnes-Butterworths, London.
4. Lewin, L., D. C. Chang, and E. F. Kuester (1977). *Electromagnetic Waves and Curved Structures*, Peter Peregrinus Ltd, on behalf of IEE, London.
5. Sporleder, F. and H.-G. Unger (1979). *Waveguide Tapers, Transitions and Couplers*, Peter Peregrinus Ltd, on behalf of IEE, London.

Historical Introduction

From a historical viewpoint, although dielectric guides have been studied since the early years of this century (Hondros and Debye, 1910) the interest in optical applications has only come about during the last two decades. The achievement of lasing action in semiconductors by three groups in 1962 (Hall et al., 1962; Nathan et al., 1962; Quist et al., 1962) was soon followed by the realization that guidance in the p–n junction plane was closely associated with this phenomenon (Yariv and Leite, 1963; Bond et al., 1963). Improved understanding of the nature of this guiding led eventually to the development of the heterostructure laser in the late 1960's (Kressel and Nelson, 1969; Hayashi et al., 1969; Panish et al., 1969; Alferov et al., 1969) with the breakthrough of c.w. room-temperature operation in 1970 (Hayashi et al., 1970; Alferov et al., 1970a). In a closely related field it was found in 1964 that the guiding action of a p–n junction could be used in a modulator via the electro-optic effect (Nelson and Reinhart, 1964). The development of the p–n junction modulator has since followed that of the junction laser via the steps of epitaxial GaAs films (Hall et al., 1970) and heterostructures (Reinhart and Miller, 1972).

In the area of planar passive guides, early work on thin films (D. B. Anderson, 1965; Osterberg and Smith, 1964) led eventually to the new concept of integrated optics (Miller, 1969). The problem of coupling laser beams into planar guides was solved by the prism coupler (Tien et al., 1969) and the grating coupler (Dakss et al., 1970). As regards active components for integrated optics, in addition to the electro-optic modulator noted above (and other versions in other materials) there have also been developments in acousto-optic (Kuhn et al., 1971) and magneto-optic (Tien et al., 1972) techniques. The distributed feedback (DFB) principle (Kogelnik and Shank, 1972), first demonstrated in dye lasers, has provided an important source for integrated optics in the semiconductor DFB laser (Nakamura et al., 1973), which in its heterostructure form can operate c.w. at room temperature (Nakamura et al., 1975). As it is now clear that epitaxial layers of AlGaAs can be used to produce all the elements required—sources, modulators, couplers, waveguides, and detectors—this may well be the material for true optical integration in the future. Already there have been demonstrations of laser-waveguide-modulators (Reinhart and Logan, 1974), laser-taper-coupler-modulators (Reinhart and Logan, 1975a), monolithically integrated optical repeaters (Yust et al., 1979), and frequency-multiplexed sources using monolithically integrated DFB lasers (Aiki et al., 1976). Whilst practical applications in optical communications have been relatively slow to materialize, in the area of signal processing the advantages offered by integrated optical systems, e.g. an RF spectrum analyser (Hamilton et al., 1977), ought to ensure a future for this field of research.

In the circular cylindrical waveguide, early work on modes in dielectric rods (Snitzer, 1961; Snitzer and Osterberg, 1961; Kapany and Burke, 1961) was followed by the suggestion of using optical fibres for long-distance communications (Kao and Hockham 1966; Werts, 1966; Borner, 1966). The high losses present in fibres at that time were overcome first by workers at Corning Glass Works who demonstrated a loss of only 20 dB/km at 0.6328 μm in 1970 (Kapron et al., 1970). Subsequent improvements led to a minimum loss around 2 dB/km in 1974 (French et al., 1974; Payne and Gambling, 1974), 0.47 dB/km at 1.2 μm in

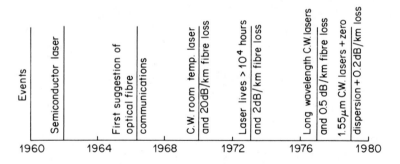

Figure (i) Important developments in optical waveguides

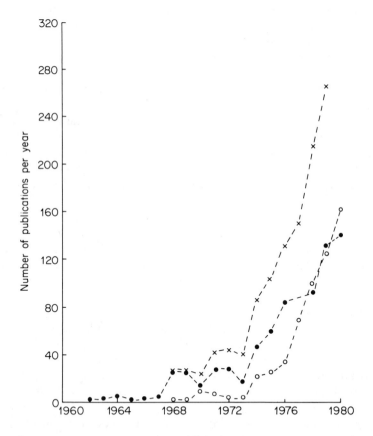

Figure (ii) Number of papers published each year in a few relevant technical journals: o—*Electronics Letters* ●—*Applied Optics* ×—*Electronics Letters* and *Applied Optics* and *Bell Systems Technical Journal*

1976 (Horiguchi and Osanai, 1976), and 0.2 dB/km at 1.55 μm in 1979 (Miya *et al.*, 1979). Although single-mode fibres offer the ultimate bandwidth for optical communications, the tight tolerances on joints, coupling to laser sources, etc. make the multimode fibre with graded refractive index profile a strong contender for some systems. The prediction of a precise form of the profile to be employed has stimulated much theoretical study, from the early suggestions of a parabolic variation with radius (Kawakami and Nishizawa, 1968) to the more sophisticated modifications of current interest (Gloge and Marcatili, 1973; Olshansky and Keck, 1976; Arnaud, 1977; Marcatili, 1977; Olshansky, 1978, 1979). A recent development has been the interest in operation at longer wavelengths than hitherto, since both attenuation and chromatic dispersion properties of high-silica fibres are minimized in the range 1.2–1.6 μm (Payne and Gambling, 1975; Gambling et al., 1979). This, in turn, has stimulated interest in new sources for use at longer wavelengths where the larger spectral widths of LED's need not be a major disadvantage if the fibre material dispersion is small.

LED's for the wavelength range 1–1.6 μm are now becoming available and are based on quaternary semiconductor materials rather than binary or ternary compounds. The quaternary $In_xGa_{1-x}As_yP_{1-y}$ has been used for high-power LED's (Goodfellow et al., 1979) and for c.w. room-temperature lasers with long operating lives (Hsieh et al., 1976; Shen et al., 1977).

This brief historical summary should suffice to show the main areas of interest in optical waveguides and to illustrate the recent rapid growth of interest in the potential applications in communications systems. The situation is expressed in an oversimplified but dramatic fashion by the graphs in Figures (i) and (ii). Figure (i) marks some of the important developments in the subject on a dateline over the last twenty years. Figure (ii) gives an indication of the increase of interest in the field by plotting numbers of published papers in a few relevant technical journals. It shows that each new breakthrough leads to a resurgence of interest until the next major problem is encountered. On the whole, however, the growth is exponential, with many major problems being solved (or at least defined) in the last few years. It would therefore appear that the broad field of optical communications has an assured future, with many field trials of high-bandwidth long-haul systems already in progress or in preparation (Senmoto and Okura, 1976; Jacobs, 1978; Midwinter, 1979a; Mogensen, 1980).

It is with this future in mind that this book has been prepared; clearly there will be a need for engineers in optical communications to have a knowledge of the principles of optical waveguides. Indeed, some parts of the text have been developed from lectures given to final-year undergraduates in electronics. The layout is such that the subjects of planar, rectangular and circular guides are dealt with in sequence, with, in each case, progress from the conducting-wall to dielectric versions, including graded-index distributions. The notation throughout is as consistent as seemed reasonably possible and it is hoped that this will lead to a simple way for the student to make meaningful comparisons between the various waveguides discussed.

The task facing the writer of a work which is to be used for teaching, reference, or background reading, is quite different from that undertaken by the author of a research monograph. Here we try to combine the two approaches by tackling simple problems immediately to give the reader confidence without too much emphasis on formality. Later it is shown that similar methods are useful in problems at the present state-of-the-art of optical waveguide research and that general statements of some overall validity may be usefully made for certain classes of waveguides. Hence it is hoped that the text could act as a stimulus for further research by directing the reader fairly quickly from the elementary concepts to the problems of current interest.

Chapter 1
Two-dimensional Conducting-wall Waveguides

In order to introduce the concepts, notation, and approaches to be used in dielectric waveguide problems later, we treat first the case of a two-dimensional waveguide whose boundaries are assumed to be perfect conductors. For simplicity, consider the waveguide to be filled with a homogeneous isotropic medium of (real) dielectric permittivity ε and magnetic permeability μ_0 (i.e. the permeability of free space). We show first that the familiar plane-polarized transverse electromagnetic (TEM) wave is a solution of the two-dimensional conducting wall guide problem and is therefore a waveguide mode. By the term 'waveguide mode' we shall mean an elementary wave characteristic of the waveguide; it propagates with well-defined phase velocity, group velocity, cross-sectional intensity distribution, and polarization.

In microwave terminology the two-dimensional guide with perfectly-conducting walls would be called the parallel strip transmission line, neglecting edge effects and problems associated with non-infinite conductivity ('skin-depth'); an approach via Maxwell's equations would then be appropriate. However, in optical terms the problem is equivalent to the situation of two plane parallel mirrors; in this case the simple concept of a ray propagating by repeated reflection from the mirror surfaces forms an alternative treatment. It will be shown that the results of this geometrical approach are the same as those resulting from the rigorous electromagnetic theory. This theme of the duality of mode and ray pictures in the treatment of waveguide problems will be stressed throughout the book, so it is important that it should be well-established by this very simple example.

The chapter concludes with the introduction of a set of normalized variables which reduce the results to a very general form easily applicable to any set of numerical parameters. This use of a consistent set of normalized variables forms a second theme running through the book. Numerical and analytical solutions to the various waveguide problems will therefore be given whenever appropriate in terms of these variables.

1.1 Transverse Electromagnetic Waves

1.1.1 Plane-polarized TEM wave in an unbounded dielectric

Consider the electric and magnetic field vectors associated with a plane-polarized wave in an unbounded isotropic medium. For a transverse electromagnetic (TEM) wave the electric field **E** and magnetic field **H** are at right-angles to each other and to the direction of propagation, taken as the z-axis. The situation is shown in Figure 1.1 where a Cartesian coordinate system has been chosen so that

$$\mathbf{E} = (E_x, 0, 0)$$
$$\mathbf{H} = (0, H_y, 0).$$

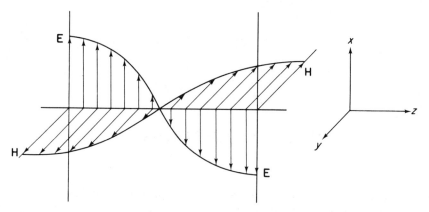

Figure 1.1 Electric and magnetic field configurations for a plane-polarized TEM wave in an unbounded dielectric

We now apply Maxwell's equations (see, for example, Stratton, 1941):

$$\nabla \times \mathbf{H} = \varepsilon \frac{d\mathbf{E}}{dt}, \quad \nabla \times \mathbf{E} = -\mu_0 \frac{d\mathbf{H}}{dt}$$

to obtain

$$\nabla \times \mathbf{H} = \left(-\frac{\partial H_y}{\partial z}, 0, \frac{\partial H_y}{\partial x}\right) = \varepsilon \left(\frac{\partial E_x}{\partial t}, 0, 0\right) \tag{1.1}$$

$$\nabla \times \mathbf{E} = \left(0, \frac{\partial E_x}{\partial z}, -\frac{\partial E_x}{\partial y}\right) = -\mu_0 \left(0, \frac{\partial H_y}{\partial t}, 0\right). \tag{1.2}$$

Assume E_x and H_y to have a periodic time- and z-dependence given by $\exp(i\beta z - i\omega t)$. Then equations (1.1) and (1.2) give

$$\beta H_y = \varepsilon \omega E_x \quad \text{and} \quad \beta E_x = \mu_0 \omega H_y$$

i.e.

$$\beta^2 = \omega^2 \varepsilon \mu_0. \tag{1.3}$$

As a consequence we have the following elementary results:

$$\frac{E_x}{H_y} = \left(\frac{\mu_o}{\varepsilon}\right)^{1/2} \tag{1.4}$$

Phase velocity: $\quad V_p = \frac{\omega}{\beta} = \frac{1}{\sqrt{\varepsilon\mu_0}} \quad \left(\equiv \text{group velocity}, \frac{\partial \omega}{\partial \beta}\right) \tag{1.5}$

Poynting vector: $\quad = \mathbf{E} \times \mathbf{H} = (0, 0, E_x H_y) \tag{1.6}$

Time-averaged power flow: $\quad = \frac{1}{2}\text{Re}(\mathbf{E} \times \mathbf{H}^*) = \frac{1}{2}|E_x|^2 \left(\frac{\varepsilon}{\mu_o}\right)^{1/2}. \tag{1.7}$

1.1.2 Parallel strip transmission line

Consider now an electromagnetic wave travelling in a medium bounded by two parallel plates assumed to have infinite conductivity. Let the plates be parallel to the y–z plane, of width d,

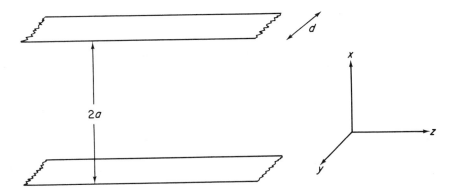

Figure 1.2 Parallel conducting planes which may guide a TEM mode

separated by distance $2a$, and of infinite length, as illustrated by Figure 1.2. (The reason for choosing $2a$ as the separation distance is to conform with later notation when we will consider guides of circular cylindrical geometry with radius a.) The boundary conditions to be obeyed by the electric and magnetic fields at the surface of each plate are then as follows (Stratton, 1941):

$$\varepsilon(E)_{\text{normal}} = \varepsilon E_x = s \text{ (surface charge density)}$$
$$(E)_{\text{tangent}} = (E_y \text{ and } E_z) = 0$$
$$(H)_{\text{normal}} = H_x = 0$$
$$(H)_{\text{tangent}} = H_y \text{ (say)} = K \text{ (current/unit width)}.$$

From the way these equations have been formulated it is clear that the TEM fields $(E_x, 0, 0)$, $(0, H_y, 0)$ satisfy these conditions. The results are then summarized as follows:

Current in strips $= I = dK = dH_y$ (1.8)

Voltage between strips $= V = 2aE_x$ (1.9)

Characteristic impedance $= Z_0 = \dfrac{V}{I} = \dfrac{2aE_x}{dH_y} = \dfrac{2a}{d}\left(\dfrac{\mu_0}{\varepsilon}\right)^{1/2}$ (1.10)

Instantaneous power $= \displaystyle\int_0^{2a}\int_0^d E_x H_y \,dx\,dy = VI = Z_0 I^2.$ (1.11)

In fact it is easily shown that the field patterns of principal waves on transmission lines of arbitrary cross-section are also TEM modes (Huxley, 1947; Collin, 1960) and the results (1.8)–(1.11) may be generalized to permit transmission lines to be discussed either in terms of electromagnetic theory or in the language of circuit theory.

The TEM wave is not the only one which can be propagated along parallel strip transmission lines; other modes are also possible, as will become evident in subsequent sections.

1.2 Ray–Mode Connections

It is appropriate at this stage to introduce some concepts of ray optics into our discussion. We therefore re-interpret the two-dimensional guide with conducting walls in terms of two parallel plane mirrors separated by distance $2a$. In this case it is sensible to refer to the

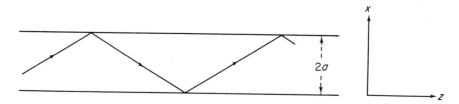

Figure 1.3 Zig-zag ray path between two parallel plane mirrors

refractive index n_1 of the medium between the mirrors, rather than the dielectric permittivity ε; the two are related by $\varepsilon = n_1^2 \varepsilon_0$ where ε_0 is the permittivity of free space. The situation is illustrated by Figure 1.3 where a 'zig-zag' ray path is also shown; the ray propagates by repeated reflection at the surface of each mirror in turn, the ray path between reflections taking the form of straight line segments since the medium is homogeneous and isotropic.

In order to see the connection between the zig-zag ray model and the electromagnetic mode picture which was used in previous sections, consider what happens when a ray is incident at angle θ on the surface of one mirror, as shown in Figure 1.4. From Snell's law, we know that

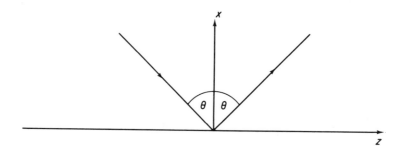

Figure 1.4 Incident and reflected rays at the surface of one mirror

the incident and reflected rays lie in the same plane (the x–z plane) and that the angle of reflection is also θ.

There are 2 cases to consider:

(a) Electric vector perpendicular to plane of incidence; $\mathbf{E} = (0, E_y, 0)$ (corresponds to TE mode—Transverse Electric).
(b) Electric vector parallel to plane of incidence; $\mathbf{H} = (0, H_y, 0)$ (corresponds to TM mode—Transverse Magnetic).

The analysis is similar in both cases; we treat here TE modes/rays. Define $k = \omega(\mu_0 \varepsilon_0)^{1/2} \equiv 2\pi/\lambda$, where λ = wavelength in free space. The electric field is then given by:

Incident ray: $\qquad E_{y_i} = E_0 \exp[-i\omega t - ikn_1 x \cos\theta + ikn_1 z \sin\theta]$ (1.12)

Reflected ray: $\qquad E_{y_r} = -E_0 \exp[-i\omega t + ikn_1 x \cos\theta + ikn_1 z \sin\theta]$ (1.13)

Total field: $\qquad E_{y_i} + E_{y_r} = E_0 e^{-i\omega t + ikn_1 z \sin\theta} \, 2i \sin(kn_1 x \cos\theta).$ (1.14)

Equation (1.14) links ray quantities (n_1, θ) with mode concepts (β, q), for we define

$$\beta = kn_1 \sin\theta$$
$$q = kn_1 \cos\theta = (k^2 n_1^2 - \beta^2)^{1/2}.$$

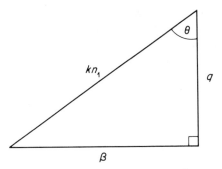

Figure 1.5 Localized plane wave diagram in two dimensions

Hence

$$\text{total field} \propto e^{-i\omega t + i\beta z} \sin(qx). \tag{1.15}$$

We shall see later that equation (1.15) is already in the form of a normal mode of the waveguide; for now it is sufficient to note that the quantities β, q, kn_1, and θ may be related geometrically by the localized plane wave diagram, Figure 1.5. Furthermore, denoting the phase velocity (ω/β) by V_p and the velocity of light in vacuo $(\mu_0 \varepsilon_0)^{1/2}$ by c, we have the following conventional definitions:

$$\beta = \text{longitudinal propagation constant} \tag{1.16}$$

$$\lambda_g = \frac{2\pi}{\beta} = \text{guide wavelength} \tag{1.17}$$

$$\frac{\lambda}{\lambda_g} = \frac{V_p}{c} = \text{equivalent index, } n \tag{1.18}$$

$$q = \text{transverse wave number.} \tag{1.19}$$

1.2.1 Ray treatment of two parallel plane mirrors

We now take account of the existence of the second mirror at $x = 2a$ in the zig-zag ray model of Figure 1.3. This implies that values of the longitudinal propagation constant β for guided rays (or equivalently the angle of incidence θ) are not unlimited but in fact must satisfy a quantization condition. To see this we follow a ray argument first used explicitly for dielectric slab waveguides by Tien and Ulrich (1970) and for metallic walls by Gandrud (1971), but which in its most general form derives from the work of Keller and Rubinow (1960). We follow one ray path on Figure 1.3 and assume an observer moving along the z-axis who sees only the transverse motion, i.e. in the x-direction. Then for a self-consistent picture the sum of all the phase shifts which occur in going from one boundary at $x = 0$ up to the second one at $x = 2a$ and back again must add to an integral multiple of 2π. This requirement, often termed the 'transverse resonance condition', must apply to all phase shifts incurred in this round-trip of the guide, including those on reflection at both walls. If the phase shift on reflection is denoted by $-\delta_\perp$, it follows from equations (1.12) and (1.13) at $x = 0$ that:

$$\frac{E_{yr}}{E_{yi}} = -1 = e^{-i\delta_\perp}$$

i.e.

$$\delta_\perp = \pi. \tag{1.20}$$

Also, the transverse phase change in traversing the guide twice is given from equations (1.12) and (1.13) as $4kn_1 a \cos\theta$. Hence, adding on the phase shifts at each wall, the transverse phase resonance condition becomes:

$$4akn_1 \cos\theta - 2\pi = 2N\pi \quad (N = 0, 1, 2, \ldots). \tag{1.21}$$

From the definition of q given above ($q = kn_1 \cos\theta$) it follows that:

$$2aq = (N+1)\pi. \tag{1.22}$$

In other words, we have derived a quantization condition on the longitudinal propagation constant β:

$$\beta^2 = k^2 n_1^2 - \left(\frac{(N+1)\pi}{2a}\right)^2. \tag{1.23}$$

This gives a characteristic waveguide plot of β versus k as shown in Figure 1.6 for the lowest-order modes. For comparison, Figure 1.6 also shows the plot for TEM modes of the two-

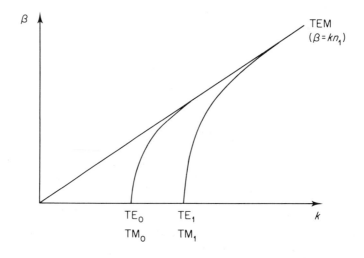

Figure 1.6 Propagation constant β versus wave number k (schematic) for low-order modes/rays confined by two parallel plane mirrors

dimensional conducting-wall guide. We see that whilst TEM modes can have $\beta = 0$ when $k = 0$, for TE (and similarly TM) modes there are *cut-off* values of k (equivalently, angular frequency ω) corresponding to $\beta = 0$. For frequencies below cut-off, β becomes imaginary and the wave is evanescent. From (1.23) the cut-off values k_c are given by:

$$k_c n_1 = \frac{(N+1)\pi}{2a}. \tag{1.24}$$

In the case of TM modes the analysis follows the same course as that for TE, with the boundary conditions now leading to a phase shift $\delta_\parallel = 0$ or 2π in place of the result (1.20). Once again, however, all the results (1.22)–(1.24) still hold true for the TM case.

1.2.2 Mode treatment of two parallel mirrors

For transverse electric (TE) modes, $\mathbf{E} = (0, E_y, 0)$ where, from equation (1.15):

$$E_y = A e^{-i\omega t + i\beta z} \sin(qx) \tag{1.25}$$
$$q^2 = k^2 n_1^2 - \beta^2 \tag{1.26}$$

and A is a constant in equation (1.25). It follows from Maxwell's equations that:

$$\mathbf{H} = (H_x, 0, H_z)$$

where

$$H_x = -\frac{\beta}{\omega \mu_0} A e^{-i\omega t + i\beta z} \sin(qx) \tag{1.27}$$

$$H_z = -\frac{iq}{\omega \mu_0} A e^{-i\omega t + i\beta z} \cos(qx). \tag{1.28}$$

The boundary conditions are that $E_y = 0$ on the mirror surfaces ($x = 0$ and $x = 2a$); the other boundary condition ($H_x = 0$ at $x = 0$ and $x = 2a$) merely duplicates this requirement. Since we have constructed the field in (1.25) in such a way that $E_y = 0$ at $x = 0$, the remaining condition yields:

$$\sin(2aq) = 0$$
$$2aq = N'\pi \quad (N' = 1, 2, \ldots). \tag{1.29}$$

Equation (1.29) is identical with (1.22) provided $N' = N + 1$. Hence the ray and mode pictures are equivalent descriptions and yield the same eigenvalue equation in this simple case. Some typical field distributions for the lowest order TE modes are shown in Figure 1.7.

For transverse magnetic (TM) modes there are only three nonzero field components: H_y, E_x, and E_z. In analogy with equations (1.25), (1.27), (1.28), those components are given by:

$$H_y = B e^{-i\omega t + i\beta z} \cos(qx) \tag{1.30}$$

$$E_x = \frac{\beta}{\omega \varepsilon} B e^{-i\omega t + i\beta z} \cos(qx) \tag{1.31}$$

$$E_z = -\frac{iq}{\omega \varepsilon} B e^{-i\omega t + i\beta z} \sin(qx) \tag{1.32}$$

where B is a constant and q is again given by equation (1.26). The boundary conditions in this case are that $E_z = 0$ at $x = 0$ and $x = 2a$, so that once again an eigenvalue equation of the form (1.29) (or equivalently (1.22)) is obtained.

1.2.3 Normalized variables

Although Figure 1.6 provides a useful description of the results for two parallel plane mirrors, the detailed structure still depends on the refractive index n_1 and guide width $2a$. In order to render the results independent of these parameters and express them in the most general form (and to anticipate future results on dielectric guides) we define the normalized variables u, v, b:

$$u^2 = a^2(k^2 n_1^2 - \beta^2) \tag{1.33}$$
$$v^2 = a^2 k^2 n_1^2 \tag{1.34}$$
$$b = 1 - \frac{u^2}{v^2}. \tag{1.35}$$

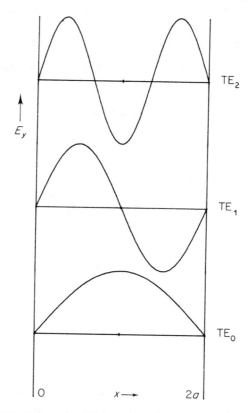

Figure 1.7 Schematic field distributions for some low-order modes confined between two perfectly conducting parallel planes

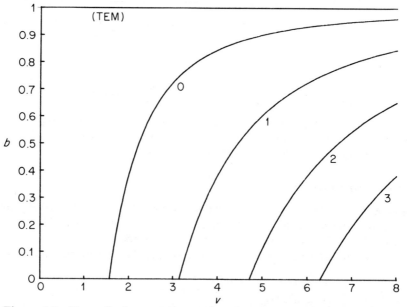

Figure 1.8 Normalized propagation constant b versus normalized frequency v for conducting-wall two-dimensional waveguide; labelling parameter gives the value of mode number N.

In general, the quantity b is a normalized propagation constant, and v is the normalized frequency. It is convenient therefore to express the result (1.29) in terms of b and v:

$$b = 1 - \left(\frac{N'\pi}{2v}\right)^2. \tag{1.36}$$

Hence we give in Figure 1.8 plots of b versus v for comparison later with similar plots for other waveguides (e.g. dielectric slab, fibres, graded-index guides, etc.). These plots form the most general version of the numerical results of a waveguide dispersion relation (eigenvalue equation).

We close this chapter by noting that the case considered here of a waveguide with walls of infinite conductivity (perfect metal) is a somewhat academic example chosen to illustrate the main concepts and notation to be employed in later chapters. However, the subject of guides with walls of finite conductivity is an important one not only in the microwave field but also at optical wavelengths. Hence we will return to this subject in discussing 'lossy' and 'hollow' waveguides in future chapters.

Chapter 2
Three-layer Dielectric Slab Waveguides

The dielectric slab waveguide has been discussed many times in the literature of integrated optics and semiconductor lasers, as well as in the conventional microwave context. The reasons for including a fairly extensive account of this structure here are (i) to emphasize again the duality of geometric-optics and modal approaches, (ii) to unify the notation via normalized variables, and (iii) to extend the conventional treatment of real waveguides to include the important classes of structures which become possible when the dielectric permittivity is considered as a complex quantity.

In order to set the scene for the geometric optics treatment we begin by considering the standard problem of reflection and refraction at an interface between two isotropic dielectric media. This yields Snell's laws, Fresnel's laws for reflection and transmission coefficients, and the expressions for phase shift on reflection of a plane wave at a dielectric interface. This leads naturally to a derivation of the eigenvalue equation for a three-layer dielectric slab via the condition of transverse phase-shift resonance discussed earlier for a simpler problem in Chapter 1. The electromagnetic mode treatment of the dielectric slab is also given and shown to yield an identical eigenvalue equation. There follow sections on the mode numbers and cut-off values, normalization in terms of power flow, and the normalized variables applied to a weakly-guiding symmetric slab. Finally the complex nature of the dielectric permittivity is introduced and the discussion widened to allow inclusion of the important topics of semiconductor injection lasers and metal-clad waveguides for integrated optics applications.

2.1 Reflection and Refraction at a Plane Dielectric Interface

We shall be concerned here with the derivation of Snell's laws of refraction and reflection, Fresnel's laws for transmission and reflection, the critical angle and the Brewster angle, and appropriate expressions for the phase shift on reflection. As mentioned above, this latter quantity is needed for the ray optics treatment of the dielectric slab waveguide. Consider therefore a plane interface ($x = 0$) in the y–z plane between two media characterized by ε_1, μ_0 and ε_2, μ_0. Consider a plane wave incident in the x–z plane from medium 1 on this interface. Then assume the incident, reflected, and refracted rays are coplanar, i.e. planes of constant phase are normal to the plane of incidence. The situation is indicated schematically in Figure 2.1.

Denoting the incident, reflected, and transmitted waves by $\mathbf{E}_i$, $\mathbf{E}_r$, and $\mathbf{E}_t$, respectively, we then have the following equations:

Incident wave: $\mathbf{E}_i = \mathbf{E}_1 e^{-i\omega t} \exp[ikn_1(-x\cos\theta_1 + z\sin\theta_1)]$ (2.1)

Reflected wave: $\mathbf{E}_r = \mathbf{E}_3 e^{-i\omega t} \exp[ikn_1(x\cos\theta_3 + z\sin\theta_3)]$ (2.2)

Transmitted wave: $\mathbf{E}_t = \mathbf{E}_2 e^{-i\omega t} \exp[ikn_2(-x\cos\theta_2 + z\sin\theta_2)]$ (2.3)

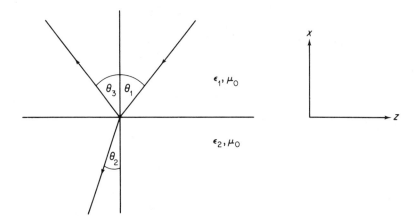

Figure 2.1 Incident, reflected, and transmitted rays at a plane interface between two dielectric media

where $k = \omega(\mu_0\varepsilon_0)^{1/2} = 2\pi/\lambda$, n_1 and n_2 are the refractive indices in media 1 and 2 ($\varepsilon_j = n_j^2 \varepsilon_0$; $j = 1, 2$), and $\theta_1, \theta_2, \theta_3$ are the angles of incidence, refraction, and reflection, respectively, as shown in Figure 2.1

2.1.1 Snell's laws

At the dielectric interface $x = 0$ the usual boundary conditions of continuity of tangential electric and magnetic fields must be satisfied. Hence a necessary condition for this is that the arguments of the exponents at $x = 0$ in equations (2.1)–(2.3) should be equal. It follows that

$$\theta_3 = \theta_1 \tag{2.4}$$

and

$$\frac{\sin\theta_2}{\sin\theta_1} = \frac{n_1}{n_2} \equiv \left(\frac{\varepsilon_1}{\varepsilon_2}\right)^{1/2}. \tag{2.5}$$

Equations (2.4) and (2.5), together with our assumption that the incident, refracted, and reflected rays are coplanar (which may also be proved from first principles (Stratton, 1941)) are Snell's laws. To proceed further, we note that there are two cases to be considered: (a) $\mathbf{E}_1$ normal to the plane of incidence, and (b) $\mathbf{E}_1$ in the plane of incidence.

2.1.2 Fresnel's laws

Consider first the case (a) of $\mathbf{E}_1$ normal to the plane of incidence: $\mathbf{E}_1 = (0, E_{1y}, 0)$. The boundary conditions at the interface are, then, first that the electric field should be continuous:

$$E_{1y} + E_{3y} = E_{2y} \tag{2.6}$$

and secondly that the tangential magnetic field should be continuous

$$-n_1 \cos\theta_1 E_{1y} + n_1 \cos\theta_1 E_{3y} = -n_2 \cos\theta_2 E_{2y}. \tag{2.7}$$

From (2.6) and (2.7), solving for E_{3y} and E_{2y} in terms of E_{1y},

$$E_{3y} = E_{1y} \left(\frac{n_1 \cos\theta_1 - n_2 \cos\theta_2}{n_1 \cos\theta_1 + n_2 \cos\theta_2} \right) \qquad (2.8)$$

$$E_{2y} = E_{1y} \left(\frac{2n_1 \cos\theta_1}{n_1 \cos\theta_1 + n_2 \cos\theta_2} \right). \qquad (2.9)$$

Using equation (2.5), these equations become (for $\theta_1 \neq 0$):

$$E_{3y} = E_{1y} \frac{\sin(\theta_2 - \theta_1)}{\sin(\theta_2 + \theta_1)} \qquad (2.10)$$

$$E_{2y} = E_{1y} \frac{2 \sin\theta_2 \cos\theta_1}{\sin(\theta_2 + \theta_1)}. \qquad (2.11)$$

Equations (2.10) and (2.11) relate the reflected and transmitted field amplitudes to the incident field. We may now derive expressions for the power reflection and transmission coefficients, using the fact that the mean power flow is given by $\frac{1}{2}(\mathbf{E} \wedge \mathbf{H}^*)$.

It follows therefore in this case that the mean energy flow normal to the dielectric interface is

$$S_x = \left| \frac{i}{2\mu_0 \omega} E_y \frac{\partial E_y^*}{\partial x} \right|.$$

Hence using equations (2.1)–(2.3) we have the following expressions for the incident, reflected, and transmitted powers normal to the interface:

Incident power:
$$S_{ix} = \frac{kn_1 \cos\theta_1}{2\mu_0 \omega} |E_{1y}|^2$$

Reflected power:
$$S_{rx} = \frac{kn_1 \cos\theta_1}{2\mu_0 \omega} |E_{3y}|^2$$

Transmitted power:
$$S_{tx} = \frac{kn_2 \cos\theta_2}{2\mu_0 \omega} |E_{2y}|^2.$$

It follows that the power reflection and transmission coefficients are given from (2.10), (2.11) by

Reflection coefficient:
$$R_\perp = \frac{S_{rx}}{S_{ix}} = \left| \frac{E_{3y}}{E_{1y}} \right|^2 = \frac{\sin^2(\theta_2 - \theta_1)}{\sin^2(\theta_2 + \theta_1)} \qquad (2.12)$$

Transmission coefficient:
$$T_\perp = \frac{S_{tx}}{S_{ix}} = \left| \frac{E_{2y}}{E_{1y}} \right|^2 \frac{n_2 \cos\theta_2}{n_1 \cos\theta_1} = \frac{\sin 2\theta_1 \sin 2\theta_2}{\sin^2(\theta_1 + \theta_2)}. \qquad (2.13)$$

Equations (2.12) and (2.13) are Fresnel's laws of reflection and transmission for a plane wave incident on a dielectric interface with the electric field vector normal to the plane of incidence.

We now derive the analogous results for the electric vector $\mathbf{E}_1$ in the plane of incidence, i.e. $\mathbf{H}_1 = (0, H_{1y}, 0)$. The boundary conditions are (i) tangential magnetic field continuous:

$$H_{1y} + H_{3y} = H_{2y} \qquad (2.14)$$

and (ii) tangential electric field continuous:

$$\frac{kn_1 \cos\theta_1}{\varepsilon_1} H_{1y} - \frac{kn_1 \cos\theta_1}{\varepsilon_1} H_{3y} = \frac{kn_2 \cos\theta_2}{\varepsilon_2} H_{2y}. \qquad (2.15)$$

From (2.14) and (2.15), solving for H_{3y} and H_{2y} in terms of H_{1y},

$$H_{3y} = H_{1y} \left(\frac{\frac{\cos\theta_1}{n_1} - \frac{\cos\theta_2}{n_2}}{\frac{\cos\theta_1}{n_1} + \frac{\cos\theta_2}{n_2}} \right) \tag{2.16}$$

$$H_{2y} = H_{1y} \left(\frac{\frac{2\cos\theta_1}{n_1}}{\frac{\cos\theta_1}{n_1} + \frac{\cos\theta_2}{n_2}} \right) \tag{2.17}$$

Using equation (2.15) these equations become (for $\theta_1 \neq 0$):

$$H_{3y} = H_{1y} \frac{\tan(\theta_1 - \theta_2)}{\tan(\theta_1 + \theta_2)} \tag{2.18}$$

$$H_{2y} = H_{1y} \frac{2\sin\theta_1 \cos\theta_1}{\sin(\theta_1 + \theta_2)\cos(\theta_1 - \theta_2)}. \tag{2.19}$$

The reflection and transmission coefficients now follow in an analogous manner to that for the previously-considered polarization. In this case the mean energy flow normal to the interface is

$$S_x = \left| \frac{i}{2\varepsilon\omega} H_y \frac{\partial H_y^*}{\partial x} \right|.$$

Hence we obtain:

Incident power:
$$S_{ix} = \frac{kn_1 \cos\theta_1}{2\varepsilon_1 \omega} |H_{1y}|^2$$

Reflected power:
$$S_{rx} = \frac{kn_1 \cos\theta_1}{2\varepsilon_1 \omega} |H_{3y}|^2$$

Transmitted power:
$$S_{tx} = \frac{kn_2 \cos\theta_2}{2\varepsilon_2 \omega} |H_{2y}|^2.$$

It follows from (2.18) and (2.19) that the required coefficients are given by

Reflection coefficient:
$$R_\| = \left|\frac{H_{3y}}{H_{1y}}\right|^2 = \frac{\tan^2(\theta_1 - \theta_2)}{\tan^2(\theta_1 + \theta_2)} \tag{2.20}$$

Transmission coefficient:
$$T_\| = \frac{\varepsilon_1 n_2 \cos\theta_2}{\varepsilon_2 n_1 \cos\theta_1} \left|\frac{H_{2y}}{H_{1y}}\right|^2 = \frac{\sin 2\theta_1 \sin 2\theta_2}{\sin^2(\theta_1 + \theta_2)\cos^2(\theta_1 - \theta_2)}. \tag{2.21}$$

2.1.3 Normal incidence

Equations (2.20) and (2.21) are Fresnel's laws for reflection and refraction of a plane wave with the electric vector in the plane of incidence, and are analogous to equations (2.12) and (2.13) for the opposite polarization. All these equations are only valid, however, for non-normal incidence, i.e. $\theta_1 \neq 0$. For the special case $\theta_1 = 0$ we must return to the boundary conditions once more to treat the situation of normal incidence separately. For the electric vector normal to the plane of incidence it follows from equation (2.8) that

$$E_{3y} = \left(\frac{n_1 - n_2}{n_1 + n_2}\right) E_{1y}.$$

Similarly, for the electric vector in the plane of incidence, equation (2.16) yields

$$H_{3y} = \left(\frac{n_2 - n_1}{n_2 + n_1}\right) H_{1y}.$$

Hence we find that the expressions for the power reflection coefficient for normal incidence are the same for both polarizations and are given by

$$R_\perp = R_\parallel = \left(\frac{n_1 - n_2}{n_1 + n_2}\right)^2. \tag{2.22}$$

Turning now to the transmitted wave for normal incidence, equation (2.9) gives

$$E_{2y} = \left(\frac{2n_1}{n_1 + n_2}\right) E_{1y}$$

so that the transmission coefficient is given by

$$T_\perp = \frac{n_2}{n_1} \left|\frac{E_{2y}}{E_{1y}}\right|^2 = \frac{4 n_1 n_2}{(n_1 + n_2)^2}. \tag{2.23}$$

For the other polarization, equation (2.17) gives

$$H_{2y} = \left(\frac{2n_2}{n_1 + n_2}\right) H_{1y}$$

and the corresponding coefficient becomes

$$T_\parallel = \frac{n_1}{n_2} \left|\frac{H_{2y}}{H_{1y}}\right|^2 = \frac{4 n_1 n_2}{(n_1 + n_2)^2}. \tag{2.24}$$

As in the case of the reflection coefficient for normal incidence (equation (2.22)) we see that the transmission coefficient for this situation is the same for each polarization. Equations (2.22), (2.23), and (2.24) are the Fresnel reflection and transmission results for normal incidence at a dielectric interface.

2.1.4 Brewster angle

Consider the possibility that for some angle of incidence the power reflection coefficient might be zero. Inspection of the appropriate equations (2.12) and (2.20) shows that this can only occur for the case of the electric vector lying in the plane of incidence (equation (2.20)) and for

$$\theta_1 + \theta_2 = \frac{\pi}{2}$$

i.e.

$$\sin \theta_2 = \cos \theta_1.$$

From (2.5) it follows that

$$\tan \theta_1 = \frac{n_2}{n_1}. \tag{2.25}$$

The angle defined by (2.25) is the Brewster angle, for which the reflection coefficient for waves polarized with the electric vector lying in the plane of incidence is zero. The physical mechanism responsible for the phenomenon of the Brewster angle is associated with the interaction of the radiation with electrons in the second dielectric medium (Born and Wolf, 1970; Marcuse, 1972). Since we may think of the interaction in terms of the vibrations of oscillating electric dipoles, it follows that for transverse waves the direction of oscillation is always normal to the direction of the ray considered. Thus the transmitted wave in region 2 has associated with it a set of electrons whose vibrations are in the plane of incidence and normal to the direction of the ray; such a situation would normally give rise to a re-radiated reflected ray in medium 1. However, since contributions to this ray for an oscillating electric dipole can only occur in a direction perpendicular to that of the oscillation, in the case $\theta_1 + \theta_2 = \pi/2$ there is no contribution to the reflected ray.

2.1.5 Critical angle

Consider now the possibility that for some angle of incidence the power transmission coefficient might be zero. For this situation equations (2.13) and (2.21) show that we require

$$\theta_2 = \frac{\pi}{2}.$$

From (2.5) it follows that

$$\sin \theta_1 = \frac{n_2}{n_1}. \tag{2.26}$$

The angle defined by (2.26) is the critical angle for total internal reflection of waves of either polarization at a dielectric interface.

2.1.6 Phase shift on total internal reflection

We derive next the expressions for phase shift on reflection at a dielectric interface which are needed for the subsequent treatment of waveguide problems. We are therefore concerned only with the situation of total internal reflection, i.e. angles of incidence in excess of the critical angle:

$$\sin \theta_1 > \frac{n_2}{n_1}.$$

From equation (2.5), it follows that the angle of transmission is complex and is given by

$$\cos \theta_2 = (1 - \sin^2 \theta_2)^{1/2} = \frac{i}{n_2} (n_1^2 \sin^2 \theta_1 - n_2^2)^{1/2}. \tag{2.27}$$

From this result it is a simple matter to show that the transmitted wave is a (decaying) evanescent wave normal to the interface. For the phase of the transmitted wave is given from (2.3) as

$$kn_2(-x \cos \theta_2 + z \sin \theta_2) = k(-ix(n_1^2 \sin^2 \theta_1 - n_2^2)^{1/2} + zn_1 \sin \theta_1).$$

In other words, the transmitted electric field is

$$\mathbf{E}_t = \mathbf{E}_2 e^{-i\omega t + i\beta z + \alpha x} \tag{2.28}$$

where

$$\beta = kn_1 \sin\theta_1$$

$$\alpha = kn_1 \left(\sin^2\theta_1 - \frac{n_2^2}{n_1^2}\right)^{1/2}.$$

Since α is a real quantity and is positive (since the field vanishes as $x \to -\infty$) the field in (2.28) represents a decaying wave in the $-x$ direction.

For the reflected wave, equations (2.8) and (2.27) yield for the electric field normal to the plane of incidence:

$$E_{3y} = E_{1y}\left[\frac{n_1\cos\theta_1 - i(n_1^2\sin^2\theta_1 - n_2^2)^{1/2}}{n_1\cos\theta_1 + i(n_1^2\sin^2\theta_1 - n_2^2)^{1/2}}\right]. \tag{2.29}$$

We may write (2.29) in terms of a phase shift $\delta_\perp$:

$$E_{3y} = E_{1y}e^{-i\delta_\perp}. \tag{2.30}$$

Now, if

$$\frac{a-ib}{a+ib} = e^{-i\delta},$$

as in the form of equation obtained by equating (2.29) and (2.30), then

$$\tan\frac{\delta}{2} = \frac{e^{i\delta/2} - e^{-i\delta/2}}{i(e^{i\delta/2} + e^{-i\delta/2})} = \frac{1 - e^{-i\delta}}{i(1 + e^{-i\delta})} = \frac{b}{a}. \tag{2.31}$$

Hence, using (2.29) and (2.31),

$$\delta_\perp = 2\tan^{-1}\left[\frac{(\sin^2\theta_1 - n_2^2/n_1^2)^{1/2}}{\cos\theta_1}\right]. \tag{2.32}$$

Similarly, for polarization with the electric field in the plane of incidence, equations (2.16) and (2.27) yield:

$$H_{3y} = H_{1y}\left[\frac{n_2^2\cos\theta_1 - in_1^2(\sin^2\theta_1 - n_2^2/n_1^2)^{1/2}}{n_2^2\cos\theta_1 + in_1^2(\sin^2\theta_1 - n_2^2/n_1^2)^{1/2}}\right]. \tag{2.33}$$

Using (2.33) and (2.31) it follows that the phase shift in this case is

$$\delta_\| = 2\tan^{-1}\left[\frac{(\sin^2\theta_1 - n_2^2/n_1^2)^{1/2}}{(n_2^2/n_1^2)\cos\theta_1}\right]. \tag{2.34}$$

Equations (2.32) and (2.34) give expressions for the phase shift on total internal reflection for the two polarizations, and will be used in subsequent sections to deal with waveguide problems from the viewpoint of geometric optics.

2.1.7 The Goos–Haenchen shift

Thus far we have dealt only with the properties of rays, in the sense of plane waves, on reflection and refraction at a dielectric interface. However, in any practical situation, including waveguides of various types, it is often more realistic to speak in terms of a ray in the sense of the axis of a beam of light with some well-defined cross-section. With this concept it follows that there is a lateral shift of the reflected ray with respect to the incident ray, which is termed the Goos–Haenchen shift. This phenomenon occurs because, if one thinks of the incident beam as composed of a set of plane waves, each elementary wave will have a slightly different angle of incidence. Since the phase change on reflection depends on the angle of incidence (as shown in equations (2.32) and (2.34)), the reflected beam will not be a perfect reconstitution of the incident beam at the point of incidence. The situation is indicated in Figure 2.2, where the lateral shift of the beam is designated $2z_s$.

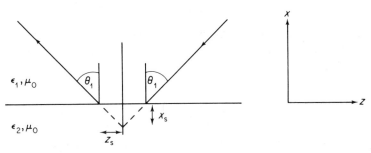

Figure 2.2 Goos–Haenchen shift of a ray on reflection at a dielectric interface

To calculate the value of the Goos–Haenchen shift consider a simple model of the spatial wave-packet or beam of incident light, consisting of two plane waves incident at slightly different angles $\theta_1 \pm \Delta\theta_1$, where $\Delta\theta_1 \ll \theta_1$. It follows that the field of the incident wave-packet is given (adding the two component plane waves represented as in equation (2.1), for example) by

$$\mathbf{E}_i \simeq \mathbf{E}_1 e^{-i\omega t} \exp\left[ikn_1(-x\cos\theta_1 + z\sin\theta_1)\right]$$
$$\times \left[\exp(ikn_1 z \cos\theta_1 \Delta\theta_1) + \exp(-ikn_1 z \cos\theta_1 \Delta\theta_1)\right]$$
$$= \mathbf{E}_1 e^{-i\omega t} \exp(ikn_1(-x\cos\theta_1 + z\sin\theta_1)) 2\cos(kn_1 z \cos\theta_1 \Delta\theta_1) \quad (2.35)$$

where we have retained terms only up to order $\Delta\theta_1$. After total internal reflection at the interface the reflected wave-packet is given in terms of the phase shift of each component plane wave, written as $\delta(\theta_1 \pm \Delta\theta_1)$ to emphasize its dependence on angle of incidence, as

$$\mathbf{E}_r = \mathbf{E}_1 e^{-i\omega t} \exp\left[ikn_1(x\cos\theta_1 + z\sin\theta_1)\right]$$
$$\times \{\exp[ikn_1 z \cos\theta_1 \Delta\theta_1 - i\delta(\theta_1 + \Delta\theta_1)] + \exp[-ikn_1 z \cos\theta_1 \Delta\theta_1 - i\delta(\theta_1 - \Delta\theta_1)]\}. \quad (2.36)$$

Retaining terms only to order $\Delta\theta_1$ in the Taylor series expansion of δ, we find

$$\delta(\theta_1 \pm \Delta\theta_1) \simeq \delta(\theta_1) \pm \Delta\theta_1 \frac{\partial \delta}{\partial \theta_1}.$$

Hence equation (2.36) becomes

$$\mathbf{E}_r = \mathbf{E}_1 e^{-i\omega t} \exp\left[ikn_1(x\cos\theta_1 + z\sin\theta_1)\right] \exp[-i\delta(\theta_1)]$$
$$\times 2\cos\left[kn_1 z \cos\theta_1 \Delta\theta_1 - \Delta\theta_1 \frac{\partial \delta}{\partial \theta_1}\right]. \quad (2.37)$$

Comparison of equations (2.35) and (2.37) for the incident and reflected wave-packets shows that there has been a lateral shift of the beam axis on reflection given by $2z_s$ where

$$z_s = \frac{1}{2kn_1 \cos\theta_1} \frac{\partial \delta}{\partial \theta_1}. \quad (2.38)$$

It follows, using equations (2.32) and (2.34), that the shifts $z_{s\perp}$ and $z_{s\|}$ for the two polarizations are

$$z_{s\perp} = \frac{\tan\theta_1}{k(n_1^2 \sin^2\theta_1 - n_2^2)^{1/2}} \qquad (2.39)$$

$$z_{s\|} = \frac{n_2^2 \tan\theta_1}{k(n_1^2 \sin^2\theta_1 - n_2^2)^{1/2}(n_1^2 \sin^2\theta_1 - n_2^2 \cos^2\theta_1)}. \qquad (2.40)$$

Note from Figure 2.2 that it is sometimes more convenient to describe the Goos–Haenchen shift in terms of the apparent penetration depth of the ray, denoted by x_s on the figure, and related to z_s by

$$x_s = z_s \cot\theta_1. \qquad (2.41)$$

According to equations (2.39) and (2.40) the lateral shift becomes infinite at the critical angle $\sin^{-1}(n_2/n_1)$ and at grazing incidence ($\theta_1 = \pi/2$). However, the derivation of these equations may be somewhat dubious for angles of incidence close to these values. For a more detailed discussion of this point and related considerations, the reader is referred to the extensive literature which exists on the subject of the Goos–Haenchen shift. General references will be found in an article by Lotsch (1968) and in Section 5.2 of Arnaud's book (1976). The application of the effect in dielectric waveguides is discussed by Kapany and Burke (1972) and by Kogelnik and Weber (1974). The extensions to anisotropic dielectric media were given by Kogelnik et al. (1973) and Ramaswamy (1974).

2.1.8 Summary and numerical example

We summarize here the principal results of Section 2.1 which will be required in later sections, and conclude with a numerical example of the use of these results. For a plane dielectric interface Snell's laws state (i) that the incident, reflected, and transmitted rays are coplanar, (ii) the angle of reflection is the same as the angle of incidence, and (iii) the angles of transmission and incidence are connected by the familiar form of equation (2.5). There are two cases of polarization to be considered, namely the electric field either normal or parallel to the plane of incidence. For the former case the Fresnel reflection and transmission coefficients are given by equations (2.12) and (2.13), respectively, whilst for the latter the relevant equations are (2.20) and (2.21). These relationships all involve the angles of incidence and transmission and are valid only for non-normal incidence. For normal incidence the two polarizations yield identical results for the Fresnel coefficients, and these are given by equations (2.22) (for reflection) and (2.24) (for transmission). In all these cases the sum of reflection and transmission coefficients is unity for a plane dielectric interface between two media of real dielectric permittivities. The reflection coefficient can only be zero for waves polarized with the electric vector in the plane of incidence and incident at the Brewster angle defined by equation (2.25). The transmission coefficient is zero for both polarizations for incidence at the critical angle defined by equation (2.26). For angles greater than the critical angle, total internal reflection occurs and there is only an evanescent wave in the second dielectric medium. This wave decays faster with increasing angle of incidence according to equation (2.28). The reflected wave suffers a phase change in this situation, the appropriate expressions being given by (2.32) and (2.34) for the two polarizations. For a beam of finite cross section there is a corresponding lateral shift of the beam-axis upon reflection—the Goos–Haenchen shift—given by equations (2.38)–(2.40).

As a numerical example of these results, consider an interface between the III–V semiconductor gallium arsenide (GaAs) and air. GaAs is an appropriate material to choose since its technology of growth and characterization has been stimulated by the discovery of lasing action in 1962. It now forms the basis of many types of injection laser and integrated optics component and we shall frequently refer to this material and its associated alloys for numerical examples in future sections. In Figure 2.1, therefore, material 1 will be GaAs ($n_1 = 3.6$) and material 2 air ($n_2 = 1$). Figure 2.3 shows plots of the Fresnel reflection

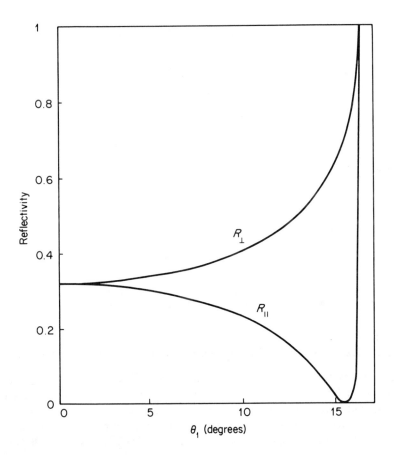

Figure 2.3 Reflectivities of a GaAs/air interface for two polarizations as functions of the angle of incidence θ_1. (From Adams and Cross (1971). Reproduced by permission of the IEE)

coefficient versus angle of incidence θ_1 for the two polarizations from equations (2.12) and (2.20) (Adams and Cross, 1971a). At $\theta_1 = 0$ the value for both polarizations is given by equation (2.22) as 0.32. For the electric field in the plane of incidence, the reflection coefficient first decreases to zero at the Brewster angle given by (2.25) as 15.5°, and then increases to unity at the critical angle which is given from (2.26) as 16.1° for this interface. For the other polarization the reflection coefficient increases monotonically with increasing angle of incidence. Figure 2.4 shows the variation of phase shift in total internal reflection with angle of incidence for waves of each polarization. Both polarizations yield a phase shift of zero at the critical angle and π at grazing incidence. However, between these values the two curves are widely separated, with the curve for polarization with electric vector parallel to the plane of incidence approaching close to π over a substantial range of angles of incidence.

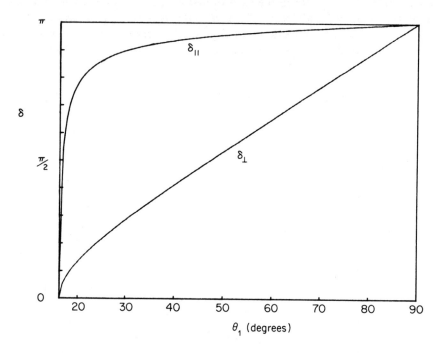

Figure 2.4 Phase shifts at a GaAs/air interface as functions of the angle of incidence θ_1

2.2 Ray Treatment of the Dielectric Slab Waveguide: Bound Rays

We are now in a position to deal with the three-layer dielectric slab waveguide from the viewpoint of geometric optics. Consider a uniform isotropic dielectric of refractive index n_1, thickness $2a$, sandwiched between two semi-infinite layers of (lower) index n_2, n_3. The situation is illustrated in Figure 2.5, which shows the zig-zag path of a ray travelling at an angle ϕ to the axis of the guide (taken as the z-axis). For definiteness, consider the situation

$$n_1 > n_2 \geqslant n_3 \tag{2.42}$$

which yields an asymmetric guiding structure for $n_2 \neq n_3$ and a symmetric guide when $n_2 = n_3$. Since we have in this structure two dielectric interfaces of the type discussed in Section 2.1, it is clear that the behaviour of a ray in the guide will depend crucially on the value of the angle ϕ. In Figure 2.5 we show the case for which ϕ is such that the angle of incidence of the ray exceeds the critical angle at each interface. From Section 2.1.5, with $\phi = \pi/2 - \theta_1$, it follows that we are dealing with the case

$$\cos \phi \geqslant \frac{n_2}{n_1} \geqslant \frac{n_3}{n_1}. \tag{2.43}$$

For such angles ϕ, the ray suffers total internal reflection at each interface and hence the light is trapped within the central layer of the structure. This corresponds to a guided mode of the

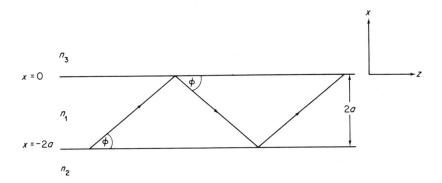

Figure 2.5 Zig-zag path of a ray in a dielectric slab waveguide

waveguide. From the considerations of Section 1.3, we know that if the waves propagate in the z-direction as exp (iβz), then the longitudinal propagation constant β is related to the ray angle ϕ by

$$\beta = n_1 k \cos \phi. \tag{2.44}$$

Hence it follows from (2.43) and (2.44) that the propagation constant β is constrained for guided modes by:

$$n_1 k \geqslant \beta \geqslant n_2 k \geqslant n_3 k, \tag{2.45}$$

2.2.1 Eigenvalue equation

The minimum value of β for guided propagation, i.e. the cut-off condition for the mode, is seen to be $n_2 k$. However, β and ϕ are only allowed to take discrete values between their respective limits, not a continuum as implied so far (cf. Section 1.2); this is due to phase information which is neglected in the discussion of this simple picture. To correct this omission, we invoke again the condition for transverse phase resonance which was used previously in Section 1.2. According to this condition (Tien and Ulrich, 1970; Tien, 1971), the transverse phase shift occurring in one complete zig-zag must add up to a multiple of 2π. Since the geometrical path length for one such round trip is $4a \sin \phi$, and recalling that phase changes occurring on reflection at the boundaries must also be included, the phase resonance condition becomes

$$4akn_1 \sin \phi - \delta_{12} - \delta_{13} = 2N\pi \quad (N = 0, 1, 2, \ldots) \tag{2.46}$$

where δ_{12}, δ_{13} denote the phase shifts occurring at interfaces 1–2 and 1–3, respectively. For a ray whose electric vector is normal to the plane of incidence (corresponding to a TE mode), equation (2.32) yields:

$$\delta_{12\perp} = 2 \tan^{-1} \left[\frac{(\cos^2 \phi - (n_2/n_1)^2)^{1/2}}{\sin \phi} \right]$$

$$= 2 \tan^{-1} \left[\left(\frac{\beta^2 - n_2^2 k^2}{n_1^2 k^2 - \beta^2} \right)^{1/2} \right]$$

where equation (2.44) has been used to introduce β and k. A similar expression holds for $\delta_{13\perp}$

at the 1–3 interface for this polarization. We define

$$p^2 = \beta^2 - n_2^2 k^2$$
$$q^2 = n_1^2 k^2 - \beta^2$$
$$r^2 = \beta^2 - n_3^2 k^2.$$

Using these results in equation (2.46) yields

$$4aq - 2\tan^{-1}\left(\frac{p}{q}\right) - 2\tan^{-1}\left(\frac{r}{q}\right) = 2N\pi \quad (N = 0, 1, 2, \ldots)$$

which may be written more conveniently as:

$$\tan(2aq - N\pi) = \frac{(p+r)q}{q^2 - pr} \quad (N = 0, 1, 2, \ldots). \tag{2.47}$$

Equation (2.47) is the TE eigenvalue equation for rays in the three-layer dielectric slab.
Similarly for rays of the opposite polarization, using the above definitions, equation (2.34) gives

$$\delta_{12\parallel} = 2\tan^{-1}\left[\left(\frac{n_1}{n_2}\right)^2 \frac{p}{q}\right]$$

which, together with the analogous result for the 1–3 interface, enables the resonance condition to be written

$$\tan(2aq - N\pi) = \frac{(n_3^2 p + n_2^2 r) n_1^2 q}{n_2^2 n_3^2 q^2 - n_1^4 pr} \quad (N = 0, 1, 2, \ldots). \tag{2.48}$$

Equation (2.48) is the TM eigenvalue equation for the three-layer dielectric slab. Although analytic approximations are possible for the solution of equations (2.47) and (2.48) (see, for example, Marcuse, 1972, 1974; Lotspeich, 1975; Miyagi and Nishida, 1979, 1979a), recourse must generally be made to numerical techniques. The form of the results is schematically illustrated in Figure 2.6, in the form of β–k plots for the guided modes. The β-values permitted for guided modes are bounded by the lines $\beta = kn_1$ and $\beta = kn_2$ on the plot, according to equation (2.45). Each different value of N in equations (2.47) and (2.48) corresponds to a different β–k plot, corresponding to a guided mode.

2.2.2 Effective width

The derivation of the eigenvalue equation given above was established first by Tien and Ulrich (1970). Slightly different treatments, still from a ray point of view, have been given by Marcuse (1974) and Love and Snyder (1976). Clearly, the solution of the eigenvalue equation will yield the phase velocity and, by differentiation, the group velocity. However, this latter quantity may be found more directly from consideration of the Goos–Haenchen effect occurring at the waveguide boundaries (Kogelnik and Weber, 1974). To see this, recall from Section 2.1.7 that associated with reflection at a dielectric interface there is an apparent ray penetration depth. Physically this corresponds to an evanescent field in the material of lower refractive index. If we include these apparent depths and their associated lateral ray shifts in the zig-zag ray model, the situation becomes as illustrated in Figure 2.7. At the 1–2 interface the lateral shift is denoted by z_{s12} and is given by an expression of the form (2.38), whilst the apparent penetration depth is denoted by x_{s12} and is given by an expression like (2.41). Similar

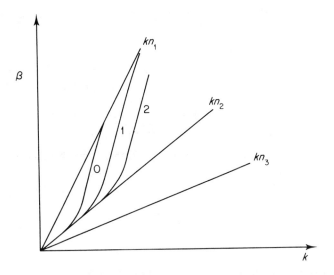

Figure 2.6 Propagation constant β versus wave number k (schematic) for low-order modes/rays of the dielectric slab waveguide

considerations apply to the 1–3 interface. Hence we may think of the zig-zag ray as propagating in a waveguide of effective width w given by

$$w = 2a + x_{s12} + x_{s13}. \tag{2.49}$$

Using equations (2.39) and (2.41) for the apparent penetration depth for polarization with electric vector normal to the x–z plane, together with the definitions of p and r given above, equation (2.49) becomes

$$w_\perp = 2a + \frac{1}{p} + \frac{1}{r}. \tag{2.50}$$

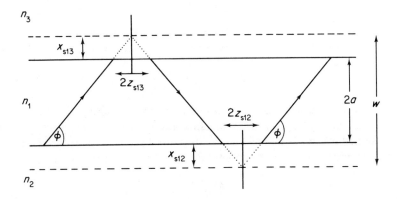

Figure 2.7 Modified zig-zag ray path in a dielectric waveguide, including the effects of the Goos–Haenchen shifts

Similarly, for waves polarized with electric vector in the $x-z$ plane, equations (2.40), (2.41), and (2.49) yield:

$$w_\| = 2a + \frac{1}{p\left(\dfrac{\beta^2}{n_1^2 k^2} + \dfrac{\beta^2}{n_2^2 k^2} - 1\right)} + \frac{1}{r\left(\dfrac{\beta^2}{n_1^2 k^2} + \dfrac{\beta^2}{n_3^2 k^2} - 1\right)}$$

$$= 2a + \frac{n_1^2 n_2^2 (p^2 + q^2)}{p(p^2 n_1^4 + q^2 n_2^4)} + \frac{n_1^2 n_3^2 (r^2 + q^2)}{r(r^2 n_1^4 + q^2 n_3^4)}. \tag{2.51}$$

Intuitively, it may be thought that this effective width w gives a measure of the depth of electromagnetic field spreading in the waveguide. This is indeed the case, as we shall see when we consider the analysis of the three-layer slab from an electromagnetic mode viewpoint.

2.2.3 Group velocity

For the present, however, we are concerned with the derivation of the group velocity. For this purpose it is important to realize that the Goos–Haenchen shift will also have a time delay associated with the time the ray apparently spends in the lower index material. Following an analogous argument to that in Section 2.1.7 for a spatial wave-packet, in this case including the temporal aspect, it follows that this time delay $2\tau_s$ is given by considering two plane waves with slightly different angular frequencies $\omega \pm \Delta\omega$. Hence we find τ_s is given by

$$\tau_s = -\frac{1}{2}\frac{\partial \delta}{\partial \omega} = -\frac{1}{2c}\frac{\partial \delta}{\partial k} \tag{2.52}$$

where δ is again the phase shift on total reflection and c is the velocity of light in free space. The time which the ray spends in crossing the guide width $2a$ is

$$\tau_a = \frac{2a\, m_1}{c \sin \phi} \tag{2.53}$$

where m_1 is the group index of dielectric 1:

$$m_1 = n_1 - \lambda \frac{dn_1}{d\lambda} \equiv n_1 + \omega \frac{dn_1}{d\omega} \equiv n_1 + k \frac{dn_1}{dk} \tag{2.54}$$

and where we have used the fact that the group velocity in medium 1 is appropriate here, rather than the phase velocity c/n_1. Therefore the total time involved τ_w for the ray to cross its apparent guide-width w is given by

$$\tau_w = \tau_a + \tau_{s12} + \tau_{s13} \tag{2.55}$$

where τ_{s12}, τ_{s13} are the delays incurred on reflection at interfaces 1–2 and 1–3, respectively. During this time the ray has travelled a distance along the axis of the guide given by $w \cot \phi$, where w is defined in equation (2.49). Now, the overall group velocity V_g of the zig-zag ray in the waveguide is given by the ratio of the total axial distance $w \cot \phi$ to the total time taken τ_w:

$$V_g = \frac{2a \cot \phi + z_{s12} + z_{s13}}{\tau_a + \tau_{s12} + \tau_{s13}}. \tag{2.56}$$

We will now prove that this result is the same as that obtained by differentiating the eigenvalue equation for the guide. Using this equation in its original form (2.46), and differentiating with respect to β (allowing n_1 to be a function of frequency ω),

$$4a \sin \phi \left(n_1 + k\frac{dn_1}{dk}\right)\frac{\partial k}{\partial \beta} + 4akn_1 \cos \phi \frac{d\phi}{d\beta} - \frac{\partial}{\partial \beta}(\delta_{12} + \delta_{13}) - \frac{\partial k}{\partial \beta}\frac{\partial}{\partial k}(\delta_{12} + \delta_{13}) = 0$$

or, in terms of frequency ω and the group index m_1, defined in (2.54),

$$\frac{4am_1 \sin \phi}{c}\frac{\partial \omega}{\partial \beta} + 4akn_1 \cos \phi \frac{d\phi}{d\beta} - \frac{\partial}{\partial \beta}(\delta_{12} + \delta_{13}) - \frac{\partial \omega}{\partial \beta}\frac{\partial}{\partial \omega}(\delta_{12} + \delta_{13}) = 0. \quad (2.57)$$

We may eliminate $d\phi/d\beta$ by differentiating (2.44):

$$\frac{d\phi}{d\beta} = \frac{\frac{m_1 \cos \phi}{c}\frac{\partial \omega}{\partial \beta} - 1}{n_1 k \sin \phi}.$$

Hence equation (2.57) yields an expression for $\partial \omega/\partial \beta$ (or, equivalently, group velocity V_g) which may be written, after a little manipulation, as

$$V_g = \frac{\partial \omega}{\partial \beta} = \frac{2a \cot \phi + \frac{1}{2}\frac{\partial \delta_{12}}{\partial \beta} + \frac{1}{2}\frac{\partial \delta_{13}}{\partial \beta}}{\frac{2am_1}{c \sin \phi} - \frac{1}{2}\frac{\partial \delta_{12}}{\partial \omega} - \frac{1}{2}\frac{\partial \delta_{13}}{\partial \omega}}. \quad (2.58)$$

Inspection of equation (2.58) shows that the second and third terms of the numerator are in fact z_{s12} and z_{s13} (from the definitions of (2.38) and (2.44)), whilst the terms in the numerator are, respectively, τ_a, τ_{s12}, and τ_{s13} (from (2.53) and (2.52)). Hence we have proved that equation (2.56) for the group velocity, derived purely from physical arguments based on the zig-zag ray model, is identical with (2.58) which was derived by differentiating the eigenvalue equation for the guide.

2.2.4 Losses

We can push the ray treatment a little further by using it to treat the case of a dielectric slab waveguide composed of lossy media. In this case the media possess complex dielectric permittivities, and we may express this by replacing the real refractive index $n_j (j = 1, 2, 3)$ used hitherto by a complex refractive index $n_j + iK_j$. The quantity K_j is sometimes termed the extinction coefficient and is related to the usual attenuation coefficient per unit length α_j by

$$\alpha_j = -2K_j k \quad (2.59)$$

where k is again the wave number. We assume that the attenuation α_j is due to absorption of radiation in region j of the guide, and the problem to be addressed is the calculation of the total attenuation suffered by a bound ray propagating in the guide. Clearly this attenuation stems from three sources:

(i) absorption in the core region 1,
(ii) transmission into medium 2 on reflection at the 1–2 interface, and
(iii) transmission into medium 3 on reflection at the 1–3 interface.

The losses (ii) and (iii) occur because total internal reflection is strictly only possible at an interface between lossless dielectrics; the presence of loss implies a complex wave transmission coefficient and a (real) nonzero *power* transmission coefficient. It follows that the attenuation coefficient of a ray taking a path of the form shown in Figure 2.7, defined as the relative power loss per unit length, is given as a sum of α_1 and the contributions from transmission on partial reflection at the interfaces. In one round trip of the zig-zag path the relative power loss is given by $(1 - R_{12}R_{13})$ where the R's are the power reflection coefficients defined as in Section 2.1.2 with appropriate inclusion of the effects of complex permittivities. Since this loss occurs whilst the ray travels in the guide of effective width w, it follows that the axial distance travelled is $2w \cot \phi$. Hence the ray attenuation coefficient α is given by (Reisinger, 1973; Ulrich and Prettl, 1973):

$$\alpha = \alpha_1 + \frac{1 - R_{12}R_{13}}{2w \cot \phi}. \tag{2.60}$$

Useful analytical results for the attenuation coefficient may be obtained from (2.60) in two important cases: (a) a hollow dielectric guide with metallic walls, and (b) a relatively low-loss weakly-guiding dielectric slab.

(a) Hollow dielectric guide with metallic walls

We consider the special case (Gandrud, 1971), $K_1 = 0, n_2 = n_3, K_2 = K_3$; i.e. a symmetric guide with metallic walls and a lossless dielectric core. In other words, we return here to the subject of guides with walls of finite conductivity, which was briefly introduced at the end of Chapter 1. For the wave polarized with electric vector normal to the plane of incidence the ratio of reflected to incident waves is given from (2.8) and (2.27) by $(\phi = \pi/2 - \theta_1)$ (Born and Wolf, 1970):

$$\frac{E_{3y}}{E_{1y}} = \frac{n_1 \sin \phi - i[n_1^2 \cos^2 \phi - (n_2 + iK_2)^2]^{1/2}}{n_1 \sin \phi + i[n_1^2 \cos^2 \phi - (n_2 + iK_2)^2]^{1/2}}. \tag{2.61}$$

Since we are dealing with metallic walls, it follows that K_2 is usually large, and in fact for most combinations of dielectric core and metal claddings at optical frequencies (Gandrud, 1971):

$$n_1^2 \cos^2 \phi \ll |n_2^2 - K_2^2|. \tag{2.62}$$

In this case equation (2.61) simplifies and we obtain for the reflection coefficient, from (2.12),

$$R_{12\perp} \simeq \frac{(n_1 \sin \phi - n_2)^2 + K_2^2}{(n_1 \sin \phi + n_2)^2 + K_2^2}. \tag{2.63}$$

The analogous result for the opposite polarization is

$$R_{12\parallel} \simeq \frac{(n_2 \sin \phi - n_1)^2 + K_2^2 \sin^2 \phi}{(n_2 \sin \phi + n_1)^2 + K_2^2 \sin^2 \phi}. \tag{2.64}$$

For most metallic boundaries the Goos–Haenchen shifts may be neglected and we may replace w in (2.60) by the guide width $2a$. The attenuation calculation is then straightforward, via the use of equation (2.63) or (2.64).

A particularly simple result is obtained for the TE_0 mode in an overmoded guide. By the term 'overmoded' here we refer to one which can support many modes, so that its width is much larger than the wavelength. More specifically, if we assume $2an_1k \gg \pi$, and we use the fact that the real part of the propagation constant is likely to be very little affected by the wall conductivity being finite, then we can utilize a result from Section 1.2 (equation (1.22) with $q = n_1 k \sin \phi$, $N = 0$) to obtain:

$$\sin \phi \ll 1.$$

Adoption of this approximation in equations (2.60) and (2.63), and ignoring the Goos–Haenchen shifts, yields for the attenuation coefficient (Gandrud, 1971):

$$\alpha_{TE_0} \simeq \frac{\pi^2 n_2}{2a^3 k^2 n_1 (n_2^2 + K_2^2)}. \tag{2.65}$$

Equation (2.65) represents a useful approximation for the attenuation coefficient of the TE_0 mode in a hollow dielectric guide with metallic walls. Although this result was derived here purely from a ray treatment, we shall see later (Section 2.5) that the same form of equation can be obtained also by the standard microwave frequency approximation, i.e. the application of perturbation theory to the boundary conditions for the formal electromagnetic mode treatment.

(b) Relatively low-loss weakly-guiding dielectric slab

Consider now the situation which is, in a sense, the opposite to that of (a) above, i.e. the relatively low-loss slab guide:

$$K_j \ll n_j (j = 1, 2, 3).$$

The guiding action is still dominated by the real refractive index effect, so equation (2.42) also applies. In this situation Snell's law (equation (2.5)) yields

$$(n_2 + iK_2)\cos\theta_2 \simeq i(n_1^2 \sin^2\theta_1 - n_2^2)^{1/2} + \frac{(n_2 K_2 - n_1 K_1 \sin^2\theta_1)}{(n_1^2 \sin^2\theta_1 - n_2^2)^{1/2}}. \tag{2.66}$$

Using this result in the expression (2.8) for ratio of reflected to incident plane waves polarized with electric vector normal to the plane of incidence yields the following result for the power reflection coefficient (in terms of the angle ϕ of Figure 2.7) (Snyder and Love, 1975):

$$R_{12\perp} \simeq 1 - \frac{4n_1 n_2^2 \sin\phi (K_2/n_2 - K_1/n_1)}{(n_1^2 - n_2^2)(n_1^2 \cos^2\phi - n_2^2)^{1/2}}. \tag{2.67}$$

The corresponding result for plane waves of the opposite polarization from (2.16) becomes:

$$R_{12\parallel} \simeq 1 - \frac{4n_1 n_2^2 \sin\phi (K_2/n_2 - K_1/n_1)(2n_1^2 \cos^2\phi - n_2^2)}{(n_1^2 - n_2^2)(n_1^2 \cos^2\phi - n_2^2)^{1/2}(n_1^2 \cos^2\phi - n_2^2 \sin^2\phi)}. \tag{2.68}$$

In deriving equations (2.67) and (2.68) we have assumed

$$|K_2/n_2 - K_1/n_1| \ll |n_1^2 \cos^2\phi - n_2^2|,$$

and have only retained terms up to a linear power of $K_j (j = 1, 2)$. The results for the 1–3 interface are obtained simply by replacing the suffix '2' wherever it appears in (2.67) and (2.68) by '3'. The calculation of modal attenuation is now simply a matter of using these expressions for reflection coefficient, together with the appropriate result for effective width w from (2.50) or (2.51), in equation (2.60).

The result for attenuation is especially simple if we consider the case of a 'weakly-guiding' dielectric slab. By the term 'weakly-guiding' (Gloge, 1971) we mean that the refractive indices of the three layers are not too dissimilar, or to be specific:

$$(n_1 - n_j) \ll n_1 \quad (j = 2, 3). \tag{2.69}$$

For this special case, which will be discussed in more detail later in the chapter, both polarizations yield the same results to a good degree of approximation. Thus, of the equations derived earlier in this section, we find that (2.48) reduces to (2.47) for the eigenvalue equation, (2.51) becomes identical with (2.50) for the effective width, and similarly (2.68) reduces to be the same as (2.67) for the reflection coefficient. For this situation of approximate TE/TM degeneracy, re-writing (2.67) in terms of p, q, r as defined in Section 2.2.1, equation (2.60) becomes:

$$\alpha \simeq \frac{\dfrac{2q^2(\alpha_2-\alpha_1)}{p(p^2+q^2)} + \dfrac{2q^2(\alpha_3-\alpha_1)}{r(r^2+q^2)}}{\left(2a+\dfrac{1}{p}+\dfrac{1}{r}\right)} + \alpha_1$$

$$= \frac{\left(2a+\dfrac{r}{q^2+r^2}+\dfrac{p}{p^2+q^2}\right)\alpha_1 + \left(\dfrac{q^2}{p(p^2+q^2)}\right)\alpha_2 + \left(\dfrac{q^2}{r(r^2+q^2)}\right)\alpha_3}{\left(2a+\dfrac{1}{p}+\dfrac{1}{r}\right)}. \quad (2.70)$$

Equation (2.70) is the simplest form of the expression for attenuation of a ray in an asymmetric low-loss weakly-guiding dielectric slab guide. We shall see later that the coefficients of $\alpha_j (j=1, 2, 3)$ occurring in (2.70) may be interpreted simply as the relative proportions of power travelling in each region j.

2.3 Electromagnetic Mode Treatment of the Dielectric Slab Waveguide: Guided Modes

In this section we will be concerned with finding the guided modes of the slab waveguide directly from Maxwell's equations. We will also show that the eigenvalue equation which results is the same as that derived by the ray treatment in the previous section (equations (2.47) and (2.48)). Once again we consider the (lossless) asymmetric dielectric slab shown in Figures 2.5 and 2.7. Writing Maxwell's equations in terms of the refractive index $n_j(j=1, 2, 3)$ of the three layers, and assuming the magnetic permeability is everywhere the same as that of free space μ_0, we have:

$$\nabla \times \mathbf{H} = n_j^2 \varepsilon_0 \frac{d\mathbf{E}}{dt} \quad (2.71)$$

$$\nabla \times \mathbf{E} = -\mu_0 \frac{d\mathbf{H}}{dt} \quad (2.72)$$

$$\nabla \cdot \mathbf{E} = 0 \quad (2.73)$$

$$\nabla \cdot \mathbf{H} = 0 \quad (2.74)$$

If we apply the curl operator to equation (2.72):

$$\nabla \times (\nabla \times \mathbf{E}) = -\mu_0 \nabla \times \left(\frac{d\mathbf{H}}{dt}\right) = -\mu_0 n_j^2 \varepsilon_0 \frac{d^2\mathbf{E}}{dt^2} \quad (2.75)$$

where equation (2.71) has been used to eliminate H. To simplify (2.75) further, we use the vector identity

$$\nabla \times (\nabla \times \mathbf{A}) = \nabla(\nabla \cdot \mathbf{A}) - \nabla^2 \mathbf{A}$$

where A is any vector. With the aid of this result and equation (2.73), we obtain from (2.75):

$$\nabla^2 \mathbf{E} = \mu_0 \varepsilon_0 n_j^2 \frac{d^2\mathbf{E}}{dt^2}.$$

Writing this in the more familiar notation (assuming $\mathbf{E} \propto \exp(-i\omega t)$) with $k^2 = \omega^2 \mu_0 \varepsilon_0$:

$$\nabla^2 \mathbf{E} + k^2 n_j^2 \mathbf{E} = 0 \tag{2.76}$$

which is the familiar wave equation for a uniform dielectric with refractive index n_j. We may simplify this result by noting that since we are only concerned with a two-dimensional situation all derivatives with respect to y will be zero. This is equivalent to assuming the guide extends to infinity in the positive and negative y-directions, so that the field distributions of the modes are uniform throughout the y-direction. If we further assume a z-dependence of the form $\exp(i\beta z)$, with β as the longitudinal propagation constant, (2.76) may be re-written for the three regions of the guide:

Region 3:
$$\frac{\partial^2 \mathbf{E}_3}{\partial x^2} - r^2 \mathbf{E}_3 = 0 \tag{2.77a}$$

Region 1:
$$\frac{\partial^2 \mathbf{E}_1}{\partial x^2} + q^2 \mathbf{E}_1 = 0 \tag{2.77b}$$

Region 2:
$$\frac{\partial^2 \mathbf{E}_2}{\partial x^2} - p^2 \mathbf{E}_2 = 0 \tag{2.77c}$$

where the previous notation of Section 2.2 has been used:

$$q^2 = n_1^2 k^2 - \beta^2, \quad p^2 = \beta^2 - n_2^2 k^2, \quad r^2 = \beta^2 - n_3^2 k^2.$$

Similar forms of the wave equation in the three regions may easily be derived for the magnetic field $\mathbf{H}$ from Maxwell's equations (2.71)–(2.74).

2.3.1 Transverse electric (TE) guided modes

Our assumption $\partial/\partial y = 0$ above implies (via Maxwell's equations (2.71)–(2.74)) that the only nonzero field components for TE modes are E_y, H_x, and H_z. For convenience we will omit the time- and z-dependent factor $\exp(i\beta z - i\omega t)$, so that equation (2.72) yields:

$$H_x = -\frac{\beta}{\omega \mu_0} E_y \tag{2.78}$$

$$H_z = -\frac{i}{\omega \mu_0} \frac{\partial E_y}{\partial x}. \tag{2.79}$$

Equations (2.78) and (2.79) therefore express the two nonzero magnetic field components in terms of the single nonzero electric field component E_y, which itself is given by the solution of wave equations in each region; these equations are of the form of (2.77a–c). The other requirement to be satisfied by these field components is that the tangential components E_y, H_z should be continuous at the interfaces 1–2 and 1–3 between the dielectric layers. Let us choose the origin of the x-axis at the 1–3 interface, so that the 1–2 interface is $x = -2a$ (see Figure 2.5).

For guided modes we require that the power be confined largely to the central layer of the guide, i.e. region 1. The form of equations (2.77a–c) then implies that this requirement will be satisfied for an oscillatory solution in region 1 ($q^2 \geq 0$) with evanescent 'tails' in the cladding regions 2 and 3 ($p^2, r^2 \geq 0$). Combining these conditions on β, we find we have derived inequalities identical to (2.45):

$$n_1 k \geq \beta \geq n_2 k \geq n_3 k.$$

From the above considerations we may immediately write down the solutions for E_y (omitting the t- and z-dependence) in the three regions for a guided mode:

$$E_y = \begin{cases} A e^{-rx} & x \geq 0 \quad (2.80a) \\ A \cos qx + B \sin qx & 0 \geq x \geq -2a \quad (2.80b) \\ (A \cos 2aq - B \sin 2aq) e^{p(x+2a)} & -2a \geq x. \quad (2.80c) \end{cases}$$

The form of equations (2.80a–c) has been chosen so that the requirement of continuity of E_y at $x = 0$ and $x = -2a$ is already satisfied explicitly. To complete the boundary requirements, it merely remains to ensure continuity of H_z. This component is given, from (2.79), as:

$$H_z = \frac{-i}{\omega \mu_0} \begin{cases} -rA e^{-rx} & x \geq 0 \quad (2.81a) \\ q(-A \sin qx + B \cos qx) & 0 \geq x \geq -2a \quad (2.81b) \\ p(A \cos 2aq - B \sin 2aq) e^{p(x+2a)} & -2a \geq x. \quad (2.81c) \end{cases}$$

The continuity condition yields the two equations

at $x = 0$: $\quad -rA = qB$

at $x = -2a$: $\quad q(A \cos 2aq + B \sin 2aq) = p(A \cos 2aq - B \sin 2aq).$

Eliminating the ratio A/B from these equations yields

$$\tan(2aq) = \frac{q(p+r)}{q^2 - pr} \quad (2.82)$$

which is the eigenvalue equation for TE modes. It is identical with equation (2.47), which was derived by the techniques of geometrical optics.

2.3.2 Transverse magnetic (TM) guided modes

For this polarization the only nonzero field components are H_y, E_x, and E_z. Again omitting the factor $\exp(i\beta z - i\omega t)$, equation (2.71) yields:

$$E_x = \frac{\beta}{\omega n_j^2 \varepsilon_0} H_y \quad (j = 1, 2, 3) \quad (2.83)$$

$$E_z = \frac{i}{\omega n_j^2 \varepsilon_0} \frac{\partial H_y}{\partial x} \quad (j = 1, 2, 3). \quad (2.84)$$

These equations relate the electric field components E_x, E_z to the only nonzero magnetic field component H_y which itself is a solution of wave equations in the three regions similar to equations (2.77a–c). As in the case of guided TE modes discussed above, simple considerations of the wave equation lead again to the condition (2.45) on the longitudinal propagation constant β. Similarly, it follows that the solution for H_y may be written:

$$H_y = \begin{cases} C e^{-rx} & x \geq 0 \quad (2.85a) \\ C \cos qx + D \sin qx & 0 \geq x \geq -2a \quad (2.85b) \\ (C \cos 2aq - D \sin 2aq) e^{p(x+2a)} & -2a \geq x. \quad (2.85c) \end{cases}$$

From (2.84) it follows that E_z is given by

$$E_z = \frac{i}{\omega\varepsilon_0} \begin{cases} \dfrac{-rC}{n_3^2} e^{-rx} & x \geqslant 0 \quad\quad (2.86a) \\ \dfrac{q}{n_1^2}(-C\sin qx + D\cos qx) & 0 \geqslant x \geqslant -2a \quad\quad (2.86b) \\ \dfrac{p}{n_2^2}(C\cos 2aq - D\sin 2aq)e^{p(x+2a)} & -2a \geqslant x. \quad\quad (2.86c) \end{cases}$$

Continuity of E_z at $x = 0$ and $x = -2a$ gives:

$$\frac{-rC}{n_3^2} = \frac{qD}{n_1^2}$$

and

$$\frac{q}{n_1^2}(C\sin 2aq + D\cos 2aq) = \frac{p}{n_2^2}(C\cos 2aq - D\sin 2aq).$$

Eliminating the ratio C/D from these equations yields

$$\tan(2aq) = \frac{(n_3^2 p + n_2^2 r)n_1^2 q}{n_2^2 n_3^2 q^2 - n_1^4 pr} \quad\quad (2.87)$$

which is the eigenvalue equation for TM modes, and is identical with (2.48) derived by a ray approach.

2.3.3 Mode numbers and cut-offs

The notation TE_N (and similarly TM_N) is used here to refer to a mode possessing N nodes in the field distribution (cf. the notation of Section 1.2 and Figure 1.7 in the case of conducting-wall guides). The value of N is obtained by taking the argument of the tangent in the eigenvalue equation (2.82) or (2.87) to be $(2aq - N\pi)$, as was the case in (2.47) or (2.48). The cut-off condition is always given by

$$\beta = kn_2$$

which corresponds in the ray picture to loss of total internal reflection, and in the modal analysis to loss of optical confinement and field-spreading throughout region 2.

With the aid of the two definitions given above we may find expressions for the cut-off frequencies for the TE and TM guided modes of the slab waveguide. For the TE case, equation (2.82) yields (substituting the appropriate expressions for p, q, r at cut-off):

$$\tan(2ak_c(n_1^2 - n_2^2)^{1/2} - N\pi) = \left(\frac{n_2^2 - n_3^2}{n_1^2 - n_2^2}\right)^{1/2}$$

where k_c is the wave number corresponding to cut-off. If we define the *normalized frequency*, v:

$$v = ak(n_1^2 - n_2^2)^{1/2} \quad\quad (2.88)$$

then the cut-off value v_c may be written from above as

TE: $\quad\quad v_c = \tfrac{1}{2}\tan^{-1}\left[\left(\dfrac{n_2^2 - n_3^2}{n_1^2 - n_2^2}\right)^{1/2}\right] + \dfrac{N\pi}{2} \quad\quad (2.89)$

where $\tan^{-1}$ is restricted to the range $0-\pi/2$. This relation may also be used as a method of counting *the number of guided TE modes*. If we allow for the first mode being designated $N = 0$, then for a general normalized frequency v equation (2.89) yields for the number M of guided modes:

TE:
$$M = \left\{\frac{1}{\pi}\left(2v - \tan^{-1}\left[\left(\frac{n_2^2 - n_3^2}{n_1^2 - n_2^2}\right)^{1/2}\right]\right)\right\}_{\text{int}} \quad (2.90)$$

where the subscript 'int' indicates the next largest integer.

The corresponding results for TM modes are as follows; for the cut-off frequency (from (2.87)):

TM:
$$v_c = \tfrac{1}{2}\tan^{-1}\left[\left(\frac{n_1}{n_3}\right)^2\left(\frac{n_2^2 - n_3^2}{n_1^2 - n_3^2}\right)^{1/2}\right] + \frac{N\pi}{2} \quad (2.91)$$

and for the number M of guided modes:

TM:
$$M = \left\{\frac{1}{\pi}\left(2v - \tan^{-1}\left[\left(\frac{n_1}{n_3}\right)^2\left(\frac{n_2^2 - n_3^2}{n_1^2 - n_2^2}\right)^{1/2}\right]\right)\right\}_{\text{int}}. \quad (2.92)$$

2.3.4 Normalization in terms of power flow

The time-averaged power flow, P, in the guide is given by the integral over the guide cross-section of the z-component of the Poynting vector (S_z):

$$P = \int_{-\infty}^{\infty} S_z \, dx = \tfrac{1}{2}\int_{-\infty}^{\infty} \text{Re}(\mathbf{E} \times \mathbf{H}^*)_z \, dx. \quad (2.93)$$

For TE modes, S_z is given by

$$S_z = -\tfrac{1}{2}E_y H_x^* = \frac{\beta}{2\omega\mu_0}|E_y|^2 \quad (2.94)$$

where the complex conjugate of equation (2.78) has been used. The integral in equation (2.93) must be split into three parts corresponding to the three layers. If we denote the power in each region by $P_j (j = 1, 2, 3)$, then we obtain for TE modes (using (2.80a–c)):

$$P_3 = \left(\frac{\beta}{2\omega\mu_0}\right)\frac{A^2}{2r} \quad (2.95a)$$

$$P_1 = \left(\frac{\beta}{2\omega\mu_0}\right)\frac{A^2}{2}\left(\frac{q^2 + r^2}{q^2}\right)\left[2a + \frac{p}{q^2 + p^2} + \frac{r}{q^2 + r^2}\right] \quad (2.95b)$$

$$P_2 = \left(\frac{\beta}{2\omega\mu_0}\right)\frac{A^2}{2p}\left(\frac{q^2 + r^2}{p^2 + q^2}\right). \quad (2.95c)$$

In deriving equations (2.95b) and (2.95c) in this simple form the boundary condition relating A and B has been used from Section 2.3.1, together with the eigenvalue equation (2.82). It follows that the total power P is given by

$$P = P_1 + P_2 + P_3 = \left(\frac{\beta}{2\omega\mu_0}\right)\frac{A^2}{2}\left(\frac{q^2 + r^2}{q^2}\right)\left[2a + \frac{1}{p} + \frac{1}{r}\right]. \quad (2.96)$$

The final term in (2.96) is recognized as the quantity $w_\perp$, termed an 'effective guide width' in Section 2.2.2, equation (2.50), where the expression was derived from a ray argument. On the basis of the modal analysis here, it is easily seen from (2.80a–c) that this quantity is the distance between points in regions 2 and 3 at which the field intensities have fallen to $1/e$ of their values at the respective interfaces, i.e. $w_\perp$ is indeed a measure of the field-spreading into the cladding layers, as asserted at the end of Section 2.2.2.

We may use equation (2.96) to normalize the modal fields of (2.80a–c) in terms of the power flow P:

$$A^2 \equiv \left(\frac{qB}{r}\right)^2 = \frac{P}{\left(\dfrac{\beta}{2\omega\mu_0}\right)\left(\dfrac{q^2+r^2}{2q^2}\right)\left(2a+\dfrac{1}{p}+\dfrac{1}{r}\right)}. \tag{2.97}$$

Furthermore, equations (2.95) and (2.96) may be used to obtain relatively simple expressions for the ratios (i) of power in the core (P_{core}) to total power P, and (ii) of power in the cladding layers (P_{clad}) to P (W. W. Anderson, 1965):

$$\frac{P_{\text{core}}}{P} = \frac{2a + \dfrac{p}{q^2+p^2} + \dfrac{r}{q^2+r^2}}{2a + \dfrac{1}{p} + \dfrac{1}{r}} \tag{2.98}$$

$$\frac{P_{\text{clad}}}{P} = \frac{\dfrac{q^2}{p(q^2+p^2)} + \dfrac{q^2}{r(q^2+r^2)}}{2a + \dfrac{1}{p} + \dfrac{1}{r}}. \tag{2.99}$$

From the preceding equations it is easily seen that the ratios P_j/P ($j = 1, 2, 3$) are the same as the coefficients of the losses α_j in the attenuation for a ray in a weakly-guiding lossy waveguide as given in equation (2.70). In other words, (2.70) may be re-written as:

$$\alpha = \frac{P_1\alpha_1 + P_2\alpha_2 + P_3\alpha_3}{P}.$$

This result may also be proved directly from the eigenvalue equation by a perturbation technique.

For TM modes, the z-component of the Poynting vector is given by

$$S_z = \tfrac{1}{2} H_y^* E_x = \frac{\beta}{2\omega n_j^2 \varepsilon_0}|H_y|^2 \quad (j = 1, 2, 3) \tag{2.100}$$

where equation (2.83) has been used. In this case the power confined in each region, P_j, is given, from (2.85a–c), by

$$P_3 = \left(\frac{\beta}{2\omega n_3^2 \varepsilon_0}\right)\frac{C^2}{2r} \tag{2.101a}$$

$$P_1 = \left(\frac{\beta}{2\omega n_1^2 \varepsilon_0}\right)\frac{C^2}{2}\left(\frac{q^2 n_3^4 + r^2 n_1^4}{q^2 n_3^4}\right)\left[2a + \frac{pn_1^2 n_2^2}{q^2 n_2^4 + p^2 n_1^4} + \frac{rn_1^2 n_3^2}{q^2 n_3^4 + r^2 n_1^4}\right] \tag{2.101b}$$

$$P_2 = \left(\frac{\beta}{2\omega n_2^2 \varepsilon_0}\right)\frac{C^2}{2p}\left(\frac{n_2^4}{n_3^4}\right)\left(\frac{q^2 n_3^4 + r^2 n_1^4}{q^2 n_2^4 + p^2 n_1^4}\right) \tag{2.101c}$$

where the C/D relationship from Section 2.3.2 and the eigenvalue equation (2.87) have been used. It follows that the total power P is given by

$$P = P_1 + P_2 + P_3$$
$$= \left(\frac{\beta}{2\omega\varepsilon_0}\right)\frac{C^2}{2}\left(\frac{q^2n_3^4 + r^2n_1^4}{q^2n_1^2n_3^4}\right)\left[2a + \frac{n_1^2n_2^2(p^2+q^2)}{p(q^2n_2^4 + p^2n_1^4)} + \frac{n_1^2n_3^2(q^2+r^2)}{r(q^2n_3^4 + r^2n_1^4)}\right].$$
(2.102)

As in the TE case, here the final term in (2.102) is recognized as the effective guide width $w_\parallel$ defined for TM modes in equation (2.51).

Equation (2.102) may be used to normalize the modal fields of (2.85a–c) in terms of the power flow P:

$$C^2 \equiv \left(\frac{qn_3^2 D}{rn_1^2}\right)^2 = \frac{P}{\left(\dfrac{\beta}{2\omega\varepsilon_0}\right)\left(\dfrac{q^2n_3^4 + r^2n_1^4}{2q^2n_1^2n_3^4}\right)\left[2a + \dfrac{n_1^2n_2^2(p^2+q^2)}{p(q^2n_2^4 + p^2n_1^4)} + \dfrac{n_1^2n_3^2(q^2+r^2)}{r(q^2n_3^4 + r^2n_1^4)}\right]}.$$
(2.103)

The corresponding ratios of power in the core (P_{core}) and power in the cladding (P_{clad}) to P are given by

$$\frac{P_{core}}{P} = \frac{2a + \dfrac{pn_1^2n_2^2}{q^2n_2^4 + p^2n_1^4} + \dfrac{rn_1^2n_3^2}{q^2n_3^4 + r^2n_1^4}}{2a + \dfrac{n_1^2n_2^2(q^2+p^2)}{p(q^2n_2^4 + p^2n_1^4)} + \dfrac{n_1^2n_3^2(q^2+r^2)}{r(q^2n_3^4 + r^2n_1^4)}}$$
(2.104)

$$\frac{P_{clad}}{P} = \frac{\dfrac{q^2n_1^2n_2^2}{p(q^2n_2^4 + p^2n_1^4)} + \dfrac{q^2n_1^2n_3^2}{r(q^2n_3^4 + r^2n_1^4)}}{2a + \dfrac{n_1^2n_2^2(q^2+p^2)}{p(q^2n_2^4 + p^2n_1^4)} + \dfrac{n_1^2n_3^2(q^2+r^2)}{r(q^2n_3^4 + r^2n_1^4)}}.$$
(2.105)

In conclusion, it is worth noting that the expressions for total power flow P from equations (2.96) and (2.102) may be cast into particularly simple forms in terms of the effective guide width w and the peak values of the field distributions. For the TE case, if the maximum value of E_y with respect to x is designated E_{ym}, then it is a simple matter to show that this occurs at

$$x = \frac{1}{q}\tan^{-1}\left(\frac{-r}{q}\right)$$

and that the value of E_{ym} is given by

$$\frac{E_{ym}}{A} = \frac{(q^2+r^2)^{1/2}}{q}.$$
(2.106)

Hence the total power flow P for TE modes may be simply represented, combining equations (2.96), (2.50), and (2.106), as (Kogelnik, 1975):

$$P = \frac{\beta}{4\omega\mu_0}E_{ym}^2 w_\perp \equiv \frac{E_{ym} H_{xm}}{4} w_\perp$$
(2.107)

where the symbol H_{xm} indicates the maximum value of H_x, and equation (2.78) has been used.

Similarly, for TM modes the maximum value H_{ym} of H_y occurs for

$$x = \frac{1}{q}\tan^{-1}\left(\frac{-rn_1^2}{qn_3^2}\right)$$

and is given by

$$\frac{H_{ym}}{C} = \frac{(q^2 n_3^4 + r^2 n_1^4)^{1/2}}{qn_3^2}. \tag{2.108}$$

Hence, from equations (2.102), (2.51), and (2.108), the total power flow P for TM modes may be simply expressed as (Kogelnik, 1975):

$$P = \frac{\beta}{4\omega\varepsilon_0 n_1^2} H_{ym}^2 w_\| \equiv \frac{E_{xm} H_{ym}}{4} w_\| \tag{2.109}$$

where E_{xm} is the maximum value of E_x, and equation (2.83) has been employed.

The relative simplicity of equations (2.107) and (2.109) for the power flows, and their obvious physical interpretation, is an elegant demonstration of the power of the effective width concept.

2.3.5 Weakly-guiding symmetric slab waveguide

In this subsection we shall show how the preceding results may be simplified in the case of a weakly-guiding symmetric slab guide, and give examples of the use of these results. The results will be expressed in terms of general normalized variables of the type introduced in Section 1.2. The assumption of a symmetric guide means that $n_3 = n_2$ in the structure of Figure 2.5. The term 'weakly-guiding' has already been encountered in Section 2.2.4, and is used to imply that n_1 and n_2 have reasonably similar values when compared with the magnitude of n_1, i.e. (2.69) becomes in the symmetric case:

$$n_1 - n_2 \ll n_1.$$

As we noted previously, this condition implies that the TE and TM modes become approximately degenerate, so that for the rest of this subsection we need only use the results for the TE case derived above. We shall see in future chapters that this situation provides at least a zeroth-order approximation for the meridional rays in a circular cylindrical dielectric guide. That is, if we consider a circular cylinder cut along a plane through the axis, the resulting waveguide structure will appear, in the plane of the cut, as a symmetrical slab guide, so that the meridional rays have at least a superficial resemblance to the rays in such a slab.

Consider first the eigenvalue equation for guided modes/rays. For the weakly-guiding symmetric slab, equations (2.82) and (2.87) reduce to

$$\tan(2aq) = \frac{2pq}{q^2 - p^2}. \tag{2.110}$$

To express this result in the most general form, we define the normalized variables v, u, and b; v is the normalized frequency already introduced in (2.88). Note that the definition of v here is half that of Kogelnik and Ramaswamy (1974); this is to agree with later notation for circular cylindrical waveguides. The remaining variables u and b are given by (Kogelnik and Ramaswamy, 1974; Gloge, 1971):

$$u^2 = a^2(k^2 n_1^2 - \beta^2) \equiv a^2 q^2 \tag{2.111}$$

$$b = 1 - \frac{u^2}{v^2} \equiv \frac{\beta^2 - k^2 n_2^2}{k^2(n_1^2 - n_2^2)} \equiv \frac{a^2 p^2}{v^2}. \tag{2.112}$$

Using u or b as the dependent variable, and v as the independent variable, equation (2.110) becomes either

$$\tan(2u) = \frac{2u(v^2 - u^2)^{1/2}}{2u^2 - v^2} \tag{2.113}$$

or

$$2v(1-b)^{1/2} = \tan^{-1}\left(\frac{2b^{1/2}(1-b)^{1/2}}{1-2b}\right) + N\pi \quad (N = 0, 1, 2, \ldots) \tag{2.114}$$

where N is the mode number. Equation (2.114) is the eigenvalue equation for the symmetric slab, expressed in terms of the normalized variables b and v, which permits all such guides to be described by a universal graph of b versus v, with one curve for each mode. A numerical solution yields plots such as those in Figure 2.8. In this notation, the cut-off condition is

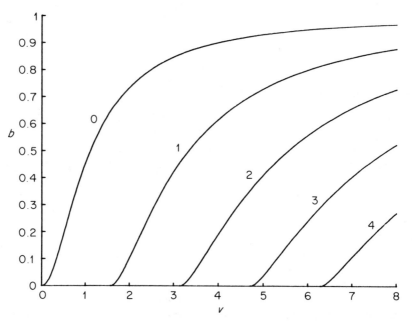

Figure 2.8 Normalized propagation constant b versus normalized frequency v for symmetric three-layer dielectric slab waveguide; labelling parameter gives the value of mode number N corresponding to the number of nodes in the field distributions

simply $b = 0$, whilst far from cut-off $b \to 1$. Hence for each mode the associated b–v curve begins at $b = 0$ and rises asymptotically to 1. The values of v for cut-off are given from equations (2.89) and (2.91) as

$$v_c = \frac{N\pi}{2} \quad (N = 0, 1, 2, \ldots) \tag{2.115}$$

where N is the mode number. The corresponding expression for the number M of propagating modes in a guide of normalized frequency v is given from (2.90) and (2.92) as

$$M = \left(\frac{2v}{\pi}\right)_{\text{int}}. \tag{2.116}$$

The parameter b defined in (2.112) above has a particularly simple interpretation as follows:

$$b = \frac{(\beta - kn_2)(\beta + kn_2)}{k^2(n_1 - n_2)(n_1 + n_2)} \simeq \frac{\beta - kn_2}{k(n_1 - n_2)}$$

since $k(n_1 - n_2) \ll k(n_1 + n_2) \simeq (\beta + kn_2) \gg (\beta - kn_2)$. Hence (Gloge, 1971)

$$\frac{\beta}{k} \simeq n_2 + b(n_1 - n_2). \tag{2.117}$$

We see that b is an effective propagation constant, since for the weakly-guiding situation there is a linear relationship (2.117) between β and b. Thus the b–v curves of Figure 2.8 are simply interpreted in terms of the effective index (β/k) versus wave number k for any given set of waveguide parameters a, n_1, n_2.

The effective guide width w for the symmetric weakly-guiding slab is given from equations (2.50) or (2.51) as

$$w = 2\left(a + \frac{1}{p}\right).$$

This may be normalized (Kogelnik and Ramaswamy, 1974) by multiplying both sides by $k(n_1^2 - n_2^2)^{1/2}$; denoting the normalized width by W, we find:

$$W = 2(v + b^{-1/2}). \tag{2.118}$$

Plots of W versus v for several modes are given in Figure 2.9. For large v, the effects associated

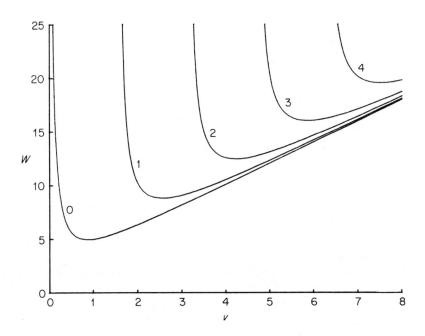

Figure 2.9 Normalized effective width W, defined in equation (2.118), as a function of normalized frequency v for some low-order modes of the symmetric slab waveguide

with the Goos–Haenchen shift and associated field penetration into the cladding layers become relatively unimportant and W tends to the asymptotic expression $2(v+1)$. For values of v close to cut-off of each mode, on the other hand, the field can spread substantially into the cladding layers so that W shows a characteristic sharp increase close to the cut-off of each mode ($b \to 0$ in (2.118)). Hence there is a fairly broad minimum in the W–v plot for each mode. For the lowest-order mode ($N = 0$) the absolute minimum occurs for $v = 0.865$, when $W = 4.93$.

The group velocity, V_g, and the group delay (ray transit time = length/V_g) may be found by differentiating (2.117):

$$\frac{d\beta}{dk} = m_2 + (n_1 - n_2)\frac{d(bk)}{dk} + b\frac{d(n_1 - n_2)}{dk} \tag{2.119}$$

where $m_2 \equiv (n_2 + k dn_2/dk)$ is the group index of the cladding, defined in an analogous way to that of dielectric 1 in equation (2.54). The group delay τ_g is then given, with L as the guide length, by

$$\tau_g = \frac{L}{V_g} = \frac{L}{c}\frac{d\beta}{dk} \tag{2.120}$$

where c is the speed of light in vacuo. In equation (2.119), the first term (m_2) is the same for all modes and characterizes the material dispersion effects. The second term on the r.h.s. is the principal mode-dependent term and is usually called the waveguide dispersion. The final term expresses the effects of different material dispersions in core and cladding materials and is related to the so-called 'profile dispersion' effect which is of importance in graded-index guides: this term was first included explicitly in group delay calculations by Arnaud (1974). However, for most materials commonly employed in slab waveguides this term is relatively small, and usually entirely dominated by the waveguide term. In addition, for many materials $k dn/dk \ll n$, so that m_2 in (2.119) may be replaced by n_2. If products of order $(n_1 - n_2)k dn/dk$ may also be ignored, we obtain the much-simplified result (Gloge, 1971):

$$\frac{d\beta}{dk} \simeq n_2 + (n_1 - n_2)\frac{d(vb)}{dv}. \tag{2.121}$$

Hence the waveguide dispersion is governed by the simple expression $d(vb)/dv$. For the weakly-guiding symmetric slab, db/dv is easily found by differentiating (2.114) (Kogelnik and Ramaswamy, 1974):

$$\frac{db}{dv} = \frac{2(1-b)}{(v+b^{-1/2})} \equiv \frac{4(1-b)}{W} \tag{2.122}$$

where equation (2.118) for W has been used.

Equation (2.122) may be employed to generate curves of $d(vb)/dv$ versus v of the type shown in Figure 2.10 for a number of modes. These curves contain all the waveguide dispersion information in a normalized form and may easily be used to find the group delay τ_g for any given slab guide, via equations (2.120) and (2.121). It is of interest to compare these results with those which would be obtained from the over-simplified zig-zag ray model of Figure 2.5. This model is deficient in neglecting the Goos–Haenchen shifts at the core–cladding interfaces; the group velocity may therefore be computed from equation (2.56) by omitting the Goos–Haenchen terms z_{s12}, z_{s13}, τ_{s12}, τ_{s13}, or (equivalently) by differentiating the eigenvalue equation (1.36) for the conducting-wall guide:

$$\frac{db}{dv} = \frac{2(1-b)}{v} \tag{2.123}$$

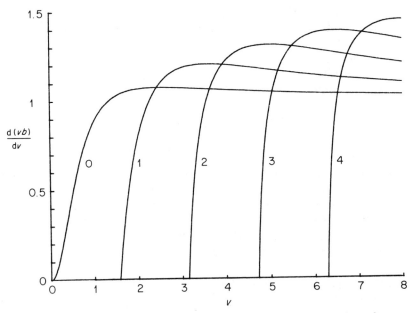

Figure 2.10 Normalized dispersion parameter d(vb)/dv as a function of normalized frequency v for some low-order modes of the symmetric slab waveguide

which is, of course, identical to replacing W by $2v$ in (2.122). Plots of the waveguide dispersion parameter d(vb)/dv for this over-simplified model are given in Figure 2.11. As we anticipate from our earlier discussion of the deficiencies of this model, the curves are in strong

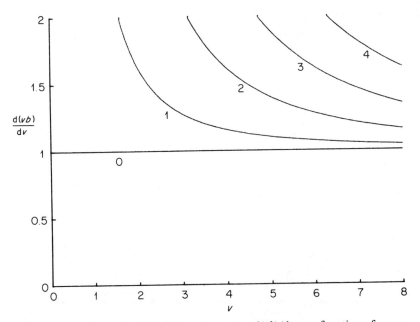

Figure 2.11 Normalized dispersion parameter d(vb)/dv as a function of normalized frequency v for a simplified waveguide model omitting the Goos–Haenchen shifts

disagreement with those of Figure 2.10 in the region of each mode cut-off, but tend asymptotically to resemble the corresponding curves far away from cut-off. Comparison of Figures 2.10 and 2.11 demonstrates in a particularly dramatic manner the importance of including the Goos–Haenchen shifts in a ray treatment of slab waveguides. Inspection of (2.122) and Figure 2.10 also shows that the difference in transit times $\Delta\tau$ between the fastest and slowest rays in a multimode slab guide can be of order

$$\Delta\tau \simeq \frac{L}{c}(n_1 - n_2). \tag{2.124}$$

Hence a multimode guide could have the effect of elongating a short pulse (in the time domain) of radiation transmitted over a length of the guide, as a result of this delay difference. This effect is of considerable importance in the use of dielectric waveguides for long-distance optical communications, since it imposes a limitation on the bandwidth and repeater separation to be employed in such systems. Although the waveguides for this application would usually be circularly symmetric, rather than the slabs considered here, we shall see later that results for pulse-spreading of the order of (2.124) still apply for homogeneous-core optical fibres.

As regards the proportion of power flow occurring in the core and cladding regions, equations (2.98) and (2.104) reduce for the weakly-guiding symmetric case to

$$\frac{P_{\text{core}}}{P} = \frac{2(v + b^{1/2})}{W} \tag{2.125}$$

and (2.99) and (2.105) become

$$\frac{P_{\text{clad}}}{P} = \frac{2(1-b)}{Wb^{1/2}}. \tag{2.126}$$

Plots of these quantities versus v are given in Figure 2.12. Far from cut-off of each mode, we

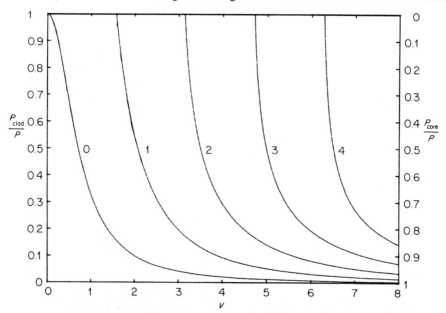

Figure 2.12 Power ratios P_{core}/P and P_{clad}/P, from equations (2.125) and (2.126), versus normalized frequency v for the symmetric slab waveguide

see that the power is concentrated in the core region. As cut-off is approached, the power spreads out into the cladding regions until guidance is lost completely at cut-off. For a slightly lossy waveguide we may use the theory of Section 2.2.4, together with our identification of the power ratios, to express the modal attenuation coefficient α as

$$\alpha = \alpha_1 \frac{P_{\text{core}}}{P} + \alpha_2 \frac{P_{\text{clad}}}{P} \qquad (2.127)$$

where α_1, α_2 are the losses in core and cladding, respectively. An alternative notation which is frequently used (especially in the laser literature) is to define a 'radiation confinement factor' $\Gamma = P_{\text{core}}/P$ (Hayashi et al., 1971) so that equation (2.127) becomes

$$\alpha = \alpha_1 \Gamma + \alpha_2 (1 - \Gamma). \qquad (2.128)$$

This result is of great practical importance in calculating threshold currents for semiconductor injection lasers.

Finally in this subsection, we prove an interesting and useful relationship between phase velocity, group velocity, and power confinement factor, neglecting material dispersion effects. The phase velocity V_p is defined as

$$V_p = \frac{ck}{\beta}. \qquad (2.129)$$

The group velocity V_g is defined as

$$V_g = c \frac{dk}{d\beta} \qquad (2.130)$$

where the derivative is conveniently found from (2.112):

$$\left(\frac{\beta}{k}\right)^2 = n_2^2 + b(n_1^2 - n_2^2) \qquad (2.131)$$

$$\frac{d\beta}{dk} = \frac{\beta}{k} + \frac{vk(n_1^2 - n_2^2)}{2\beta} \frac{db}{dv}$$

$$= \frac{\beta}{k} + \frac{vk(n_1^2 - n_2^2)}{\beta} \frac{2(1-b)}{W} \qquad (2.132)$$

where (2.122) has been used for db/dv. Hence, from (2.129)–(2.132),

$$\frac{c^2}{V_p V_g} = \frac{\beta}{k} \frac{d\beta}{dk} = bn_1^2 + n_2^2(1-b) + \frac{v(n_1^2 - n_2^2)2(1-b)}{W}$$

$$= n_1^2 \frac{2(v + b^{1/2})}{W} + n_2^2 \frac{2(1-b)}{Wb^{1/2}}.$$

We immediately recognize the coefficients of n_1^2 and n_2^2, from (2.125) and (2.126), as P_{core}/P and P_{clad}/P, respectively, so that in terms of the confinement factor Γ we find

$$\frac{c^2}{V_p V_g} = n_1^2 \Gamma + n_2^2 (1 - \Gamma). \qquad (2.133)$$

Equation (2.133) forms a neat counterpart to (2.128) as representing useful relationships between waveguide properties which are amenable to measurement and/or calculation (see also Buus, 1980). However, equation (2.133) also represents a special case of a general result for dielectric waveguides of many different types, and we will return to this class of relationships in due course.

It should be emphasized once more that all the results of this subsection have been derived for the weakly-guiding symmetric waveguide. Extensions to the general asymmetric slab in terms of normalized variables are fairly obvious but somewhat more tedious. In particular, a normalized parameter to represent the dielectric asymmetry has been introduced by W. W. Anderson (1965) and Kogelnik and Ramaswamy (1974). The latter authors have given normalized plots of b versus v and W versus v for TE and TM modes for a range of asymmetries, thus extending the results of Figures 2.8 and 2.9 here. The analogous extension of Figure 2.12 for P_{core}/P versus v to the asymmetric case was also given recently (Adams, 1977).

2.3.6 Waveguiding properties of heterostructure injection lasers

In order to demonstrate the application of the preceding theory of slab waveguides, we take the semiconductor injection laser as an example. This device has an assured future as a source for optical communications systems using optical fibres as the transmission medium.

Lasing action in semiconductors was first reported in 1962 (Hall *et al.*, 1962; Nathan *et al.*, 1962; Quist *et al.*, 1962). At that time the device used consisted of gallium arsenide (typical dimensions 500 μm × 100 μm × 100 μm) incorporating a p–n junction, which could be made to emit laser radiation at a wavelength of about 0.9 μm by the application of a strong forward bias. Laser operation in the early days was usually at low temperatures, since very large currents were required to achieve lasing at room temperature. The room temperature threshold current density was then of the order of 10^5 A cm^{-2} which produced strong heating of the junction region and necessitated pulsed operation in order to allow time for heat dissipation.

Early in the history of the injection laser, it was realized that guidance of the emitted radiation in the junction plane played an important role in achieving efficient laser action (Yariv and Leite 1963; Bond *et al.*, 1963). At that time the mechanism giving rise to the change of dielectric permittivity at the p–n junction was uncertain. Whilst some (Yariv and Leite, 1963) argued that a real guiding action occurred via refractive index changes associated with a free-carrier plasma effect, others (Lasher, 1963; Hall and Olechna, 1963; McWhorter *et al.*, 1963) cited the change in imaginary part of the dielectric constant due to the existence of a region with gain sandwiched between two layers of lossy material. For a review of these mechanisms and a guide to the early literature the reader is referred to an article on the theory of the homostructure laser (Adams and Landsberg, 1969). Further study of this effect led to an improved understanding of the waveguiding action and to the realization that a real refractive index variation was desirable to achieve improved confinement of the recombination radiation to the active layer of the device. Following the suggestions of Kroemer (1963) and Alferov and Kazarinov (1963), the use of heterojunctions (rather than a single p–n junction) achieved this objective, resulting in dramatic decreases of current density required to achieve lasing action (Kressel and Nelson, 1969; Hayashi *et al.*, 1969; Panish *et al.*, 1969; Alferov *et al.*, 1969).

The light-emitting region and adjacent layers of the injection laser structure are shown schematically in Figure 2.13. The current is shown as a vertical with the various layers of the device in the horizontal plane. The light-emitting region, termed the 'active region', is the thin layer of width $2a$ which forms the central part of the 'sandwich' structure illustrated. Although the early lasers were usually fabricated by a diffusion process of the p-type dopant (usually zinc) into the n-type GaAs substrate, great improvements have been achieved by the

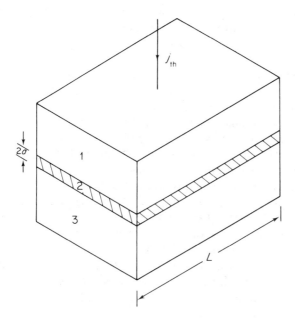

Figure 2.13 Schematic three-layer structure of the semiconductor injection laser. (This figure appeared in Vol. 4, pp. 273–280 of *Optics and Laser Technology* published by IPC Science and Technology Press Ltd, Guildford, Surrey, UK and is by M. J. Adams)

use of a liquid phase epitaxy (l.p.e.) technique. This process involves the epitaxial growth of successive layers of specified material on the n-type GaAs substrate. These layers can be of GaAs doped to the required level in the case of 'homostructure' devices, or of the alloy $Al_xGa_{1-x}As$ for the newer heterostructure lasers. The choice of $Al_xGa_{1-x}As$ was made on account of the near equality between lattice constants of the compounds GaAs and AlAs (the values are 565.35 pm and 563.90 pm respectively).

The interpretation of the three layers of Figure 2.13 is given for different structures in Table 2.1. In addition to the three regions shown, heterostructure lasers usually incorporate additional layers. On the n-side there is the GaAs substrate on which all the other layers have been grown, and on the p-side a further (optional) thin layer of GaAs may be added to improve the thermal and electrical contacting properties of the device. To complete the structure, contacts are added by evaporation or alloying, and the devices are cut to the desired dimensions. The partially reflecting mirrors which form the laser cavity are achieved simply

Table 2.1 Interpretation of the three layers of Figure 2.13 for various injection laser structures

Structure	1	2	3
Homostructure	p^+-GaAs	p-GaAs	n-GaAs
Single heterostructure	p^+-$Al_xGa_{1-x}As$	p-GaAs	n-GaAs
Double heterostructure	p^+-$Al_xGa_{1-x}As$	p-GaAs	n-$Al_xGa_{1-x}As$

by cleaving the crystal along a crystal plane (usually the (110) plane when the epitaxial layers are in the (100) plane). The refractive index of GaAs at the lasing wavelength is about 3.6, so that a simple GaAs–air interface gives a reflectivity of about 32 per cent, which is sufficient to form the laser cavity.

Under operating conditions, a forward bias is applied to the laser which causes electrons to be injected from the n-type material (region 3 of Table 2.1) into the p-type active region 2. Thus electrons and holes are both present simultaneously in the active region and, under favourable conditions, recombination takes place. The energy difference of the recombining carriers can be dissipated in several ways, e.g. to photons, phonons, or to other carriers by electron collision (Auger) effects. The successful operation of the injection laser depends principally on the fact that the great majority of this recombination traffic is radiative. It has been estimated that the efficiency of radiative recombination in GaAs can approach 100 per cent at low temperatures and is still very high at room temperature.

The epitaxial layers of the heterostructure laser help to improve its performance in two ways: by confining the carriers to the active layer and simultaneously confining the electromagnetic radiation to this region. The carrier confinement is achieved by potential 'steps' introduced at the layer interfaces, which confine both electrons and holes to the central layer. At the same time the material and doping changes at these interfaces alter the optical properties (refractive index and absorption coefficient) of the laser medium in such a way that a dielectric slab waveguide is produced.

The use of $Al_xGa_{1-x}As$ as the passive layers results in a refractive index discontinuity at the interfaces of typically 5–10 per cent, thus providing effective guidance and reducing the losses suffered by the laser radiation. The consequence of this reduced optical loss and improved carrier confinement is seen most strikingly in the improvement of threshold current density. Figure 2.14 shows the progress of this quantity achieved during the years 1962–1972, and illustrates the marked improvement occurring with the introduction of double heterostructure devices around 1970. Typical threshold current densities for these lasers are about $1000 \, A/cm^2$ or even less.

We may apply the preceding theory developed for the slab waveguide to the problem of designing a double heterostructure laser optimized for minimum threshold current density. To do this we must first derive the threshold condition for amplification of radiation within the Fabry–Perot cavity formed by the cleaved facets of the device. Consider the general cavity structure of Figure 2.15, consisting of two mirrors, reflectivity R_1, R_2, separated by a cavity length L parallel to the z-axis. If the propagation constant of a wave in the cavity is β and the nett gain per unit length is G, then the variation of the wave is described by $\exp(i\beta z + Gz)$. The condition for nett amplification is that after one round trip of the cavity the wave shall have the same form, i.e.

$$R_1 R_2 \exp(2i\beta L + 2GL) = 1.$$

Taking the phase and amplitude parts of this equation, and noting that β may be written as $\beta = 2\pi/\lambda_g$, where λ_g is the guide wavelength, we obtain

$$L = N_L \frac{\lambda_g}{2} \tag{2.134}$$

(where N_L is an integer) and

$$G = \frac{1}{2L} \ln\left(\frac{1}{R_1 R_2}\right). \tag{2.135}$$

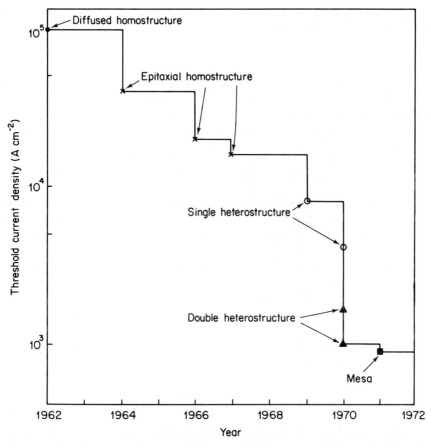

Figure 2.14 Progress with room-temperature threshold current density of injection lasers during the years 1962–1972. (This figure appeared in Vol. 4, pp. 273–280 of *Optics and Laser Technology* published by IPC Science and Technology Press Ltd, Guildford, Surrey, UK and is by M. J. Adams)

Equation (2.134) is the familiar result that the length should contain an integer number of half-wavelengths, whilst equation (2.135) is the threshold condition that the nett modal gain G should equal the end-loss through the mirrors. Now we use the expression for the modal

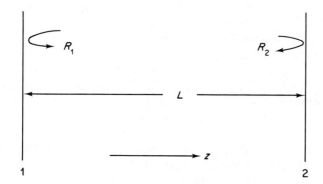

Figure 2.15 Schematic Fabry–Perot cavity structure

gain (i.e. negative attenuation) derived in Section 2.3.5 above for the symmetric slab guide, and expressed in equation (2.128). This may be re-written here (noting that $G = -\alpha$):

$$G = -\alpha_1 \Gamma - \alpha_2(1 - \Gamma) \tag{2.136}$$

where α_1, α_2 are the losses in core and cladding regions, respectively, and Γ is the radiation confinement parameter (sometimes also termed the 'filling factor'). For G to be a positive quantity, of course, there must be gain in the core region rather than loss. If this gain per unit length is $g \ (\equiv -\alpha_1)$ then equations (2.135) and (2.136) may be combined to give

$$g\Gamma = \alpha_2(1 - \Gamma) + \frac{1}{2L}\ln\left(\frac{1}{R_1 R_2}\right) \tag{2.137}$$

which is the conventional threshold relationship for the injection laser. It states that the gain experienced by that part of the mode occupying the active core region is exactly balanced by the loss suffered by the proportion of the mode in the lossy cladding regions, together with the end-loss through the cavity mirrors. The expressions derived in the previous subsection for the filling-factor Γ (e.g. equation (2.125)) may be used in (2.137) to calculate the gain required at lasing threshold.

In order to calculate the current density j required to produce a specified gain per unit length, g, in the active region, we need the gain–current relationship. The precise form of this depends on the details of the material properties of the active region. A general relationship (see Adams and Landsberg, 1969, for a detailed derivation) is as follows (Lasher, 1963; Yariv and Leite, 1963; Lasher and Stern, 1964):

$$j = \frac{8\pi e 2 a n_1^2 \Delta v}{\eta_i \lambda^2} \gamma g \tag{2.138}$$

where e is the electron charge, η_i is the internal quantum efficiency, λ is the wavelength of emission, Δv is the linewidth of the spontaneous emission, n_1 is the refractive index, and $2a$ is the guide width as defined previously. The remaining quantity γ in (2.138) is a 'demerit' factor introduced to account for line shape and temperature effects on the emission and is defined as

$$\gamma = \frac{R_{sp}}{h\Delta v r_{st}(\lambda)} \tag{2.139}$$

where h is Planck's constant, R_{sp} is the total spontaneous emission rate (summed over all wavelengths), and $r_{st}(\lambda)$ is the stimulated emission rate at the wavelength λ corresponding to peak gain g. The factor γ may be calculated for various empirical line shapes (Adams and Landsberg, 1969) or for appropriate models of the recombination transitions responsible for radiative emission. The results for the latter alternative can yield complicated expressions for γ, involving in general a further dependence on g, so that the simple linear dependence of (2.138) is lost. However, for a simple model involving band tails with an exponential density-of-states (Adams, 1969), the factor $\Delta v \gamma$ is rendered independent of g and the simple linear relationship is recovered:

$$j = C2ag \tag{2.140}$$

where C is a constant independent of guide half-width a and gain g. The form of (2.140) has also some experimental confirmation (Hakki, 1973; Hakki and Paoli, 1975) and further theoretical justification (Stern, 1973) for low doping levels.

Equations (2.137) and (2.140) may be combined to permit the explicit calculation of threshold current density j_{th}:

$$j_{th} = C2a\left[\alpha_2\frac{(1-\Gamma)}{\Gamma} + \frac{1}{2L\Gamma}\ln\left(\frac{1}{R_1R_2}\right)\right]. \quad (2.141)$$

For a double heterostructure laser the end-losses usually dominate over the cladding loss α_2. This is because (i) α_2 is largely due to free-carrier absorption which is of order $10\,\text{cm}^{-1}$ or less, and (ii) Γ is usually reasonably close to 1 for a well-designed structure, so that little power penetrates the cladding. In this situation equation (2.141) may be simplified to the form

$$j_{th} \simeq \frac{C2a}{2L\Gamma}\ln\left(\frac{1}{R_1R_2}\right) = D\left(\frac{Wv}{v+b^{1/2}}\right) \quad (2.142)$$

where D is independent of waveguide parameters, and equation (2.125) has been used for Γ. The variation of the threshold parameter v/Γ, as given in equation (2.142), is shown as a function of v for the lowest-order modes in Figure 2.16. Each curve approaches a linear

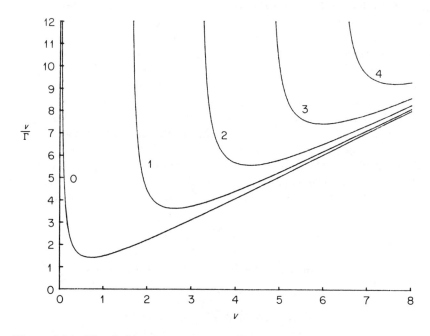

Figure 2.16 Threshold parameter v/Γ as a function of v for symmetric double heterostructure lasers

dependence on v asymptotically; near cut-off, however, there is a fairly broad minimum, with a sharp increase of threshold as cut-off is approached. These features are simply interpreted as follows: far away from cut-off, the fields are well confined to the central active layer and the threshold simply increases linearly with the width of this layer. Nearer cut-off the fields spread into the passive regions and this feature dominates the behaviour (Kressel et al., 1971) until at cut-off the fields spread uniformly throughout all space, Γ approaches zero, and the threshold becomes very large. It follows that for each mode there is an optimum v-value for minimum threshold; for the zeroth-order mode this value is $v = 0.71$ (Unger, 1971). Using the definition of v, equation (2.88), this yields a simple relationship between refractive indices, n_1, n_2, wavelength λ, and guide width $2a$:

$$\frac{2a}{\lambda} = \frac{0.227}{(n_1^2 - n_2^2)^{1/2}}.$$

Using typical values for a GaAs laser with 30 per cent Al content in the passive layers: $n_1 = 3.6$, $n_2 = 3.4$, $\lambda = 0.85 \,\mu\text{m}$, this yields an optimum active layer width of $0.16 \,\mu\text{m}$. It should be noted that to ensure single-mode operation the width should satisfy the following condition (from (2.115)):

$$\frac{2a}{\lambda} \leqslant \frac{0.5}{(n_1^2 - n_2^2)^{1/2}}$$

i.e. for the above numerical example, we must have $2a \leqslant 0.36 \,\mu\text{m}$ for operation in the zero-order mode.

Calculations of the filling-factor Γ for specific sets of numerical parameters have been presented by many authors (Hayashi et al., 1971; Casey et al., 1973; Hakki and Paoli, 1975; Dumke, 1975; Butler and Kressel, 1977); the generalized form presented in Figure 2.12 was given first for the zeroth order mode in injection lasers by the present author (Adams, 1977). Corresponding threshold calculations along the lines indicated above have also been performed with various degrees of sophistication and levels of experimental agreement (Thompson and Kirkby, 1973; Dyment et al., 1974; Kressel and Ettenberg, 1976; Nash et al., 1976; Casey, 1978).

Although minimization of threshold current represents one desirable aim in the design of heterostructure lasers, a secondary requirement has been the achievement of small angular spread of the emitted radiation. This has been achieved by the use of very narrow active layers (Kressel et al., 1971; Selway and Goodwin, 1972) so that the radiation is permitted to penetrate some distance into the passive layers, i.e. the filling-factor is small. For this situation it is useful to derive an approximate expression for the filling-factor of the lowest-order mode (Dumke, 1975). From equation (2.114) we have, for the zeroth-order mode:

$$\tan[v(1-b)^{1/2}] = \frac{b^{1/2}}{(1-b)^{1/2}}.$$

Hence, for very small v, a useful approximation for the eigenvalue is given by (Marcuse, 1974):

$$b \simeq v^2.$$

Using this approximation in the expression for Γ (equation 2.125),

$$\Gamma \simeq \frac{2v^2}{1+v^2}. \tag{2.143}$$

This expression is a reasonable approximation for v less than about 0.4; similar approximations have been derived by Dumke (1975) and Botez (1978). Figure 2.17 shows a plot of the accurate Γ versus v with, for comparison, the approximation (2.143) in the range $0 < v < 0.5$. It is clear that approximations of this type are useful for interpreting experimental data on devices with narrow active regions and/or low Al content in the passive layers (Kressel and Ettenberg, 1976). A general discussion of more approximations of this kind has been given by Botez (1978a), who has also derived analytic approximations for threshold currents (Botez, 1979).

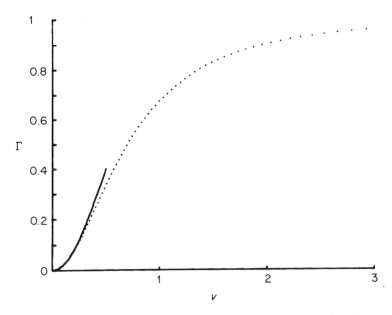

Figure 2.17 Filling-factor Γ versus normalized frequency v for the symmetric dielectric slab waveguide (lowest-order mode) Solid line, approximation from equation (2.143); dotted line, numerical solution

2.4 Effects of Loss and Gain

Thus far we have considered only guided modes in real dielectric waveguides, with the exception of some allowance for lossy media in Section 2.2.4. However, it is also possible to have wave-guidance in media whose real refractive index is the same, with the guiding action produced by changes in the imaginary part of the dielectric permittivity. This may be intuitively understood by thinking of a medium with gain sandwiched between two layers with loss, all three having the same refractive index. It is then clear that the central layer is more transparent than the outer layers and hence will guide an electromagnetic wave. The guidance occurs physically via two effects: (i) the gain continually creates radiation to replace that lost to the outer layers, and (ii) the change in imaginary part of dielectric permittivity across the layers implies a finite reflectivity at the boundaries. Although such a wave exhibits local regions of growing or decaying behaviour, it is still reasonable to speak of a guided mode of the waveguide since the overall attenuation (in the sense of (2.127), for example) can in principle be zero. However, we may on occasion with rather less precision also use the term 'guided mode' even for situations where the wave is subject to attenuation or loss.

2.4.1 Mode stability

To consider gain-guidance in more detail we define the complex dielectric permittivity ε_j of layer j (as in Section 2.2.4) as

$$\varepsilon_j = (n_j + iK_j)^2. \tag{2.144}$$

For simplicity let us consider only symmetric slab waveguides, so that we may define a normalized frequency v in analogy with the real case (equation (2.88)):

$$v^2 = a^2 k^2 (\varepsilon_1 - \varepsilon_2) \tag{2.145}$$

where a is again the guide half-width and k the wave number. With this definition the preceding theory for the real dielectric slab guide may be taken directly over for the complex case, on the understanding that all the relevant parameters now assume complex values. In particular, the eigenvalue equation for TE modes (equation (2.113) or (2.114) in normalized coordinates) still applies with complex b and u. A similar equation for TM modes may be written in terms of normalized variables and the new definitions of $\varepsilon_1, \varepsilon_2$; its derivation from (2.87) is straightforward and will not be given here.

Our definitions of complex permittivity and normalized frequency v above allow four possible combinations of relative values for refractive indices n_1, n_2 and extinction coefficients K_1, K_2. These may be listed as

(a) $n_1 \geqslant n_2$; $K_1 \geqslant K_2$
(b) $n_1 \geqslant n_2$; $K_1 \leqslant K_2$
(c) $n_1 \leqslant n_2$; $K_1 \geqslant K_2$
(d) $n_1 \leqslant n_2$; $K_1 \leqslant K_2$.

Strictly speaking all four possibilities will support guided modes (Marcuse, 1972). However, it is important to distinguish between 'guided' modes and 'stable' modes; a stable mode is one which is unaffected by perturbation. Although an unstable mode is a perfectly acceptable solution of Maxwell's equations and the boundary conditions for the system, if displaced slightly from the axis of the waveguide it will not return to its original configuration. Marcuse (1970, 1972) has shown by numerical computation that solutions exist for all four possibilities (a)–(d) listed above. However, not all of these solutions yield stable modes and the most general criterion for stability is the existence of a gain maximum on the axis of the guide, i.e. as in cases (a) and (c) of the above list. Whilst this condition is easily proved for guidance in a square-law medium (see Chapter 3) no simple proof exists for the dielectric slab guide considered here. Note also that for certain situations, e.g. case (b) with $n_1 > n_2 \gg K_2 > K_1$, this criterion is somewhat academic, since the real index guidance clearly dominates strongly over the gain anti-guidance. In this case we expect that the effect of a perturbation will only be felt after a very long distance of propagation in the guide, so that for many applications the requirement of gain maximum on-axis is unnecessarily restrictive. An analogous result of a somewhat more intuitive nature has been given by Schlosser (1973); we will derive this result now.

Define first another normalized variable w (Gloge, 1971):

$$w = vb^{1/2}. \tag{2.146}$$

This parameter should not be confused with the effective width of the slab guide defined in equation (2.49). Here w gives the variation of the field distributions in the outer layers of the guide as $\exp(-w|x|/a)$ (cf. equation (2.80)). It is complex-valued with real and imaginary parts w_r and w_i, respectively, the former governing the field decay and the latter determining the phase fronts. Since we consider only guided modes at present, it is clear that w_r must be positive in order to ensure the field's decay to zero as x tends to infinity (either positive or negative). In addition, for the mode to have stable characteristics it is desirable that the phase fronts have positive curvature in the direction of propagation of the mode (the negative z-direction throughout the present work). In other words the phase fronts represent a wave expanding about a point which advances in the direction of propagation. For this requirement to be satisfied it is necessary that w_i also be positive.

With the above definition of w, equation (2.113) may be re-written in two simpler forms corresponding to odd and even order modes:

Even modes: $\qquad\qquad\qquad w = u \tan u \qquad\qquad\qquad$ (2.147a)

Odd modes: $\qquad\qquad\qquad w = -u \cot u.\qquad\qquad\qquad$ (2.147b)

These equations are valid for the case of TE modes. Similar results hold for the TM case but will not be dealt with here in the interests of brevity; for the weakly-guiding low-loss case the results are identical. Now, writing $u = u_r + iu_i$, separating the real and imaginary parts of equation (2.147a), and using the conditions $w_r > 0$, $w_i > 0$, we find:

$$u_r \sin 2u_r - u_i \sinh 2u_i > 0 \qquad (2.148)$$
$$u_i \sin 2u_r + u_r \sinh 2u_i > 0. \qquad (2.149)$$

Multiplying (2.148) by u_r, (2.149) by u_i, and adding, it is easily seen that

$$\sin 2u_r > 0.$$

Since $u_i \sinh u_i$ is always positive, it follows from (2.148) that

$$u_r > 0.$$

Equation (2.149) now implies that

$$u_i > 0.$$

If we combine these two latter inequalities with the original conditions on w_r, w_i, we find

$$u_r u_i + w_r w_i > 0. \qquad (2.150)$$

Using the definition of w given in (2.146), together with that of b in (2.112), we conclude that

$$\mathrm{Im}(v^2) = \mathrm{Im}(w^2 + u^2) = 2(u_r u_i + w_r w_i) > 0 \qquad (2.151)$$

where Im signifies the imaginary part of the quantity in brackets.

Using the definitions (2.145) for v and (2.144) for ε_j, the inequality (2.151) can be written in the form (Schlosser, 1973):

$$n_1 K_1 - n_2 K_2 > 0. \qquad (2.152)$$

Equation (2.152) represents a necessary (but not sufficient) condition for stability of a guided mode in a slab waveguide of complex permittivity. The same result is found via an analogous argument for the odd modes from equation (2.147b). It provides a criterion for examining the four possible combinations of values of n_1, n_2, K_1, K_2 listed as (a)–(d) at the beginning of this subsection. The result may be reduced to particularly simple forms in the following cases:

(i) $K_1 = K_2$: Equation (2.152) now reads $n_1 > n_2$, which is the well-known condition (2.42) for real index guiding in a symmetric slab.

(ii) $n_1 = n_2$: equation (2.152) yields $K_1 > K_2$ which is the same as Marcuse's condition for mode stability discussed earlier in this subsection.

(iii) *Weak-guidance:* $n_1 = n_2 + \Delta n$, where $\Delta n \ll n_1, n_2$. In this case (2.152) becomes

$$\frac{\Delta K}{K_1} + \frac{\Delta n}{n_2} > 0 \qquad (2.153)$$

where $\Delta K = K_1 - K_2$. This is a particularly interesting case, since it shows that either gain-guided or index-guided modes can be stable *even in the presence of anti-guidance due to the opposite effect* provided the guiding action is strong enough. This is one way around the over-restrictive nature of the condition $K_1 > K_2$ discussed earlier in this subsection.

2.4.2 Mode cut-offs

From the remarks concerning the condition for a guided mode which were made above in Sections 2.3.3 and 2.4.1, it should be clear that this condition corresponds to the requirement of a field decaying continuously as we move away in the direction normal to the guide axis. In the limiting case therefore the cut-off condition is $w_r = 0$ or equivalently $\text{Re}(b) = 0$, where Re denotes the real part of the quantity in brackets. Applying this condition to the eigenvalue equations (2.147a) and (2.147b) for even and odd TE modes yields:

Even modes: $\qquad\qquad u_r \sin 2u_r - u_i \sinh 2u_i = 0 \qquad\qquad$ (2.154a)

Odd modes: $\qquad\qquad u_r \sin 2u_r + u_i \sinh 2u_i = 0.\qquad\qquad$ (2.154b)

If the real and imaginary parts of v (as defined in (2.145)) are denoted v_r and v_i, respectively, then the cut-off condition also produces the simple result

$$u_r u_i = v_r v_i. \qquad (2.155)$$

If we recast the eigenvalue equations (2.147a) and (2.147b) in a slightly different form:

Even modes: $\qquad\qquad w = v \sin u \qquad\qquad$ (2.156a)

Odd modes: $\qquad\qquad w = -v \cos u \qquad\qquad$ (2.156b)

then the cut-off condition, using (2.155) and (2.156) may be re-written:

Even modes: $\qquad\qquad v_r^2 = u_r u_i \cot u_r \tanh u_i \qquad\qquad$ (2.157a)

$\qquad\qquad\qquad\qquad v_i^2 = u_r u_i \tan u_r \coth u_i \qquad\qquad$ (2.157b)

Odd modes: $\qquad\qquad v_r^2 = -u_r u_i \tan u_r \tanh u_i \qquad\qquad$ (2.158a)

$\qquad\qquad\qquad\qquad v_i^2 = -u_r u_i \cot u_r \coth u_i. \qquad\qquad$ (2.158b)

Equations (2.154) and (2.157) or (2.158) may be solved numerically to find values of (v_r, v_i) pairs at which mode cut-offs occur in the slab guide of complex permittivity. One method of solution is to choose a value of u_r and solve equation (2.154a) or (2.154b), as appropriate, for u_i; equation (2.157) or (2.158) then yields the (v_r, v_i) values. The results are perhaps best expressed as mode-lines on a plot of modulus versus phase of v, as calculated by Suematsu and Yamada (1974) and Buus (1977). Figure 2.18 shows a plot of this kind for the first 5 modes of a slab waveguide (Buus, 1977). Along the vertical axis ($\arg(v) = 0$) the intercepts coincide with the v-values for mode cut-off in the case of real index-guidance (cf. Figure 2.8 and equation (2.115)). Note that the zero-order mode has no cut-off for values of $\arg(v)$ less than $\pi/4$, i.e. for v^2 lying in the first quadrant. This situation corresponds to a real index-guidance situation assisting the gain-guidance of the mode. For values of $\arg(v)$ greater than $\pi/4$, however, the lowest-order mode has a cut-off, indicating that in this case there is an index-antiguiding effect, i.e. an index depression, competing with the gain-guidance mechanism (Schlosser, 1973; Suematsu and Yamada, 1974; Buus, 1977). The line $\arg(v) = \pi/4$ corresponds to the case where there is no index discontinuity ($n_1 = n_2$) and pure gain-guiding provides the wave confinement mechanism. The line $\arg(v) = \pi/2$ is a region where no guided modes exist and may correspond to an inverted structure ($n_2 > n_1$) which supports only leaky modes. Further discussion of these hollow waveguides will be deferred to the appropriate place below.

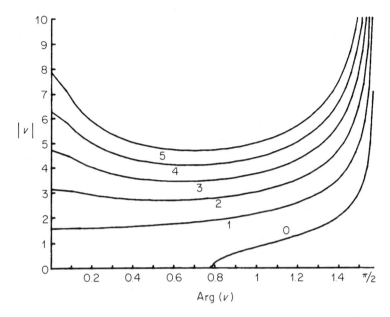

Figure 2.18 Mode cut-offs for the first six modes of symmetric slab waveguides of complex permittivity. (From Buus (1977). *Proc. 7th European Microwave Conference, Copenhagen, September, 1977*. pp. 29–33. Reproduced by permission of Microwave Exhibitions and Publishers Ltd)

2.4.3 Pure gain-guidance

One important case where normalized results of a rather general nature can be obtained for the dielectric slab of complex permittivity is in the situation of pure gain-guidance with no index discontinuity. If we consider only the case of relatively low loss or gain values so that $K_j \ll n_j (j = 1, 2)$ then TE and TM modes are again described by the same eigenvalue equation. The normalized frequency v in this situation becomes:

$$v = \frac{|v|(1+i)}{\sqrt{2}}. \tag{2.159}$$

To solve the eigenvalue equation for this case, it is convenient to re-cast equations (2.147) or (2.156) into the equivalent forms:

Even modes:
$$v = \frac{u}{\cos u} \tag{2.160a}$$

Odd modes:
$$v = \frac{u}{\sin u}. \tag{2.160b}$$

Combining equations (2.159) and (2.160) yields simple relations between u_r and u_i:

Even modes:
$$\frac{u_r - u_i}{u_r + u_i} = \tan u_r \tanh u_i \tag{2.161a}$$

Odd modes:
$$\frac{u_r - u_i}{u_r + u_i} = -\cot u_r \tanh u_i. \tag{2.161b}$$

Once again numerical solution of these equations is easily effected by choosing a value of u_r and computing values of u_i and $|v|$. The results may be expressed, as for the real dielectric slab waveguide, in terms of general plots of the normalized propagation constant b (equation (2.112)) versus $|v|$. In this case, of course, b is complex and its real and imaginary parts, denoted by b_r and b_i, are shown plotted as functions of $|v|$ in Figures 2.19 and 2.20 respectively for the first few modes (Schlosser, 1973).

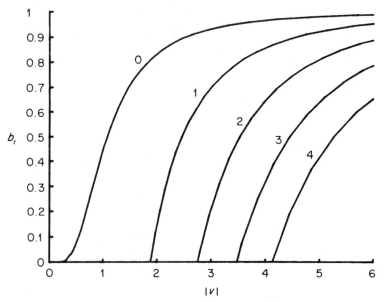

Figure 2.19 Real part b_r of the normalized propagation constant versus $|v|$ for the lowest five modes of a gain-guiding dielectric slab. (From Schlosser (1973). Copyright 1973, American Telephone and Telegraph Company, reprinted by permission)

The physical interpretation of these results for a normalized propagation constant is aided by expressing the actual propagation constant β in terms of b. For the relatively low loss and gain case considered here, the definition of b yields:

$$b = \frac{\beta^2 - \varepsilon_2 k^2}{(\varepsilon_1 - \varepsilon_2)k^2} \simeq \frac{\beta/k - (n_2 + iK_2)}{i(K_1 - K_2)}$$

where the definition of ε_j in equation (2.144) has been used together with the fact that $k(\varepsilon_1 - \varepsilon_2) \ll k(\varepsilon_1 + \varepsilon_2) \simeq (\beta + \varepsilon_2^{1/2} k) \gg (\beta - \varepsilon_2^{1/2} k)$. Hence the analogous gain-guiding result to equation (2.117) for the index guiding case becomes (Schlosser, 1973):

$$\frac{\beta}{k} \simeq n_2 - b_i(K_1 - K_2) + i[K_2 + b_r(K_1 - K_2)]. \tag{2.162}$$

Hence the real part of β is given by the variation of b_i, and the imaginary part of β which governs the gain or attenuation of the mode depends only on b_r. Close to cut-off the value of b_r is very small and hence the attenuation is strongly dependent on the loss in the cladding medium, K_2. As we move away from cut-off, b_r increases (Figure 2.19), so that the gain in the core region is then of relatively more importance in determining the overall modal attenuation. Asymptotically b_r approaches 1, as in the real index-guiding case, and then the gain in the core region dominates.

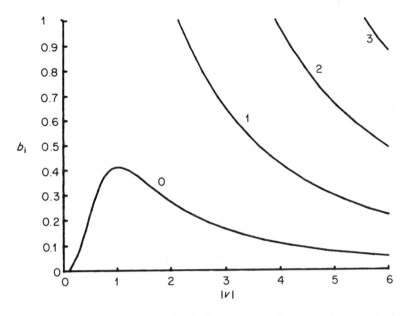

Figure 2.20 Imaginary part b_i of the normalized propagation constant versus $|v|$ for the lowest four modes of a gain-guiding dielectric slab. (From Schlosser (1973). Copyright 1973, American Telephone and Telegraph Company, reprinted by permission)

The real part of β is largely determined by the refractive index $n_2 = n_1$, since we have limited the discussion to the case $K_j \ll n_j$. Hence the importance of the parameter b_i is limited only to the regions close to cut-off of modes other than the zero-order mode.

2.4.4 Application to stripe-geometry lasers

Much of the stimulus for the study of gain-induced guiding in slab waveguides has come from the study of stripe-geometry injection lasers. The stripe-geometry configuration, which is shown schematically in Figure 2.21, was originally introduced (Dyment, 1967) in an effort to overcome problems associated with filamentary lasing in homostructure devices. This phenomenon consists of the formation of one or more lasing filaments along the length of the cavity thus giving rise to a near-field distribution at the facet which takes the appearance of a series of bright spots along the junction plane. It was thought that a stripe contact would confine the active area in the junction plane to dimensions where only one filament could occur, with a consequent improvement in the controllability of the near-field and other laser characteristics.

In the earliest lasers of this type the stripe contact was defined by an oxide barrier as indicated in Figure 2.21. This structure also had the advantage of producing better conduction of the heat produced in the active region to the heat sink, so that homostructure lasers could be run c.w. at higher ambient temperature than hitherto (Dyment and D'Asaro, 1967). The structure is also used in heterostructure lasers (Ripper et al., 1971). Further developments have involved the definition of the stripe contact by methods of proton bombardment (Dyment et al., 1972), doping profiles (Yonezu et al., 1973), oxygen implantation (Blum et al., 1975), and other techniques. For a more comprehensive guide to the details of these structures the interested reader is referred to review articles by D'Asaro (1973) and Selway (1976).

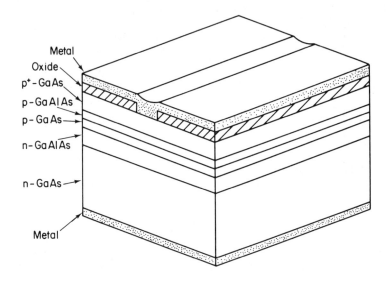

Figure 2.21 The stripe-geometry injection laser structure. (This figure appeared in Vol. 4, pp. 273–280 of *Optics and Laser Technology* published by IPC Science and Technology Press Ltd, Guildford, Surrey, UK and is by M. J. Adams)

Contrary to the original viewpoint mentioned above of the stripe contact acting as a filament selection mechanism, more recent study (Cross and Adams, 1972, Schlosser, 1973; Suematsu and Yamada, 1973; Nash, 1973; Hakki, 1973; Cook and Nash, 1975; Kirkby et al., 1977) has shown that this structure can produce its own waveguiding effects in the plane of the junctions. Although some stripe fabrication techniques, e.g. in the case of the buried heterostructure (Tsukada, 1974) and mesa-stripe lasers (Tsukada et al., 1972), deliberately introduce real refractive index waveguides in the junction plane, the other structures mentioned above rely at least in part on the gain-guidance mechanism. If we consider the oxide-insulated structure of Figure 2.21, for example, it is clear that in the region under the stripe, and possibly for some distance on either side due to current spreading (Hakki, 1973, 1975), there is a region of gain; elsewhere along the junction plane the remaining area will be lossy. The refractive index of these regions with gain and loss will be, to a first approximation, the same. In this case we have the situation of gain-guidance described in Section 2.4.3 with the stripe width equal to $2a$. We will calculate the lasing threshold condition for this case and investigate its dependence on the laser parameters. For the moment we shall ignore the waveguide effect normal to the heterostructure layers which was considered in Section 2.3.6. We shall see in a later chapter that this assumption is justified if the effective index method of analysis for rectangular guides is adopted, provided the refractive index used in the present section is regarded as an effective value resulting from the solution of the waveguide problem of Section 2.3.6.

For the situation of pure gain-guidance, equation (2.162) gives an expression for the nett gain, G, of the mode in terms of the imaginary part of the propagation constant:

$$\frac{G}{2k} = \text{Im}\left(\frac{\beta}{k}\right) = K_2(1-b_r) + K_1 b_r. \qquad (2.163)$$

It follows from the definition of attenuation coefficient α_j ($j = 1, 2$) in terms of extinction coefficient K_j, equation (2.59), that

$$G = gb_r - \alpha_2(1 - b_r) \qquad (2.164)$$

where g is the gain per unit length in region 1 of the waveguide. We now invoke the lasing threshold condition of nett modal gain equal to end-loss through the cavity mirrors, as given by equation (2.135), to obtain:

$$gb_r = \alpha_2(1 - b_r) + \frac{1}{2L} \ln\left(\frac{1}{R_1 R_2}\right). \qquad (2.165)$$

This result is the threshold condition for lasing in a cavity of length L with facet reflectivities R_1, R_2 where the field distribution is confined in one transverse direction by a gain-induced guiding action. It is analogous to equation (2.137) for a real index-guidance mechanism with the confinement parameter ('filling factor') Γ, here replaced by b_r. Since the same physical interpretation may be given to the individual terms of (2.165) as was given to those of (2.137), it follows for this situation of pure gain-guidance only that we may take

$$\Gamma = b_r. \qquad (2.166)$$

This result may also be proved by the more usual evaluation of the filling factor Γ as in Section 2.3.4 (Buus, 1977).

The threshold current density, j_{th}, may now be calculated from (2.165) in a similar way to the calculation in Section 2.3.6 provided that the gain–current relationship is known. For the simple case of a linear relationship as in equation (2.140), which is justified for low doping levels (Stern, 1973), the result becomes

$$j_{th} = E\left[\alpha_2\left(\frac{|v|(1-b_r)}{b_r}\right) + \left(\frac{|v|}{b_r}\right)\frac{1}{2L}\ln\left(\frac{1}{R_1 R_2}\right)\right] \qquad (2.167)$$

where E is a constant, and the remaining expression contains only the cladding loss, end loss, and normalized waveguide parameters. The functions $(|v|(1 - b_r)/b_r)$ and $(|v|/b_r)$ clearly contain all the relevant waveguide information, and these are therefore plotted as functions of $|v|$ in Figure 2.22 for the zero-order mode.

The reader will recall that in Section 2.3.6 it was possible to proceed further for the case of a real index guide by comparing the magnitudes of end loss and cladding loss and disregarding the latter as being dominated by the former for the heterostructure laser. An analogous process is not possible here since, in general, for GaAs/air reflectivities of about 0.3, cavity lengths L of order 300–500 μm, the end loss is around 20–30 cm^{-1}, whilst the cladding loss α_2 can be dominated by interband absorption which may be of order 10–100 cm^{-1}. Hence it is perhaps preferable to discuss the general nature of results to be expected for threshold current density than to examine for example the minimum value of $(|v|/b_r)$ which occurs at $|v| = 1.29$. Hence we note that if only one mode propagates in the guide $|v| < 1.88$ and from Figure 2.22 it follows that the threshold current density will increase quite strongly with decreasing stripe width $(2a)$, independent of the relative magnitudes of end and cladding losses. This result is in good agreement with the experimental observations on stripe-geometry lasers (Schlosser, 1973; Hakki, 1973; Blum et al., 1975). To see if the values of gain, loss, and stripe width are reasonably realistic consider the value of $|v|$ at cut-off of the first-order mode which is given by

$$|v| = a(kn(g + \alpha_2))^{1/2} = 1.88 \qquad (2.168)$$

where equations (2.59), (2.144), and (2.145) have been used. Then for a 10 μm stripe ($a = 5\,\mu$m) at the GaAs emission wavelength of 0.9 μm, equation (2.168) gives $(g + \alpha_2) \simeq 56$ cm^{-1}, a value entirely realistic in terms of the gain and loss of some stripe lasers.

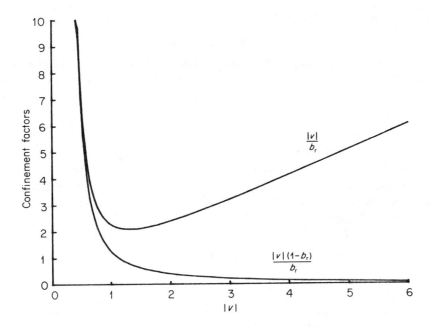

Figure 2.22 Confinement factors $|v|/b_r$ and $|v|(1-b_r)/b_r$, appearing in equation (2.167), versus $|v|$ for gain-induced guidance in a symmetric dielectric slab (lowest-order mode)

However, the reader should not conclude from the above discussion that this provides a complete description of the waveguiding properties of stripe geometry lasers. There are in fact many oversimplifications, which have been the subject of much further study of the phenomenon (Nash, 1973; Hakki, 1973; Cook and Nash, 1975; Hakki, 1975; Kirkby et al., 1977). The principal omission was the neglect of any real index discontinuity as playing a part in the optical confinement. In fact since the gain in the waveguide 'core' is produced by injected carriers, there is a related effect on the refractive index via the Kramers–Kronig relation (W. W. Anderson, 1965; Adams and Cross, 1971; Thompson, 1972; Cross and Adams, 1974). The effect of injected carriers is to decrease the refractive index of the core region by a small amount, thus providing a real index anti-guiding mechanism which competes with the gain-guiding effect discussed above. We can describe this effect, as in Section 2.4.1, by writing $n_1 = n_2 + \Delta n$ with $\Delta n \ll n_1, n_2$, whence:

$$v^2 \simeq 2n_2 \Delta n + 2n_2 i(K_1 - K_2). \tag{2.169}$$

We see that if there is a refractive index depression in the core ($\Delta n < 0$) then v^2 will lie in the second quadrant. It follows from Figure 2.18 that the zero-order mode now has a cut-off frequency $|v|$. There are corresponding changes in the solutions of the eigenvalue equation and in the results for threshold current density, and these have been investigated in more detail by Schlosser (1973). Other than the work of Schlosser (1973), Suematsu and Yamada (1974), and Buus (1977), the stripe-geometry laser has usually been analysed by assuming a graded variation of index and gain/loss under the stripe, rather than abrupt discontinuities in these quantities. This approach, which has some justification in view of the current-spreading effect under the stripe (Hakki, 1973, 1975), has led to simpler and more empirical treatments of the device physics; we shall return to this topic in Chapter 3.

2.5 Metal-clad and Hollow Waveguides

In Section 2.4 we introduced the topics of loss and gain in dielectric media with the aid of a complex dielectric permittivity, defined as in equation (2.144). If we consider now the use of metals in waveguides at optical frequencies then we may still retain the concept of a complex permittivity. Contrary to the common practice at microwave frequencies, it is not legitimate in the optical region to think of a metal merely as an imperfect conductor. It has been shown experimentally (Wilmot and Schineller, 1966) that the modes of a metal-clad optical waveguide are strongly affected by the dielectric properties of the metal. More recently metal-clad guides have found important applications in integrated optics (Tien, 1977) and as flexible waveguides for carbon dioxide laser radiation at 10.6 μm (Garmire et al., 1976). In addition the closely-related topic of hollow dielectric waveguides is of practical importance for waveguide lasers (Degnan, 1976). Hence in the present section we will discuss various forms of metal-clad and hollow dielectric waveguides as further examples of the slab guide.

2.5.1 Waveguide classification

As an aid to understanding the effects to be considered, Table 2.2 gives the refractive index n and extinction coefficient K of a few selected metals at the He–Ne and CO_2 laser wavelengths. The most obvious point to note first is that $|K| > n$ for all the cases listed except for chromium at 0.633 μm. This implies that for these (and many other metals) the real part of the dielectric permittivity, i.e. $(n^2 - K^2)$, is negative. Hence it follows from (2.45) that if such a metal is used as the cladding material for a waveguide, *the refractive index of the dielectric core region may be arbitrary* and there will still be a guiding action. It follows also that the propagation constant β of a mode in such a guide has an allowed range of values from a maximum value of the wave number in the core (kn_1, where n_1 = core index) down to zero. In some special cases, as we shall see, the range of propagation constants can be even larger than this. Hence metal-clad guides can usually support a larger number of modes than the equivalent dielectric guides.

A second point arising from Table 2.2 is the result that for many cases $|K| \gg n$. Thus it is frequently permissible to find the eigenvalue equation for modes in such a guide by the methods of Chapter 1, treating the cladding as a metal of real negative permittivity. Inclusion of the imaginary component of permittivity leads to improved results for the modal propagation constants and to estimates of the attenuation of each mode. This provides some justification for the procedure adopted in Section 2.2.4, case (a), for the calculation of TE_0 mode attenuation. In the present section, however, we will not make quite such restrictive assumptions concerning the magnitudes of real and imaginary parts of the complex permittivity. We are therefore interested here in deriving rather more accurate results than those of equations (1.22) and (2.65).

Consider now materials such as chromium at 0.633 μm in Table 2.2 which have a positive real part of complex permittivity. The effects of such materials used as claddings for a dielectric-core waveguide have been analysed by Rashleigh (1976); there are four cases to be considered (Batchman and McMillan, 1977). Denoting the core refractive index by n_1 and the metal permittivity by $\varepsilon = \varepsilon' + i\varepsilon''$, these four cases are:

(i) $n_1^2 > \varepsilon'$; ε'' negligible: This corresponds to the lossless dielectric slab guide considered in Sections 2.2 and 2.3. If ε'' is nonzero the modes will be attenuated and the analysis of Section 2.2.4, case (b), may apply in such a case.

(ii) $n_1^2 > \varepsilon'$; ε'' large: This lossy slab guide has received little attention since few materials seem to satisfy this condition. In principle, however, this represents a combination of

Table 2.2 Optical constants for various metals; the dielectric permittivity is given by $(n+iK)^2$, i.e. negative K corresponds to loss

Metal	Wavelength 0.633 μm			Wavelength 10.6 μm		
	n	K	Reference	n	K	Reference
Aluminium	1.2	−7	Gray (1963)	25	−67	Beattie (1957)
Chromium	3.19	−2.26	Batchman and McMillan (1977)	11	−22.8	Lenham and Treherne (1966)
Copper	0.15	−3.2	Gray (1963)	12.6	−64.3	Hass (1965)
Gold	0.15	−3.2	Gray (1963)	7.4	−53.4	Hass (1965)
Silver	0.065	−4	Gray (1963)	10.7	−69	Beattie (1957)

positive guidance due to both real index and loss as discussed in general in Section 2.4. For a symmetric guide of this type the TE mode cut-offs will be given by the region $0 < \arg(v) < \pi/4$ on Figure 2.18.

(iii) $n_1^2 < \varepsilon'$; ε'' *negligible*: This is the hollow dielectric guide which supports only leaky modes. A symmetric guide of this type would be described on Figure 2.18 as being at $\arg(v) = \pi/2$. Nonzero ε'' may correspond to $\arg(v)$ somewhat less than $\pi/2$.

(iv) $n_1^2 < \varepsilon'$; ε'' *large*: From the discussion of Section 2.3 it follows that this structure supports guided modes which are stable if ε'' is sufficiently large. Such modes have, however, been termed 'damped leaky modes' (Batchman and McMillan, 1977) since they may be thought of as leaky modes of the real guide with heavy damping resulting from the absorption in the metal. Instead of the familiar exponential tails in the cladding, the mode fields are damped oscillatory in this region. Chromium in Table 2.2 satisfies the requirements for a cladding material of this type.

In general all of the structures discussed in this subsection with positive or negative permittivity claddings must be analysed by a solution of the appropriate eigenvalue equation. This normally implies numerical solution of equation (2.47) for TE and (2.48) for TM modes, where all the quantities in these equations are replaced by their complex equivalents. There are several accounts of numerical techniques suitable for this problem available in the literature (Batchman and Rashleigh, 1972; Chang and Loh, 1972; Kaminow *et al.*, 1974). Alternatively, for negative permittivity metals the simplified model of a lossless dielectric of negative permittivity may sometimes be useful (Takano and Hamasaki, 1972; Polky and Mitchell, 1974; Fink, 1976). In a few cases of potential importance for applications, some simplifying assumptions may be made so that useful analytical approximations become possible. We shall consider some of these cases below. They may be conveniently subdivided into the subjects of asymmetric and symmetric metal-clad guides, and hollow dielectric guides, respectively.

2.5.2 Asymmetric metal-clad dielectric waveguide

Consider the asymmetric structure shown in Figure 2.23 consisting of a metal substrate surmounted by two layers of dielectric. Let the dielectric core have refractive index n_1, the dielectric cladding have index n_2, and the metal cladding have permittivity $\varepsilon_3 = (n_3 + iK_3)^2$. The eigenvalue equation for this structure is then given by equations (2.47) for TE modes or (2.48) for TM modes. The two cases can be combined into the single equation

$$4aq = 2\tan^{-1}\left(\eta_{12}\frac{p}{q}\right) + 2\tan^{-1}\left(\eta_{13}\frac{r}{q}\right) + 2N\pi \quad (N = 0, 1, 2, \ldots) \quad (2.170)$$

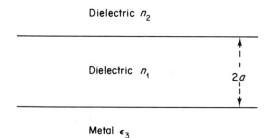

Figure 2.23 The asymmetric metal-clab waveguide structure

where the following definitions now apply:

$$p^2 = \beta^2 - n_2^2 k^2$$
$$q^2 = n_1^2 k^2 - \beta^2$$
$$r^2 = \beta^2 - \varepsilon_3 k^2$$

$$\eta_{12} = \begin{cases} 1, & \text{for TE modes} \\ \left(\dfrac{n_1}{n_2}\right)^2, & \text{for TM modes} \end{cases}$$

$$\eta_{13} = \begin{cases} 1, & \text{for TE modes} \\ \dfrac{n_1^2}{\varepsilon_3}, & \text{for TM modes.} \end{cases}$$

Note that, in general, since ε_3 is complex all the parameters p, q, r entering (2.170) are also complex quantities.

Equation (2.170) may be solved numerically for given sets of parameter values n_1, n_2, ε_3 and k (see e.g. Kaminow et al., 1974). However, a useful approximate solution can be found for the special case (Garmire and Stoll, 1972):

$$n_1 - n_2 \ll n_1 \tag{2.171a}$$
$$k|(n_1^2 - \varepsilon_3)^{1/2} \eta_{13}| \gg \mathrm{Re}(q). \tag{2.171b}$$

Equation (2.171a) is the now-familiar condition for weak-guidance on the dielectric-clad side of the waveguide; its implication here is that $\eta_{21} \simeq 1$. Equation (2.171b) represents a condition on the metal/dielectric interface whose physical implications will be seen after a solution has been obtained. If we examine the argument of the second $\tan^{-1}$ function in (2.170) in the light of condition (2.171b), we see that

$$\eta_{13}\frac{r}{q} = \eta_{13}\frac{[k^2(n_1^2 - \varepsilon_3) - q^2]^{1/2}}{q} \simeq \frac{|\eta_{13} k(n_1^2 - \varepsilon_3)^{1/2}|}{\mathrm{Re}(q)} \gg 1.$$

Hence a first approximation to equation (2.170) is obtained by taking

$$\tan^{-1}\left(\frac{\eta_{13} r}{q}\right) \simeq \frac{\pi}{2}.$$

The reader will recall that this approximation is equivalent to the condition for phase shift on reflection at the surface of a perfectly conducting metal, as stated in equation (1.20). In other words we are here assuming that there is very little field penetration into the metal cladding.

With the aid of assumptions (2.171), equation (2.170) simplifies to

$$\frac{p}{q} \simeq \tan[2aq - (N + \tfrac{1}{2})\pi].$$

Adopting our earlier definition of normalized frequency, $v = ak(n_1^2 - n_2^2)^{1/2}$, and transverse propagation constant, $u \equiv u_r + i u_i = aq$, this eigenvalue equation may be re-written (Garmire and Stoll, 1972):

$$v = \frac{u_r(-1)^N}{\sin 2u_r} \tag{2.172}$$

where we have also used the fact that $u_r \gg u_i$, which is again consistent with little field penetration into the metal. Figure 2.24 gives plots of the solutions of (2.172) for the four

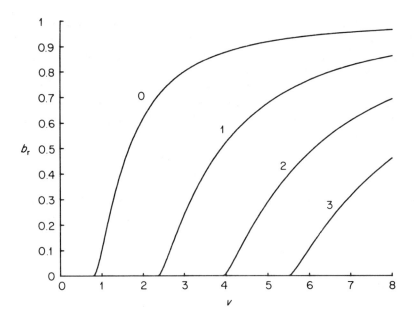

Figure 2.24 Real part b_r of the normalized propagation constant versus normalized frequency v for asymmetric metal-clad guides, calculated from numerical solution of equation (2.172) for four lowest-order modes

lowest-order modes in terms of the normalized parameter b_r (defined in the usual way, see for example Section 2.4) versus v. Clearly, as $v \to \infty$ and $\beta \to kn_1$, a simpler approximation results in the limit of (2.172):

$$\lim_{v \to \infty} (u_r) \to \frac{(N+1)\pi}{2} \qquad (2.173)$$

which corresponds to (1.22) derived for the perfect metallic surface in Chapter 1.

With the aid of (2.173) we may investigate further the physical meaning of assumption (2.171b). Since, from Table 2.2, we note that for many metals of interest

$$|K_3| \gg n_3 \quad \text{and} \quad |\varepsilon_3| \simeq |K_3^2| \gg n_1^2$$

then (2.171b) may be reduced to:

TE modes:
$$a \gg \frac{(N+1)\pi}{2k|K_3|} \qquad (2.174a)$$

TM modes:
$$a \gg \frac{(N+1)\pi|K_3|}{2kn_1^2}. \qquad (2.174b)$$

In each case we see that the condition corresponds to demanding that the waveguide thickness exceeds some critical value. This condition is clearly easier to satisfy for TE than for TM modes.

There is another problem associated with the TM case, which we have avoided up to now in this subsection. It is that for the TM_0 mode the $\tan^{-1}$ functions in (2.170) can have opposite signs (the real part of η_{13} is normally negative) so that they may sum to zero. This implies that as $\beta \to kn_1$ ($q \to 0$) there may still be a finite value of ka. In order to permit $ka \to \infty$ along the real axis it is therefore necessary to allow $\beta > n_1 k$. This phenomenon, which will be discussed in more detail in due course, corresponds to a surface plasma wave (Kaminow et al., 1974; Otto and Sohler, 1971). For this case the assumption (2.171b) is violated and all the resulting approximations fail. *The reader is therefore cautioned that the results (2.172), (2.173) and conditions (2.174) cannot apply to the TM_0 mode.*

With the above caveat in mind we proceed to the calculation of attenuation for modes in this structure, on the understanding that the TM_0 mode must be considered separately in due course. From the ray treatment of Section 2.2, we have equation (2.60) for loss, α, which may be re-written in the form:

$$\alpha = \frac{1 - R_{12} R_{13}}{2w \cot \phi} \quad (2.175)$$

where the R's are power reflection coefficients at the two walls, w is the effective waveguide width, and ϕ is the angle of propagation of the ray. For the asymmetric metal-clad structure under consideration, it follows that

$$R_{12} \simeq 1$$

$$\cot \phi = \frac{\beta}{q} \simeq \frac{akn_1}{u_r}$$

$$w \simeq 2a + \frac{1}{p} = \frac{a}{u_r}(2u_r - \tan 2u_r).$$

The reflectivity R_{13} at the metal/dielectric interface may be found for TE modes from equation (2.8), and for TM modes from (2.16).

Combining these results we find:

$$R_{13} = \frac{(n_1 \sin \phi - \text{Re}[\eta_{13}(\varepsilon_3 - n_1^2)^{1/2}])^2 + (\text{Im}[\eta_{13}(\varepsilon_3 - n_1^2)^{1/2}])^2}{(n_1 \sin \phi + \text{Re}[\eta_{13}(\varepsilon_3 - n_1^2)^{1/2}])^2 + (\text{Im}[\eta_{13}(\varepsilon_3 - n_1^2)^{1/2}])^2}.$$

Hence

$$1 - R_{12} R_{13} \simeq 4 n_1 \sin \phi \, \text{Re}\left(\frac{1}{\eta_{13}(\varepsilon_3 - n_1^2)^{1/2}}\right).$$

Using the above results in equation (2.175), we obtain finally (Garmire and Stoll, 1972):

$$\alpha \simeq \frac{-2u_r^3}{a^3 k^2 (2u_r - \tan 2u_r) n_1} \text{Im}\left(\frac{1}{\eta_{13}(n_1^2 - \varepsilon_3)^{1/2}}\right). \quad (2.176)$$

This equation is easily divided into a mode-dependent part (γ) which can be written in normalized variables and a material-dependent part (κ):

$$\frac{\alpha}{k} = \gamma \kappa \quad (2.177a)$$

$$\gamma = \frac{u_r^3}{v^3 (2u_r - \tan 2u_r)} \quad (2.177b)$$

$$\kappa = \frac{-2(n_1^2 - n_2^2)^{3/2}}{n_1} \text{Im}\left(\frac{1}{\eta_{13}(n_1^2 - \varepsilon_3)^{1/2}}\right). \quad (2.177c)$$

The mode-dependent part of the attenuation coefficient (γ) as given in (2.177b) is plotted versus the normalized frequency v for the four lowest-order modes in Figure 2.25. Apart from the maximum in each curve occurring close to the cut-off values ($v = (N + \frac{1}{2})\pi/2$ for

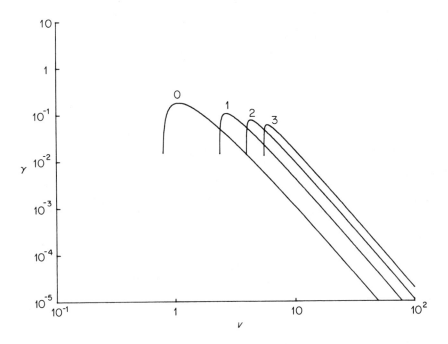

Figure 2.25 Mode-dependent loss function γ (equation (2.177b)) versus normalized frequency v for asymmetric metal-clad guides

$N = 0, 1, 2, \ldots$) the results show the dominance of the $1/v^3$ term in the limit of large v. This is easily seen by taking the limiting case of the eigenvalue equation (2.173) and substituting in (2.176).

Let us now turn to the special case of the TM_0 mode. If we permit $\beta > n_1 k$ as $ka \to \infty$, then equation (2.170) may be re-written in full ($N = 0$, $\varepsilon_3' < 0$, $|\varepsilon_3'| \gg |\varepsilon_3''|$) as:

$$2ak \simeq \frac{k}{(\beta^2 - k^2 n_1^2)^{1/2}} \left\{ \tanh^{-1} \left[\frac{n_1^2}{|\varepsilon_3'|} \left(\frac{\beta^2 - \varepsilon_3' k^2}{\beta^2 - n_1^2 k^2} \right)^{1/2} \right] \right.$$

$$\left. - \tanh^{-1} \left[\frac{n_1^2}{n_2^2} \left(\frac{\beta^2 - n_2^2 k^2}{\beta^2 - n_1^2 k^2} \right)^{1/2} \right] \right\}. \quad (2.178)$$

For $ak \to \infty$ on the real axis, the argument of the first $\tanh^{-1}$ function must approach unity. The corresponding value of β is given by:

$$\left(\frac{\beta}{k} \right)^2 = \frac{n_1^2 \varepsilon_3'}{n_1^2 + \varepsilon_3'}. \quad (2.179)$$

It follows from (2.179) that β/k is real for $|\varepsilon_3'| > n_1^2$ (since then the numerator and denominator of (2.179) are less than zero because $\varepsilon_3' < 0$), and it is imaginary for $|\varepsilon_3'| < n_1^2$. For $|\varepsilon_3'| = n_1^2$, $\beta/k \to \infty$, i.e. the phase velocity of the wave goes to zero.

In fact the above argument is also applicable for the case when ε_3' is replaced by ε_3 in equations (2.178) and (2.179). The $\tanh^{-1}$ function in the complex plane is such that as the argument approaches $1 + i0$ the imaginary part is indeterminate but finite, whilst the real part is infinite. Hence we find from (2.179) with ε_3' replaced by ε_3, $|\varepsilon_3''| \ll |\varepsilon_3'|$, and $|\varepsilon_3'| > n_1^2$, for

$\beta = \beta_r + i\beta_i$ (Kaminow et al., 1974):

$$\frac{\beta_r}{k} = \left(\frac{n_1^2 \varepsilon_3'}{n_1^2 + \varepsilon_3'}\right)^{1/2} \qquad (2.180a)$$

$$\frac{\beta_i \beta_r}{k^2} = \frac{1}{2}\frac{n_1^4 \varepsilon_3''}{(n_1^2 + \varepsilon_3')^2}. \qquad (2.180b)$$

We note that these results for β_r and β_i as $ka \to \infty$ are independent of ka. The fact that $\beta > n_1 k$ results in an evanescent wave on the core side of the metal–dielectric interface, so that the TM_0 mode has its field distribution largely confined to this interface.

A further property of the TM_0 mode is that as $ka \to 0$ the wave continues to propagate, i.e. there is no cut-off frequency. The condition $ka = 0$ occurs for equal arguments of the $\tanh^{-1}$ function in (2.170). This yields (Otto and Sohler, 1971; Kaminow et al., 1974):

$$\left(\frac{\beta}{k}\right)^2 = \frac{n_2^2 \varepsilon_3}{n_2^2 + \varepsilon_3}. \qquad (2.181)$$

For all other modes of the asymmetric metal-clad guide cut-off occurs at the usual condition $\beta = kn_2$.

Field distributions for a few low-order TE and TM modes are shown in Figure 2.26. The surface-wave characteristics of the TM_0 mode are clearly seen whilst all other field distributions have features somewhat similar to those in dielectric guides. For TE modes there is little field penetration into the metal, whereas for TM modes there is a substantially greater portion of the field inside the metal and the losses are correspondingly higher. This latter point may be easily seen from our approximate results for loss in (2.177). For the case $|\varepsilon_3''| \ll |\varepsilon_3'|, |\varepsilon_3'| > n_1^2, \varepsilon_3' < 0$, we may take the ratio of κ in (2.177c) for TM and TE modes:

$$\frac{\kappa_{TM}}{\kappa_{TE}} \simeq \frac{2n_1^2 - \varepsilon_3'}{n_1^2}. \qquad (2.182)$$

Clearly, this ratio always exceeds 2, and for most metal/dielectric combinations it is of order 10 or larger.

The principal application envisaged for metal-clad guides of the type discussed here lies in the field of integrated optics. This term, first used in 1969 (Miller), is used to cover all activities in which conventional or novel optical devices are fabricated in a miniaturized form on a planar substrate. An important aim is the production of integrated circuits made up of such devices connected by waveguides in order to perform various forms of optical signal processing. A recent comprehensive review of the subject has been given by Tien (1977), to which the reader is referred for more details.

Since some components, e.g. modulators, require metal electrodes in close proximity to the guided wave in an integrated optics circuit, the subject of metal-clad guides has been studied extensively by workers in the field (Suematsu et al., 1972; Garmire and Stoll, 1972; Takano and Hamasaki, 1972; Reisinger, 1973; Kaminow et al., 1974; Polky and Mitchell, 1974; Nosu and Hamasaki, 1976; Batchman and McMillan, 1977). Measurements of the real part of the propagation constant, β_r, have been made using a prism-film coupler (Tien et al., 1969) on Al–LiF–air guides (Otto and Sohler, 1971) and on metal–polymer–glass and metal–polymer–air systems (Kaminow et al., 1974; Tien et al., 1975) using Al, Ag, and Au as the metal claddings. In general the agreement with calculated results of the type indicated above is very good. Loss measurements on these guides have also been made (Suematsu et al., 1972; Reisinger, 1973a; Kaminow et al., 1974; Tien et al., 1975) and again there is good

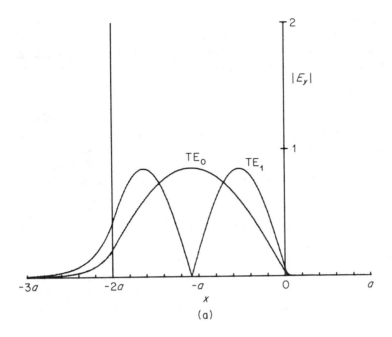

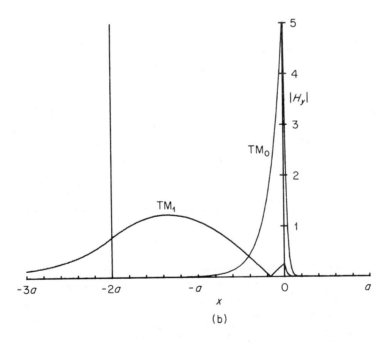

Figure 2.26 (a) Field amplitude distributions $|E_y|$ for TE_0 and TE_1 modes. Parameters are $a = 1$ μm, $\lambda = 0.633$ μm, $n_1^2 = 2.3$, $n_2^2 = 2.25$, $\varepsilon_3' = -16$, $\varepsilon_3'' = -0.54$. (b) Field amplitude distributions $|H_y|$ for TM_0 and TM_1 modes. Note that the vertical axis is calibrated in the same units as for (a)

agreement with the calculated results as regards the variations of loss with mode number, polarization, and waveguide thickness. The loss is lowest in the TE_0 mode and for silver-clad polymer guides about 3 μm thick at He–Ne wavelength the loss of this mode can be less than 1 dB/cm (Kaminow et al., 1974; Tien et al., 1975). Measurements on positive-permittivity guides have also been performed using chromium and germanium claddings on polystyrene (Batchman and McMillan, 1977).

The heavy attenuation of the TM_0 mode and the fact that losses of other TM modes are typically at least ten times greater than those of corresponding TE modes has suggested the possibility of using metal-clad guides as mode analysers (Suematsu et al., 1972; Garmire and Stoll, 1972). These could either take the form of simple mode filters transmitting the TE_0 mode and cutting off TM_o and higher-order modes, or be used as part of a novel modulator. By using an anisotropic dielectric the polarization of a given mode could be switched with an applied voltage by using the electro-optic effect (Garmire and Stoll, 1972); since the TE mode is transmitted much more than the TM mode this could result in a variable loss, or amplitude modulator. Another application lies in insulating high-index from relatively low-index dielectrics in a hybrid integrated optical circuit. High-index materials are used in heterostructure lasers, electro-optic modulators and photodetectors, whilst glass or organic compounds have traditionally been used for passive waveguide components. It has been proposed therefore (Tien et al., 1975) that hybrid circuits could be formed by covering parts of a high-index substrate with a layer of silver so that active devices are grown on the substrate and passive components fabricated on the silver.

2.5.3 Symmetric metal-clad guide

We turn now to the symmetric structure shown in Figure 2.27 consisting of a layer of dielectric, refractive index n_1, thickness $2a$, between two identical metals of permittivity

$$\varepsilon_2 = \varepsilon'_2 + i\varepsilon''_2 = (n_2 + iK_2)^2.$$

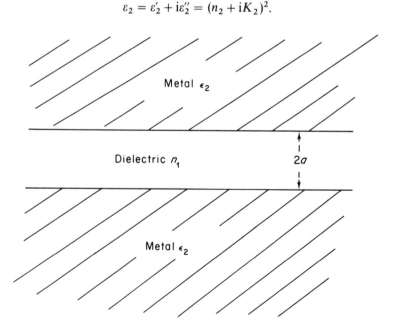

Figure 2.27 The symmetric metal-clad waveguide structure

Again we limit attention to the situation where the real part of the metal permittivity is negative, i.e. $\varepsilon_2' < 0$. In this case the eigenvalue equation (2.170) becomes:

$$2aq = 2\tan^{-1}\left(\eta_{12}\frac{p}{q}\right) + N\pi \quad (N = 0, 1, 2, \dots) \tag{2.183}$$

where $q^2 = n_1^2 k^2 - \beta^2$, $p^2 = \beta^2 - \varepsilon_2 k^2$, and

$$\eta_{12} = \begin{cases} 1, & \text{for TE modes} \\ \dfrac{n_1^2}{\varepsilon_2}, & \text{for TM modes.} \end{cases}$$

As for the case of the asymmetric metal-clad guide discussed in the previous subsection, the symmetric structure TM_0 mode has to be dealt with separately. In addition, for the present structure the TM_1 mode also exhibits surface wave behaviour. The TM_0 and TM_1 modes are respectively the symmetric and antisymmetric combinations of the surface waves at the two metal/dielectric interfaces.

For all modes other than TM_0 and TM_1 we may obtain useful approximate solutions of (2.183) as $ka \to \infty$. For this purpose the trigonometric identity

$$\tan^{-1}(x) = \frac{\pi}{2} - \tan^{-1}\left(\frac{1}{x}\right)$$

may be used in (2.183), so that for $ka \to \infty$ and $\beta \to n_1 k$, we obtain:

$$2a(n_1^2 k^2 - \beta^2)^{1/2} = (N+1)\pi - 2\tan^{-1}\left[\frac{(n_1^2 k^2 - \beta^2)^{1/2}}{\eta_{12}(\beta^2 - \varepsilon_2 k^2)^{1/2}}\right]$$

$$\simeq (N+1)\pi - \frac{2(n_1^2 k^2 - \beta^2)^{1/2}}{\eta_{12} k(n_1^2 - \varepsilon_2)^{1/2}}. \tag{2.184}$$

It follows after a little algebra that for situations of interest where $|\beta_r| \gg |\beta_i|$ (Kaminow et al., 1974):

$$\beta_r \simeq n_1 k\left\{1 - \frac{\pi^2}{2n_1^2}\left(\frac{N+1}{2ak}\right)^2\left[1 - \frac{2}{ak}\text{Re}\left(\frac{1}{\eta_{12}(n_1^2 - \varepsilon_2)^{1/2}}\right)\right]\right\} \tag{2.185}$$

$$\beta_i \simeq \left(\frac{N+1}{2}\right)^2 \frac{\pi^2}{a^3 k^2 n_1}\text{Im}\left(\frac{1}{\eta_{12}(n_1^2 - \varepsilon_2)^{1/2}}\right). \tag{2.186}$$

We note that if ak is sufficiently large that the final term on the r.h.s. of (2.185) can be ignored, then the simpler approximation of (2.173) (or, equivalently, (1.22)) is obtained again:

$$\lim_{ak \to \infty}(u_r) \to \left(\frac{N+1}{2}\right)\pi.$$

Similarly, if this limit is used in equation (2.176) for the *asymmetric* metal-clad dielectric waveguide we find that

$$\beta_i \equiv -\frac{\alpha}{2} \simeq \left(\frac{N+1}{2}\right)^2 \frac{\pi^2}{2a^3 k^2 n_1}\text{Im}\left(\frac{1}{\eta_{13}(n_1^2 - \varepsilon_3)^{1/2}}\right). \tag{2.187}$$

Comparing (2.186) and (2.187) we see that the asymptotic loss formula for the symmetric metal-clad guide is twice that for the asymmetric structure with a single metallic layer.

Note also that further simplification of (2.186) for the TE_0 mode with $|\varepsilon_2| \gg n_1^2$ yields

$$\alpha_{TE_0} = -2\beta_{iTE_0} \simeq \frac{\pi^2}{2a^3 k^2 n_1}\left(\frac{n_2}{n_2^2 + K_2^2}\right)$$

which is identical with our earlier approximation (2.65), obtained by the techniques of geometric optics.

For the special case of the TM_0 mode the arguments given in the previous subsection still hold true. In particular the arguments of the two $\tanh^{-1}$ functions in (2.178) are now equal, and $\varepsilon_3, \varepsilon_3', \varepsilon_3''$ are replaced by $\varepsilon_2, \varepsilon_2', \varepsilon_2''$, respectively. It follows that the results as $ak \to \infty$ for β_r, β_i given in (2.180) still apply with the appropriate change of subscript. For the TM_1 mode equation (2.183) becomes for $\varepsilon_2' < 0$, $|\varepsilon_2'| \gg |\varepsilon_2''|$:

$$2ak = \frac{2k}{(\beta^2 - n_1^2 k^2)^{1/2}}\left\{\tanh^{-1}\left[\frac{n_1^2}{|\varepsilon_2'|}\left(\frac{\beta^2 - \varepsilon_2' k^2}{\beta^2 - n_1^2 k^2}\right)^{1/2}\right] - \frac{i\pi}{2}\right\}$$

$$= \frac{2k}{(\beta^2 - n_1^2 k^2)^{1/2}} \tanh^{-1}\left[\frac{|\varepsilon_2'|}{n_1^2}\left(\frac{\beta^2 - n_1^2 k^2}{\beta^2 - \varepsilon_2' k^2}\right)^{1/2}\right] \quad (2.188)$$

where the identity $\tanh^{-1}(x) - i\pi/2 = \tanh^{-1}(1/x)$ has been used. In this case we can find explicitly the waveguide thickness for which $\beta \to n_1 k$. It is given from (2.188) by

$$ak = \frac{|\varepsilon_2'|}{n_1^2 (n_1^2 - \varepsilon_2')^{1/2}}. \quad (2.189)$$

As $ka \to \infty$, we again need the condition that the argument of the $\tanh^{-1}$ function in (2.188) should be unity. Hence equation (2.179) with ε_3' replaced by ε_2' is again true in this limit. A similar extension to the complex domain is again possible (Kaminow et al., 1974) so that (2.180) holds also for the TM_1 mode of the symmetric metal-clad guide with the appropriate change of subscript. The only difference between the TM_0 and TM_1 modes in this structure is the sign of the imaginary part of q; the field amplitudes are the same. As for the case of the asymmetric metal-clad guide, the TM_1 mode is not cut-off and continues to propagate as $ka \to 0$. In this limit, equation (2.188) shows that (Kaminow et al., 1974):

$$\frac{\beta}{k} \to \frac{n_1^2}{ak\varepsilon_2}.$$

Measurements on symmetrical metal-clad waveguides of the type discussed here are naturally difficult. For example it is impractical to use the sliding prism method to measure β' and the attenuation, since insufficient optical power is coupled out through the metal layers. Hence the only reliable measurement technique is that of total transmission loss through guides of different lengths. Measurements of this type have been made on metal-clad optical strip lines (Yamamoto et al., 1975). However, since the optical strip line described by Yamamoto et al. is strictly a three-dimensional structure (consisting of a partially-clad slab guide) we postpone more detailed discussion to the appropriate place in Chapter 6.

At far-infrared wavelengths, the parallel-plate waveguide forms a potentially useful transmission medium (Nishihara et al., 1974; Garmire et al., 1976, 1976a). The application envisaged would be for flexible waveguides capable of delivering CO_2 laser radiation to remote targets for use in welding, cutting, surgery, or other such operations. Since optical fibres which are transparent in the infra-red wavelength region are not available as yet the hollow metal waveguide would seem to have an important future in these applications. Let us estimate the loss at 10.6 μm in an aluminium-clad hollow guide (i.e. the core is air or vacuum with $n_1 = 1$) of width $2a = 0.5$ mm. Using the Al data from Table 2.2, equation (2.186) gives for the TE$_0$ mode:

$$\alpha_{TE_0} = -2\beta_{iTE_0} = 0.02 \text{ dB/m}.$$

Experimental measurements on real aluminium guides (Garmire et al., 1976a) do not give quite such remarkably low attenuations as this figure would indicate. Since commercially available Al does not have as large a reflectivity as that of a freshly evaporated film (corresponding to the values of Table 2.2), the values for n_2 and K_2 must be amended somewhat to obtain a more realistic answer. However, measurements indicate that transmission through 1 m lengths can be of order 80–90 per cent (Garmire et al., 1976a) which is quite adequate for some applications. In addition the bending loss of such guides is also extremely low (95 per cent transmission per radian of bend) and practically independent of waveguide thickness, bend radius, and wavelengths (Garmire, 1976; Garmire et al., 1977; Krammer, 1977), so that the range of applications may include those where radiation is required to be piped around sharp corners.

2.5.4 Hollow dielectric guide

In this subsection we shall be concerned with the dielectric slab waveguide, of the form shown earlier in Figure 2.5, with the difference here that the refractive index of the central core layer is considered lower than those of the cladding layers, i.e. $n_1 < n_2, n_3$. For simplicity we restrict attention to the symmetric case $n_2 = n_3$ which is the one of most practical interest, although the theory is easily extended to include asymmetric hollow guides. Although the nomenclature implies that the central layer is composed of air or vacuum ($n_1 = 1$), this is by no means essential and we will retain a general index n_1 throughout.

From our discussion of bound rays in Section 2.2 and guided modes in Section 2.3 for the conventional slab guide ($n_1 > n_2$) it is clear that the present structure will not support rays or modes of these types. On the ray picture, total internal reflection is never attained so that every ray loses a portion of its energy as a real transmitted ray at every contact with the core–cladding interface. Hence the hollow waveguide is obviously lossy. From the electromagnetic mode description the guide is found to support only leaky waves, that is those with oscillatory fields throughout the core and cladding regions and no purely evanescent behaviour as in the case of guided modes. Leaky waves can in fact exist also on the conventional slab with $n_1 > n_2$ where they form the analytic continuation of the guided mode spectrum below cut-off. In the latter context the leaky waves should be distinguished also from the continuum of radiation modes of the slab waveguide below cut-off. Since a comprehensive account of the radiation modes and leaky modes has been given by Marcuse (1974) and Vassallo (1979) we have deliberately excluded them from our discussion of the conventional dielectric slab and restricted attention to guided modes.

Returning to the hollow slab waveguide we will begin with a summary of the formal TE mode analysis and later illustrate some approximate results to be obtained from the zig-zag ray model. The modal fields are again described by equations (2.80) for the TE modes and (2.85) for the TM modes, provided all the quantities appearing in these equations are treated as complex rather than real. Similarly the eigenvalue equations are given by (2.82) and (2.87), respectively. Restricting attention to the TE modes and changing to *complex* normalized variables, equations (2.113) and (2.114) will still hold for the hollow waveguide. Now the normalized frequency v is given in this case by

$$v^2 = a^2 k^2 (n_1^2 - n_2^2) = -|v|^2, \text{ since } n_1 < n_2.$$

Hence if we re-cast the normalized TE eigenvalue equation into the form (2.160), we find:

Even modes: $$i|v| = \frac{u}{\cos u} \qquad (2.190a)$$

Odd modes: $$i|v| = \frac{u}{\sin u}. \qquad (2.190b)$$

Writing $u = u_r + iu_i$, separating real and imaginary parts of (2.190), and eliminating $|v|$ yields:

Even modes: $$\frac{u_r}{u_i} = \tan u_r \tanh u_i \qquad (2.191a)$$

Odd modes: $$\frac{u_r}{u_i} = -\cot u_r \tanh u_i. \qquad (2.191b)$$

These equations are now in a particularly convenient form for numerical solution; choice of a value of u_r and solution for u_i may be used to find $|v|$. The results may be expressed as usual in terms of normalized propagation constant $b \equiv b_r + ib_i$:

$$b_r = 1 + \frac{(u_r^2 - u_i^2)}{|v|^2} \qquad (2.192)$$

$$b_i = \frac{2 u_r u_i}{|v|^2}. \qquad (2.193)$$

Computed results for b_r and b_i versus $|v|$ for a few low-order modes of the hollow dielectric waveguide are shown (solid lines) in Figures 2.28 and 2.29, respectively. Note that in this case a value of b_r equal to zero does not imply cut-off of the mode; the modes are all below cut-off in the usual definition. If we define w as in (2.146), then conventional guided mode cut-off is given by $w_r = 0$, whereas $b_r = 0$ for the hollow guide coincides with $w_r = w_i$. In the limit of large $|v|$, $b_r \to 1$ corresponding to $w_i \gg w_r$ (Marcuse, 1974).

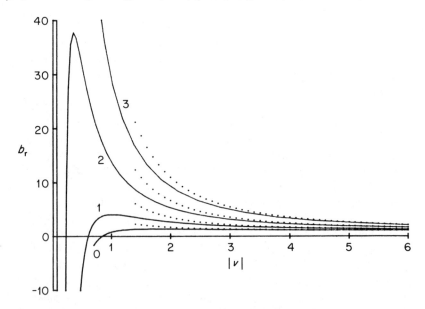

Figure 2.28 Real part b_r of the normalized propagation constant versus $|v|$ for a symmetric hollow dielectric waveguide (TE modes). Solid lines, numerical solution; dotted lines, approximation given by equation (2.194)

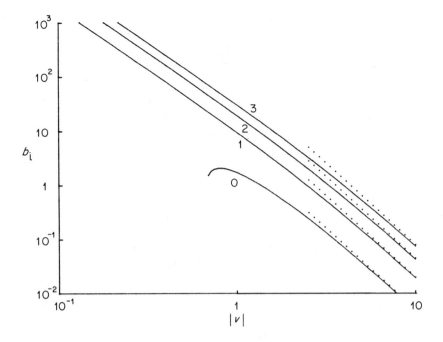

Figure 2.29 Imaginary part b_i of the normalized propagation constant versus $|v|$ for a symmetric hollow dielectric waveguide (TE modes). Solid lines, numerical solution; dotted lines, approximation given by equation (2.197)

Let us turn now to the ray description of the hollow waveguide (Marcuse, 1972a; Ulrich and Prettl, 1973). From the zig-zag ray model of Figure 2.5 with partial reflection at each contact with the core–cladding interface, equation (2.46) yields a quantization condition on ray angle ϕ. However, since we are interested naturally in cases of relatively low loss we consider the case of nearly grazing incidence (ϕ very small). In the limit of phase shift $\delta_{12} \to \pi$ a useful approximation is found from (2.46):

$$2ak_1 \sin \phi \simeq (N+1)\pi \quad (N = 0, 1, 2, \dots).$$

In terms of our normalized variables this result assumes the familiar form ($u_r \gg u_i$):

$$u_r \simeq \frac{(N+1)\pi}{2}$$

$$b_r \simeq 1 + \left(\frac{(N+1)\pi}{2|v|}\right)^2. \tag{2.194}$$

The values of b_r calculated from this equation are shown as broken lines on Figure 2.28; for large $|v|$ they approach the numerical solutions asymptotically.

It remains to calculate the loss, α, of a given mode from the loss equation (2.175) with equal reflectivities $R_{13} = R_{12}$. For TE modes, equation (2.8) yields the result for R_{12} in terms of propagation constant β:

$$R_{12} = \left[\frac{(k^2 n_1^2 - \beta^2)^{1/2} - (k^2 n_2^2 - \beta^2)^{1/2}}{(k^2 n_1^2 - \beta^2)^{1/2} + (k^2 n_2^2 - \beta^2)^{1/2}}\right]^2$$

where, for low-loss rays (ϕ small), $(k^2n_1^2 - \beta^2)^{1/2} \ll (k^2n_2^2 - \beta^2)^{1/2}$. Using this approximation it follows that

$$1 - R_{12\perp}^2 \simeq 8\left(\frac{k^2n_1^2 - \beta^2}{k^2n_2^2 - \beta^2}\right)^{1/2}. \tag{2.195}$$

For partial reflections at the core–cladding interface, the Goos–Haenchen shift is zero, so that the effective width $w_\perp$ of the guide, defined as in (2.49), is simply $2a$. Hence we find for TE modes:

$$\alpha = \frac{1 - R_{12\perp}^2}{2w_\perp \cot\phi} \simeq \frac{2(k^2n_1^2 - \beta^2)}{a\beta(k^2n_2^2 - \beta^2)^{1/2}}. \tag{2.196}$$

We may express this result in terms of the normalized parameter b_i for comparison with the computed results of Figure 2.29. Noting that $\alpha = -2\beta_i$, it follows from (2.193) that

$$b_i = -\frac{2\beta_r\beta_i a^2}{|v|^2} \simeq \frac{2(k^2n_1^2 - \beta_r^2)a}{|v|^2(k^2n_2^2 - \beta_r^2)^{1/2}}.$$

Using our earlier ray theory result for β_r and expressing everything in terms of normalized variables:

$$b_i \simeq \frac{(N+1)^2\pi^2}{2|v|^3}. \tag{2.197}$$

The values of b_i calculated from this equation are shown as a function of $|v|$ on Figure 2.29 (broken lines); asymptotically these results coincide with the curves computed numerically from the modal description.

Equation (2.197) illustrates the a^{-3} dependence of loss for large a which was found also for the asymmetric and symmetric metal-clad guides (equations (2.177) and (2.186), respectively). An experimental study of the properties of leaky waves in hollow planar dielectric guides has been reported by Ulrich and Prettl (1973). Using a fluorescent dye between two plates of fused quartz, the attenuation of the guide was measured as a function of thickness $2a$; good agreement was observed with the predicted a^{-3} dependence.

The principal application at present for hollow dielectric waveguides is in the field of waveguide lasers. In the case of gas lasers the gas is confined inside a tube and excited by an electric discharge. The gain thus induced and the optimum gas pressure are found to vary approximately inversely with tube diameter. Hence the use of a hollow waveguide to confine the gas can assist in confining also the laser radiation and increasing the gain. Since the waveguide loss varies as a^{-3} there is frequently an optimum waveguide thickness for a given waveguide laser configuration. A similar situation occurs in dye lasers which are optically pumped and absorb most of the pump radiation near the walls of the tube containing the dye. Hence here also the gain is greatest near the tube walls and the use of a hollow waveguide structure can improve the efficiency of lasing action. The first waveguide laser (Smith, 1971) used a glass capillary to confine the lasing medium, in this case a He–Ne gas mixture. We shall consider the circular hollow waveguide in more detail in Chapter 7. Since 1971 waveguide lasers have been fabricated also with rectangular and planar guides, and details of many practical structures have been summarized by Degnan (1976) in a recent and thorough review. One particular virtue of the planar configuration wich we have dealt with here is its compatibility with the distributed feedback (DFB) principle for laser operation. This feature, proposed originally by Marcuse (1972a), has been studied further for use in a DFB hollow waveguide CO_2 laser (Miles and Grow, 1978, 1979).

2.6 Multilayer Waveguides

We consider next the topic of multilayer waveguides as an extension of the three-layer dielectric slab guide. In particular the general modal theory of four-layer planar guides will be discussed, together with some applications in integrated optics and semiconductor lasers. Later the symmertic five-layer slab theory will be outlined, including the case of high-index outer cladding layers—the so-called W-guide, where the distribution of refractive index across the five layers can resemble the shape of the letter W. The subject of multilayer guides of more than five layers will not be considered. However, multilayer guides with a periodic distribution of index (Bragg waveguides) have been studied both theoretically (Fox, 1974; Yeh and Yariv, 1976; Yeh et al., 1977) and experimentally (Cho et al., 1977; Yeh et al., 1978), and have been successfully used in semiconductor laser structures (Shellan et al., 1978; Dupuis and Dapkus, 1978).

2.6.1 Four-layer asymmetric slab theory

Consider the general four-layer waveguide whose geometry and dielectric distribution are shown in Figure 2.30. Region 1 is taken as the primary region of electromagnetic confinement, although for some cases there is also confinement to region 2. Region 1 has width $2a$ and region 2 has width $2d$; the origin of the x-axis is taken as the boundary between regions 1 and 2. The situation illustrated corresponds to a real guide with $n_1 \geqslant n_2 \geqslant n_3 \geqslant n_4$. However, the formalism to be developed will be sufficiently general to include other possible distributions, e.g. $n_1 \geqslant n_3 \geqslant n_2 \geqslant n_4$, and the possibility of a complex dielectric permittivity in one or more layers is not excluded. For guided modes in the structure shown there are clearly two cases to be considered for the propagation constant β, viz (A) $kn_2 \geqslant \beta \geqslant kn_3$, (B) $kn_1 \geqslant \beta \geqslant kn_2$. Later we will also discuss the case of leaky waves $kn_1 \geqslant kn_3 \geqslant \beta \geqslant kn_2$.

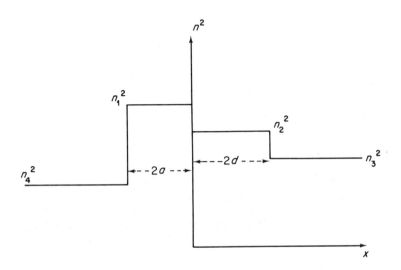

Figure 2.30 The four-layer asymmetric slab structure

A. $kn_2 \geqslant \beta \geqslant kn_3$

For this case the TE and TM modal fields may be expressed as:

$$E_y, H_y = \begin{cases} (-A \sin 2ah_1 + B \cos 2ah_1)e^{h_4(x+2a)}, & x \leqslant -2a & (2.198a) \\ A \sin h_1 x + B \cos h_1 x, & -2a \leqslant x \leqslant 0 & (2.198b) \\ \dfrac{B \cos (h_2 x + \chi)}{\cos \chi}, & 0 \leqslant x \leqslant 2d & (2.198c) \\ \dfrac{B \cos (2dh_2 + \chi)}{\cos \chi} e^{h_3(2d-x)} & x \geqslant 2d & (2.198d) \end{cases}$$

where

$$\left.\begin{aligned} h_1^2 &= k^2 n_1^2 - \beta^2 \\ h_2^2 &= k^2 n_2^2 - \beta^2 \\ h_3^2 &= \beta^2 - k^2 n_3^2 \\ h_4^2 &= \beta^2 - k^2 n_4^2 \end{aligned}\right\}. \tag{2.199}$$

The boundary conditions yield:

at $x = -2a$: $\quad h_4(B \cos 2ah_1 - A \sin 2ah_1) = \dfrac{h_1}{\eta_{14}} (A \cos 2ah_1 + B \sin 2ah_1)$

at $x = 0$: $\quad \dfrac{h_1}{\eta_{12}} A = -h_2 B \tan \chi$

at $x = 2d$: $\quad \dfrac{h_2}{\eta_{23}} \sin (2dh_2 + \chi) = h_3 \cos (2dh_2 + \chi)$

where

$$\eta_{ij} = \begin{cases} 1, & \text{for TE modes} \\ \dfrac{n_i^2}{n_j^2}, & \text{for TM modes.} \end{cases}$$

Eliminating the ratio A/B from the boundary equations at $x = -2a$ and $x = 0$ leaves:

$$2ah_1 = N\pi + \tan^{-1}\left(\eta_{14}\dfrac{h_4}{h_1}\right) + \tan^{-1}\left(\eta_{12}\dfrac{h_2}{h_1} \tan \chi\right) \quad (N = 0, 1, 2, \dots). \tag{2.200}$$

From the boundary condition at $x = 2d$, we have for χ:

$$\chi = \tan^{-1}\left(\eta_{23}\dfrac{h_3}{h_2}\right) - 2dh_2. \tag{2.201}$$

Combining equations (2.200) and (2.201), the eigenvalue equation for the asymmetric four-layer structure is given by (Yamamoto et al., 1975a):

$$2ah_1 = N\pi + \tan^{-1}\left(\eta_{14}\dfrac{h_4}{h_1}\right) + \tan^{-1}\left\{\eta_{12}\dfrac{h_2}{h_1} \tan\left[\tan^{-1}\left(\eta_{23}\dfrac{h_3}{h_2}\right) - 2dh_2\right]\right\}$$

$$(N = 0, 1, 2, \dots) \tag{2.202}$$

where the inverse tangent functions take values between 0 and π. It is perhaps worth noting that equation (2.202) may also be derived by a ray approach, when the second and third terms on the r.h.s. may be interpreted in terms of the phase shifts on reflection at the boundaries $x = -2a$ and $x = 0$, respectively (Tien et al., 1973; Sohler, 1973; Sun and Muller, 1977).

B. $kn_1 \geqslant \beta \geqslant kn_2$

For this case the modal fields may be written, in analogy with (2.198), as:

$$E_y, H_y = \begin{cases} (-A \sin 2ah_1 + B \cos 2ah_1)e^{h_4(x+2a)}, & x \leqslant -2a & (2.203a) \\ A \sin h_1 x + B \cos h_1 x & -2a \leqslant x, \leqslant 0 & (2.203b) \\ \dfrac{B \cosh (h_2'' x + \chi)}{\cosh \chi}, & 0 \leqslant x \leqslant 2d & (2.203c) \\ \dfrac{B \cosh (2dh_2'' + \chi)}{\cosh \chi} e^{h_3(2d-x)}, & x \geqslant 2d & (2.203d) \end{cases}$$

where h_1, h_3, h_4 are defined as in (2.199), and $h_2''^2 = \beta^2 - k^2 n_2^2$. Applying the boundary conditions at the three dielectric interfaces and eliminating A/B and χ as in case A, the eigenvalue equation is found to be

$$2ah_1 = N\pi + \tan^{-1}\left(\eta_{14}\frac{h_4}{h_1}\right) + \tan^{-1}\left\{\eta_{12}\frac{h_2''}{h_1} \tanh\left[\tanh^{-1}\left(\eta_{23}\frac{h_3}{h_2''}\right) + 2dh_2''\right]\right\}$$

$(N = 0, 1, 2, \ldots)$. (2.204)

Note that equation (2.204) may also be derived directly from case A by replacing h_2 by ih_2'' in equation (2.202).

Numerical solutions of the eigenvalue equations (2.202) and (2.204) for real four-layer dielectric waveguides are easily obtained (Smith, 1968; Tien et al., 1973; Sun and Muller, 1977; Cherny et al., 1979). The results have been checked against measurements of mode index (β/k) on garnet layers with a high level of agreement (Sun and Muller, 1977). However, rather than giving plots of propagation constant for these structures we will proceed to consider some applications of the theory and to develop some useful approximations for special cases.

2.6.2 Applications of the four-layer slab

A simple example of four-layer slab applications is furnished by the large optical cavity (LOC) laser (Lockwood et al., 1970; Kressel et al., 1971a). This structure differs from the conventional double heterostructure laser described in Section 2.3 in that the p–n junction is displaced from one of the hetero-interfaces into the central region. The four-layer structure thus created exhibits gain only in the p-layer of the region between the heterojunctions as a consequence of the dominance of electron injection in GaAs. Since the band-gap of p-type GaAs is smaller than that of n-type, the refractive index of this layer will also be slightly increased. In summary the regions of Figure 2.30 are identified for the LOC laser as:

region 1: p-type GaAs
region 2: n-type GaAs
region 3: n-type $Al_x Ga_{1-x} As$
region 4: p-type $Al_x Ga_{1-x} As$.

The original objective of this structure was to achieve large optical confinement regions (layers 1 and 2) for high-power output; since the structure was usually multimode, optical power was more evenly distributed at the cavity facets than for single-mode lasers and this resulted in a higher limit for catastrophic mirror damage (Lockwood et al., 1970). However, the structure was also produced with a narrow n-GaAs region (2) resulting in a low-threshold c.w. LOC laser (Kressel et al., 1971a); a detailed theoretical analysis of these structures was made by Butler and Kressel (1972). Single-mode operation of the LOC structure with an optical cavity width of 2 μm (and hence high power and narrow output beam divergence) was achieved with $a/d = 2$ (Figure 2.30) by Paoli et al. (1973). Further theoretical and experimental work has concentrated on achieving high-power single-mode output from a large-cavity four-layer structure including small amounts of Al in the n-region (region 2 of Figure 2.30) (Kirkby and Thompson, 1972; Hakki and Hwang, 1974; Krupka, 1975).

Four-layer slab theory also lies at the heart of the optical waveguide lens (Southwell, 1977) and the taper coupler used to transfer guided radiation from one layer to another. The use of a smooth taper as the transition region between two layers of different refractive index was first demonstrated in organo-silicone compounds (Tien et al., 1973). More recently semiconductor lasers with intracavity taper-coupled passive waveguide sections have been fabricated by liquid phase epitaxy (Reinhart and Logan, 1975; Logan and Reinhart, 1975). This basic structure has also been incorporated as part of an intracavity frequency-modulated heterostructure laser (Reinhart and Logan, 1975a) making use of the linear electro-optic effect. Devices such as these show good promise for monolithically-integrated optical circuits based on $Al_xGa_{1-x}As$ alloys.

Metal-clad waveguides have also been investigated in a four-layer configuration (Polky and Mitchell, 1974; Yamamoto et al., 1975a; Rashleigh, 1976a, 1976b). For such a structure, layer 3 of Figure 2.30 is metallic, characterized by $\varepsilon_3 = \varepsilon_3' + i\varepsilon_3''$ rather than n_3^2, and layer 2 is a dielectric buffer layer. One function of this buffer layer is to reduce the attenuation coefficients of the modes in a metal-clad waveguide whilst retaining the mode-selective properties. We may investigate this behaviour by an approximate analysis of the modal attenuation (Yamamoto et al., 1975a). For large thickness $2a$ of the core region 1 we may again use the trigonometric identity

$$\tan^{-1}(x) = \frac{\pi}{2} - \tan^{-1}\left(\frac{1}{x}\right)$$

so that equation (2.204) assumes the approximate form:

$$2ah_1 \simeq (N+1)\pi - \tan^{-1}\left(\frac{h_1}{\eta_{14}h_4}\right)$$

$$- \tan^{-1}\left\{\frac{h_1}{\eta_{12}h_2'' \tanh\left[\tanh^{-1}\left(\frac{\eta_{23}h_3}{h_2''}\right) + 2dh_2''\right]}\right\}. \quad (2.205)$$

The zero-order solution for h_1 is given by the familiar expression

$$h_1^{(0)} \simeq \frac{(N+1)\pi}{2a}. \quad (2.206)$$

Hence, for large values of a, (2.205) may be further approximated by taking the first term in the expansion of each $\tan^{-1}$ function; this yields a first-order approximation for h_1:

$$h_1^{(1)} \simeq \frac{(N+1)\pi}{\left\{2a + \dfrac{1}{\eta_{14}h_4} + \dfrac{1}{\eta_{12}h_2''\tanh[\tanh^{-1}(\eta_{23}h_3/h_2'') + 2dh_2'']}\right\}}$$

$$\simeq \frac{(N+1)\pi}{2a}\left\{1 - \frac{1}{2a\eta_{14}h_4} - \frac{1}{2a\eta_{12}h_2''\tanh[\tanh^{-1}(\eta_{23}h_3/h_2'') + 2dh_2'']}\right\} \quad (2.207)$$

where h_2'', h_3, and h_4 on the r.h.s. of (2.207) are to be interpreted as their zero-order approximations expressed in terms of $h_1^{(0)}$ from (2.206). In general h_3 is now a complex quantity, defined in terms of ε_3, with real and imaginary parts denoted by h_3', h_3'', respectively. It follows that $h_1^{(1)}$ is also complex with real and imaginary parts $h_1'^{(1)}$ and $h_1''^{(1)}$. Hence the power attenuation α is given from (2.199) as

$$\alpha = -2\beta_i \simeq \frac{2h_1''^{(1)} h_1'^{(1)}}{\beta_r} \simeq \frac{2(N+1)\pi h_1''^{(1)}}{2akn_1}. \tag{2.208}$$

Using (2.207) to find an expression for $h_1''^{(1)}$ in (2.208) yields (Yamamoto et al., 1975a):

$$\alpha \simeq \frac{2(N+1)^2 \pi^2}{(2a)^3 kn_1} \left(\frac{\eta_{23}}{\eta_{12}}\right) \left[\frac{4h_3''}{\zeta_1 \exp(4dh_2'') + \zeta_2 \exp(-4dh_2'') + \zeta_3} \right] \tag{2.209}$$

where

$$\zeta_1 = \eta_{23}^2 (h_3'^2 + h_3''^2) + h_2''^2 + 2\eta_{23} h_3' h_2'' \tag{2.210a}$$
$$\zeta_2 = \eta_{23}^2 (h_3'^2 + h_3''^2) + h_2''^2 - 2\eta_{23} h_3' h_2'' \tag{2.210b}$$
$$\zeta_3 = 2[\eta_{23}^2 (h_3'^2 + h_3''^2) - h_2''^2] \tag{2.210c}$$

and

$$h_2''^2 \simeq k^2(n_1^2 - n_2^2) - \frac{(N+1)^2 \pi^2}{4a^2}$$

$$h_3'^2 \simeq k^2(n_1^2 - \varepsilon_3') - \frac{(N+1)^2 \pi^2}{4a^2}$$

$$h_3'' \simeq -\frac{k^2 \varepsilon_3''}{2h_3'} \qquad \eta_{23} = \begin{cases} 1 & \text{(TE)} \\ \dfrac{n_2^2}{\varepsilon_3'} & \text{(TM)}. \end{cases}$$

Equation (2.209) gives a first-order approximation for the attenuation coefficient of the Nth mode (TE or TM) of a four-layer metal-clad waveguide incorporating a dielectric buffer layer of thickness $2d$. In the limit $d \to \infty$ the loss tends to zero and the structure reverts to the conventional three-layer dielectric slab waveguide. In the limit $d \to 0$ the loss tends to the expression (2.187) obtained for the asymmetric metal-clad guide, provided $|\varepsilon_3'| \gg |\varepsilon_3''|$. Notice also that (2.209) retains the $(N+1)^2/a^3$ dependence exhibited by the symmetric and asymmetric metal-clad guides ((2.186) and (2.187), respectively) and by the hollow dielectric waveguide (equation (2.197)).

Let us examine the variation of α in equation (2.209) with variation of dielectric buffer-layer thickness $2d$. For the TE modes, where $\eta_{ij} = 1$, the behaviour of α is monotonically decreasing with increasing values of $2d$; for small thicknesses the attenuation is similar to that for the corresponding modes of the asymmetric metal-clad guide, whilst for larger thicknesses an exponential decay occurs:

$$\alpha \propto \frac{(N+1)^2}{k^2(2a)^3} \exp(-4dh_2'').$$

For the TM modes, on the other hand, if the metal cladding has a negative real permittivity ($\varepsilon_3' < 0$) then η_{23} is negative and the behaviour of α may exhibit a maximum for one value of thickness $2d$. This may be understood in (2.209) for the case when $\zeta_2 > \zeta_1$; the value of $2d$ at which the maximum α occurs is found by equating to zero the derivative of (2.209) with respect to d. It is given by (Yamamoto et al., 1975a):

$$(2d)_m = \frac{1}{4h_2''} \ln\left(\frac{\zeta_2}{\zeta_1}\right). \qquad (2.211)$$

The occurrence of this attenuation peak has been confirmed by numerical solutions of the eigenvalue equation (2.204) (Polky and Mitchell, 1974; Yamamoto et al., 1975a; Rashleigh, 1976a) which also demonstrate that (2.211) is a good approximation for the position of the peak. The phenomenon has been attributed to a resonant coupling of the TM modes to the lossy surface plasma wave discussed for metallic claddings in Section 2.5.

In the region close to the attenuation peak, the approximation (2.209) for α no longer holds since the approximate form (2.205) of the eigenvalue equation is not appropriate. At the peak half the phase shift on internal reflection is not $\pi/2$, as in (2.205), but zero. In this case the zero-order approximation for h_1 is

$$h_1^{(0)} \simeq \frac{(N+\tfrac{1}{2})\pi}{2a}.$$

Following a similar argument in approximating (2.204) to that used previously we find that in the vicinity of the attenuation maximum the loss α_m is given to first order by (Yamamoto et al., 1975a):

$$\alpha_m \simeq \frac{n_1}{ak\,\varepsilon_3'}\left[\frac{4h_3''\,h_2''^2}{\zeta_1 \exp(4dh_2'') + \zeta_2 \exp(-4dh_2'') + \zeta_3}\right] \qquad (2.212)$$

where the ζ's are defined as in (2.210) and the h's are given by:

$$h_2''^2 \simeq k^2(n_1^2 - n_2^2) - \frac{(N+\tfrac{1}{2})^2 \pi^2}{4a^2}$$

$$h'^2 \simeq k^2(n_1^2 - \varepsilon_3') - \frac{(N+\tfrac{1}{2})^2 \pi^2}{4a^2}$$

$$h_3'' \simeq -\frac{k^2 \varepsilon_3''}{2h_3'}.$$

From the form of equation (2.212) it is clear that the earlier estimate for the position $(2d)_m$ of the attenuation peak is given as before by equation (2.211). Numerical solutions of (2.204) for the TM modes close to the resonant attenuation verify that (2.212) is a valid approximation (Yamamoto et al., 1975a).

For buffer layer thicknesses close to that given by (2.211) the TE and TM attenuations for the same order mode can differ by the order of 1000 dB/cm (Polky and Mitchell, 1974). Hence the structure considered here shows merit for use as a polarizer. A second application which might be envisaged would be an absorption modulator utilizing the steep slope of the TM attenuation curve for $2d > (2d)_m$. Such a device should achieve a high extinction ratio for a small change of refractive index. A further interesting property of the structure which may also prove useful for device applications is the fact that TE_N and TM_N modes can have the same phase velocity for appropriate choices of core and buffer layer thicknesses $2a$ and $2d$ (Rashleigh, 1976b). Values for these parameters may be found by equating the corresponding TE and TM eigenvalue equations; numerical solution is relatively simple for the case $|\varepsilon_3'| \gg |\varepsilon_3''|$. Rib waveguide polarizers utilizing the buffer-layer effect discussed above have recently been demonstrated (Reinhart et al., 1980).

The treatment of the metal-clad four-layer waveguide given above introduced the concept of a complex transverse propagation parameter h_3. A somewhat analogous situation occurs if we consider the dielectric four-layer waveguide of Figure 2.30 with $n_3 \geq n_2$. In this case the low-index buffer layer implies that modes can exist for which $kn_3 \geq \beta \geq kn_2$, i.e. in a zero-order approximation h_3 will be purely imaginary and denoted by ih_3''. Then the derivation leading to equation (2.209) for the loss of such a mode still holds with appropriate redefinition of h_3'', and $h_3' = 0$. The result becomes therefore

$$\alpha \simeq \frac{2(N+1)^2 \pi^2}{(2a)^3 kn_1} \left(\frac{\eta_{23}}{\eta_{12}}\right) \left[\frac{2h_3''}{(\eta_{23}{}^2 h_3''{}^2 + h_2''{}^2)\cosh(4dh_2'') + (\eta_{23}{}^2 h_3''{}^2 - h_2''{}^2)}\right] \quad (2.213)$$

where

$$h_3''{}^2 = \frac{(N+1)^2 \pi^2}{4a^2} - k^2(n_1{}^2 - n_3{}^2); \quad \eta_{23} = \begin{cases} 1 & \text{(TE)} \\ \dfrac{n_2{}^2}{n_3{}^2} & \text{(TM)} \end{cases}$$

and the other symbols have their previous definitions. Equation (2.213) exhibits a monotonic decrease of α with increasing buffer-layer thickness $2d$. For TE modes at large thicknesses the result assumes a particularly simple form:

$$\alpha_{TE} \simeq \frac{2(N+1)^2 \pi^2}{(2a)^3 kn_1} \left(\frac{4h_3''}{k^2(n_3{}^2 - n_2{}^2)}\right) \exp(-4dh_2'').$$

The field distributions of these leaky waves have oscillatory forms in the core region (1), evanescent decay in the buffer layer (2), and a further oscillatory behaviour in the high-index cladding (3) where the wave radiates outwards from the guiding structure. A practical situation where such a structure occurs is the case of silicon nitride films grown on silicon and separated from the substrate by a buffer layer of SiO_2; such structures show good promise for waveguide applications in integrated optics (Stutius and Streifer, 1977). Another example of the phenomenon is a heterostructure laser where the cladding layers of the simple three-layer structure are no longer infinitely thick as assumed in the discussion of this device in Section 2.3. In practice these layers are of order 1 μm and are bounded by substrate and contact layers of GaAs in the GaAs/AlGaAs laser. This means that field penetration can occur through the low-index AlGaAs layers into the outer GaAs layers (Casey and Panish, 1975). Normally this penetration results in loss, as described by (2.213), which implies increased threshold currents and is therefore undesirable; however, in c.w. lasers thin AlGaAs layers are of assistance in providing efficient heat-sinking (the thermal resistivity of AlGaAs greatly exceeds that of GaAs). Hence optimum values of the thickness $2d$ exist for these devices and detailed numerical analyses of the loss in such structures have been made (Butler *et al.*, 1975; Butler and Wang, 1976; Streifer *et al.*, 1976).

This same effect of field penetration through the low-index buffer layer has been used also to design a leaky wave laser (Scifres *et al.*, 1976). In this application the buffer layer is made very thin ($\sim 0.1\,\mu$m) as compared with the core region ($\sim 1.2\,\mu$m) so that the radiation is strongly coupled into the substrate. The mode then emerges from the substrate at a well-defined angle with a high degree of collimation (about 2°)—a feature not normally associated with semiconductor injection lasers. The angle θ which the emergent beam makes with the z-axis is easily seen to be given by

$$\sin\theta = \left(n_3{}^2 - \frac{\beta^2}{k^2}\right)^{1/2}.$$

The loss penalty paid for this collimated beam in the leaky wave laser is an increase of threshold current density of about 30 per cent over that of a conventional double heterostructure.

A structure which is closely related to the one just described is that in which the buffer layer (layer 2 of Figure 2.30) is a metal whose real part permittivity ε'_2 is negative. The loss of a mode in such a structure may then be approximately described by equation (2.213) with the factor n_2^2 in all the definitions replaced by ε'_2, provided the imaginary part of the permittivity is negligibly small. The modes of this structure with a thin metallic cladding in region 3 have been analysed numerically by Rashleigh (1976c). As in other metal-clad guides, the TM modes can couple strongly to the surface plasma waves occurring at both metal–dielectric interfaces. A cut-off polarizer of this type has been fabricated (Rollke and Sohler, 1977) by depositing a thin (~ 50 Å) silver stripe on the surface of a guide. The effect of the thin metal cladding is to couple the TE_0 mode out into the substrate (layer 4 of Figure 2.30) whilst permitting the TM_0 mode to propagate with relatively low loss. Extinction ratios as high as 18 dB were reported with silver layers on a glass waveguide. A further analysis of this structure by Moshkun et al. (1978) has shown that the TE_0 modal losses exhibit a peak as the buffer layer thickness is varied.

2.6.3 Five-layer symmetric slab theory

The five-layer symmetric slab waveguide is shown schematically in Figure 2.31. It consists of a

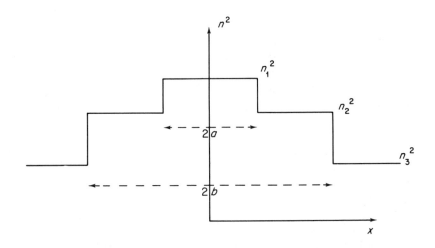

Figure 2.31 The five-layer symmetric slab structure

central layer, refractive index n_1, thickness $2a$, embedded in cladding layers of index n_2, overall thickness $2b$,* which are in turn sandwiched between outer claddings of index n_3. For the structure considered here, $n_1 \geqslant n_2 \geqslant n_3$; in the next subsection we shall consider the W-guide for which $n_1 \geqslant n_3 > n_2$. The five-layer guide is symmetric about the midpoint of the core region which is accordingly designated as $x = 0$. As in the case of the asymmetric four layer guide of Section 2.6.1 there are two cases for the guided modes: (A) $kn_2 \geqslant \beta \geqslant kn_3$, (B) $kn_1 \geqslant \beta \geqslant kn_2$. For simplicity, and since our main interest lies in the zero-order mode of this structure, we will consider only even-order modes. The five-layer symmetric guide with anisotropic dielectric permittivity has been considered by Nelson and McKenna (1967).

* Note here that the b should not be confused with the normalized propagation constant used elsewhere.

A. $kn_2 \geq \beta \geq kn_3$

The field distributions for even-order modes are written:

$$E_y, H_y = \begin{cases} A\cos h_1 x, & 0 \leq |x| \leq a \quad (2.214a) \\ \dfrac{A\cos h_1 a \cos(h_2|x|+\chi)}{\cos(h_2 a+\chi)}, & a \leq |x| \leq b \quad (2.214b) \\ \dfrac{A\cos h_1 a \cos(h_2 b+\chi)}{\cos(h_2 a+\chi)} e^{h_3(b-|x|)}, & |x| \geq b \quad (2.214c) \end{cases}$$

where

$$\left. \begin{array}{l} h_1^2 = k^2 n_1^2 - \beta^2 \\ h_2^2 = k^2 n_2^2 - \beta^2 \\ h_3^2 = \beta^2 - k^2 n_3^2 \end{array} \right\} \quad (2.215)$$

The boundary conditions yield:

at $x = a$: $\quad \dfrac{h_1}{\eta_{12}} \tan h_1 a = h_2 \tan(h_2 a + \chi)$

at $x = b$: $\quad \dfrac{h_2}{\eta_{23}} \tan(h_2 b + \chi) = h_3$.

Eliminating χ, we find:

$$h_1 a = M\pi + \tan^{-1}\left\{ \eta_{12} \dfrac{h_2}{h_1} \tan\left[\tan^{-1}\left(\eta_{23}\dfrac{h_3}{h_2} \right) - h_2(b-a) \right] \right\}$$
$$(M = 0, 1, 2, \ldots) \quad (2.216)$$

where

$$\eta_{ij} = \begin{cases} 1 & \text{for TE modes} \\ \dfrac{n_i^2}{n_j^2} & \text{for TM modes.} \end{cases}$$

Note that the index M in the eigenvalue equation (2.216) gives only the even-order modes, i.e., $N = 2M$ in the notation previously used.

B. $kn_1 \geq \beta \geq kn_2$

In this case the even-order field distributions become:

$$E_y, H_y = \begin{cases} A\cos h_1 x, & 0 \leq |x| \leq a \quad (2.217a) \\ \dfrac{A\cos h_1 a \cosh(h_2''|x|+\chi)}{\cosh(h_2'' a+\chi)}, & 0 \leq |x| \leq b \quad (2.217b) \\ \dfrac{A\cos h_1 a \cosh(h_2'' b+\chi)}{\cosh(h_2'' a+\chi)} e^{h_3(b-|x|)}, & |x| \geq b \quad (2.217c) \end{cases}$$

where h_1, h_3 are defined as in (2.215) and $h_2''^2 = \beta^2 - k^2 n_2^2$.
The corresponding eigenvalue equation is:

$$h_1 a = M\pi + \tan^{-1}\left\{ \eta_{12} \dfrac{h_2''}{h_1} \tanh\left[\tanh^{-1}\left(\eta_{23}\dfrac{h_3}{h_2''} \right) + h_2''(b-a) \right] \right\} \quad (M = 0, 1, 2, \ldots).$$

$$(2.218)$$

The eigenvalue equations (2.216) and (2.218) for even-order modes in a five-layer symmetric slab may be conveniently re-written in terms of normalized variables for the weakly-guiding situation $\eta_{ij} \simeq 1$ for which TE and TM modes are approximately degenerate. We may define, in analogy with the earlier normalized variables,

$$v^2 = a^2 k^2 (n_1^2 - n_3^2) \tag{2.219}$$
$$u^2 = a^2 (k^2 n_1^2 - \beta^2) \equiv a^2 h_1^2 \tag{2.220}$$
$$w^2 = a^2 (\beta^2 - k^2 n_3^2) = v^2 - u^2 \equiv a^2 h_3^2 \tag{2.221}$$
$$t^2 = a^2 (k^2 n_2^2 - \beta^2) = u^2 - v^2 c^2 \equiv a^2 h_2^2 \tag{2.222a}$$
$$t''^2 = a^2 (\beta^2 - k^2 n_2^2) = v^2 c^2 - u^2 \equiv a^2 h_2''^2 \tag{2.222b}$$

where

$$c^2 = \frac{n_1^2 - n_2^2}{n_1^2 - n_3^2} \simeq \frac{n_1 - n_2}{n_1 - n_3}. \tag{2.223}$$

Using these definitions, (2.216) becomes

$$u = M\pi + \tan^{-1}\left\{\frac{t}{u} \tan\left[\tan^{-1}\left(\frac{w}{t}\right) - t\left(\frac{b}{a} - 1\right)\right]\right\} \tag{2.224}$$

and (2.218) becomes ($t \to it''$ in (2.224)):

$$u = M\pi + \tan^{-1}\left\{\frac{t''}{u} \tanh\left[\tanh^{-1}\left(\frac{w}{t''}\right) + t''\left(\frac{b}{a} - 1\right)\right]\right\}. \tag{2.225}$$

Equations (2.224) and (2.225) are the eigenvalue equations for the weakly-guiding symmetric five-layer waveguide in the cases (A) $c \leqslant u/v \leqslant 1$ and (B) $0 \leqslant u/v \leqslant c$, respectively. They may easily be solved numerically to yield, say, u as a function of v for various values of the two parameters c and b/a.

A fairly simple application of the five-layer symmetric slab theory occurs in separate confinement heterostructure (SCH) lasers (Casey et al., 1974). In these devices, also termed localized gain region (LGR) lasers (Thompson and Kirkby, 1973), the layers are fabricated in such a way as to confine the region where electron–hole recombination occurs to the central layer (1) of Figure 2.31. Layers 1, 2, 3 are, respectively GaAs, $Al_y Ga_{1-y} As$, and $Al_x Ga_{1-x} As$ ($x > y$) so that the optical field is allowed to spread into the whole of layers 1 and 2 for appropriate values of alloy compositions x and y (related to the parameter c defined in (2.223)). The advantage of this structure is that the confinement of radiation to the recombination region can be made superior to that for the conventional double heterostructure (DH) laser, so that there is more efficient pumping of the optical field by the recombination process. We may demonstrate this effect by calculating the confinement factor Γ defined, as in Section 2.3.5, as the ratio of power confined in the core to the total power. For the conventional DH laser an expression for $\Gamma (\equiv P_{\text{core}}/P)$ has already been given in equation (2.125). For the SCH structure, the definition of Γ is

$$\Gamma = \frac{\int_0^a |E_y|^2 dx}{\int_0^\infty |E_y|^2 dx} \tag{2.226}$$

where we have considered only TE modes, assuming approximate TE/TM degeneracy. Using the field distributions given in (2.214) and (2.217), and expressing the results in terms of the normalized variables defined in (2.219)–(2.223), a straightforward calculation yields:

A. $c \leqslant u/v \leqslant 1$:

$$\Gamma = \frac{u + \sin u \cos u}{u + \sin u \cos u \left(1 - \frac{u^2}{t^2}\right) + u\left(\cos^2 u + \frac{u^2}{t^2}\sin^2 u\right)\left(\frac{b}{a} - 1 + \frac{1}{w}\right)} \quad (2.227a)$$

B. $0 \leqslant u/v \leqslant c$:

$$\Gamma = \frac{u + \sin u \cos u}{u + \sin u \cos u \left(1 + \frac{u^2}{t''^2}\right) + u\left(\cos^2 u - \frac{u^2}{t''^2}\sin^2 u\right)\left(\frac{b}{a} - 1 + \frac{1}{w}\right)}. \quad (2.227b)$$

Examples of the variation of Γ with v for the lowest order mode from equations (2.227a, b) obtained by numerical solution of (2.224) and (2.225) are given in Figure 2.32 (solid lines) for

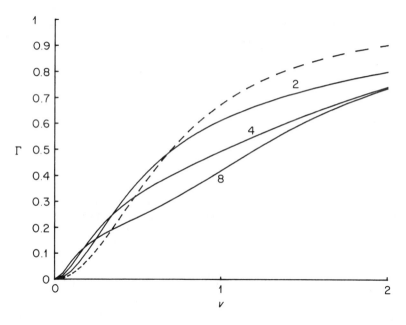

Figure 2.32 Confinement factor Γ versus v for the lowest-order mode of the symmetric five-layer waveguide (solid lines) with $c^2 = \frac{1}{3}$ (defined as in equation (2.223)) and $b/a = 2, 4, 8$; labelling parameter gives the value of b/a. The broken line gives the variation of Γ with v for the equivalent three-layer slab

$c^2 = \frac{1}{3}$ and $b/a = 2, 4,$ and 8. The broken line gives the variation of Γ for the equivalent three-layer symmetric slab ($b/a = 1$). For small v the values of Γ obtained with the five-layer structure are larger than those for the equivalent three-layer slab. Since in a semiconductor laser the threshold current density is inversely dependent on Γ (see Section 2.3.6) the result is a lower threshold for the SCH laser as compared with the equivalent DH device. Although the effect, as shown in Figure 2.32, is small, it is sufficient to produce worthwhile improvements; it may be augmented in practice by a superlinear gain–current relationship (Thompson and Kirkby, 1973) rather than the linear assumption used in equation (2.140). Using the SCH structure, threshold current densities as low as 575 A/cm^2 have been achieved (Thompson and Kirkby, 1973a; Thompson et al., 1976).

A further advantage offered by the SCH structure is that it may be used in devices where it is desired to prevent current carriers from reaching a region where nonradiative recombination may occur, whilst permitting the optical field to penetrate this region. An example of such a situation occurs in the distributed feedback (DFB) laser (Kogelnik and Shank, 1971, 1972; Wang, 1973) where the feedback corrugations may introduce nonradiative recombination centres. In this case use of the SCH structure has eliminated the problem and resulted in room temperature c.w. DFB semiconductor lasers (Casey et al., 1975; Aiki et al., 1975; Nakamura et al., 1975).

2.6.4 The 'W' guide

A refractive index distribution resembling the shape of the letter W can be obtained in a five-layer slab with $n_1 \geqslant n_3 > n_2$ (see Figure 2.31) i.e. $c^2 > 1$ in (2.223). This structure has interesting properties with respect to mode cut-offs, confinement factors, and mode filter characteristics (Ohtaka et al., 1974). The guided modes of the structure may be described by the formalism developed for case B of the previous subsection; there are also leaky modes of interest for which the transverse propagation constant h_3 assumes complex values. Let us consider first the condition for cut-off of the first-order mode in order to examine the range of v-values for which the zero-order mode propagates alone. Restricting attention to the weakly-guiding case and using the normalized variables defined in (2.219)–(2.223), a simple extension of the results of the previous subsection to odd-order modes yields for the eigenvalue equation:

$$u = (M' + \tfrac{1}{2})\pi + \tan^{-1}\left\{\frac{t''}{u}\tanh\left[\tanh^{-1}\left(\frac{w}{t''}\right) + t''\left(\frac{b}{a} - 1\right)\right]\right\} \quad (M' = 0, 1, 2, \ldots) \tag{2.228}$$

where M' is related to N, the mode number, by $N = 2M' + 1$. At cut-off, $w = 0$, $u = v_c$, and $t''^2 = v_c^2(c^2 - 1)$, so that equation (2.228) becomes (Kawakami and Nishida, 1974):

$$-\cot v_c = (c^2 - 1)^{1/2} \tanh\left[v_c(c^2 - 1)^{1/2}\left(\frac{b}{a} - 1\right)\right]. \tag{2.229}$$

Plots of the cut-off values v_c for the first-order mode versus the ratio b/a for various values of c^2 are shown in Figure 2.33. For $b/a = 1$ the cut-off is given by $v_c = \pi/2$ as for the three-layer slab, but with increasing values of b/a and c the cut-off is shifted to higher values. This is a potentially useful feature for obtaining single-mode guides with larger dimensions and/or refractive index differences than is possible in the three-layer waveguide. There is, however, a limit on the values of cut-off which can be obtained; for $b/a \to \infty$ equation (2.229) gives

$$\cot v_c = -(c^2 - 1)^{1/2}$$

so that the maximum value of cut-off which can be achieved as $c \to \infty$ is $v_c = \pi$. In principle therefore W-guides can be made with core thicknesses twice those of conventional symmetric three-layer structures for single-mode operation with the same difference of refractive index between core and outer cladding.

The confinement factor Γ for the zero-order mode in the W-guide is also of interest; this is given by equation (2.227b) for $0 \leqslant u \leqslant v$, where u is the solution of the eigenvalue equation (2.225). Computed results for Γ versus v of the lowest-order mode are given in Figure 2.34 (solid lines) for $b/a = 1.5$ and various values of the parameter c; the corresponding results for the symmetric three-layer slab ($b/a = 1$) from equation (2.125) are also shown as a broken

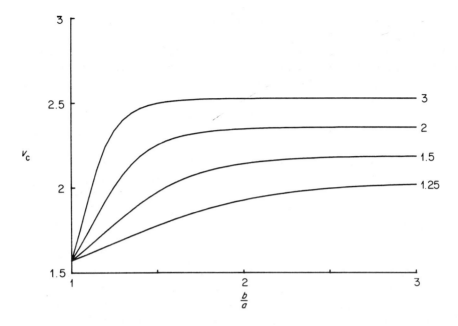

Figure 2.33 Cut-off values v_c of the first-order mode of the W waveguide as a function of layer thickness ratio b/a. Labelling parameter gives the value of c^2 defined as in (2.223)

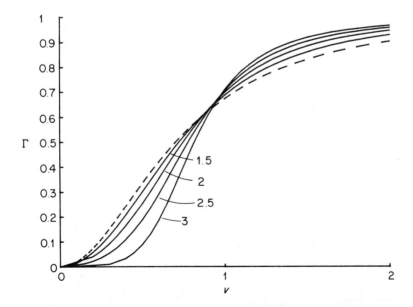

Figure 2.34 Confinement factor Γ versus v for the lowest-order mode of the W-guide (solid lines) for $b/a = 1.5$; labelling parameter gives the value of c^2 defined as in equation (2.223). The broken line gives the variation of Γ with v for the equivalent three-layer slab

line for comparison. At low v the three-layer slab gives better mode confinement, corresponding to the fact that close to $w = 0$ there is a considerable field penetration in layer 2 of the W-guide. However, for values of v above about 1 the W-guide gives better confinement, the effect increasing with increasing values of c; for this region of operation the mode of the W-guide is characterized by a strong decay of the field distribution in layer 2. If we recall the earlier result of an extended range of single-mode operation it becomes clear that this effect of improved confinement at higher v-values in the W-guide can be an attractive feature of the structure, for example in cases where the optical field is to be excluded from lossy outer cladding layers.

Although the W-guide offers an increased range of single-mode operation as compared with the symmetric three-layer slab, it should be remembered that the first-order and higher-order modes may still propagate below cut-off as leaky or quasi-guided modes of relatively low loss (Kawakami and Nishida, 1974; Suematsu and Furuya, 1975). It is therefore important to estimate the loss of a given mode below cut-off in order to find if the mode filter characteristics of the W-guide are physically realizable. We make such an analysis using the eigenvalue equations (2.225) and (2.228) as starting points. For the leaky modes of interest a zero-order approximation implies, as in the case of the four-layer analysis of Section 2.6.2, replacing h_3 by ih'_3, i.e. in normalized variables, replacing w by iw'' where $w''^2 = a^2(k^2 n_3^2 - \beta^2)$, so that there is an oscillatory field distribution in the outer cladding layer. Using the approximations of Section 2.6.2, the zero-order approximations for u are given by:

even-order modes (from (2.225)): $\quad u^{(0)} = \dfrac{(2M+1)\pi}{2}$ \hfill (2.230a)

odd-order modes (from (2.228)): $\quad u^{(0)} = \dfrac{(2M'+2)\pi}{2}.$ \hfill (2.230b)

In each case, making the change to the conventional mode number N, we find the familiar result:

$$u^{(0)} = (N+1)\frac{\pi}{2}. \tag{2.230c}$$

The corresponding first-order approximations are:

even-order modes: $\quad u^{(1)} = \dfrac{(2M+1)\pi}{2}\left\{1 - \dfrac{\dfrac{iw''}{t''} + \tanh\left[t''\left(\dfrac{b}{a}-1\right)\right]}{t'' + iw'' \tanh\left[t''\left(\dfrac{b}{a}-1\right)\right]}\right\},$ \hfill (2.231a)

odd-order modes: $\quad u^{(1)} = \dfrac{(2M'+2)\pi}{2}\left\{1 - \dfrac{\dfrac{iw''}{t''} + \tanh\left[t''\left(\dfrac{b}{a}-1\right)\right]}{t'' + iw'' \tanh\left[t''\left(\dfrac{b}{a}-1\right)\right]}\right\},$ \hfill (2.231b)

Combining both these results in terms of the usual mode number N ($= 2M, 2M'+1$) and following a similar argument to that of Section 2.6.2, we find for the loss α:

$$\alpha \simeq \frac{(N+1)^2 \pi^2}{a^2 k n_1}\left\{\frac{w''}{v^2(c^2-1)\cosh[2t''(b/a-1)] + (t''^2 - w''^2)}\right\} \tag{2.232}$$

where

$$w''^2 = \frac{(N+1)^2\pi^2}{4} - v^2; \quad t''^2 = v^2c^2 - \frac{(N+1)^2\pi^2}{4}.$$

Equation (2.232) gives the first-order approximation for the attenuation of the Nth order mode of the W-guide; it is equal to twice the loss for the equivalent mode in the asymmetric four-layer guide with low-index buffer layer, as given by equation (2.213) (TE: $2d = b - a$). A derivation of an equivalent loss formula by the methods of ray optics for the W-guide has been given by Love and Winkler (1977).

To obtain a useful mode filter from the W-guide structure it is important to choose the parameters such that, for example, the lowest-order mode is guided whilst the higher-order modes are leaky waves with high losses. For the attenuation of these modes to be large, equation (2.232) says that the parameters c^2 and b/a should be small, i.e. close to 1. However, under these conditions one no longer obtains the desirable properties of (i) extended v-ranges for single-mode operation and (ii) improved zero-order mode confinement, as discussed above. We may conclude therefore that the W-guide may offer some advantages in these respects over conventional three-layer guides but only for rather restricted ranges of the design parameters v, c^2, and b/a.

A structure with rather more tolerance in the design parameters as regards achieving desirable mode filter characteristics is obtained for a W-guide with $n_1 = n_3$. In this case all the modes are of the leaky type but there can be large changes in the loss of the various order modes. For this case equation (2.232) reduces, for low-order modes ($t'' \gg w''; t''(b/a - 1) \gg 1$) to the simple expression (Suematsu and Furuya, 1975):

$$\alpha \simeq \frac{(N+1)^3\pi^3}{a^4 k^3 n_1 (n_1^2 - n_2^2)} \exp\left[-2ak(n_1^2 - n_2^2)^{1/2}\left(\frac{b}{a} - 1\right)\right]. \tag{2.233}$$

Since this equation gives the variation of loss as $(N+1)^3$ for low-order leaky modes the strong mode filter action is clearly achieved in a W-structure with outer cladding refractive index the same as that of the core.

Chapter 3
Two-dimensional Parabolic-index Media

A large proportion of the remainder of this book deals with waveguides whose refractive index is inhomogeneous in the direction(s) normal to the waveguide axis, i.e. graded-index guides. This grading occurs as a result of the fabrication process, e.g. diffusion or ion implantation in planar guides, and may produce waveguides with characteristics unique to the specific variation of refractive index obtained. It is therefore appropriate that this chapter should deal with the simplest form of symmetric refractive index variation $n(x)$, that obeying the parabolic law:

$$n^2(x) = n_1^2 \left[1 - 2\Delta \left(\frac{x}{a} \right)^2 \right] \tag{3.1}$$

where x is the direction normal to the axis of propagation (z), n_1 is the refractive index on-axis, i.e. at the guide centre, and Δ and a are two parameters governing the index variation which are to be defined in due course. It is to be noted that equation (3.1) gives in fact a parabolic variation of $n^2(x)$ rather than $n(x)$, i.e. dielectric permittivity rather than refractive index. However, provided x is not too large, so that $2\Delta(x/a)^2 \ll 1$, we also have a parabolic variation of index:

$$n(x) \simeq n_1 \left(1 - \Delta \left(\frac{x}{a} \right)^2 \right). \tag{3.2}$$

Equations (3.1) and (3.2) may be taken as first approximations to any symmetric dielectric distribution which may be expanded in a Taylor series about the axis $(x = 0)$ of the waveguide. A particular case which is often considered is that of a graded-index core ($|x| \leq a$) surrounded by a cladding of uniform refractive index n_2. The situation is illustrated in Figure 3.1 for the case of no index discontinuity at the boundary $|x| = a$, i.e. $n(a) = n_2$. From the figure the definition of the parameter Δ is clear:

$$2\Delta = \frac{n_1^2 - n_2^2}{n_1^2}. \tag{3.3}$$

In the weakly-guiding approximation $n_1 - n_2 \ll n_1, n_2$, equation (3.3) reduces to

$$\Delta \simeq \frac{n_1 - n_2}{n_1}$$

which is the relative core–cladding index difference of a slab waveguide (cf. Chapter 2). These definitions of Δ and a are useful (although not essential) in that one may speak of a normalized frequency v defined by analogy with the slab waveguide case as:

$$v^2 = a^2 k^2 (n_1^2 - n_2^2) = a^2 k^2 2\Delta n_1^2 \tag{3.4}$$

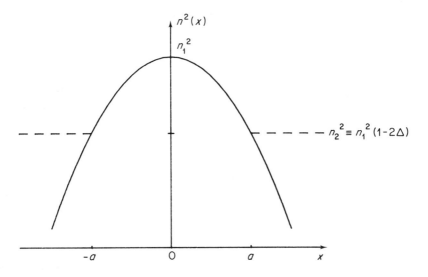

Figure 3.1 The parabolic dielectric profile. Solid line, infinitely-extended parabolic profile; broken line, cladded parabolic profile

where $k = 2\pi/\lambda$. The use of equation (3.4) facilitates comparisons of results for the parabolic index medium with those for the slab and other waveguides. However, a word of caution is necessary for the variation of refractive index discussed here: the parabolic variation of (3.1) and (3.2) yields modes whose characteristics will only resemble those of the cladded parabolic variation (Figure 3.1, broken line) when the field is tightly confined to the centre of the core. For increasing field penetration of the cladding layers the approximation furnished by (3.1) and (3.2) will become progressively worse until at cut-off the approximate field distributions will not bear much resemblance to those of the cladded-parabolic profile. With this warning in mind we will restrict attention here to the infinitely-extended parabolic-index medium as characterized by equations (3.1) and (3.2), whilst retaining the definitions (3.3) and (3.4) for the purposes of comparison with results of other index profiles.

The discussion commences with a ray treatment of parabolic-index media including a derivation of the ray trajectories and the eigenvalue equation. The electromagnetic mode treatment of the scalar wave equation is then given and the results compared with those obtained from the ray approach. Results are given in terms of the now-familiar normalized variables wherever possible. Finally the extension to media including loss or gain variations is considered and the theory is applied to analyses of stripe-geometry injection lasers where it forms a powerful tool for diagnosis of the dielectric variations present in these devices.

3.1 Ray Treatment of Parabolic-Index Media

This topic was the subject of considerable study during the mid-1960's, when its application to the analogous case of lens waveguides was under discussion (Marcatili, 1964; Miller, 1965; Tien et al., 1965; Gordon, 1966). Here we begin with a somewhat novel derivation of the equations of motion of a ray in a parabolic-index medium via the local-plane-wave method. Solutions of these equations yield the ray trajectories and demonstrate the focusing action of the medium; the ray transit time is also found. In order to derive an eigenvalue equation for the allowed values of propagation constant, the phase resonance condition is invoked; this in turn necessitates a discussion of the phase change occurring at a caustic. Finally in this section the transit time is found directly from the eigenvalue equation and is shown to be identical to that derived from the ray trajectory.

3.1.1 Local plane wave derivation of ray equations

The ray equations may be derived from the scalar wave equation by solving for the surfaces of constant phase and using the fact that the rays are normal to these surfaces (see Section 8.1). They may also be derived from Fermat's principle or from the Hamilton equations by analogy with classical mechanics. For a survey of these approaches the reader is referred to the books by Marcuse (1972) and Arnaud (1976). Since we are here primarily concerned with the predictions of the ray equations as applied to parabolic-index media we will give a rather informal derivation based on the decomposition of the local wave vector (Gloge and Marcatili, 1973). This approach, which has already been briefly referred to in the discussion of Figure 1.5 in Chapter 1, although lacking some of the rigour of other methods has the advantage of being easily interpreted physically and of giving the constants of motion in terms of wave parameters already known.

Figure 3.2(a) shows the decomposition of the local wave vector $kn(x)$ in a two-dimensional Cartesian coordinate system at the point (x, z). We are here assuming (since this is to be a ray

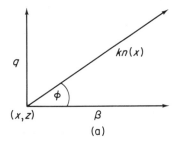 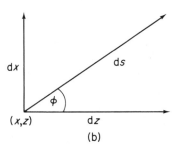

Figure 3.2 (a) Decomposition of the local plane wave vector into its x and z components. (b) Decomposition of an element of ray path into its x and z components

description) that all linear dimensions are greatly in excess of the wavelength λ and in particular that the refractive index $n(x)$ varies sufficiently slowly with x that it may be considered effectively constant on the scale of λ. Hence the z and x components of the local wave vector are β and q, respectively, where

$$q^2 = k^2 n^2(x) - \beta^2.$$

A ray which propagates at the point (x, z) has the local direction of $kn(x)$ at this point and the decomposition into its elementary x and z components of motion are shown in Figure 3.2(b). The line element along the ray is given by

$$ds^2 = dx^2 + dz^2.$$

Comparison of Figures 3.2(a) and (b) then yields the results:

$$\frac{ds}{dx} = \frac{kn(x)}{(k^2 n^2(x) - \beta^2)^{1/2}} \tag{3.5}$$

$$\frac{dz}{dx} = \frac{\beta}{(k^2 n^2(x) - \beta^2)^{1/2}} \tag{3.6}$$

which are the ray equations for propagation in a two-dimensional medium of refractive index $n(x)$. Note that the constant of motion is β which has been associated with the conventional definition of longitudinal propagation constant.

From the form of equations (3.5) and (3.6) it is clear that there may exist values of x for which $dx = q = 0$, i.e. $\beta = kn(x)$. These values correspond to the turning points or caustics between which the parameter q is real, corresponding to a real ray; for values of x outside this region q is imaginary (corresponding to an evanescent wave). Hence the ray path is constrained to lie between values of x for which $\beta = kn(x)$. It follows also from (3.5) and (3.6) that the transit time τ of a ray in a length L of such a medium is given by

$$\tau = \frac{L}{c} \frac{\oint n(x) \, ds}{\oint dz} \tag{3.7}$$

where c is the velocity of light in vacuo, and the symbol $\oint$ indicates that the integration is taken over a complete ray period (assumed much less than L). In equation (3.7), $n(x)\,ds$ represents an element of optical path length along the ray rather than the geometrical path length ds. We are here implicitly neglecting material dispersion effects since a more accurate treatment would use as an element of optical path length the expression $m\,ds$ where m is the group index $(n - \lambda \, dn/d\lambda)$.

3.1.2 The ray trajectory and transit time

We may apply equation (3.6) directly to the index distribution (3.1) in order to find the ray trajectory in x–z space. The differential equation then reads:

$$\frac{dz}{dx} = \frac{\beta}{[k^2 n_1^2 - k^2 n_1^2 2\Delta(x/a)^2 - \beta^2]^{1/2}} = \frac{a\beta}{[u^2 - v^2(x/a)^2]^{1/2}} \tag{3.8}$$

where the definition of v^2 in (3.4) has been used together with the conventional definition (2.111) of the normalized variable u. If the boundary condition is taken as $x = x_0$ at $z = 0$, integration of (3.8) gives

$$\frac{zv}{a^2 \beta} = \sin^{-1}\left(\frac{vx}{ua}\right) - \sin^{-1}\left(\frac{vx_0}{ua}\right) \tag{3.9}$$

whence we see that the ray path is periodic with period $2\pi a^2 \beta/v$, and the maximum amplitude of oscillations is ua/v. To cast (3.9) into a more conventional form, define $\tan \phi_0$ as the initial ray slope (dx/dz) at $x = x_0$, $z = 0$:

$$a\beta \tan \phi_0 = \left[u^2 - v^2 \left(\frac{x_0}{a}\right)^2\right]^{1/2}. \tag{3.10}$$

Rearrangement of equation (3.9) then yields

$$x = x_0 \cos\left(\frac{zv}{a^2 \beta}\right) + \frac{a^2 \beta}{v} \tan \phi_0 \sin\left(\frac{zv}{a^2 \beta}\right) \tag{3.11}$$

which gives the displacement x as a function of axial distance z for a ray whose initial position and slope are x_0 and $\tan \phi_0$.

Since we are primarily concerned with rays which do not stray too far from the axis (so that the parabolic-index medium may serve as a model for a weakly-guiding cladded-parabolic guide) we will in practice be dealing with the *paraxial approximation* (the ray angle ϕ of

Figure 3.2 must be small). For this case $\beta \simeq kn_1$ and equation (3.11) reduces finally to the form (Miller, 1965):

$$x \simeq x_0 \cos\left((2\Delta)^{1/2}\frac{z}{a}\right) + \frac{a\phi_0}{\sqrt{2\Delta}} \sin\left((2\Delta)^{1/2}\frac{z}{a}\right). \tag{3.12}$$

Note that within the limits of the paraxial approximation equation (3.12) states that all rays have the same period. Plots of some ray trajectories calculated from equation (3.12) are given in Figure 3.3. The periodic nature of the ray paths gives rise to a focusing effect in parabolic-index media; the images will occur at multiples of the ray period $2\pi a/(2\Delta)^{1/2}$. Hence a length L of such a medium acts as a lens; Figure 3.4 illustrates this property by showing the path of a

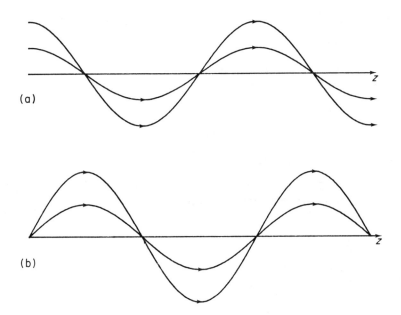

Figure 3.3 Ray trajectories in the parabolic-index medium, according to equation (3.12): (a) parallel rays at input ($\phi_0 = 0$); (b) coincident rays at input on axis ($x_0 = 0$)

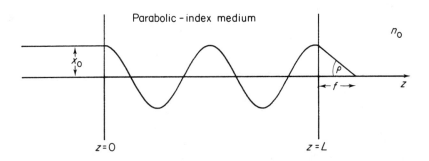

Figure 3.4 The lens effect of a length L of parabolic index medium, focal length f

ray which enters the medium parallel to the z axis ($\phi_0 = 0$). It follows that the ray leaves the medium after length L at a displacement x given by

$$x = x_0 \cos\left((2\Delta)^{1/2}\frac{L}{a}\right).$$

If the medium outside the parabolic-index region has refractive index n_0, then the ray leaves at an angle ρ (see Figure 3.4) given in the paraxial approximation by

$$\rho \simeq -\frac{n_1(2\Delta)^{1/2}x_0}{n_0 a}\sin\left((2\Delta)^{1/2}\frac{L}{a}\right).$$

It follows that the focal length f of the parabolic-index medium of length L is given by the distance from $z = L$ at which the ray crosses the z-axis, i.e. (Miller, 1965):

$$f \simeq \left|\frac{x}{\rho}\right| \simeq \frac{n_0 a}{n_1(2\Delta)^{1/2}}\cot\left((2\Delta)^{1/2}\frac{L}{a}\right). \tag{3.13}$$

Note that the focal length as given in the paraxial approximation by equation (3.13) is independent of the specific ray input displacement x_0 and the parabolic-index medium therefore acts as a true lens for rays which do not travel too far from the axis.

The ray transit time for the refractive index profile of equation (3.1) may be found via the expression in equation (3.7). The denominator of this expression is given (in complete generality, without using the paraxial approximation) from (3.9) by taking a ray half-period from $x_0 = -ua/v$ to $x = ua/v$:

$$\oint dz = \frac{a^2 \beta \pi}{v}. \tag{3.14}$$

Similarly, the numerator of (3.7) may be found by direct integration of (3.5) over a ray half-period:

$$\oint n(x)\,ds = \int_{-ua/v}^{ua/v}\frac{kn^2(x)\,dx}{(k^2n^2(x)-\beta^2)^{1/2}}$$

$$= \int_{-ua/v}^{ua/v}\frac{akn_1^2(1-2\Delta(x/a)^2)\,dx}{[u^2-v^2(x/a)^2]^{1/2}}$$

$$= \frac{a^2 k n_1^2 \pi}{v}\left(1 - \frac{\Delta u^2}{v^2}\right). \tag{3.15}$$

Hence the ray transit time from (3.7) becomes

$$\tau = \frac{L}{c}\left(\frac{kn_1^2}{\beta}\right)\left(1 - \frac{\Delta u^2}{v^2}\right). \tag{3.16}$$

Note that within the limits of the paraxial approximation $\beta \simeq kn_1$ and equation (3.16) gives the ray transit time τ as simply Ln_1/c, i.e. the same value for all rays. This result may be understood physically by considering the ray trajectories of Figure 3.3. The geometrical ray path lengths are seen to be different for each ray, those with greater displacement from the axis having a correspondingly greater path length. However, for much of this length these rays travel in a medium of lower refractive index; hence they travel faster than the other rays which do not suffer such large lateral displacements. It follows that the overall transit time of all rays may be the same provided the index profile is chosen appropriately. The parabolic profile of equation (3.1) possesses this property in the paraxial approximation, as we have just proved. We will return to this important feature of the parabolic-index medium for a more detailed investigation in the course of the modal treatment in Section 3.2.

3.1.3 Phase shift at a caustic

In order to find an eigenvalue equation for the parabolic-index medium using a ray treatment we may use the transverse phase resonance condition already utilized for the waveguides of Chapters 1 and 2. However, in this case the situation is somewhat more complicated since it is necessary to find the phase shift occurring at the caustics $x = \pm au/v$ which define the limits of the ray's displacement from the z-axis. All rays of the same β are tangential to the caustic, as may be easily verified from equation (3.6) for any refractive index profile $n(x)$. In order to calculate the phase change occurring at the caustic we may use the general expressions (2.32) and (2.34) derived for the phase shift at a dielectric interface, provided that we approximate the continuous curve $n(x)$ as a staircase curve and take the limit of vanishing index steps and grazing incidence (Hocker and Burns, 1975; Marcuse, 1976); we will now give the details of this argument.

Consider a ray immediately before reaching its caustic as travelling in a medium of index $n(x_c - \Delta x)$, where the position of the caustic occurs at $x = x_c$; on the opposite side of the caustic the refractive index will be $n(x_c + \Delta x)$, where Δx is assumed small. Now the expressions for phase shifts $\delta_\perp$ and $\delta_\parallel$ for the electric field normal and parallel to the plane of incidence, respectively, are given from (2.32) and (2.34) with $\beta = n(x)k \sin \theta_1 \, (= n(x)k \cos \phi$ in Figure 3.2) as:

$$\delta_\perp = 2 \tan^{-1} \left\{ \left[\frac{(\beta/k)^2 - n^2(x_c + \Delta x)}{n^2(x_c - \Delta x) - (\beta/k)^2} \right]^{1/2} \right\} \quad (3.17)$$

$$\delta_\parallel = 2 \tan^{-1} \left\{ \frac{n^2(x_c - \Delta x)}{n^2(x_c + \Delta x)} \left[\frac{(\beta/k)^2 - n^2(x_c + \Delta x)}{n^2(x_c - \Delta x) - (\beta/k)^2} \right]^{1/2} \right\}. \quad (3.18)$$

The limit we require to find for grazing incidence and continuous $n(x)$ is given by $\Delta x \to 0$, when the numerator and denominator of the $\tan^{-1}$ functions in (3.17) and (3.18) both go to zero. To find the limit we may use either a Taylor series expansion of $n(x)$ about x_c (Hocker and Burns, 1975) or l'Hospital's rule; the latter approach will be used here. It gives:

$$\tan^2 \left(\frac{\delta_\perp}{2} \right) = \lim_{\Delta x \to 0} \left\{ \frac{n(x_c + \Delta x)(dn/dx)_{x = x_c + \Delta x}}{n(x_c - \Delta x)(dn/dx)_{x = x_c - \Delta x}} \right\}$$

$$= 1.$$

A similar result holds for $\delta_\parallel$, so that we obtain:

$$\delta_\perp = \delta_\parallel = \frac{\pi}{2} \quad (3.19)$$

i.e. the phase shift (as defined in equation (2.30)) at a caustic is $\pi/2$ for each polarization. Alternative derivations of this result, usually involving somewhat more difficult arguments than the one given above, may be found elsewhere (see, for example, Keller, 1956; Keller and Rubinow, 1960).

3.1.4 The eigenvalue equation

We are now in a position to derive the eigenvalue equation for the parabolic-index medium by considering the total transverse phase change in a ray period. We have already noted in

Chapters 1 and 2 that for a bound ray the total transverse phase change in one round trip of a guide must be an integral multiple of 2π. The phase change must include the shifts at the caustics dealt with in the previous subsection, together with the phase change suffered by the ray in traversing the optical path length of one period. This latter quantity (denoted by Φ) may be found by integrating the transverse wave vector component q of Figure 3.2(a) over a complete ray period:

$$\Phi = \oint q \, dx = 2 \int_{-x_c}^{x_c} (k^2 n^2(x) - \beta^2)^{1/2} \, dx. \tag{3.20}$$

The phase resonance condition then reads, both for the TE and TM cases:

$$\Phi - 2\delta = 2N\pi. \tag{3.21}$$

Using equation (3.1) in (3.20) we find

$$\Phi = 2 \int_{-ua/v}^{ua/v} \left[u^2 - v^2 \left(\frac{x}{a}\right)^2 \right]^{1/2} \frac{dx}{a} = \frac{\pi u^2}{v}.$$

Hence equation (3.21) with δ from (3.19) gives the eigenvalue equation for the parabolic-index medium in the form:

$$u^2 = v(2N+1) \quad (N = 0, 1, 2, \ldots). \tag{3.22}$$

The ray transit time τ may be found directly by differentiation of the eigenvalue equation (3.22), since $\tau = (L/c)(d\beta/dk)$. From the definition of u in (2.111) we have:

$$u \frac{du}{dk} = a^2 \left(n_1^2 k - \beta \frac{d\beta}{dk} \right)$$

and from the definition of v in (3.4) it follows that

$$u \frac{du}{dv} = \frac{ka^2}{v} \left(n_1^2 k - \beta \frac{d\beta}{dk} \right). \tag{3.23}$$

From equation (3.22) we have the result that

$$u \frac{du}{dv} = \frac{(2N+1)}{2} = \frac{u^2}{2v}. \tag{3.24}$$

Hence using equations (3.23) and (3.24) we find:

$$\tau = \left(\frac{L}{c}\right) \frac{d\beta}{dk} = \left(\frac{L}{c}\right) \frac{1}{\beta} \left(n_1^2 k - \frac{u^2}{2ka^2} \right) = \left(\frac{L}{c}\right) \frac{n_1^2 k}{\beta} \left(1 - \frac{\Delta u^2}{v^2} \right) \tag{3.25}$$

which confirms our earlier result (3.16) obtained from the integration of the ray equations.

3.2 Electromagnetic Mode Treatment of Parabolic-index Media

The ray approach of the previous section was based implicitly on the scalar wave equation

$$\nabla^2 \psi + k^2 n^2(x) \psi = 0 \tag{3.26}$$

where ψ represents any field component. In the present section we shall be concerned with a somewhat more careful analysis, with Maxwell's equations as a starting point, in which the polarization properties of the normal modes will be retained. We are therefore concerned with the derivation of rather more accurate forms of the wave equation than (3.26). However, we shall show that in the weakly-guiding situation the approximation represented by (3.26) is reasonably accurate and in the limiting case the results of the mode treatment reduce to those obtained by the ray analysis of the previous section.

3.2.1 The vector wave equations for general graded-index media

For a general graded-index medium whose refractive index is $n(x)$, Maxwell's equations may be written (Marcuse, 1972):

$$\nabla \times \mathbf{H} = n^2(x)\varepsilon_0 \frac{d\mathbf{E}}{dt} \tag{3.27}$$

$$\nabla \times \mathbf{E} = -\mu_0 \frac{d\mathbf{H}}{dt} \tag{3.28}$$

$$\nabla \cdot (n^2(x)\varepsilon_0 \mathbf{E}) = 0 \tag{3.29}$$

$$\nabla \cdot \mathbf{H} = 0. \tag{3.30}$$

If we apply the curl operator to equation (3.28), we find:

$$\nabla \times (\nabla \times \mathbf{E}) = -\mu_0 \nabla \times \frac{d\mathbf{H}}{dt} = -\mu_0 n^2(x)\varepsilon_0 \frac{d^2\mathbf{E}}{dt^2} \tag{3.31}$$

where equation (3.27) was used to eliminate $\mathbf{H}$. Using the vector identity

$$\nabla \times (\nabla \times \mathbf{E}) = \nabla(\nabla \cdot \mathbf{E}) - \nabla^2 \mathbf{E}$$

together with equation (3.29), equation (3.31) becomes:

$$\nabla^2 \mathbf{E} + \nabla \left(\frac{\mathbf{E} \cdot \nabla n^2(x)}{n^2(x)} \right) + k^2 n^2(x) \mathbf{E} = 0 \tag{3.32}$$

where the time dependence $\exp(-i\omega t)$ of $\mathbf{E}$ has been assumed and the conventional notation $k^2 = \omega^2 \mu_0 \varepsilon_0$ has been used. Equation (3.32) is the vector wave equation satisfied by the electric field $\mathbf{E}$.

To derive the corresponding wave equation for the magnetic field, first apply the curl operator to equation (3.27) to find:

$$\nabla \times (\nabla \times \mathbf{H}) = n^2(x)\varepsilon_0 \left(\nabla \times \frac{d\mathbf{E}}{dt} \right) + \varepsilon_0 \left(\nabla n^2(x) \times \frac{d\mathbf{E}}{dt} \right)$$

$$= -n^2(x)\varepsilon_0 \mu_0 \frac{d^2\mathbf{H}}{dt^2} + \frac{1}{n^2(x)} (\nabla n^2(x)) \times (\nabla \times \mathbf{H}) \tag{3.33}$$

where equations (3.27) and (3.28) have been used to eliminate $\mathbf{E}$. In this case, since $\nabla \cdot \mathbf{H} = 0$ from (3.30), use of the same vector identity as before yields:

$$\nabla^2 \mathbf{H} + \frac{1}{n^2(x)} (\nabla n^2(x)) \times (\nabla \times \mathbf{H}) + k^2 n^2(x) \mathbf{H} = 0 \tag{3.34}$$

which is the vector wave equation for $\mathbf{H}$.

Next we use the fact that $n(x)$ is a function of transverse coordinate x only; this enables equations (3.32) and (3.34) to be expressed as scalar wave equations in the components E_x and

H_x, respectively. From (3.34), assuming a z-dependence as $\exp(i\beta z)$ and with $d/dy = 0$, we find:

$$\frac{d^2 H_x}{dx^2} + (k^2 n^2(x) - \beta^2) H_x = 0. \tag{3.35}$$

From (3.32) on the other hand, it follows that:

$$\frac{d^2 E_x}{dx^2} + \frac{d}{dx}\left[\frac{E_x}{n^2(x)} \frac{dn^2(x)}{dx}\right] + (k^2 n^2(x) - \beta^2) E_x = 0. \tag{3.36}$$

In order to eliminate the (dE_x/dx) term in (3.36) we introduce the transformation (Ghatak and Kraus, 1974; Kirchhoff, 1972):

$$E_x = \frac{\psi}{n(x)}. \tag{3.37}$$

Substitution of (3.37) into (3.36) results in the scalar wave equation for ψ:

$$\frac{d^2 \psi}{dx^2} + \left[\frac{1}{2n^2(x)} \frac{d^2(n^2(x))}{dx^2} - \frac{3}{4n^4(x)}\left(\frac{d(n^2(x))}{dx}\right)^2 + k^2 n^2(x) - \beta^2\right]\psi = 0. \tag{3.38}$$

Hence we have derived two scalar wave equations (3.35) and (3.38) for the field components H_x and E_x (via ψ) in a medium whose refractive index varies only with the transverse coordinate x.

3.2.2 TE solutions for parabolic-index media

For the TE modes of the general graded-index medium, take $E_x = 0$ and assume H_x is known as the solution of equation (3.35). The remaining field components are then given from (3.27) and (3.28) as:

$$E_z = H_y = 0 \tag{3.39}$$

$$E_y = -\frac{\omega \mu_0}{\beta} H_x \tag{3.40}$$

$$H_z = \frac{i}{\beta} \frac{dH_x}{dx}. \tag{3.41}$$

For the special case of the parabolic-index medium where $n^2(x)$ is given by (3.1), equation (3.35) becomes

$$\frac{d^2 H_x}{dx^2} + \left[k^2 n_1^2 \left(1 - 2\Delta\left(\frac{x}{a}\right)^2\right) - \beta^2\right] H_x = 0. \tag{3.42}$$

As a trial solution, set

$$H_x(x) = X(x) \exp\left(-\frac{x^2}{w_0^2}\right) \tag{3.43}$$

so that equation (3.42) becomes:

$$\frac{d^2 X}{dx^2} - \frac{4x}{w_0^2} \frac{dX}{dx} - \frac{2}{w_0^2}\left(1 - \frac{2x^2}{w_0^2}\right) X + \left[k^2 n_1^2 \left(1 - 2\Delta\left(\frac{x}{a}\right)^2\right) - \beta^2\right] X = 0. \tag{3.44}$$

The x^2-terms in this equation may be eliminated by an appropriate definition of the 'beam-waist', w_0, of the Gaussian in (3.43):

$$w_0^2 = \frac{2a}{kn_1(2\Delta)^{1/2}} = \frac{2a^2}{v}. \tag{3.45}$$

The equation for X is now transformed to a more familiar form by writing

$$x' = \frac{x\sqrt{2}}{w_0}.$$

With this transformation and the definition (3.45), equation (3.44) becomes:

$$\frac{d^2X}{dx'^2} - 2x'\frac{dX}{dx'} + \left[(k^2n_1^2 - \beta^2)\frac{w_0^2}{2} - 1\right]X = 0. \tag{3.46}$$

The solutions of (3.46) are the Hermite polynomials $H_N(x')$ (see, for example, Abramowitz and Stegun, 1964) where N and β are related by

$$2N = (k^2n_1^2 - \beta^2)\frac{w_0^2}{2} - 1 \quad (N = 0, 1, 2, \ldots). \tag{3.47}$$

The solution for the TE modes of the parabolic medium is thus seen to be given from (3.43) as Hermite–Gaussian functions:

$$H_x = \frac{2^{1/4}}{\pi^{1/4}(2^N N! w_0)^{1/2}} H_N\left(\frac{x\sqrt{2}}{w_0}\right) \exp\left(\frac{-x^2}{w_0^2}\right) \quad (N = 0, 1, 2, \ldots) \tag{3.48}$$

where the constant has been chosen so that the modes are normalized according to

$$\int_{-\infty}^{\infty} |H_x|^2 \, dx = 1. \tag{3.49}$$

The result (3.48) may be expressed in terms of v rather than w_0 by using (3.45):

$$H_x = \frac{v^{1/4}}{\pi^{1/4}(2^N N! a)^{1/2}} H_N\left(\frac{v^{1/2}x}{a}\right) \exp\left(-\frac{vx^2}{2a^2}\right) \quad (N = 0, 1, 2, \ldots). \tag{3.50}$$

Field distributions for some of the low-order Hermite–Gaussian modes are shown in Figure 3.5.

The eigenvalue equation for these modes is given by (3.47) which may be re-cast in terms of normalized variables as

$$u^2 = v(2N+1) \quad (N = 0, 1, 2, \ldots) \tag{3.51}$$

which is identical with the result (3.22) of the ray analysis. In terms of the normalized propagation constant b, defined in (2.112), this result becomes

$$b = 1 - \left(\frac{2N+1}{v}\right) \quad (N = 0, 1, 2, \ldots). \tag{3.52}$$

Plots of b versus v calculated from (3.52) for a few low-order modes are given in Figure 3.6. Note that the condition $b = 0$ ($\beta = kn_2$) occurs for values of v given by (Marcuse, 1973):

$$v_c = 2N+1 \quad (N = 0, 1, 2, \ldots) \tag{3.53}$$

but this does *not* correspond to a definition of waveguide cut-off in the usual sense. At values of v given by (3.53) the modal fields will not possess the characteristics of modes at cut-off as

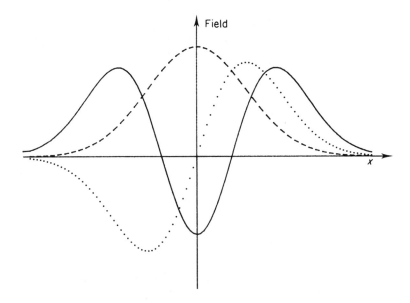

Figure 3.5 Field distributions for some low-order Hermite–Gaussian modes. Dashed line, $N = 0$; dotted line, $N = 1$; solid line, $N = 2$

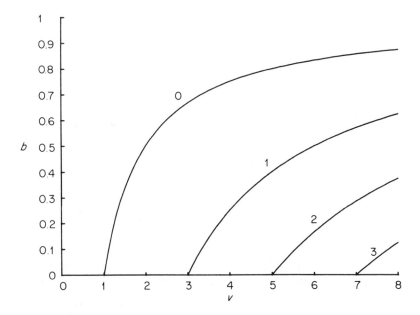

Figure 3.6 Normalized propagation constant b versus normalized frequency v for the parabolic-index medium, calculated from equation (3.52); labelling parameter gives the value of mode number N

in, say, the three-layer slab guide of Chapter 2 where the field distributions become constant in the cladding layers so that power is distributed uniformly throughout all space. For the Hermite–Gaussian modes the fields will still possess the characteristics of guided modes at all v-values, since the refractive-index distribution of (3.1) is unbounded in the x-direction.

3.2.3 TM solutions for parabolic-index media

In this case we assume $H_x = 0$ and suppose E_x can be found as the solution of equation (3.38) with the aid of the transformation (3.37). The remaining field components are then given from (3.27) and (3.28) as:

$$E_y = H_z = 0 \tag{3.54}$$

$$E_z = \frac{i}{\beta n^2(x)} \frac{d}{dx}\left(n^2(x)E_x\right) \tag{3.55}$$

$$H_y = \frac{\omega n^2(x)\varepsilon_0}{\beta} E_x. \tag{3.56}$$

For the case of parabolic-index media, equation (3.38) may be solved by expanding terms in $n^2(x)$ as a power series and neglecting terms of order $\Delta^2(x/a)^4$ and higher. The equation then becomes

$$\frac{d^2\psi}{dx^2} + \left[\left(k^2 n_1^2 - \beta^2 - \frac{2\Delta}{a^2}\right) - \left(\frac{k^2 n_1^2 2\Delta}{a^2} + \frac{16\Delta^2}{a^4}\right)x^2\right]\psi = 0. \tag{3.57}$$

The similarity of the form of this equation to that of (3.42) suggests a similar solution, and we find (Ghatak and Kraus, 1974):

$$E_x = \frac{\psi}{n(x)} = \frac{2^{1/4}}{n(x)\pi^{1/4}(2^N N! w_0)^{1/2}} H_N\left(\frac{x\sqrt{2}}{w_0}\right)\exp\left(-\frac{x^2}{w_0^2}\right) \quad (N = 0, 1, 2, \ldots) \tag{3.58}$$

where the beam waist w_0 is now given by

$$w_0^2 = \frac{2}{\left(\dfrac{k^2 n_1^2 2\Delta}{a^2} + \dfrac{16\Delta^2}{a^4}\right)^{1/2}}. \tag{3.59}$$

The corresponding eigenvalue equation is

$$\beta^2 = k^2 n_1^2 - \frac{2\Delta}{a^2} - (2N+1)\left(\frac{k^2 n_1^2 2\Delta}{a^2} + \frac{16\Delta^2}{a^4}\right)^{1/2} \quad (N = 0, 1, 2, \ldots). \tag{3.60}$$

Note that for $2\Delta \ll k^2 n_1^2 a^2$ the TM and TE solutions become similar so that in the limit equations (3.58)–(3.60) reduce to the forms (3.50), (3.45), and (3.47), respectively.

3.2.4 Group delay and radiation confinement factor

To compare the expressions for propagation constant and hence group delay which result for TE and TM modes from the expressions derived above, we may expand equations (3.47) and (3.60), respectively, in ascending powers of $\Delta^{1/2}$. Retaining terms up to order Δ, equation (3.47) yields:

TE: $$\beta \simeq kn_1 - \frac{(2\Delta)^{1/2}}{a}(N + \tfrac{1}{2}) - \frac{\Delta}{a^2 k n_1}(N + \tfrac{1}{2})^2 \tag{3.61}$$

and equation (3.60) gives (Marcuse, 1973a):

TM: $$\beta \simeq kn_1 - \frac{(2\Delta)^{1/2}}{a}(N+\tfrac{1}{2}) - \frac{\Delta}{a^2 kn_1}[(N+\tfrac{1}{2})^2 + 1]. \quad (3.62)$$

Differentiating each of these equations, we find for the group delay τ:

TE: $$\tau \simeq \frac{Ln_1}{c}\left[1 + \frac{\Delta}{a^2 k^2 n_1^2}(N+\tfrac{1}{2})^2\right] \quad (3.63)$$

TM: $$\tau \simeq \frac{Ln_1}{c}\left\{1 + \frac{\Delta}{a^2 k^2 n_1^2}\left[(N+\tfrac{1}{2})^2 + 1\right]\right\}. \quad (3.64)$$

It follows from equations (3.63) and (3.64) that

(i) other than for a few low-order modes the group delay is very similar for TE and TM modes,
(ii) for small $(\Delta/a^2 k^2 n_1^2)$ the TE and TM modes are quasi-degenerate,
(iii) for small $(\Delta/a^2 k^2 n_1^2)$ the group delay is essentially independent of mode number, i.e. the same for all modes.

This latter result (iii) can be re-interpreted for the weakly-guiding case (quasi-degeneracy of TE and TM modes) by using the normalized equation (3.52). From (3.52) it follows that the normalized dispersion parameter $d(vb)/dv$ defined in (2.121) is given by:

$$\frac{d(vb)}{dv} = 1 \quad \text{(all } N\text{)} \quad (3.65)$$

i.e. to this level of approximation the group delay is the same for all modes.

Turning to the radiation confinement factor Γ, defined (as in Chapter 2) as the ratio of power confined to the core to total power, this is given in the limit of small Δ (TE/TM degeneracy) by:

$$\Gamma = \frac{v^{1/2}}{\pi^{1/2} 2^N N! a} \int_{-a}^{a} \left[H_N\left(\frac{v^{1/2} x}{a}\right)\right]^2 \exp\left(-\frac{vx^2}{a^2}\right) dx \quad (N = 0, 1, 2, \ldots) \quad (3.66)$$

where equation (3.50) has been used for the normalized fields. The first few Hermite polynomials are given by

$$H_0(y) = 1; \quad H_1(y) = 2y;$$
$$H_2(y) = 4y^2 - 2; \quad \text{etc.}$$

It follows that values of Γ may always be evaluated in terms of the error function $\text{erf}(v^{1/2})$ (see, for example, Abramowitz and Stegun, 1964). For the two lowest-order modes we find:

$N = 0$: $$\Gamma = \text{erf}(v^{1/2}) \quad (3.67)$$

$N = 1$: $$\Gamma = \text{erf}(v^{1/2}) - \frac{2v^{1/2}}{\pi^{1/2}} e^{-v}. \quad (3.68)$$

Plots of these as functions of v are given in Figure 3.7. Note that values have only been shown for v greater than the mode cut-offs v_c given by equation (3.53). Since this is an artificial cut-off condition, as already discussed in Section 3.2.2, the confinement factor does not go to zero at these cut-off values and great caution should be exercised in applying these results to real waveguides of parabolic index variation. Results which are suited for cladded-parabolic

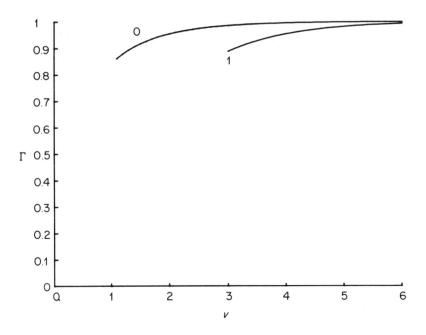

Figure 3.7 Radiation confinement factor Γ versus v for the $N = 0$ and $N = 1$ modes of the parabolic-index medium, calculated from equations (3.67) and (3.68)

guides (Figure 3.1, broken line) will be calculated and discussed by approximate methods in Chapter 5.

3.3 Variation of Loss or Gain

Consider now the more general case of a dielectric where, in addition to the parabolic index variation as given by equation (3.2), there is also a parabolic variation of the loss or gain in the medium. Let us denote this latter variation by $g(x)$, defined such that

$$g(x) = g_1\left[1 - \delta\left(\frac{x}{a}\right)^2\right] \qquad (3.69)$$

where g_1 is the gain on-axis ($x = 0$) and δ is a parameter which measures the strength of the gain-guiding action in much the same way as Δ characterizes the index-guidance. Note that equation (3.69) could equally well describe a loss variation by simply using a negative value for g_1; similarly a distribution of gain with a minimum on-axis can be obtained for a negative δ. Gain variations of the form given in equation (3.69) occur, for example, in gas lasers (Kogelnik, 1965) and (as a first-order approximation—see Section 2.4) along the junction plane of stripe-geometry semiconductor lasers (Nash, 1973). It is worth noting that in these examples no assumptions may be made as regards the value of δ, although $\Delta \ll 1$ for the index variation may sometimes be justified. However, g_1/k is usually a great deal smaller than the index n_1 so that a small amount of index-guidance is equivalent to quite a strong gain-guiding effect.

In view of the remarks made above, we may define the complex permittivity $\varepsilon(x)$, for the case $\Delta \ll \delta$ and $g_1/k \ll n_1$, as follows:

$$\varepsilon(x) = \left(n(x) + \frac{ig(x)}{2k}\right)^2$$

$$\simeq n_1\left(n_1 + \frac{ig_1}{k}\right) - n_1\left[2\Delta n_1 + \frac{i\delta g_1}{k}\right]\left(\frac{x}{a}\right)^2. \quad (3.70)$$

This gives a general parabolic variation of the complex permittivity $\varepsilon(x)$ which we will analyse in more detail below. In order to reduce the complexity of the algebraic expressions, let us define

$$\varepsilon(x) = \varepsilon_1 - b_0^2 x^2 \quad (3.71)$$

where

$$\varepsilon_1 = n_1^2 + \frac{in_1 g_1}{k} \quad (3.72)$$

$$b_0^2 = \frac{2\Delta n_1^2}{a^2} + \frac{in_1 \delta g_1}{ka^2}. \quad (3.73)$$

We are now ready to find the general modes of media characterized by equation (3.71); in analogy with the real permittivity case already discussed, the modes can be expressed in terms of Hermite–Gaussian functions of complex argument. The case of off-axis Gaussian beam modes is worth including in the analysis, since this leads to a simple proof of the mode stability criterion mentioned earlier (Section 2.4). An interpretation of the mode parameters in terms of Gaussian beam waist and phase front radius of curvature leads to simple formulae for analysis of stripe-geometry laser observations. Finally, the case of pure gain-guidance (no index variation) will be discussed in analogy with the analysis made for dielectric slab guides in Section 2.4.3.

3.3.1 General beam modes

We take as a starting point the scalar wave equation with $d/dy = 0$;

$$\frac{d^2 H_x}{dx^2} + \frac{d^2 H_x}{dz^2} + k^2 \varepsilon(x) H_x = 0 \quad (3.74)$$

which we have shown in Section 3.2 to be accurate for TE modes and a reasonable approximation in certain circumstances for TM modes. To include the general off-axis beam solutions we substitute (Kogelnik, 1965):

$$H_x = A(x, z) e^{i\beta z} \quad (3.75)$$

and assume that A is sufficiently slowly-varying with z that the second derivative may be ignored. Then (3.74) becomes

$$\frac{d^2 A}{dx^2} + 2i\beta \frac{dA}{dz} + (k^2 \varepsilon(x) - \beta^2) A = 0. \quad (3.76)$$

To find a general astigmatic Gaussian solution, we can write

$$A = B(x, z) \exp\left[-iP(z) + i\frac{Q(z)x^2}{2}\right] \quad (3.77)$$

where P and Q are complex phase and beam parameters, respectively. With this substitution, equation (3.77) takes the form:

$$\frac{d^2B}{dx^2} + 2iQx\frac{dB}{dx} + 2i\beta\frac{dB}{dz} + iQB - Q^2x^2B + 2\beta\frac{dP}{dz}B - \beta x^2\frac{dQ}{dz}B$$
$$+ (k^2\varepsilon_1 - \beta^2 - k^2b_0^2x^2)B = 0. \qquad (3.78)$$

An equation for $Q(z)$ may be found from (3.78) by equating terms in x^2:

$$Q^2 + \beta\frac{dQ}{dz} + k^2b_0^2 = 0. \qquad (3.79)$$

The remaining equation for B and P is:

$$\frac{d^2B}{dx^2} + 2iQx\frac{dB}{dx} + 2i\beta\frac{dB}{dz} + iQB + (k^2\varepsilon_1 - \beta^2)B + 2\beta\frac{dP}{dz}B = 0. \qquad (3.80)$$

The change of variables (Casperson, 1976):

$$x' = c(z)x; \quad z' = z \qquad (3.81)$$

where $c(z)$ is to be determined, reduces (3.80) to:

$$c^2\frac{d^2B}{dx'^2} + 2ix'\left[Q + \frac{\beta}{c}\frac{dc}{dz'}\right]\frac{dB}{dx'} + 2i\beta\frac{dB}{dz'} + (iQ + k^2\varepsilon_1 - \beta^2)B + 2\beta\frac{dP}{dz'}B = 0. \qquad (3.82)$$

This equation is now separable into a relationship for the complex phase P:

$$\frac{dP}{dz'} = \frac{Nc^2}{\beta} - \frac{(iQ + k^2\varepsilon_1 - \beta^2)}{2\beta} \qquad (3.83)$$

and an equation for B:

$$\frac{d^2B}{dx'^2} - 2x'\left[-\frac{iQ}{c^2} - \frac{i\beta}{c^3}\frac{dc}{dz'}\right]\frac{dB}{dx'} + \frac{2i\beta}{c^2}\frac{dB}{dz'} + 2NB = 0. \qquad (3.84)$$

Now equation (3.84) can be made to resemble the Hermite differential equation by judicious choice of the parameter c; equating the square bracket in (3.84) to 1, we have (Casperson, 1976):

$$-iQ - \frac{i\beta}{c}\frac{dc}{dz'} = c^2. \qquad (3.85)$$

Equation (3.84) is now satisfied by the Hermite polynomials of complex argument:

$$B = H_N(x'). \qquad (3.86)$$

The full solution for the higher-order beam modes is given by solving (3.79) for $Q(z)$, (3.85) for $c(z)$, and (3.83) for $P(z)$. An extension of the arguments given here to cases of three-dimensional z-dependent permittivity, including linear terms in x and y, has been given by Casperson (1976).

3.3.2 Mode stability

In the equilibrium case $(d/dz = 0)$ the solution of equation (3.79) yields for the beam parameter Q_m:

$$Q_m = \pm ikb_0. \qquad (3.87)$$

In the general case, however, equation (3.79) may be solved by the substitution (Kogelnik, 1965):

$$Q = \frac{\beta}{r}\frac{dr}{dz} \tag{3.88}$$

when (3.79) becomes

$$\frac{d^2r}{dz^2} + \frac{k^2 b_0^2}{\beta^2} r = 0. \tag{3.89}$$

The general solution of (3.89) is

$$r = g e^{i\gamma z} + h e^{-i\gamma z} \tag{3.90}$$

where

$$\gamma = \frac{k b_0}{\beta}. \tag{3.91}$$

The solution (3.90) is the same as that for the ray displacement in a parabolic medium, which was given for the real dielectric case in equation (3.11). For real dielectric media, the displacement of the beam centre obeys the usual ray equation; for the complex case considered here, the interpretation of r in (3.89) is not so clear as in the real case.

The general solution for Q, using (3.88) and (3.90) is:

$$Q = i\gamma\beta\left(\frac{g - h e^{-2i\gamma z}}{g + h e^{-2i\gamma z}}\right)$$

which may be written in terms of Q_0, the value at $z = 0$, as:

$$Q = ikb_0\left[\frac{(i\gamma\beta + Q_0) - (i\gamma\beta - Q_0)e^{-2i\gamma z}}{(i\gamma\beta + Q_0) + (i\gamma\beta - Q_0)e^{-2i\gamma z}}\right]. \tag{3.92}$$

We can use equation (3.92) to study the stability of the beam modes by comparing the asymptotic results from this expression with the equilibrium value given in (3.87). Bearing in mind that we are dealing with waves travelling in the $-z$ direction (from (3.75)), the asymptotic limits are given by:

$Im(\gamma) > 0$: $Q(-\infty) = ikb_0$ (3.93a)
$Im(\gamma) < 0$: $Q(-\infty) = -ikb_0$. (3.93b)

From (3.77) and (3.87) it is clear that for a guided wave, whose field amplitude is to decay as x tends to $\pm\infty$, then

$Re(\gamma) > 0$:   choose $+$ sign in (3.87)
$Re(\gamma) < 0$:   choose $-$ sign in (3.87).

Combining these results with those of (3.93), we find that for $Q(-\infty) = Q_m$ the necessary condition is (Ganiel and Silberberg 1975):

$$\text{Sign}(Re(\gamma)) = \text{Sign}(Im(\gamma)). \tag{3.94}$$

It follows that provided equation (3.94) is satisfied a beam which is perturbed from equilibrium will ultimately return to the on-axis position, i.e. the mode is stable against perturbations in the sense discussed earlier, in Section 2.4.1. Since we are concerned principally with the case of relatively small gain or loss, it follows that $\beta_r \gg \beta_i$ (subscripts r and i indicating real and imaginary parts) so that (3.94) may equally well be written

$$\text{Sign}(\text{Re}(b_0)) = \text{Sign}(\text{Im}(b_0))$$

or

$$\text{Im}(b_0^2) > 0. \tag{3.95}$$

Application of (3.95) to the definition of b_0^2 given in (3.73) (under the condition stated) gives the result that for mode stability

$$\delta g_1 > 0. \tag{3.96}$$

We have thus proved that a gain maximum on-axis is necessary for stable modes; although guided solutions exist for the case of a gain minimum at $x = 0$, such modes are not stable and may diverge from the axis in response to a perturbation. This result may also be proved from an expansion of the perturbed beam in terms of the normal modes (Marcuse, 1970, 1972). In spite of these alternative proofs the condition for mode stability in media containing gain profiles has been the subject of recent controversy; a summary and relevant references will be found in a letter by Casperson and Ganiel (1977).

3.3.3 Phase front curvature

For the real dielectric medium the identification of the complex beam parameter Q is simple: comparison of equations (3.43) and (3.75) shows that Q is equivalent to $2i/w_0^2$, where w_0 is the Gaussian spot size (which may now be a function of z). For the general complex medium, on the other hand, Q may be interpreted in terms of w_0 and a phase front radius of curvature R (Kogelnik, 1965):

$$Q = \frac{kn_1}{R} + \frac{2i}{w_0^2}. \tag{3.97}$$

The definition of R in (3.97) implies that for the equilibrium case the surfaces of constant phase are cylindrical with radius R. To see this, note first that from (3.97) and (3.77) these surfaces are given by

$$\beta_r z + \frac{x^2}{2} \text{Re}(Q) = \text{constant} = C. \tag{3.98}$$

Now consider the equation of a circle in the x–z plane centred at $x = 0, z = z_0$, with radius R; for small angles ($x/R \ll 1$) this is:

$$z - z_0 \simeq R - \frac{x^2}{2R}. \tag{3.99}$$

Equations (3.98) and (3.99) may be made equivalent if we make the identifications:

$$\frac{C}{\beta_r} = R + z_0; \quad \text{Re}(Q) = \frac{\beta_r}{R} \simeq \frac{kn_1}{R}.$$

Using the equilibrium value of Q from (3.87), it is clear that the phase fronts are cylindrical with radius of curvature R given by (Cook and Nash, 1975):

$$R = \frac{n_1}{\mp \text{Im}(b_0)} \tag{3.100}$$

where the upper and lower signs correspond respectively to positive and negative values of

$\mathrm{Re}(b_0)$. The corresponding value of spot size w_0 is given from (3.87) and (3.97), with the same sign convention, as:

$$w_0{}^2 = \frac{2}{\pm k \mathrm{Re}(b_0)}. \tag{3.101}$$

Equations (3.100) and (3.101) have been used recently in analyses of the guiding mechanism associated with the stripe geometry in heterostructure lasers (Cook and Nash, 1975; Kirkby et al., 1977). As discussed in Section 2.4.4, the problem here is to elucidate the relative roles played by real index and gain-guidance effects; in fact, all four possibilities of gain minimum or maximum on-axis together with index minimum or maximum on-axis may occur in these devices, i.e. Δ and δ may each take positive or negative signs. Fortunately, however, the values of w_0 and R are measurable from observations of the near- and far-field distributions, and these quantities may be used to calculate the corresponding values of Δ and δ.

From equation (3.73) we have:

$$\frac{2\Delta n_1{}^2}{a^2} = \mathrm{Re}(b_0{}^2) = (\mathrm{Re}(b_0))^2 - (\mathrm{Im}(b_0))^2$$

so that the definitions of (3.100) and (3.101) yield:

$$\Delta n_1 = \frac{a^2}{2n_1}\left[\left(\frac{2}{kw_0{}^2}\right)^2 - \left(\frac{n_1}{R}\right)^2\right]. \tag{3.102}$$

Similarly, the imaginary part of (3.73) gives;

$$\frac{n_1 \delta g_1}{ka^2} = \mathrm{Im}(b_0{}^2) = 2\mathrm{Re}(b_0)\mathrm{Im}(b_0)$$

so that (3.100) and (3.101) yield:

$$\delta g_1 = -\frac{4}{R}\left(\frac{a}{w_0}\right)^2. \tag{3.103}$$

If we identify a with the stripe half-width, then equations (3.102) and (3.103) may be used to find values for the index and gain differences between the centre and edges of the stripe, from measurements of w_0 and R. In particular, it is clear that the sign of R gives the corresponding sign of the gain difference δg_1. Since we have considered waves travelling in the $-z$ direction, we see that positive and negative values of δg_1 correspond respectively to concave and convex phase fronts as viewed along the direction of propagation (Kirkby et al., 1977). The corresponding sign of the index difference Δn_1 is determined from (3.102) in terms of the relative magnitudes of R and w_0. Experimental results analysed in this manner indicate that for narrow-stripe ($\simeq 10\,\mu\mathrm{m}$) lasers the guidance is dominated by the gain-guiding mechanism ($\delta > 0$) (Cook and Nash, 1975), whereas in wider stripes ($\gtrsim 20\,\mu\mathrm{m}$) the dominant effect is a real index guide ($\Delta > 0$, $\delta < 0$). This latter result is thought to be a self-focusing effect resulting from spatial hole-burning in the transverse distribution of current carriers under the stripe (Kirkby et al., 1977).

3.3.4 Equilibrium solutions

Let us return to the solution of the wave equation (3.74) in the equilibrium on-axis case ($d/dz = 0$). Then the value of Q is given by (3.87) and we find from equation (3.85):

$$c^2 = \pm kb_0. \tag{3.104}$$

It follows that $P = 0$ is a solution of (3.83) and the final expression for $A(x)$ from (3.77), (3.81), (3.86), and (3.87) is:

$$A = H_N(cx)\exp\left(\mp \frac{kb_0 x^2}{2}\right) \quad (N = 0, 1, 2, \ldots) \tag{3.105}$$

where the upper and lower signs correspond again to positive and negative values of $\text{Re}(b_0)$. As a check on this result, consider the real dielectric where, from (3.73), $b_0^2 = 2\Delta n_1^2/a^2$. It follows that (3.105) becomes:

Real dielectric:
$$A = H_N\left(\left(\frac{kn_1}{a}\right)^{1/2}(2\Delta)^{1/4}x\right)\exp\left[-\frac{kn_1(2\Delta)^{1/2}x^2}{2a}\right]$$
$$(N = 0, 1, 2, \ldots) \tag{3.106}$$

which is in agreement with equations (3.48) and (3.50), except that in (3.106) the normalization coefficient has been omitted.

The eigenvalue equation for the complex dielectric on-axis beam is given by (3.83), (3.87), and (3.104) as:

$$\beta^2 = k^2\varepsilon_1 \mp kb_0(2N+1) \quad (N = 0, 1, 2, \ldots). \tag{3.107}$$

Once again this result may be checked for the real dielectric medium ($\varepsilon_1 = n_1^2$), where we find:

$$\beta^2 = k^2 n_1^2 - \frac{kn_1(2\Delta)^{1/2}}{a}(2N+1) \quad (N = 0, 1, 2, \ldots) \tag{3.108}$$

which is identical with equation (3.47).

The general complex results (3.105) and (3.107) may be cast into more familiar forms by defining a complex normalized frequency v as:

$$v = \pm a^2 k b_0 \tag{3.109}$$

whence we find

$$A = H_N\left(\frac{v^{1/2}x}{a}\right)\exp\left(-\frac{vx^2}{2a^2}\right) \quad (N = 0, 1, 2, \ldots) \tag{3.110}$$

which is the complex analogue of equation (3.50). Defining also a complex u:

$$u^2 = a^2(k^2\varepsilon_1 - \beta^2) \tag{3.111}$$

equation (3.107) becomes:

$$u^2 = v(2N+1) \quad (N = 0, 1, 2, \ldots) \tag{3.112}$$

which corresponds to the real equation (3.51). Similarly, in terms of the complex propagation constant b, defined in (2.112), we have the result

$$b = 1 - \left(\frac{2N+1}{v}\right). \tag{3.113}$$

If we define the modulus and phase of v as $|v|$ and θ, respectively, then the condition $b_r = 0$ (subscript r indicating real part), which corresponds to cut-off in cladded waveguides, becomes:

$$|v| = (2N+1)\cos\theta. \tag{3.114}$$

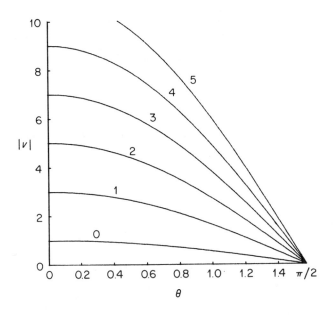

Figure 3.8 Solutions of the equation $b_r = 0$ for the first six modes of the complex parabolic-index medium, calculated from equation (3.114)

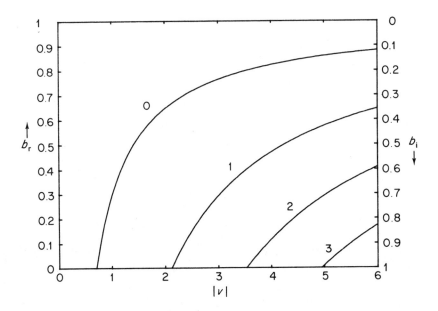

Figure 3.9 Real and imaginary parts of the complex propagation constant b versus $|v|$ for the parabolic medium calculated from equation (3.117); labelling parameter gives the value of mode number N. Left ordinate: b_r. Right ordinate: b_i

Plots of $|v|$ versus θ from equation (3.114) are given in Figure 3.8. In view of our frequent warnings concerning the noncorrespondence of $b_r = 0$ to cut-off for the infinitely-extended parabolic-permittivity medium, it is worth noting that all the results of the present section hold only for modes whose $(|v|, \theta)$ values are well above the corresponding mode-lines on Figure 3.8.

We will here apply the general results to the case of pure gain-guidance, i.e. no index variation, which as mentioned above may well be realized physically in narrow stripe-geometry lasers. For this situation, in analogy with Section 2.4.3, we find:

$$v = \frac{|v|(1+i)}{\sqrt{2}} \tag{3.115}$$

or in terms of the gain difference δg_1 from equation (3.73) with $\Delta = 0$ (Nash, 1973):

$$\mathrm{Re}(b_0) = \mathrm{Im}(b_0) = \frac{1}{a}\left(\frac{n_1 \delta g_1}{2k}\right)^{1/2}. \tag{3.116}$$

Using (3.115) in (3.112), and expressing the result in terms of the real and imaginary parts, b_r and b_i, of the complex propagation constant b:

$$b_r = 1 - \frac{(2N+1)}{|v|\sqrt{2}} \tag{3.117a}$$

$$b_i = \frac{(2N+1)}{|v|\sqrt{2}}. \tag{3.117b}$$

Results of b_r and b_i versus $|v|$ computed from (3.117) are given in Figure 3.9 for the lowest four modes. They may be compared with the equivalent curves for pure gain-guidance in homogeneous-core three-layer slab waveguides which are shown in Figures 2.19 and 2.20.

Chapter 4
Other Graded-index Two-dimensional Guides

The discussion of parabolic-index media given in Chapter 3 has provided an introduction to the subject of graded-index guides in two dimensions. However, the parabolic variation of dielectric profile suffers from two disadvantages: it is only applicable to symmetric guides and it does not include cladding effects, in particular the behaviour of modes near cut-off. In practice, many of the planar waveguides used in integrated optics are strongly asymmetric, frequently being formed by modifying the dielectric properties of the surface layer of a dielectric material, so that the cladding on one side of the guide is air. Thus we see the need to consider asymmetric cladded waveguides whose dielectric profiles are graded as a result of the fabrication process. In the present chapter therefore we will concentrate on profiles for which the scalar wave equation possesses analytic solutions. Chapter 5 will deal with approximate methods of solution which may be applied to arbitrary profiles.

We shall consider first the case of strongly asymmetric profiles such as may occur at the surface of a diffused guide where the external medium is air. The results of the parabolic profile have been applied to this situation by Standley and Ramaswamy (1974), by demanding that the fields vanish at the surface, i.e. only odd-order modes of the symmetric parabolic-index medium are permitted (see, for example, Figure 3.5, $N = 1$). This is a neat solution and we shall apply it to the exponential and linear dielectric profiles in this chapter. Both these profiles possess analytic modal solutions and both are also amenable to the approximations of geometric optics. The effects of surface dielectric coatings which reduce the asymmetry and change the guide characteristics will also be dealt with. Both profiles may also be solved in the symmetric case where the index decreases equally on either side of the axis.

A further profile which yields exact solutions is the so-called Epstein-layer model (Epstein, 1930) and this is also discussed in the present chapter. This model, originally proposed for calculations of radio-wave reflection from the ionosphere, has also many features in common with the Pöschl–Teller potential in quantum mechanics (Pöschl and Teller, 1933). This emphasizes the similarity of the scalar wave equation to Schrödinger's equation and the equivalence in mathematical terms of dielectric profile and potential well as regards the solution of these equations. The advantage of the Epstein-layer model lies in its continuous nature, making boundary problems unnecessary, and in its ability to represent a wide range of dielectric profiles from perfectly symmetric to strongly asymmetric.

4.1 The Exponential Profile

The practice established in earlier chapters of treating first the ray approach and then the modal theory will be continued with regard to this profile. Comparison between solutions obtained by these methods shows good agreement over a wide range of normalized frequency v. We deal first with the strongly asymmetric case referred to above, then with the effects of a surface coating, and finally with the symmetric exponential profile. In order that the treatment does not become too repetitive or tedious we will deal in somewhat less depth with the profiles of this chapter than in the previous two chapters. Further details of TM mode solutions which are omitted here have been given by Love and Ghatak (1979).

4.1.1 Ray treatment of the strongly asymmetric profile

The strongly asymmetric exponential profile is shown in Figure 4.1. At the surface ($x = 0$) the index is n_1, in the bulk of the material far from the surface the value is n_2, and the outside

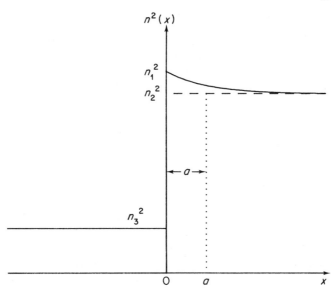

Figure 4.1 The strongly-asymmetric exponential dielectric profile

medium has index n_3 (typically $n_3 = 1$, corresponding to air). For $x > 0$ the profile is given by:

$$n^2(x) = n_1^2[1 - 2\Delta(1 - e^{-x/a})] \tag{4.1}$$

where, as in earlier work,

$$n_1^2 - n_2^2 = 2\Delta n_1^2. \tag{4.2}$$

In order to find the eigenvalue equation for this profile, we use again the transverse phase resonance condition in the form:

$$2\int_0^{x_c} (k^2 n^2(x) - \beta^2)^{1/2} dx - \delta_{12} - \delta_{13} = 2N\pi \tag{4.3}$$

where δ_{12}, δ_{13} are the phase shifts at the caustic x_c and the 1–3 interface, respectively. From Section 3.1, equation (3.19), we have $\delta_{12} = \pi/2$. For δ_{13} the general expressions (2.32) and (2.34) apply for transverse and parallel polarization, respectively, with n_2 replaced by n_3 and $\beta = n_1 k \sin \theta_1$, i.e.

$$\delta_\perp = 2 \tan^{-1}\left[\frac{((\beta/k)^2 - n_3^2)^{1/2}}{(n_1^2 - (\beta/k)^2)^{1/2}}\right] \tag{4.4a}$$

$$\delta_\parallel = 2\tan^{-1}\left[\frac{n_1^2((\beta/k)^2 - n_3^2)^{1/2}}{n_3^2(n_1^2 - (\beta/k)^2)^{1/2}}\right]. \quad (4.4b)$$

Now for the case of strong dielectric asymmetry considered here, $2\Delta n_1^2 = (n_1^2 - n_2^2) \ll (n_1^2 - n_3^2)$, and for bound rays $n_1 k \geqslant \beta \geqslant n_2 k$. It follows that the arguments of the $\tan^{-1}$ functions in (4.4) are very large, and to a good level of approximation we have (Hocker and Burns, 1975):

$$\delta_\parallel \simeq \delta_\perp \simeq \pi. \quad (4.5)$$

Using the usual normalized variables $v^2 = 2\Delta n_1^2 k^2 a^2$, $w^2 = a^2(\beta^2 - n_2^2 k^2)$, equation (4.3) becomes:

$$2\int_0^{x_c} (v^2 e^{-x/a} - w^2)^{1/2} \frac{dx}{a} = (2N + \tfrac{3}{2})\pi \quad (4.6)$$

where $x_c = 2a\ln(v/w)$ gives the position of the caustic. The integral in (4.6) may be performed by writing $y = e^{-x/2a}$, $b = w^2/v^2$, whence:

$$2v\int_{\sqrt{b}}^1 (y^2 - b)^{1/2} \frac{2dy}{y} = [(1-b)^{1/2} - b^{1/2}\cos^{-1}(b^{1/2})]4v.$$

Hence the eigenvalue equation for normalized propagation constant b in terms of v becomes:

$$(1-b)^{1/2} - b^{1/2}\cos^{-1}(b^{1/2}) = \frac{\pi}{4v}(2N + \tfrac{3}{2}) \quad (N = 0, 1, 2, \ldots). \quad (4.7)$$

Graphical results for b as a function of v for the lowest four modes calculated from (4.7) are given in Figure 4.2 (solid lines). It also follows that the mode cut-offs are given by ($b = 0$)

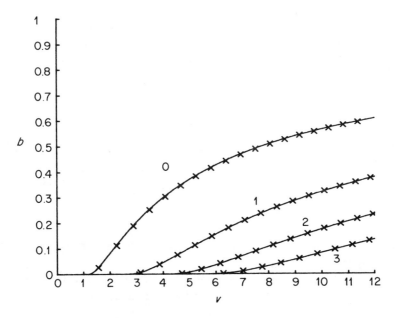

Figure 4.2 Normalized b–v plots for the strongly-asymmetric exponential profile; labelling parameter gives the value of mode number N. Solid lines, ray theory results from equation (4.7); crosses, modal theory results from equation (4.13)

(Conwell, 1975):

$$v_c = \frac{\pi}{4}(2N + \tfrac{3}{2}) \quad (N = 0, 1, 2, \ldots). \tag{4.8}$$

These results will be compared below with those obtained by an exact solution of the scalar wave equation. Similar plots of b–v curves for various modes and v-ranges have been given by Hocker and Burns (1975) and Caton (1974).

4.1.2 Mode treatment of the strongly asymmetric profile

Assuming the usual time and z-dependence of the fields the scalar wave equation for this profile may be written ($x \geq 0$):

$$\frac{d^2\psi}{dx^2} + \frac{1}{a^2}(v^2 e^{-x/a} - w^2)\psi = 0 \tag{4.9}$$

where ψ represents any transverse field component, and the familiar normalized variables have been used. If we make the substitution

$$x' = 2v \exp\left(-\frac{x}{2a}\right) \tag{4.10}$$

then equation (4.9) becomes:

$$x'^2 \frac{d^2\psi}{dx'^2} + x' \frac{d\psi}{dx'} + (x'^2 - 4w^2)\psi = 0. \tag{4.11}$$

Equation (4.11) is Bessel's differential equation (see, for example, Abramowitz and Stegun, 1964), whose solutions which behave correctly in the limit $x' \to \infty$ are the Bessel functions of the first kind $J_{2w}(x')$ (Conwell, 1973, 1974; Gubanov and Sharapov, 1970; Sharapov, 1970):

$$\psi = AJ_{2w}(2v e^{-x/2a}) \tag{4.12}$$

where A is a constant. The eigenvalue equation may now be found by matching the field distribution in (4.12) for $x \geq 0$ with that appropriate to the region $x \leq 0$. Since we are concerned with the case of strong asymmetry, a simple approximate form of this boundary condition is obtained by demanding that at $x = 0$ the fields vanish:

$$J_{2w}(2v) = 0. \tag{4.13}$$

Solutions of equation (4.13), when numbered consecutively starting from 0, can provide plots of w versus v for each mode (Carruthers et al., 1974). The solutions of (4.13) obtained from Bessel function tables (Abramowitz and Stegun, 1964) are shown as crosses on the plots of Figure 4.2; there is clearly excellent agreement with the ray solutions obtained from (4.7).

The cut-off values of v for the modes of this profile are obtained by setting $w = 0$ in (4.13):

$$J_0(2v_c) = 0. \tag{4.14}$$

The solutions of this equation for the first four modes are compared with those from the ray theory (equation (4.8)) in Table 4.1.

The generally high level of agreement may also be confirmed by taking the asymptotic form of the Bessel function in (4.14) (Abramowitz and Stegun, 1964):

$$J_0(2v_c) \sim \left(\frac{1}{\pi v_c}\right)^{1/2} \cos\left(2v_c - \frac{\pi}{4}\right) \tag{4.15}$$

whence the solution (4.8) is again found (Carruthers et al., 1974).

Table 4.1 Mode cut-offs v_c for the strongly asymmetric exponential profile

Mode number	0	1	2	3
Modal solution from (4.14)	1.202	2.760	4.327	5.896
Ray solution from (4.8)	1.178	2.749	4.320	5.890

Plots of the field distributions for the two lowest order modes at $v = 5.54$ are given in Figure 4.3. The results, which were calculated from (4.12) with the constant A evaluated for equal powers, show the weak field-confinement characteristics of the exponential profile.

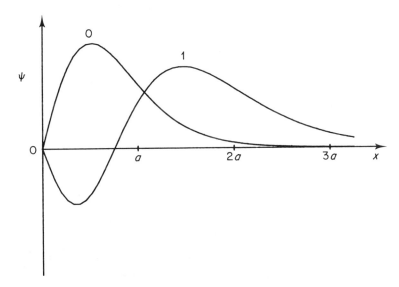

Figure 4.3 Field distributions for the two lowest-order modes of the strongly-asymmetric exponential profile; labelling parameter gives the value of mode number N

Analogous results for the case where the imaginary part of the dielectric permittivity has a strongly asymmetric exponential behaviour have been derived by Lee et al. (1978).

4.1.3 The effects of a dielectric cover

Consider now the case when there is no longer strong asymmetry, and the detailed effects of the external medium (index n_3) must be included. For simplicity we restrict attention to polarization with the electric field normal to the plane of incidence (TE modes). Let us define an asymmetry parameter c (Kogelnik and Ramaswamy, 1974; Hocker, 1976; Johnson, 1977) as follows:

$$c^2 = \frac{n_2^2 - n_3^2}{n_1^2 - n_2^2} \tag{4.16}$$

where n_1, n_2, and n_3 are as defined in Figure 4.1. Returning to the ray treatment of Section 4.1.1, the analysis proceeds as before until the definition of phase shift δ_{13} at the 1–3 interface.

For the case of a dielectric cover, equation (4.4a) gives:

$$\delta_{13} = 2 \tan^{-1}\left[\frac{(b+c^2)^{1/2}}{(1-b)^{1/2}}\right] \quad \text{(TE)}. \tag{4.17}$$

As a consequence, the eigenvalue equation (4.3) becomes:

$$[(1-b)^{1/2} - b^{1/2} \cos^{-1}(b^{1/2})]4v = \pi(2N+\tfrac{1}{2}) + 2\tan^{-1}\left[\frac{(b+c^2)^{1/2}}{(1-b)^{1/2}}\right]$$
$$(N = 0, 1, 2, \ldots). \tag{4.18}$$

Plots of b versus v obtained from (4.18) for the first three modes and various values of c are given in Figure 4.4; further curves of this type for different c-values have been presented by

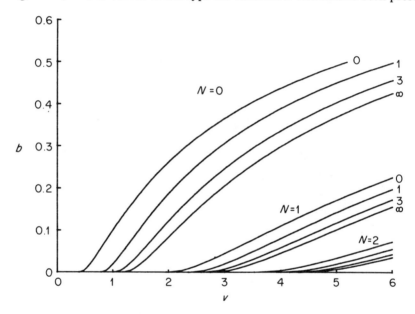

Figure 4.4 Normalized b–v plots for the asymmetric exponential profile; labelling parameter on each curve within a modal group gives the value of the asymmetry parameter c, defined as in equation (4.16)

Hocker (1976). The cut-offs are given from (4.18) as:

$$v_c = \frac{\pi}{4}(2N+\tfrac{1}{2}) + \tfrac{1}{2}\tan^{-1}(c) \quad (N = 0, 1, 2, \ldots). \tag{4.19}$$

These results may be checked with the modal treatment. For $x \geq 0$ the solution (4.12) applies; in the region $x \leq 0$ we take the usual exponentially-damped solution:

$$\psi = B \exp\left[(w^2 + c^2 v^2)^{1/2} \frac{x}{a}\right]. \tag{4.20}$$

Matching the fields (4.12), (4.20) and their derivatives at $x = 0$ yields the modal form of the TE eigenvalue equation (Conwell, 1973, 1974).

$$\frac{J_{2w-1}(2v) - J_{2w+1}(2v)}{2J_{2w}(2v)} = -\frac{(w^2 + c^2 v^2)^{1/2}}{v} \tag{4.21}$$

where the derivative J'_{2w} has been expressed in terms of J_{2w-1} and J_{2w+1}. At cut-off this equation becomes:

$$\frac{J_1(2v_c)}{J_0(2v_c)} = c. \qquad (4.22)$$

Using again the asymptotic form (4.15) for J_0 and the corresponding result for J_1, we recover equation (4.19) for the cut-off values of v.

The modal analysis outlined here has been extended to include TM modes and anisotropy by Conwell (1973, 1974) and to include a finite thickness cover layer by Weller and Giallorenzi (1975). General plots of b–v curves for the lowest-order modes obtained from numerical solution of equation (4.21) have been given by Haus and Schmidt (1976). The case of a metal cladding was considered by Findakly and Chen (1978).

4.1.4 The symmetric exponential profile

The symmetric case is shown in Figure 4.5, and is described by:

$$n^2(x) = n_1^2[1 - 2\Delta(1 - e^{-|x|/a})]. \qquad (4.23)$$

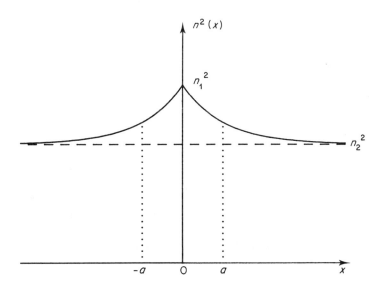

Figure 4.5 The symmetric exponential dielectric profile

For the ray theory of this profile, the analysis proceeds as in Section 4.1.1, except that here the integral in (4.3) extends from $-x_c$ to $+x_c$ and we have that $\delta_{12} = \delta_{13} = \pi/2$. It follows that the eigenvalue equation is given by

$$(1-b)^{1/2} - b^{1/2} \cos^{-1}(b^{1/2}) = \frac{\pi}{8v}(2N+1). \qquad (4.24)$$

For odd-order modes this equation reduces to (4.7), since these modes have vanishing fields at $x = 0$, corresponding to the assumption $\delta_{13} \simeq \pi$ in (4.5). Plots of the b–v curves for the first 8 modes are given in Figure 4.6 (solid lines) as calculated from equation (4.24). The cut-offs are

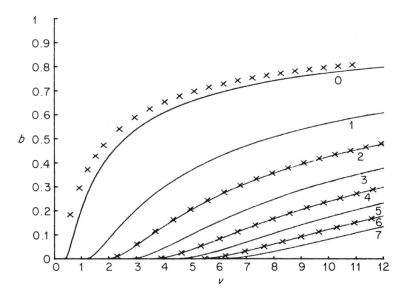

Figure 4.6 Normalized b–v plots for symmetric exponential profile; labelling parameter gives the value of mode number N. Solid lines, ray theory results from equation (4.24); crosses, modal theory results for even-order modes only from equation (4.27)

given by:

$$v_c = \frac{\pi}{8}(2N+1). \tag{4.25}$$

For $x \geq 0$ the field distributions of the symmetric exponential profile are given by (4.12); for $x \leq 0$ the appropriate solution of the scalar wave equation is (Kogelnik, 1975):

$$\psi = A J_{2w}(2v e^{x/2a}). \tag{4.26}$$

The modal eigenvalue equation is found by matching the fields in (4.12) and (4.26) and their derivatives at $x = 0$. For *odd-order modes* this reduces to equation (4.13), the solutions of which have already been discussed. For *even-order modes* the matching condition becomes

$$J'_{2w}(2v) = 0. \tag{4.27}$$

Solutions of this equation for the four lowest even-order modes are shown in Figure 4.6 as crosses. There is good agreement between these results and those of the ray theory, equation (4.24), for all except the zero-order mode. The cut-off condition for odd-order modes has already been discussed in Section 4.1.2; for the even-order modes the cut-offs are given by

$$J'_0(2v_c) = 0. \tag{4.28}$$

Table 4.2 compares the results of the ray and mode predictions for cut-off from equations (4.25) and (4.28), respectively, for the even-order modes. Once again the asymptotic form of equation (4.28) yields (4.25) for the even-order modes, although the justification for using an asymptotic result for the lowest-order mode is clearly lacking.

From the table we see the generally high level of agreement between these theories, with the exception of the results for the lowest-order mode. The reason for the failure of the ray theory for the zero-order mode is connected with the large slope of the exponential profile close to

Table 4.2 Mode cut-offs v_c for the even-order modes of the symmetric exponential profile

Mode number	0	2	4	6
Modal solution from (4.28)	0	1.916	3.508	5.087
Ray solution from (4.25)	0.393	1.963	3.534	5.105

$x = 0$. This point will be discussed in more detail in the section on WKB theory in Chapter 5.

The radiation confinement factor Γ, defined in the usual way, may be written for the symmetric exponential guide as:

$$\Gamma = 1 - \frac{\int_0^{2ve^{-1/2}} J_{2w}^2(y)\,dy/y}{\int_0^{2v} J_{2w}^2(y)\,dy/y} \tag{4.29}$$

where the field distribution (4.12) has been used, with the substitution $y = 2v\exp(-x/2a)$. The integrals in (4.29) may be evaluated in terms of series of Bessel functions (Luke, 1962):

$$\int^t J_v^2(y)\frac{dy}{y} = \frac{1}{2v}\left[J_v^2(t) + 2\sum_{k=0}^{\infty} J_{v+k}^2(t)\right] \tag{4.30}$$

and the series in (4.30) usually converges fairly rapidly for the parameter values of interest here. Plots of Γ versus v obtained from (4.29) and (4.30) for the first two modes are shown in Figure 4.7 (broken lines). For comparison the corresponding $\Gamma - v$ curves for the slab

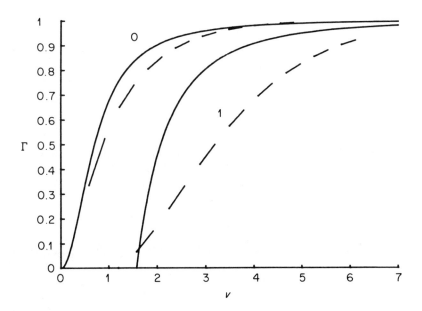

Figure 4.7 Radiation confinement factor Γ versus v for the lowest-order modes of the symmetric exponential profile (broken lines) calculated from equations (4.29) and (4.30); the solid lines give the corresponding curves for the step-index slab (see Chapter 2)

waveguide are also plotted (see Section 2.3). Clearly the exponential profile has weaker guidance properties, with the degree of optical confinement decreasing for higher-order modes.

4.2 The Linear Profile

The treatment here follows the sequence used for the exponential profile above. Once again we confine attention to TE modes for simplicity.

4.2.1 Ray treatment of the strongly asymmetric profile

Figure 4.8 shows the strongly asymmetric profile, which is described by

$$n^2(x) = \begin{cases} n_1^2[1-2\Delta(x/a)], & 0 \leqslant x \leqslant a \\ n_1^2[1-2\Delta] \equiv n_2, & x \geqslant a \end{cases} \quad (4.31)$$

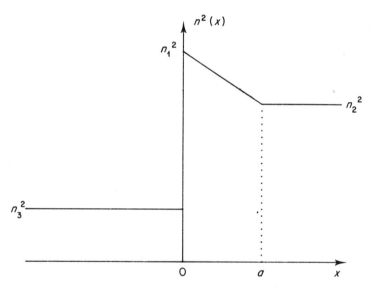

Figure 4.8 The strongly-asymmetric linear profile

where the notation of Section 4.1 has again been used. The eigenvalue equation may be found from the transverse phase resonance condition (4.3), again using $\delta_{12} = \pi/2$ and $\delta_{13} = \pi$ for the phase shifts. With the usual variables, equation (4.3) becomes

$$2 \int_0^{au^2/v^2} \left[u^2 - v^2 \left(\frac{x}{a} \right) \right]^{1/2} \frac{dx}{a} = (2N + \tfrac{3}{2})\pi \quad (N = 0, 1, 2, \ldots). \quad (4.32)$$

The integral is easily performed and we obtain the eigenvalue equation in the form (Marcuse, 1973b):

$$(1-b)^{3/2} = \frac{3\pi}{4v}(2N + \tfrac{3}{2}) \quad (N = 0, 1, 2, \ldots). \quad (4.33)$$

Plots of b–v curves calculated from (4.33) are given in Figure 4.9. The cut-offs are given as:

$$v_c = \frac{3\pi}{4}(2N + \tfrac{3}{2}) \quad (N = 0, 1, 2, \ldots). \quad (4.34)$$

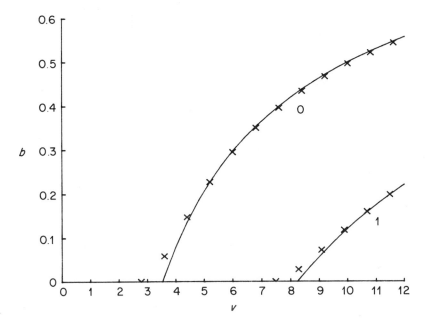

Figure 4.9 Normalized b–v plots for the strongly asymmetric linear profile; labelling parameter gives the value of mode-number N. Solid lines, ray theory results from equation (4.33); crosses, modal theory results from equation (4.42)

These results are compared with those from the exact solution of the scalar wave equation in the next subsection.

4.2.2 Mode treatment of the strongly asymmetric profile

The scalar wave equation for this profile in the region $0 \leqslant x \leqslant a$ is:

$$\frac{d^2\psi}{dx^2} + \frac{1}{a^2}\left(u^2 - v^2\frac{x}{a}\right)\psi = 0. \tag{4.35}$$

Making the change of variables

$$x' = \left(\frac{x}{a} - \frac{u^2}{v^2}\right)v^{2/3} \tag{4.36}$$

the wave equation becomes (Marcuse, 1973b; Rigrod et al., 1975; Streifer et al., 1976a):

$$\frac{d^2\psi}{dx'^2} - x'\psi = 0. \tag{4.37}$$

The solutions of this equation are the Airy functions Ai(x') and Bi(x') (see, for example, Abramowitz and Stegun, 1964) with turning points at $x' = 0$ ($x = au^2/v^2$). For $x' > 0$, Ai and Bi are linear combinations of the modified Bessel functions of order $\pm \frac{1}{3}$, i.e. $I_{\frac{1}{3}}$ and $I_{-\frac{1}{3}}$; Ai is exponentially decaying and Bi is exponentially growing. For $x' < 0$, Ai and Bi are combinations of the Bessel functions $J_{\frac{1}{3}}$ and $J_{-\frac{1}{3}}$, and both functions are oscillatory. For the Bessel and modified Bessel function representations of the Airy functions the argument is $2|x'|^{3/2}/3$. Hence the general solution of equation (4.37) may be written

$$\psi(x') = C\text{Ai}(x') + D\text{Bi}(x') \tag{4.38}$$

where C and D are constants to be evaluated from the boundary conditions at $x = 0$ ($x' = -u^2/v^{4/3}$) and $x = a$ ($x' = w^2/v^{4/3}$).

For the region $x \geqslant a$, the refractive index is a constant value (n_2) according to equation (4.31) and thus the appropriate solution for a field component ψ which vanishes as $x \to \infty$ is the familiar exponential decay:

$$\psi(x) = \psi(a)\exp\left[w\left(1 - \frac{x}{a}\right)\right]. \tag{4.39}$$

Hence the TE boundary condition at $x = a$ is given by matching the field distributions (4.38) and (4.39) and their derivatives:

$$\frac{C\,\text{Ai}'(w^2/v^{4/3}) + D\,\text{Bi}'(w^2/v^{4/3})}{C\,\text{Ai}(w^2/v^{4/3}) + D\,\text{Bi}(w^2/v^{4/3})} = -\frac{w}{v^{2/3}} \tag{4.40}$$

where the primes indicate differentiation with respect to the argument, and the relation $dx'/dx = v^{2/3}/a$ from (4.36) has been used. For the strongly asymmetric profile the boundary condition at $x = 0$ is that the field distribution (4.38) should vanish, i.e.

$$C\,\text{Ai}(-u^2/v^{4/3}) + D\,\text{Bi}(-u^2/v^{4/3}) = 0. \tag{4.41}$$

Eliminating the ratio C/D from equations (4.40) and (4.41) we obtain the modal eigenvalue equation in the form:

$$\frac{v^{2/3}\,\text{Bi}'(w^2/v^{4/3}) + w\,\text{Bi}(w^2/v^{4/3})}{v^{2/3}\,\text{Ai}'(w^2/v^{4/3}) + w\,\text{Ai}(w^2/v^{4/3})} = \frac{\text{Bi}(-u^2/v^{4/3})}{\text{Ai}(-u^2/v^{4/3})}. \tag{4.42}$$

The solutions of this equation for the first two modes are shown as crosses on the b–v plots of Figure 4.9. The agreement between ray and modal theories (equations (4.33) and (4.42), respectively) is clearly rather poor near to cut-off, but improves as v increases.

The cut-off values from the modal theory are obtained from (4.42) as the solutions of:

$$\frac{\text{Bi}(-v_c^{2/3})}{\text{Ai}(-v_c^{2/3})} = \frac{\text{Bi}'(0)}{\text{Ai}'(0)} \equiv -\sqrt{3}. \tag{4.43}$$

An asymptotic approximation of equation (4.43) may be made for purposes of comparison with the ray theory result. If we use the asymptotic forms of the Airy functions (Abramowitz and Stegun, 1964):

$$\text{Ai}(-t) \sim \frac{1}{\sqrt{\pi}\,t^{1/4}}\sin\left(\zeta + \frac{\pi}{4}\right) \tag{4.44a}$$

$$\text{Bi}(-t) \sim \frac{1}{\sqrt{\pi}\,t^{1/4}}\cos\left(\zeta + \frac{\pi}{4}\right) \tag{4.44b}$$

where $\zeta = 2t^{3/2}/3$, then (4.43) reduces to the form:

$$v_c \simeq \left(\frac{3N}{2} + \frac{7}{8}\right)\pi \quad (N = 0, 1, 2, \ldots). \tag{4.45}$$

This asymptotic result differs from the ray theory equation (4.34) in that the term $7\pi/8$ here is replaced by $9\pi/8$ in (4.34). The numerical values for cut-off v_c obtained by the modal equation (4.43), the asymptotic form (4.45), and the ray equation (4.34) may be compared for the first four modes in Table 4.3.

From the table it is clear that the asymptotic form (4.45) gives reasonable agreement with

Table 4.3 Mode cut-offs v_c for the strongly asymmetric linear profile

Mode number	0	1	2	3
Modal solution from (4.43)	2.800	7.482	12.186	16.895
Asymptotic form (4.45)	2.749	7.461	12.174	16.886
Ray solution from (4.34)	3.534	8.247	12.959	17.671

the exact result (4.43), the level of agreement improving as the mode number increases. The ray theory, on the other hand, gives a rather poor level of agreement for this profile. The reason for this is the failure of the result $\delta_{12} = \pi/2$ at cut-off.

4.2.3 The effects of a dielectric cover

Using again the asymmetry parameter c defined as in (4.16), the effects of the external medium (index n_3) may be included. For the ray theory the phase shift δ_{13} is again given by (4.17), and the eigenvalue equation becomes:

$$\frac{2v(1-b)^{3/2}}{3} = (N+\tfrac{1}{4})\pi + \tan^{-1}\left[\frac{(b+c^2)^{1/2}}{(1-b)^{1/2}}\right] \quad (N = 0, 1, 2, \ldots). \tag{4.46}$$

The cut-offs are given by:

$$v_c = \tfrac{3}{2}[(N+\tfrac{1}{4})\pi + \tan^{-1}(c)] \quad (N = 0, 1, 2, \ldots). \tag{4.47}$$

For the modal theory we proceed as in the previous subsection except that the region $x \leqslant 0$ now has the corresponding field distribution:

$$\psi(x) = \psi(0)\exp\left[\frac{x}{a}(w^2 + v^2 c^2)^{1/2}\right]. \tag{4.48}$$

Hence the TE boundary condition at $x = 0$ is given by matching the fields (4.38) and (4.48) and their derivatives:

$$\frac{C\operatorname{Ai}'(-u^2/v^{4/3}) + D\operatorname{Bi}'(-u^2/v^{4/3})}{C\operatorname{Ai}(-u^2/v^{4/3}) + D\operatorname{Bi}(-u^2/v^{4/3})} = \frac{(w^2 + v^2 c^2)^{1/2}}{v^{2/3}}. \tag{4.49}$$

Eliminating the ratio C/D from equations (4.40) and (4.49) we obtain the modal eigenvalue equation:

$$\frac{v^{2/3}\operatorname{Bi}'(w^2/v^{4/3}) + w\operatorname{Bi}(w^2/v^{4/3})}{v^{2/3}\operatorname{Ai}'(w^2/v^{4/3}) + w\operatorname{Ai}(w^2/v^{4/3})} = \frac{v^{2/3}\operatorname{Bi}'(-u^2/v^{4/3}) - (w^2 + v^2 c^2)^{1/2}\operatorname{Bi}(-u^2/v^{4/3})}{v^{2/3}\operatorname{Ai}'(-u^2/v^{4/3}) - (w^2 + v^2 c^2)^{1/2}\operatorname{Ai}(-u^2/v^{4/3})}. \tag{4.50}$$

At cut-off, the values of v_c are given by the solution of:

$$\frac{\operatorname{Bi}'(-v_c^{2/3}) - cv_c^{1/3}\operatorname{Bi}(-v_c^{2/3})}{\operatorname{Ai}'(-v_c^{2/3}) - cv_c^{1/3}\operatorname{Ai}(-v_c^{2/3})} = \frac{\operatorname{Bi}'(0)}{\operatorname{Ai}'(0)} \equiv -\sqrt{3}. \tag{4.51}$$

Once again an asymptotic form of (4.51) may be found by taking the approximate relations (4.44) and the corresponding results for Ai′, Bi′, to give:

$$\frac{2v_c}{3} \simeq (N+1/12)\pi + \tan^{-1}(cv_c^{1/3}) \quad (N = 0, 1, 2, \ldots). \tag{4.52}$$

This result may be compared with the corresponding ray theory equation (4.47). In the limits $c \to 0$ and $c \to \infty$ there is again the discrepancy of $\pi/4$ in predicted values of v_c which was noted in the previous subsection for the corresponding comparison in the strongly asymmetric case. Generalized versions of the theory given here have been applied to various structures by Marcuse (1973b), Rigrod et al. (1975), and Streifer et al. (1976a). Ray theory calculations for the linear profile have been made by Stewart et al. (1977) and Gallagher and De La Rue (1977). The effects of a metallic cladding have been analysed by Masuda et al. (1977) and the additional complications of a buffer layer were considered by Masuda and Koyama (1977).

4.2.4 The symmetric linear profile

Figure 4.10 illustrates the symmetric profile, according to:

$$n^2(x) = \begin{cases} n_1^2[1 - 2\Delta|x/a|], & |x| \leq a \\ n_1^2[1 - 2\Delta] \equiv n_2, & |x| \geq a. \end{cases} \quad (4.53)$$

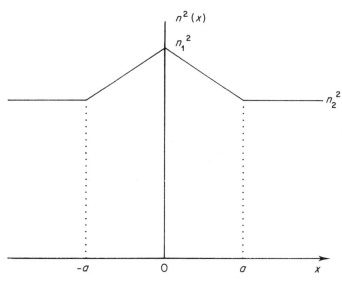

Figure 4.10 The symmetric linear profile

For the ray theory, $\delta_{12} = \delta_{13} = \pi/2$ and the eigenvalue equation is:

$$(1-b)^{3/2} = \frac{3\pi}{8v}(2N+1) \quad (N = 0, 1, 2, \ldots). \quad (4.54)$$

Note that for odd-order modes this equation reduces to the form (4.33). Plots of the b–v curves calculated from (4.54) are given in Figure 4.11 (solid lines). The cut-offs are given by

$$v_c = \frac{3\pi}{8}(2N+1) \quad (N = 0, 1, 2, \ldots). \quad (4.55)$$

Turning to the modal theory, for *odd-order modes* the solutions found in Section 4.2.2 again apply; for *even-order modes* the result (4.41) is replaced by the boundary condition that the derivatives of the fields should vanish at $x = 0$:

$$C \text{Ai}'(-u^2/v^{4/3}) + D \text{Bi}'(-u^2/v^{4/3}) = 0. \quad (4.56)$$

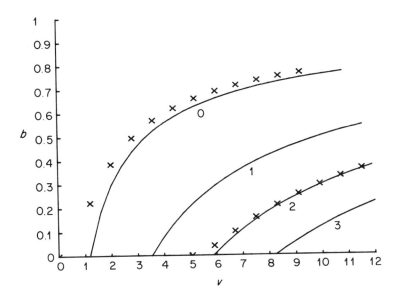

Figure 4.11 Normalized b–v plots for the symmetric linear profile; labelling parameter gives the value of mode number N. Solid lines, ray theory results from equation (4.54); crosses, modal theory results for even-order modes only from equation (4.57)

Eliminating the ratio C/D from equations (4.40) and (4.56) gives the eigenvalue equation:

$$\frac{v^{2/3}\,\text{Bi}'(w^2/v^{4/3})+w\,\text{Bi}(w^2/v^{4/3})}{v^{2/3}\,\text{Ai}'(w^2/v^{4/3})+w\,\text{Ai}(w^2/v^{4/3})} = \frac{\text{Bi}'(-u^2/v^{4/3})}{\text{Ai}'(-u^2/v^{4/3})}. \tag{4.57}$$

The solutions of this equation for the first two even-order modes are shown as crosses on the b–v plots of Figure 4.11. For the lowest order mode, the ray and modal theories are in serious disagreement, just as we found for the exponential profile in Section 4.1; for the other modes the theories disagree near the mode cut-offs, but become progressively closer as v increases.

At cut-off the values of v_c are given by the solutions of:

$$\frac{\text{Bi}'(-v_c^{2/3})}{\text{Ai}'(-v_c^{2/3})} = -\sqrt{3}. \tag{4.58}$$

The asymptotic form of this equation yields:

$$v_c \simeq \left(\frac{3N'}{2}+\frac{1}{8}\right)\pi \quad (N'=0,1,2,\ldots). \tag{4.59}$$

Note that N' as defined here corresponds to $2N$ where N is the usual mode number as used, for example, in (4.55) for the corresponding ray theory result. The values of v_c obtained from (4.58), (4.59), and (4.55) for the first four even-order modes are compared in Table 4.4.

The calculation of the radiation confinement factor Γ for the symmetric linear profile involves the integration of Airy functions:

$$\Gamma = \frac{P_1}{P_1+P_2} \tag{4.60}$$

Table 4.4 Mode cut-offs v_c for the even-order modes of the symmetric linear profile

Mode number	0	2	4	6
Modal solution from (4.58)	0	5.063	9.795	14.515
Asymptotic form (4.59)	0.393	5.105	9.817	14.530
Ray solution from (4.55)	1.178	5.890	10.603	15.315

where

$$P_1 = \int_0^a [C\,\text{Ai}(x') + D\,\text{Bi}(x')]^2 \, dx \tag{4.61}$$

$$P_2 = [C\,\text{Ai}(w^2/v^{4/3}) + D\,\text{Bi}(w^2/v^{4/3})]^2 \int_a^\infty \exp\left[2w\left(1-\frac{x}{a}\right)\right] dx. \tag{4.62}$$

The integral in (4.61) may be performed by using a special property of the Airy function (Arnaud and Mammel, 1975):

$$\int^t \text{Ai}^2(y)\,dy = t\,\text{Ai}^2(t) - \text{Ai}'^2(t). \tag{4.63}$$

Since this result applies equally to Bi^2 and AiBi, a trivial extension of (4.63) yields

$$P_1 = \frac{a}{v^{2/3}}\left[\frac{w^2}{v^{4/3}}\psi^2(w^2/v^{4/3}) + \frac{u^2}{v^{4/3}}\psi^2(-u^2/v^{4/3}) - \psi'^2(w^2/v^{4/3})\right.$$

$$\left. + \psi'^2(-u^2/v^{4/3})\right] \tag{4.64}$$

$$P_2 = \frac{a}{2w}\psi^2(w^2/v^{4/3}) \tag{4.65}$$

where ψ is defined as in (4.38). Plots of Γ versus v calculated in this way from equations (4.60), (4.64), and (4.65) are given in Figure 4.12. The curve for the lowest-order mode ($N = 0$) is rather similar to the corresponding curve for the exponential profile (Figure 4.7). It lies below the corresponding step-index curve (Figure 2.12) for v-values less than about 2.5.

4.3 The Epstein-layer Model

The most general form of the profile to be discussed is given by (Epstein, 1930):

$$n^2(x) = \varepsilon_3 + \frac{(\varepsilon_2 - \varepsilon_3)e^{2x/a}}{(1 + e^{2x/a})} + \frac{\varepsilon_1 e^{2x/a}}{(1 + e^{2x/a})^2}. \tag{4.66}$$

To agree with our earlier notation, the constants ε_1–ε_3 are chosen as follows:

$$\varepsilon_1 = 4n_1^2 - 2(n_2^2 + n_3^2) \tag{4.67a}$$

$$\varepsilon_2 = n_2^2 \tag{4.67b}$$

$$\varepsilon_3 = n_3^2. \tag{4.67c}$$

This set of definitions ensures that $n^2(0) = n_1^2$, $\lim_{x \to \infty}[n^2(x)] = n_2^2$, and $\lim_{x \to -\infty}[n^2(x)] = n_3^2$. Other choices of the constants ε_1–ε_3 are of course possible, and a further discussion of this

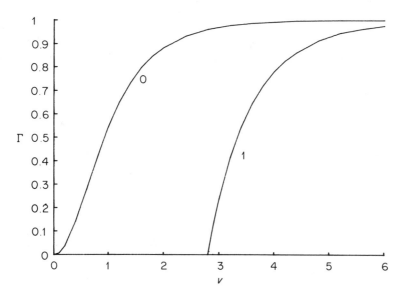

Figure 4.12 Radiation confinement parameter Γ versus v for the lowest-order modes of the symmetric linear profile calculated from equations (4.60), (4.64), and (4.65)

point has been given by Osinski (1976, 1977), including the case of complex dielectric permittivity. Using equation (4.67), together with the definitions of Δ in (4.2) and the asymmetry parameter c in (4.16), equation (4.66) assumes the general form:

$$n^2(x) = n_3^2 + 2\Delta n_1^2 \left\{ \frac{c^2 e^{2x/a}}{(1+e^{2x/a})} + \frac{2(2+c^2)e^{2x/a}}{(1+e^{2x/a})^2} \right\}. \tag{4.68}$$

The variation of $n^2(x)$ with x for a typical value of c is given in Figure 4.13; we will deal first

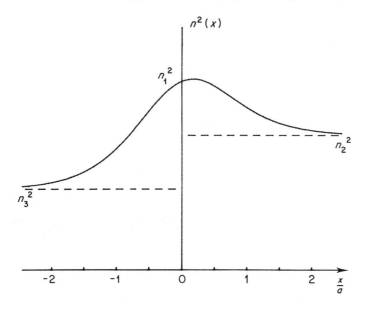

Figure 4.13 The general Epstein-layer profile

with the general asymmetric profile ($c \neq 0$) following the solution of the scalar wave equation. Later the special case of symmetry ($c = 0$) will be discussed, and the ray paths and transit times calculated. As indicated in the introductory remarks to this chapter, the Epstein-layer dielectric profile is closely related, from a mathematical point of view, to the Pöschl–Teller potential in quantum mechanics (Pöschl and Teller, 1933):

$$\frac{A}{\sin^2(\gamma x)} + \frac{B}{\cos^2(\gamma x)}.$$

The Epstein-layer is also analogous to a quantum-mechanical problem known as the Eckart model (Nelson and McKenna, 1967), a detailed account of which will be found in the standard text of Morse and Feshbach (1953).

4.3.1. Mode treatment of the general profile

The scalar wave equation for the index profile (4.68) is given in our conventional notation by:

$$a^2 \frac{d^2\psi}{dx^2} + \left[u^2 - v^2(1+c^2) + \frac{v^2 e^{2x/a}}{1+e^{2x/a}} \left(c^2 + \frac{2(2+c^2)}{1+e^{2x/a}} \right) \right] \psi = 0. \tag{4.69}$$

With the substitution $x' = e^{2x/a}$, this equation assumes the form:

$$\frac{d^2\psi}{dx'^2} + \frac{1}{x'}\frac{d\psi}{dx'} + \left[\frac{u^2 - v^2(1+c^2)}{4x'^2} + \frac{v^2 x'}{4x'^2(1+x')} \left(c^2 + \frac{2(2+c^2)}{1+x'} \right) \right] \psi = 0. \tag{4.70}$$

If we use the definitions

$$w^2 = v^2 - u^2 \tag{4.71a}$$
$$w'^2 = v^2(1+c^2) - u^2 \tag{4.71b}$$
$$d^2 - d = v^2(1+c^2/2) \tag{4.71c}$$

and make the substitution (Epstein, 1930)

$$\psi(x') = (1+x')^d x'^{w/2} f(x') \tag{4.72}$$

then (4.70) reduces to the hypergeometric equation:

$$x'(1+x')\frac{d^2 f}{dx'^2} + [x'(2d+w'+1) + (w'+1)]\frac{df}{dx'} + \left(\frac{w'}{2} + \frac{w}{2} + d\right)\left(\frac{w'}{2} - \frac{w}{2} + d\right) f = 0. \tag{4.73}$$

The solutions of equation (4.73) are hypergeometric functions (see, for example, Abramowitz and Stegun, 1964); the appropriate choice to satisfy $\psi \to 0$ as $x \to -\infty$ ($x' \to 0$) is

$$\psi = (1+x')^d x'^{w/2} F\left(\frac{w'}{2} + \frac{w}{2} + d; \frac{w'}{2} - \frac{w}{2} + d; w'+1; -x'\right). \tag{4.74}$$

For guided modes we require that $\lim_{x \to \infty} \psi(x)$ and $\lim_{x \to -\infty} \psi(x)$ are both zero. The second of these requirements is already obeyed by (4.74); in order to examine the first, we note that (Abramowitz and Stegun, 1964):

$$F\left(\frac{w'}{2} + \frac{w}{2} + d; \frac{w'}{2} - \frac{w}{2} + d; w'+1; -x'\right) = \frac{\Gamma(w'+1)}{\Gamma(w'/2 + w/2 + d)\Gamma(w'/2 - w/2 + d)}$$

$$\times \sum_{n=0}^{\infty} \frac{\Gamma(w'/2 + w/2 + d + n)\Gamma(w'/2 - w/2 + d + n)(-x')^n}{\Gamma(w'+1+n)\, n!}. \tag{4.75}$$

For the appropriate behaviour as $x \to \infty$, it is necessary that one of the arguments of the gamma functions in the denominator of (4.75) be a negative integer, $-N$. For definiteness, let us choose the case

$$\frac{w'}{2} + \frac{w}{2} + d = -N \quad (N = 0, 1, 2, \ldots). \tag{4.76}$$

In fact the alternative choice (replacing $w/2$ by $-w/2$ in (4.76)) merely duplicates the results about to be derived. Substituting for w, w', and d from (4.71) and writing the results in terms of the normalized propagation constant $b(= w^2/v^2)$ yields the explicit eigenvalue equation (Stolyarov, 1970, 1972; Osinski, 1977):

$$b = \frac{c^4 v^2}{4[(1 + 2v^2(2 + c^2))^{1/2} - (2N+1)]^2}$$
$$+ \frac{[(1 + 2v^2(2 + c^2))^{1/2} - (2N+1)]^2}{4v^2} - \frac{c^2}{2} \quad (N = 0, 1, 2, \ldots). \tag{4.77}$$

Plots of b versus v for the first three modes calculated from equation (4.77) for three values of c are shown in Figure 4.14. Note that for values of $c > 1$ there is the possibility that $b > 1$ for

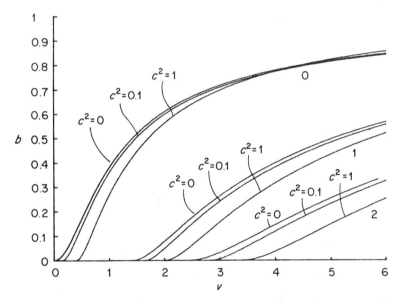

Figure 4.14 Normalized b–v plots for the general Epstein-layer profile; labelling parameter gives the value of mode number N. For each mode results for three values of the asymmetry factor c are given

part of the range of v; this is a consequence of the choice of definitions of the constants ε_1–ε_3 in equation (4.67).

For the lowest-order mode ($N = 0$) the explicit form of the field distribution is given from (4.74) as (Stolyarov, 1972; Osinski, 1977):

$$\psi(x) = \frac{e^{w'x/a}}{(1 + e^{2x/a})^{w'/2 + w/2}} \tag{4.78}$$

where w and w' may be found from (4.71) and 4.77) with $N = 0$.

4.3.2 Modal results for the symmetric profile

The symmetric Epstein-layer profile is given by $c = 0$ in equation (4.68), namely (Nelson and McKenna, 1967; Haus and Schmidt, 1976):

$$n^2(x) = n_2^2 + \frac{2\Delta n_1^2}{\cosh^2(x/a)}. \tag{4.79}$$

This function is plotted in Figure 4.15. Note that for small values of x/a, the first-order approximation of (4.79) yields

$$n^2(x) \simeq n_1^2\left[1 - 2\Delta\left(\frac{x}{a}\right)^2\right]$$

which agrees with the notation of equation (3.1). This fact provides the justification for our choice of the exponential argument $(2x/a)$ in (4.66), although it disagrees with the notation of

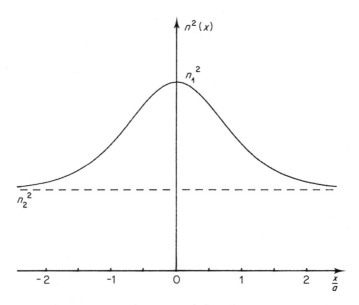

Figure 4.15 The symmetric Epstein-layer profile

earlier authors (Haus and Schmidt, 1976). Returning to the case of (4.79), the eigenvalue equation (4.77) reduces to (Kogelnik, 1975):

$$b = \left[\frac{(1+4v^2)^{1/2} - (2N+1)}{2v}\right]^2 \quad (N = 0, 1, 2, \dots) \tag{4.80}$$

the results of which were already included in Figure 4.14 for the first three modes.
The field distributions for the two lowest-order modes are given by:

$N = 0:$
$$\psi \propto \frac{1}{\cosh^w(x/a)} \tag{4.81}$$

$N = 1:$
$$\psi \propto \frac{\sinh(x/a)}{\cosh^{w+1}(x/a)}. \tag{4.82}$$

Hence the radiation confinement factor Γ is given by:

$N = 0$:
$$\Gamma = \frac{I_a(2w)}{I_\infty(2w)} \tag{4.83}$$

$N = 1$:
$$\Gamma = \frac{I_a(2w) - I_a(2w+2)}{I_\infty(2w) - I_\infty(2w+2)} \tag{4.84}$$

where

$$I_a(v) = \int_0^a \frac{dx}{\cosh^v(x/a)} \tag{4.85a}$$

and

$$I_\infty(v) = \int_0^\infty \frac{dx}{\cosh^v(x/a)}. \tag{4.85b}$$

In general the integrals in (4.85) must be performed numerically; plots of Γ versus v from (4.83) and (4.84) calculated in this way are given in Figure 4.16 (solid lines). For cases where w

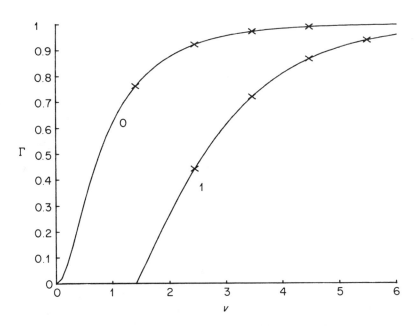

Figure 4.16 Radiation confinement parameter Γ versus v for the lowest-order modes of the symmetric Epstein-layer profile. Solid lines, numerical integration; crosses, analytic results using (4.86)

is an integer, however, an analytic solution is available (Gradshteyn and Ryzhik, 1965):

$$\int_0^t \frac{dy}{\cosh^{2m}(y)} = \frac{\sinh t}{2m-1} \left\{ \operatorname{sech}^{2m-1} t + \sum_{j=1}^{m-1} \frac{2^j(m-1)(m-2)\ldots(m-j)\operatorname{sech}^{2m-2j-1} t}{(2m-3)(2m-5)\ldots(2m-2j-1)} \right\}$$
$$(m = 2, 3, \ldots). \tag{4.86}$$

Four points calculated from this formula are shown as crosses on each of the curves for $N = 0$

and $N = 1$ in Figure 4.16, and serve as a check on the accuracy of the numerical integrations used for the solid lines.

Extensions of the theory given here for the $\mathrm{sech}^2(x/a)$ profile to situations of complex permittivity for applications in semiconductor lasers and LED's have been given by Unger (1967), Bogdankevich et al. (1969), Johnston (1971), and Zehe and Röpke (1973). Comparisons of calculated laser properties for this profile with experimental measurements have been made by Asbeck et al. (1978); the results are in good agreement for narrow-stripe-geometry lasers. Comparisons of modal and ray results for this profile have been given by Ankiewicz (1978). A modal treatment of an optical fibre with a modified hyperbolic secant index profile has been given by Silberberg and Levy (1979).

4.3.3 Ray paths for the symmetric profile

The symmetric profile of the Epstein-layer model possesses interesting properties of ray focusing and equal transit times under certain situations (Fletcher et al., 1954; Kornhauser and Yaghjian, 1967; Kawakami and Nishizawa, 1968). To investigate these effects we will determine the ray paths and transit times using the ray equations derived in Chapter 3. From equation (3.6) it follows that

$$\frac{dx}{dz} = \frac{(k^2 n^2(x) - \beta^2)^{1/2}}{\beta} = \frac{(v^2 \mathrm{sech}^2(x/a) - w^2)^{1/2}}{a\beta}. \tag{4.87}$$

The integration of equation (4.87) is easily performed and, together with the boundary condition $z = 0$ at $x = x_0$, leads to the equation for the ray path:

$$x = a \sinh^{-1}\left\{\frac{u}{w}\sin\left[\frac{zw}{a^2\beta} + \sin^{-1}\left(\frac{w \sinh(x_0/a)}{u}\right)\right]\right\}. \tag{4.88}$$

The result of equation (4.88) is a sinusoidal ray path whose period is $2\pi a^2 \beta/w$. This results in a focusing action for all rays (period independent of β) for two special cases:

(a) paraxial rays: $\beta \simeq kn_1$
(b) cladding of zero refractive index: $n_2 = 0$.

Case (a) is similar to that already discussed for the parabolic index profile in Chapter 3. However, case (b) represents a unique property of the profile

$$n^2(x) = n_1^2 \mathrm{sech}^2(x/a). \tag{4.89}$$

For this case the result (4.88) may be cast into a simple form by defining θ_0 as the initial angle the ray makes with the z-axis ($\tan \theta_0 = dx/dz$ at $x = x_0$, $z = 0$). The ray path is given by (Kornhauser and Yaghjian, 1967):

$$x = a \sinh^{-1}\left\{\frac{(1 - \cos^2\theta_0 \mathrm{sech}^2(x_0/a))^{1/2}}{\cos\theta_0 \mathrm{sech}(x_0/a)} \sin\left[\frac{z}{a} + \sin^{-1}\left(\frac{\cos\theta_0 \mathrm{sech}(x_0/a)\sinh(x_0/a)}{(1 - \cos^2\theta_0 \mathrm{sech}^2(x_0/a))}\right)\right]\right\} \tag{4.90}$$

which is periodic with period $2\pi a$.

The ray transit time for the special case of the profile of (4.89) may be evaluated from equation (3.7). In this case the ray caustic x_c is given by

$$\mathrm{sech}^2\left(\frac{x_c}{a}\right) = \frac{\beta^2}{k^2 n_1^2} \tag{4.91}$$

so that

$$z_0 = \int_0^{x_c} \frac{\beta \, dx}{(k^2 n^2(x) - \beta^2)^{1/2}} = \frac{a\pi}{2}. \tag{4.92}$$

In addition,

$$s_0 = \int_0^{x_c} \frac{k n^2(x) \, dx}{(k^2 n^2(x) - \beta^2)^{1/2}} = \int_0^{x_c} \frac{k n_1^2 \operatorname{sech}(x/a) \, dx}{(k^2 n_1^2 - \beta^2 \cosh^2(x/a))^{1/2}}. \tag{4.93}$$

The integral in (4.93) is easily reduced to a standard form by the substitution $y = \cosh^2(x/a)$, whence (Gradshteyn and Ryzhik, 1965):

$$s_0 = \frac{k n_1^2 a}{2} \int_1^{k^2 n_1^2/\beta^2} \frac{dy}{y(-\beta^2 y^2 + y(k^2 n_1^2 + \beta^2) - k^2 n_1^2)^{1/2}} = \frac{a\pi n_1}{2}. \tag{4.94}$$

It follows from equations (3.7), (4.92), and (4.94) that the ray transit time τ is given by

$$\tau = \frac{L s_0}{c z_0} = \frac{L n_1}{c} \tag{4.95}$$

which is independent of the ray constants. In other words the profile (4.89) has the property of equalizing all ray transit times (Kawakami and Nishizawa, 1968).

Chapter 5
Approximate Methods for Two-dimensional Graded-index Guides

In the previous chapter the similarities between the scalar wave equation and Schrödinger's equation were noted; the dielectric distribution corresponds to the quantum-mechanical potential, and the field components to the wave function. It is not surprising therefore that many of the techniques for approximate solution of quantum-mechanical problems may be carried over directly for application to waveguide situations. In this chapter we will be concerned in particular with three such methods—perturbation theory, WKB analysis, and the variational technique. In addition, details of other approximate methods of solution will also be considered which are particularly suited to numerical treatment. Although such methods may give remarkably accurate and detailed solutions, their main deficiency is the lack of physical insight given by these techniques. In this respect the three methods considered first in this chapter may be said to give more useful results and to provide relatively simple closed-form approximations for many situations of interest. As a means of testing the relative accuracy of the various methods we take a standard index distribution—that of the cladded parabolic profile (see, for example, Figure 3.1).

5.1 Perturbation Theory

General details of perturbation methods may be found in any book on quantum mechanics (see, for example, Schiff, 1955); here we are concerned specifically with applications to the ray equations and to the scalar wave equations for a general graded-index waveguide described by a refractive index distribution $n(x)$.

5.1.1 Perturbation solution of the ray equations

Consider the ray equation-of-motion (3.6):

$$\frac{dx}{dz} = \frac{(k^2 n^2(x) - \beta^2)^{1/2}}{\beta}. \tag{5.1}$$

Differentiation with respect to z yields

$$\frac{d^2 x}{dz^2} = \frac{k^2 n(x)}{\beta(k^2 n^2(x) - \beta^2)^{1/2}} \frac{dx}{dz} \frac{\partial n(x)}{\partial x} = \frac{k^2 n(x)}{\beta^2} \frac{\partial n(x)}{\partial x}. \tag{5.2}$$

Substituting for (k^2/β^2) from equation (5.1) yields:

$$\frac{d^2 x}{dz^2} = \left[1 + \left(\frac{dx}{dz}\right)^2\right] \frac{1}{n(x)} \frac{\partial n(x)}{\partial x}. \tag{5.3}$$

Equation (5.3) describes the variation of a ray path in the x–z plane for an arbitrary index profile $n(x)$ in a form suitable for solution by a perturbation method.

As an example to illustrate the method we choose a general symmetric polynomial index-profile:

$$n(x) = n_1 \left[1 - \frac{\alpha x^2}{2} + \frac{\gamma x^4}{4} + \ldots \right]. \tag{5.4}$$

Inserting this expression in (5.3) gives the differential equation:

$$\frac{d^2 x}{dz^2} + \left[1 + \left(\frac{dx}{dz}\right)^2 \right] (\alpha x - \gamma^* x^3 + \ldots) = 0 \tag{5.5}$$

where

$$\gamma^* = \gamma - \frac{\alpha^2}{2}. \tag{5.6}$$

For the solution of (5.5) we assume the series (Steiner, 1973):

$$x(z) = \sigma f^{(0)}(z) + \sigma^3 f^{(2)}(z) + \ldots \tag{5.7}$$

where σ is chosen as $\tan \phi$ with ϕ as the angle of incidence of the ray with the z-axis (cf. Figure 3.2). Substituting this trial solution into (5.5) and equating coefficients of corresponding powers of σ yields a set of equations defining the functions $f^{(0)}(z), f^{(2)}(z), \ldots$:

$$\frac{d^2 f^{(0)}}{dz^2} + \alpha f^{(0)} = 0 \tag{5.8}$$

$$\frac{d^2 f^{(2)}}{dz^2} + \alpha f^{(2)} = \gamma^* f^{(0)3} - \alpha f^{(0)} \left(\frac{df^{(0)}}{dz}\right)^2. \tag{5.9}$$

The solution of (5.8) with the boundary condition $x = 0, dx/dz = \tan \phi$ when $z = 0$ is given by

$$f^{(0)}(z) = \frac{1}{\alpha^{1/2}} \sin(\alpha^{1/2} z). \tag{5.10}$$

The integration of equation (5.9) is conveniently performed by the use of the Green's function $G(x, \zeta)$ (Steiner, 1973a):

$$f^{(2)}(z) = \int_0^x G(x, \zeta) \left[\gamma^* (f^{(0)}(\zeta))^3 - \alpha f^{(0)}(\zeta) \left(\frac{df^{(0)}(\zeta)}{d\zeta}\right)^2 \right] d\zeta \tag{5.11}$$

where

$$G(x, \zeta) = \frac{1}{\alpha^{1/2}} [\sin(\alpha^{1/2} x) \cos(\alpha^{1/2} \zeta) - \cos(\alpha^{1/2} x) \sin(\alpha^{1/2} \zeta)] \tag{5.12}$$

which is chosen such that at $x = \zeta$ the function vanishes and the derivative gives unity.

Using (5.10) and (5.12) in (5.11), $f^{(2)}(z)$ is given by:

$$f^{(2)}(z) = \frac{1}{8\alpha^{1/2}} \left\{ \frac{\sin(\alpha^{1/2} z)}{4} \left(\frac{9\gamma^*}{\alpha^2} - 7\right) + \frac{\sin(3\alpha^{1/2} z)}{4} \left(\frac{\gamma^*}{\alpha^2} + 1\right) \right.$$
$$\left. - \left(\frac{3\gamma^*}{\alpha^2} - 1\right) \alpha^{1/2} z \cos(\alpha^{1/2} z) \right\}. \tag{5.13}$$

Hence, using the fact that $\sigma = \phi + \phi^3/3 + \ldots$, equations (5.7), (5.10), and (5.13) yield for the ray path (Steiner, 1973):

$$x(z) = \frac{\phi}{\alpha^{1/2}}\left\{\left[1 + \frac{\phi^2}{32}\left(\frac{9\gamma^*}{\alpha^2} + \frac{11}{3}\right)\right]\sin(\alpha^{1/2}z) + \frac{\phi^2}{32}\left(\frac{\gamma^*}{\alpha^2} + 1\right)\sin(3\alpha^{1/2}z)\right.$$
$$\left. - \frac{\phi^2}{8}\left(\frac{3\gamma^*}{\alpha^2} - 1\right)\alpha^{1/2}z\cos(\alpha^{1/2}z) + O(\phi^4)\right\}. \tag{5.14}$$

Note that to order ϕ the simple periodic ray path given by $x(z) = \phi \sin(\alpha^{1/2}z)/\alpha^{1/2}$ is obtained from (5.14), in agreement with equation (3.12) for parabolic-index media with $x_0 = 0$, $\phi_0 = \phi$, $2\Delta/a^2 = \alpha$. In the general case the first and second terms on the r.h.s. of (5.14) imply a periodic ray path, whereas the third term represents a disturbance from periodicity. We see therefore that to order ϕ^3 for a polynomial profile of order 4, the ray path may be made exactly periodic if the coefficient of this third term is zero, i.e. if we choose γ^* such that

$$\frac{3\gamma^*}{\alpha^2} = 1 \tag{5.15}$$

or, using (5.6),

$$\gamma = \frac{5\alpha^2}{6}. \tag{5.16}$$

With this choice of γ, the index profile is given from (5.4) as

$$n(x) = n_1\left[1 - \frac{\alpha x^2}{2} + \frac{5\alpha^2 x^4}{24} + \cdots\right] \tag{5.17}$$

and it is easily seen that this equation represents the first terms in the expansion of $n_1 \text{sech}(\alpha^{1/2}x)$. This is in agreement with the result from Section 4.3 that this profile (cf. equation (4.89)) gives a focusing action for all rays (Fletcher et al., 1954). For this profile the ray path is given from (4.90) with $x_0 = 0$, $\theta_0 = \phi$, $a = \alpha^{-1/2}$ as:

$$x(z) = \frac{1}{\alpha^{1/2}}\sinh^{-1}[\tan\phi\sin(\alpha^{1/2}z)] \tag{5.18}$$

and it is easily confirmed that equation (5.14) gives the first terms in the expansion of (5.18).

The ray transit-time τ is also found from the perturbation solution of the ray equations. From equations (3.5)–(3.7) we have (ignoring material dispersion):

$$\tau = \frac{L}{c}\frac{\oint n(x)[1 + (dx/dz)^2]^{1/2}dz}{\oint dz} \tag{5.19}$$

where L is the length, c the velocity of light in vacuo, and the symbol $\oint$ indicates that the integration is taken over a complete ray period (assumed much less than L).

From equation (5.14) we find:

$$\left[1 + \left(\frac{dx}{dz}\right)^2\right]^{1/2} = 1 + \frac{1}{2}\left(\frac{dx}{dz}\right)^2 - \frac{1}{8}\left(\frac{dx}{dz}\right)^4 + \cdots$$
$$\simeq 1 + \frac{\phi^2}{2}\left\{\left[1 + \frac{\phi^2}{16}\left(\frac{9\gamma^*}{\alpha^2} + \frac{11}{3}\right)\right]\cos^2(\alpha^{1/2}z) + \frac{3\phi^2}{16}\left(\frac{\gamma^*}{\alpha^2} + 1\right)\cos(3\alpha^{1/2}z)\cos(\alpha^{1/2}z)\right.$$
$$\left. - \frac{\phi^2}{4}\left(\frac{3\gamma^*}{\alpha^2} - 1\right)\cos(\alpha^{1/2}z)\left[\cos(\alpha^{1/2}z) - \alpha^{1/2}z\sin(\alpha^{1/2}z)\right]\right\}$$
$$- \frac{\phi^4}{8}\cos^4(\alpha^{1/2}z). \tag{5.20}$$

Also, from (5.4) and (5.14), the index profile is given by (to order ϕ^4):

$$\frac{n(x)}{n_1} \simeq 1 - \frac{\phi^2}{2}\left\{\left[1 + \frac{\phi^2}{16}\left(\frac{9\gamma^*}{\alpha^2} + \frac{11}{3}\right)\right]\sin^2(\alpha^{1/2}z) - \frac{\phi^2}{4}\left(\frac{3\gamma^*}{\alpha^2} - 1\right)\alpha^{1/2}z\cos(\alpha^{1/2}z)\sin(\alpha^{1/2}z) \right.$$
$$\left. + \frac{\phi^2}{16}\left(\frac{\gamma^*}{\alpha^2} + 1\right)\sin(3\alpha^{1/2}z)\sin(\alpha^{1/2}z)\right\} + \frac{\gamma}{4}\frac{\phi^4}{\alpha^2}\sin^4(\alpha^{1/2}z). \quad (5.21)$$

Using these results in equation (5.19), integrating, and simplifying, we find (Steiner, 1974) to order ϕ^4:

$$\frac{\tau}{\tau_0} = 1 - \frac{3\phi^4}{32}\left(\frac{\gamma}{\alpha^2} - \frac{5}{6}\right) \quad (5.22)$$

where $\tau_0 = Ln_1/c$ is the transit time of an on-axis ray. Once again it is clear from (5.22) that if equation (5.16) holds then all rays have the same transit time to order ϕ^4; this also confirms the result of Section 4.3 where the sech-profile was shown to have the property of equalizing ray transit-times.

The perturbation solution for ray paths and transit-times given above has been extended to deal with three-dimensional graded-index guides (Steiner, 1974) and with asymmetric two-dimensional situations (Steiner, 1974a).

5.1.2 First-order perturbation theory for the scalar wave equation

The scalar wave equations for TE and TM modes in a general graded-index medium have been derived in Chapter 3. These equations are of the form:

$$H\psi(x) = \beta^2\psi(x) \quad (5.23)$$

where $\psi(x)$ is related to the appropriate field component (H_x for TE modes and $E_x n(x)$ for TM modes). In equation (5.23) the analogy with Schrodinger's equation has been made explicit by the use of the operator H. For TE modes this is given by equation (3.35):

TE: $$H = \frac{d^2}{dx^2} + k^2 n^2(x) \quad (5.24)$$

whilst for TM modes equation (3.38) yields:

TM: $$H = \frac{d^2}{dx^2} + k^2 n^2(x) + \frac{1}{2n^2(x)}\frac{d^2(n^2(x))}{dx^2} - \frac{3}{4n^4(x)}\left(\frac{d(n^2(x))}{dx}\right)^2. \quad (5.25)$$

In order to apply perturbation theory to equation (5.23) we must first show that the solutions are orthogonal. This result is most easily proved by taking a pair of eigenfunctions $\psi_N(x)$, $\psi_M(x)$ of (5.23) with corresponding eigenvalues β_N^2, β_M^2. Multiplying the wave equation for each solution by the other, integrating over the x-axis, and taking the difference, we obtain:

$$\int_{-\infty}^{\infty}(\psi_N(x)H\psi_M(x) - \psi_M(x)H\psi_N(x))\,dx = (\beta_M^2 - \beta_N^2)\int_{-\infty}^{\infty}\psi_M(x)\psi_N(x)\,dx. \quad (5.26)$$

On the l.h.s. of (5.26) the terms in H other than d^2/dx^2 are easily seen to cancel. Furthermore, the remaining d^2/dx^2 terms may be integrated by parts to yield:

$$\int_{-\infty}^{\infty}\left(\psi_N(x)\frac{d^2\psi_M(x)}{dx^2} - \psi_M(x)\frac{d^2\psi_N(x)}{dx^2}\right)dx = \left[\psi_N(x)\frac{d\psi_M(x)}{dx} - \psi_M(x)\frac{d\psi_N(x)}{dx}\right]_{-\infty}^{\infty} = 0 \quad (5.27)$$

which follows since the field distributions are required to vanish in the limit $x \to \pm\infty$ for guided modes. In fact this result holds also for radiation modes although the proof requires a somewhat more complicated argument (Marcuse, 1972).

It follows in general that the r.h.s. of equation (5.26) is also zero for $M \neq N$. For the case $M = N$ we may choose the solutions to be normalized in the usual way (see, for example, equation (3.49)), so that the general result for $\psi_N(x)$, $\psi_M(x)$ reads:

$$\int_{-\infty}^{\infty} \psi_M(x)\psi_N(x) = \delta_{MN} \tag{5.28}$$

where δ_{MN} is the Kronecker delta ($= 0$ for $M \neq N$, and 1 for $M = N$).

Let us now apply perturbation theory to equation (5.23) assuming the unperturbed eigenfunctions are $\psi_N(x)$ and the corresponding unperturbed eigenvalues are β_N^2. If the perturbed solutions are $\phi(x)$, β'^2, then we may write the perturbation as $\sigma H'$ so that the wave equation becomes:

$$(H + \sigma H')\phi(x) = \beta'^2 \phi(x). \tag{5.29}$$

The full perturbation series for these solutions may now be written (Schiff, 1955):

$$\left. \begin{aligned} \phi &= \phi^{(0)} + \sigma\phi^{(1)} + \sigma^2\phi^{(2)} + \cdots \\ \beta'^2 &= \beta^{(0)2} + \sigma\beta^{(1)2} + \sigma^2\beta^{(2)2} + \cdots \end{aligned} \right\} \tag{5.30}$$

where the various powers of the parameter σ will give the corresponding orders of perturbation. Substituting (5.30) in (5.29) and equating equivalent powers of σ:

$$H\phi^{(0)} = \beta^{(0)2}\phi^{(0)} \tag{5.31}$$

$$H\phi^{(1)} + H'\phi^{(0)} = \beta^{(0)2}\phi^{(1)} + \beta^{(1)2}\phi^{(0)} \tag{5.32}$$

$$\vdots$$

The zero-order perturbation equation (5.31) is seen to correspond to the unperturbed case, with:

$$\left. \begin{aligned} \phi^{(0)} &= \psi_N(x) \\ \beta^{(0)2} &= \beta_N^2 \end{aligned} \right\}. \tag{5.33}$$

The first-order result (5.32) may be developed further by writing $\phi^{(1)}$ as a linear superposition of the unperturbed solutions:

$$\phi^{(1)} = \sum_M a_M^{(1)} \psi_M(x). \tag{5.34}$$

Using (5.33) and (5.34) in (5.32) gives:

$$\sum_M a_M^{(1)} H\psi_M(x) + H'\psi_N(x) = \beta_N^2 \sum_M a_M^{(1)} \psi_M(x) + \beta^{(1)2}\psi_N(x). \tag{5.35}$$

Now H in the first term on the l.h.s. of (5.35) may be replaced by β_M^2; if the equation is multiplied by $\psi_P(x)$ and integrated over the entire x-axis using the orthonormalization property (5.28), then:

$$a_P^{(1)} \beta_P^2 + \int_{-\infty}^{\infty} \psi_P(x) H' \psi_N(x) dx = \beta_N^2 a_P^{(1)} + \beta^{(1)2} \delta_{PN}. \tag{5.36}$$

It follows that for the case $P = N$ the perturbation on the eigenvalues is given by

$$\beta^{(1)2} = \int_{-\infty}^{\infty} \psi_N(x) H' \psi_N(x) dx \tag{5.37}$$

whilst in the case $P \neq N$, the coefficients of the expansion for $\phi^{(1)}$ are found:

$$a_P^{(1)} = \frac{\int_{-\infty}^{\infty} \psi_P(x) H' \psi_N(x) dx}{(\beta_N^2 - \beta_P^2)} \quad (P \neq N). \tag{5.38}$$

Equations (5.37) and (5.38) represent the first-order perturbation results for the scalar wave equation. Note that all the coefficients in the expansion (5.34) are given by (5.38) except for $a_N^{(1)}$. This latter coefficient is determined by the normalization condition to be zero (Schiff, 1955). Hence to first order the results for eigenvalues and eigenfunctions are:

$$\beta'^2 = \beta_N^2 + \int_{-\infty}^{\infty} \psi_N(x) H' \psi_N(x) dx \tag{5.39}$$

$$\phi(x) = \psi_N(x) + \sum_{\substack{M \\ M \neq N}} \frac{\psi_M(x) \int_{-\infty}^{\infty} \psi_M(x) H' \psi_N(x) dx}{(\beta_N^2 - \beta_M^2)}. \tag{5.40}$$

These equations have been used by Marcuse (1973a) to investigate the effect of the $dn^2(x)/dx$ term on the TM modes of a parabolic-index medium. Taking the unperturbed solutions of the TE operator (5.24), it is relatively straightforward to calculate the perturbation resulting from the extra terms in the TM operator (5.25) for such a refractive index profile. The results confirm those already found in Chapter 3 (equations (3.62) and (3.64)) by a more direct approach.

5.1.3 Polynomial index profiles

As an example of the application of the first-order perturbation theory results, consider the index distribution described by a fourth-order symmetric polynomial:

$$n^2(x) = n_1^2 \left[1 - 2\Delta \left(\frac{x}{a}\right)^2 + 2K\Delta \left(\frac{x}{a}\right)^4 \right]. \tag{5.41}$$

If K is sufficiently small in (5.41), as might be the case for example in the series expansion of a Gaussian index profile (Lotspeich, 1976), then the fourth-order term may be regarded as a perturbation on the well-known parabolic distribution. In this case the perturbation operator H' of the previous subsection is defined as:

$$H' = 2n_1^2 k^2 K\Delta \left(\frac{x}{a}\right)^4. \tag{5.42}$$

For TE modes, the normalized unperturbed solutions of the parabolic-law are given by (3.50) as

$$\psi_N(x) = \frac{v^{1/4}}{\pi^{1/4} (2^N N! a)^{1/2}} H_N\left(\frac{v^{1/2} x}{a}\right) \exp\left(-\frac{vx^2}{2a^2}\right) \quad (N = 0, 1, 2, \ldots) \tag{5.43}$$

where H_N is a Hermite polynomial and v is defined in the usual way by (3.4). The corresponding eigenvalues β_N^2 are given by (3.51):

$$\beta_N^2 = k^2 n_1^2 - \frac{v(2N+1)}{a^2} \quad (N = 0, 1, 2, \ldots). \tag{5.44}$$

The first-order perturbation theory results (5.39) and (5.40) may now be applied to this problem. The integrals involving products of Hermite–Gaussian functions may be performed analytically following the methods discussed by Schiff (1955). For the diagonal terms, using (5.42) and (5.43), we obtain (Lotspeich, 1976):

$$\int_{-\infty}^{\infty} \psi_N(x) H' \psi_N(x) dx = \frac{3K(2N^2 + 2N + 1)}{4a^2}. \tag{5.45}$$

Hence with the first-order correction included, we may express the eigenvalues in terms of the normalized propagation constant b defined in the usual way:

$$b = \frac{a^2(\beta'^2 - k^2 n_1^2 + 2\Delta k^2 n_1^2)}{v^2}$$

$$= 1 - \frac{(2N+1)}{v} + \frac{3K(2N^2 + 2N + 1)}{4v^2}. \tag{5.46}$$

Using the strongly asymmetric approximation (cf. Chapter 4) of assuming only odd-order modes ($N = 1, 3, 5, \ldots$), Lotspeich (1976) has used the result (5.46) to calculate mode indices of a diffused optical waveguide with a Gaussian profile. Such profiles are known to occur for example in planar guides fabricated by the diffusion of metal ions into $LiNbO_3$ or $LiTaO_3$ (Standley and Ramaswamy, 1974). Calculations including higher-order terms in the Gaussian expansion have been presented by Ayant et al. (1977).

Similarly, the perturbed eigenfunctions are given by the expansion (5.34) with $\psi_N(x)$ from (5.43) and the only non zero coefficients given by (Lotspeich, 1976):

$$\left. \begin{aligned} a_{N-4}^{(1)} &= -\frac{K[N(N-1)(N-2)(N-3)]^{1/2}}{32v} \\ a_{N-2}^{(1)} &= -\frac{K(2N-1)[N(N-1)]^{1/2}}{8v} \\ a_{N+2}^{(1)} &= \frac{K(2N+3)[(N+1)(N+2)]^{1/2}}{8v} \\ a_{N+4}^{(1)} &= \frac{K[(N+1)(N+2)(N+3)(N+4)]^{1/2}}{32v} \end{aligned} \right\} \tag{5.47}$$

Using these results, Lotspeich (1976) has calculated the first-order perturbed field distributions for TE modes with $N = 1, 3$, and 5; the first-order corrections to the results for the parabolic index profile increase with increasing mode number N. Further studies of fourth-order polynomial profiles have been reported by Hashimoto (1976a) and Pratesi and Ronchi (1978).

Although the fourth-order polynomial profile (5.41) may be used as an approximation for a Gaussian distribution (as noted above), a more direct approach for the latter case is to view the Gaussian profile as a perturbation of the parabolic-index medium. Such a calculation using first-order perturbation theory has been made by Savatinova and Nadjakov (1975) who have compared the results with experimental observations on planar guides made by silver-ion exchange in glass. By comparing the theoretical and experimental mode spectra, the parameters relating to maximum index and width of the guide may be determined from such a calculation. A somewhat similar approach to the problem has been made by Kumar and Khular (1978) who treated the Gaussian profile as a perturbation of the hyperbolic secant profile (see Section 4.3). Since this profile is already a good approximation to the Gaussian shape the results of first-order perturbation theory are correspondingly improved over those of earlier treatments.

5.1.4 The cladded-parabolic profile

This profile was illustrated in Figure 3.1 (dashed line). For modes well away from cut-off the effect of the cladding layers may be viewed as a small perturbation on the results for the extended-parabolic medium. Hence to apply first-order perturbation theory to this case we may define the perturbation operator H' as:

$$H' = \begin{cases} 0, & |x| \leq a \\ 2\Delta n_1^2 k^2 \left[\left(\frac{x}{a}\right)^2 - 1 \right], & |x| \geq a. \end{cases} \quad (5.48)$$

With the aid of this definition, the first-order correction to the eigenvalues of the Nth TE mode is given from (5.37) as:

$$\beta^{(1)2} = \frac{4\Delta k^2 n_1^2 v^{1/2}}{\pi^{1/2} 2^N N! a} \int_a^\infty \left[\left(\frac{x}{a}\right)^2 - 1 \right] H_N^2\left(\frac{v^{1/2} x}{a}\right) \exp\left(-\frac{vx^2}{a^2}\right) dx. \quad (5.49)$$

Restricting attention to the lowest-order mode ($N = 0$) we find for the total normalized propagation constant b (Kumar et al., 1974; Adams, 1978):

$$b = 1 - \frac{1}{v} - \mathrm{erfc}(v^{1/2})\left(\frac{1}{2v} - 1\right) - \frac{e^{-v}}{(\pi v)^{1/2}} \quad (5.50)$$

where erfc is the complementary error function (Abramowitz and Stegun, 1964).

For modes close to cut-off the above approximation is not expected to hold, since then the perturbing effect of the cladding layers is strong (cf. introductory remarks to Chapter 3). In this case a rather more appropriate approach is to regard the unperturbed case as a homogeneous core guide with the parabolic variation as a perturbation; this situation is illustrated in Figure 5.1. We now define the perturbation operator as (Kumar et al., 1974):

$$H' = \begin{cases} -2\Delta n^2 k^2 \left(\frac{x}{a}\right)^2, & |x| \leq a \\ 0, & |x| > a. \end{cases} \quad (5.51)$$

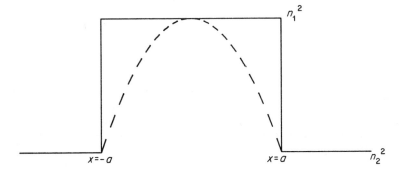

Figure 5.1 The cladded-parabolic profile, illustrating a perturbation approach for modes near cut-off. Solid line, the step-index guide; broken line: the parabolic perturbation

The unperturbed solutions are given by the symmetric slab results (see Chapter 2). Again confining attention to the lowest-order TE mode, we have:

$$\frac{va^{1/2}}{u}\left(1+\frac{1}{w}\right)^{1/2}\psi_0(x) = \begin{cases} \dfrac{\cos(ux/a)}{\cos u}, & |x| \leq a \\ \exp\left[w\left(1-\dfrac{|x|}{a}\right)\right], & |x| \geq a \end{cases} \quad (5.52)$$

where the additional factors on the l.h.s. ensure normalization according to (5.28). The normalized parameters in (5.52) are defined in the usual way, viz. $u^2 = a^2(k^2n_1^2 - \beta_0^2)$, $w^2 = v^2 - u^2$, $v^2 = a^2k^2 2\Delta n_1^2$ and are given by the solution of the eigenvalue equation for the lowest-order mode:

$$u = v \cos u. \quad (5.53)$$

With these definitions, equation (5.37) may easily be evaluated to find the first-order perturbation correction $\beta^{(1)2}$ to the eigenvalues β_0^2. If the result is expressed in terms of the correction $b^{(1)}$ to the normalized propagation constant b, then we find (Adams, 1978):

$$b^{(1)} = -\frac{1}{(1+1/w)}\left[\frac{1}{3} + \frac{\cos 2u}{2u^2} + \frac{\sin 2u}{2u}\left(1 - \frac{1}{2u^2}\right)\right]. \quad (5.54)$$

Equation (5.54) is expected to yield a better approximation in the region of mode cut-off ($v \to 0$) than the corresponding result (5.50). The two approximations represented by these equations are compared graphically in Figure 5.2. Also shown are the corresponding unperturbed step-index and parabolic solutions from equations (5.53) and (5.44), respectively (solid lines). The close to cut-off approximation, equation (5.54) (dashed line), is seen to give a

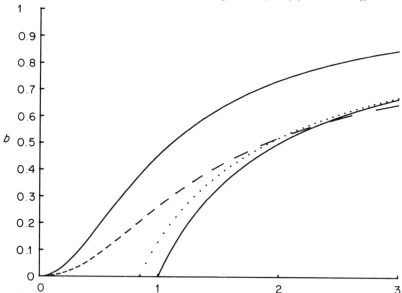

Figure 5.2 Results of first-order perturbation theory for the cladded-parabolic profile. Solid lines, step-index and extended-parabolic profiles (equations (5.53) and (5.44), respectively); dotted line, far from cut-off perturbation, according to equation (5.50); dashed line, close to cut-off perturbation, according to equation (5.54). (From Adams (1978). Reproduced by permission of Chapman and Hall)

quite marked departure from the unperturbed step index result. On the other hand, the far from cut-off approximation (5.50) (dotted line) does not produce a strong perturbation on the result for the extended parabolic medium. In fact as we shall see in Section 5.4, the two approximations taken in their corresponding regions of application (v less than and greater than 2, respectively) yield a fairly good description of the actual result for the cladded parabolic-index profile.

An extension of the perturbation approach discussed here has been made by Chua and Thomas (1977) who have calculated results for a range of different dielectric variations, all with symmetric cladding regions but different core profiles.

5.2 The WKB Approximation

This approximation, usually associated with the names of Wentzel, Kramers, Brillouin, and Jeffreys, is sometimes also known as the phase-integral method. A general discussion of the method together with a critical guide to the related literature will be found in the book by Heading (1962). In optical waveguide theory the method has found wide application, although the relevant results are usually able to be obtained by the ray optics approach. These methods are also similar in their range of applicability, both being suitable for use only when the variation of dielectric permittivity is small in distances of the order of the wavelength. However, there are cases where the ray theory may provide a better level of approximation than the usual WKB theory. As we shall see, these anomalies result from problems associated with the phase changes occurring on reflection from a dielectric discontinuity. If these phase changes are correctly accounted for, then the first-order WKB and ray methods yield identical results (Hartog and Adams, 1977).

We are concerned again with solutions of the scalar wave equation:

$$\frac{d^2\psi}{dx^2} + (k^2 n^2(x) - \beta^2)\psi = 0 \tag{5.55}$$

where ψ is related to a field component in the usual way. For a trial solution we write $\psi = \psi_0 \exp[ikS(x)]$ and expand S in powers of $1/k$:

$$S(x) = S_0(x) + \frac{1}{k}S_1(x) + \ldots \tag{5.56}$$

so that

$$\psi = \psi_0 \exp[ikS_0(x) + iS_1(x) + \ldots]. \tag{5.57}$$

Terms of higher order may be neglected provided the variation of $n(x)$ is small over a wavelength; this is the fundamental WKB assumption noted above. Using the trial solution (5.57) in the wave equation (5.55) and equating terms of similar order in k the zero- and first-order WKB approximations are obtained:

$$S_0 = \frac{1}{k} \int (k^2 n^2(x) - \beta^2)^{1/2} \, dx \tag{5.58}$$

$$S_1 = \frac{i}{2} \ln\left|\frac{dS_0}{dx}\right|. \tag{5.59}$$

From equations (5.57)–(5.59) there are therefore two possible types of solution depending on the sign of $(k^2 n^2(x) - \beta^2)$:

$$\psi(x) = \frac{\psi_0}{Q^{1/2}} \exp[\pm i \int Q \, dx], \quad k^2 n^2(x) > \beta^2 \tag{5.60}$$

$$\psi(x) = \frac{\psi_0}{P^{1/2}} \exp[\pm \int P\,dx], \qquad k^2 n^2(x) < \beta^2 \tag{5.61}$$

where
$$Q^2 = -P^2 = k^2 n^2(x) - \beta^2. \tag{5.62}$$

Suitable combinations of the solutions (5.60) and (5.61) must now be found which reproduce correctly the physical nature of guided waves, i.e. we anticipate an oscillatory behaviour in a region where $k^2 n^2(x) > \beta^2$ and an evanescent behaviour asymptotically transverse to the guide. In the region of the caustic or 'turning point' given by $kn(x) = \beta$, the approximate solutions given above will of course fail. The problem then remains to find the 'connection formulae' which describe the relationship between the forms of oscillatory and evanescent behaviour on either side of such a caustic.

5.2.1 Approximate solution near a caustic

Let the position of a turning point be given by x_c, so that $kn(x_c) = \beta$. Then in the vicinity of the turning point we may use as a first approximation a linear variation of Q^2 as defined in (5.62):
$$Q^2 = C(x_c - x) \tag{5.63}$$
where C is a constant.

Making the change of variable
$$x' = C^{1/3}(x - x_c) \tag{5.64}$$
the scalar wave equation then becomes:
$$\frac{d^2 \psi}{dx'^2} - x' \psi = 0. \tag{5.65}$$

The solutions of this equation have already been discussed in Section 4.2; two independent solutions are the Airy functions $\mathrm{Ai}(x')$ and $\mathrm{Bi}(x')$, both of which are continuous through the turning point $x' = 0$. It is clear therefore that these solutions can provide approximations for the field distributions in the vicinity of $x' = 0$ provided that their asymptotic behaviour resembles that of the WKB solutions. The asymptotic forms of the Airy functions are given by (Abramowitz and Stegun, 1964):

$$\mathrm{Ai}(x') \sim \frac{e^{-\zeta}}{2\pi^{1/2} x'^{1/4}} \tag{5.66}$$

$$\mathrm{Ai}(-x') \sim \frac{1}{\pi^{1/2} x'^{1/4}} \sin\left(\zeta + \frac{\pi}{4}\right) \tag{5.67}$$

$$\mathrm{Bi}(x') \sim \frac{e^{\zeta}}{\pi^{1/2} x'^{1/4}} \tag{5.68}$$

$$\mathrm{Bi}(-x') \sim \frac{1}{\pi^{1/2} x'^{1/4}} \cos\left(\zeta + \tfrac{\pi}{4}\right) \tag{5.69}$$

where
$$\zeta = \tfrac{2}{3} x'^{3/2}. \tag{5.70}$$

For the sake of definiteness, let us identify the region $k^2 n^2(x) > \beta^2$ with $x < x_c$ and similarly $k^2 n^2(x) < \beta^2$ with $x > x_c$. Re-writing equations (5.66) and (5.67) in terms of Q and P we find the connection formula:

$$\frac{1}{2 P^{1/2}} \exp\left(-\int_{x_c}^{x} P\,dx\right) \to \frac{1}{Q^{1/2}} \cos\left(\int_{x}^{x_c} Q\,dx - \frac{\pi}{4}\right) \tag{5.71}$$

where the arrow implies that the asymptotic solution on the l.h.s. for $x > x_c$ corresponds to the asymptotic solution on the r.h.s. for $x < x_c$. The connection formula (5.71) dictates the form of WKB solutions to be chosen from the types given in (5.60) and (5.61). Thus for an oscillatory field distribution in the region $x < x_c$ (where $k^2 n^2(x) > \beta^2$) and an evanescent solution in the region $x > x_c$ ($k^2 n^2(x) < \beta^2$), ignoring the multiplicative constants

oscillatory:
$$\psi \propto \frac{1}{Q^{1/2}} \cos\left(\int_x^{x_c} Q\,dx - \frac{\pi}{4}\right) \tag{5.72}$$

evanescent:
$$\psi \propto \frac{1}{2P^{1/2}} \exp\left(-\int_{x_c}^x P\,dx\right). \tag{5.73}$$

A further connection formula may be found from equations (5.68) and 5.69).

$$\frac{1}{P^{1/2}} \exp\left(\int_{x_c}^x P\,dx\right) \to \frac{1}{Q^{1/2}} \cos\left(\int_x^{x_c} Q\,dx + \frac{\pi}{4}\right). \tag{5.74}$$

It is clear that whilst the previous connection formula (5.71) applied to waves with a decaying behaviour as $x \to \infty$, the present case (5.74) applies to exponentially-growing waves. Hence the appropriate forms for the WKB solutions are given by:

oscillatory:
$$\psi \propto \frac{1}{Q^{1/2}} \cos\left(\int_x^{x_c} Q\,dx + \frac{\pi}{4}\right) \tag{5.75}$$

evanescent:
$$\psi \propto \frac{1}{P^{1/2}} \exp\left(\int_{x_c}^x P\,dx\right). \tag{5.76}$$

In many cases the requirement that the wave vanishes in the limit $x \to \infty$ demands that the choice of solutions is restricted to those of type (5.72) and (5.73). However, there may be cases where the superposition of both decaying and growing waves are required in a finite region of the spatial variable x, and then the forms (5.75) and (5.76) will also be required.

5.2.2 Solutions in the presence of two caustics

For the situation when two caustics occur, as in the case of guided modes in a waveguide, Figure 5.3 illustrates the behaviour. The caustics occur at x_{c1} and x_{c2} with oscillatory behaviour in the region $x_{c1} < x < x_{c2}$ ($k^2 n^2(x) > \beta^2$) and evanescent behaviour elsewhere. Hence the connection formula (5.71) must be applied at each caustic x_{c1} and x_{c2}. For matching

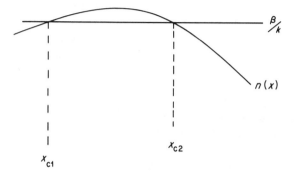

Figure 5.3 The two caustics for a mode in a general graded-index guide

a solution which decays as $x \to -\infty$ the connection formula at x_{c1} implies an oscillatory solution for $x > x_{c1}$ of the form:

$$\psi \propto \frac{1}{Q^{1/2}} \cos\left(\int_{x_{c1}}^{x} Q\,dx - \frac{\pi}{4}\right). \qquad (5.77)$$

At the other caustic, in order to match an evanescent solution which decays in the positive x-direction, the connection formula implies that for $x < x_{c2}$,

$$\psi \propto \frac{1}{Q^{1/2}} \cos\left(\int_{x}^{x_{c2}} Q\,dx - \frac{\pi}{4}\right)$$

$$= \frac{1}{Q^{1/2}} \cos\left(\int_{x_{c1}}^{x} Q\,dx - \int_{x_{c1}}^{x_{c2}} Q\,dx + \frac{\pi}{4}\right). \qquad (5.78)$$

If the two forms of oscillatory solutions given by (5.77) and (5.78) are to match smoothly, then the arguments of the cosines must be equal (up to $N\pi$, $N = 0, 1, 2, \ldots$), i.e.:

$$\int_{x_{c1}}^{x_{c2}} Q\,dx = N\pi + \frac{\pi}{2} \quad (N = 0, 1, 2, \ldots). \qquad (5.79)$$

Equation (5.79) is sometimes termed the WKB resonance condition and may be used to find the eigenvalue equation for a graded-index waveguide with N as the mode number. It is identical with the result obtained by applying the transverse phase condition in the geometric optics treatment (see, for example, equations (3.19)–(3.21) in the case of a profile symmetric about $x = 0$). For this reason throughout this book the references to calculations via WKB or geometrical optics methods are interchangeable; we have not hesitated to describe as a 'ray optics result' any equation which may equally well have been derived by the WKB method.

The result (5.79) is valid provided that the two caustics x_{c1} and x_{c2} are spaced sufficiently far apart; in the limiting case $x_{c1} \to x_{c2}$ the linear approximation (5.63) fails and should be replaced by a more appropriate form for Q^2, e.g. a quadratic dependence with a double root at the caustic x_c (see, for example, Schiff, 1955). Another case where the simple result (5.79) fails occurs in the presence of a refractive index discontinuity; this will be considered separately below. A WKB eigenvalue equation for media with gain or loss variations has been derived by Ronchi et al. (1980).

5.2.3 Solutions in the presence of an index discontinuity

Consider first the case of an index discontinuity occurring at $x = x_{c2}$ as illustrated schematically in Figure 5.4. The caustic at $x = x_{c1}$ is assumed to occur in a region of smoothly-varying refractive index as discussed previously, although the extension to include the case of two or more discontinuities is in fact trivial. Once again we assume oscillatory behaviour in the region $x_{c1} < x < x_{c2}$ ($k^2 n^2(x) > \beta^2$) and evanescent behaviour elsewhere. Hence, for a solution which decays as $x \to -\infty$, the connection formula (5.71) at x_{c1} implies a solution for $x > x_{c1}$ of the form (5.79). At $x = x_{c2}$, however, it is not appropriate to apply the connection formulae, since they were derived on the assumption of a linear form for Q^2 as in (5.63). Instead we must apply the familiar boundary conditions of field continuity at $x = x_{c2}$.

If the two refractive index values at the discontinuity x_{c2} are denoted, as in Figure 5.4, by $n(x_{c2}^-)$ and $n(x_{c2}^+)$ corresponding to approaching from $x < x_{c2}$ and $x > x_{c2}$ respectively, then the field in the region $x > x_{c2}$ chosen to vanish as $x \to \infty$ is given by

$$\psi \propto \frac{1}{P^{1/2}} \exp\left(-\int_{x_{c2}^+}^{x} P\,dx\right). \qquad (5.80)$$

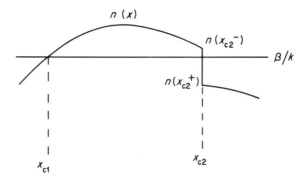

Figure 5.4 The case of an index discontinuity occurring at the caustic x_{c2}

It remains therefore to match the fields (5.77) and (5.80) and their derivatives at $x = x_{c2}$. In the case of the derivatives it would at first appear that each consists of two terms; i.e. one from the derivative of the coefficient and another from the derivative of the argument of the trigonometric or exponential function, as appropriate. In fact, on closer inspection, it turns out that this is not the case since we must be careful only to include terms up to first order in the expansion (5.56). Since the term resulting from differentiating $S_1(x)$ in this expansion would be of higher order and might be affected by higher-order terms in the original expansion then it is best *not* to differentiate this term (Heading, 1962). Hence the appropriate forms for the derivatives yield the matching condition:

$$Q^{1/2}(x_{c2}^-)\sin\left(\int_{x_{c1}}^{x_{c2}^-} Q\,dx - \frac{\pi}{4}\right) = P^{1/2}(x_{c2}^+). \tag{5.81}$$

The matching condition on the fields (5.77) and (5.80) yields:

$$\frac{1}{Q^{1/2}(x_{c2}^-)}\cos\left(\int_{x_{c1}}^{x_{c2}^-} Q\,dx - \frac{\pi}{4}\right) = \frac{1}{P^{1/2}(x_{c2}^+)}. \tag{5.82}$$

Combining equations (5.81) and (5.82) gives the eigenvalue equation for modes of order N whose caustic coincides with an index discontinuity at $x = x_{c2}$ as (Gedeon, 1974):

$$\int_{x_{c1}}^{x_{c2}^-} Q\,dx - \frac{\pi}{4} = N\pi + \tan^{-1}\left[\left(\frac{\beta^2 - k^2 n^2(x_{c2}^+)}{k^2 n^2(x_{c2}^-) - \beta^2}\right)^{1/2}\right] \quad (N = 0, 1, 2, \ldots). \tag{5.83}$$

Once again it is clear that this result is the same as that which would be given by a rigorous application of the geometrical optics theory to the situation shown in Figure 5.4. In the limiting case where the discontinuity is very large, it is a good approximation to replace the final term in (5.83) by $\pi/2$ (cf. the cases of strong dielectric asymmetry discussed in Chapter 4). Using the resulting simpler eigenvalue equation, Tien *et al.* (1974) have obtained good numerical agreement with the results of the rigorous modal theory for the case of an exponential index profile (see also Section 4.1.).

5.2.4 'Buried' modes near an index discontinuity

Consider the case illustrated in Figure 5.5 where the mode possesses two caustics x_{c1}, x_{c2} in regions of smoothly-varying index variation but with that at x_{c2} occurring in the vicinity of an index discontinuity at $x = a$, say. Oscillatory field behavior is assumed in the region

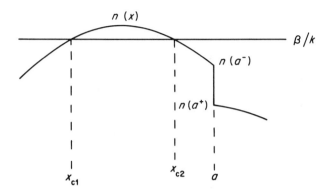

Figure 5.5 The case of an index discontinuity in the vicinity of the caustic x_{c2}

$x_{c1} < x < x_{c2}$ and evanescent behaviour elsewhere. For the boundary conditions far away from the caustics we may assume the fields decay to zero as $x \to \pm \infty$. However, there is a significant difference between this case and those considered previously in the form of field distribution permitted in the region $x_{c2} < x < a$ between the second caustic and the discontinuity. In this region there is no longer the requirement only for a decaying field distribution in the positive x-direction; hence in general it is necessary to include both decaying and growing waves (Hashimoto, 1976). Using the appropriate connection formulae (5.71) and (5.74) the field distributions in the regions of interest are given by (Shevchenko, 1974):

$$\frac{\psi}{A} = \frac{1}{Q^{1/2}} \cos\left(\int_{x_{c1}}^{x} Q\,dx - \frac{\pi}{4}\right), \quad x_{c1} < x < x_{c2} \tag{5.84}$$

$$\frac{\psi}{A} = \frac{1}{P^{1/2}} \left[\cos\left(\int_{x_{c1}}^{x_{c2}} Q\,dx\right) \exp\left(\int_{x_{c2}}^{x} P\,dx\right) + \frac{1}{2} \sin\left(\int_{x_{c1}}^{x_{c2}} Q\,dx\right) \exp\left(-\int_{x_{c2}}^{x} P\,dx\right) \right],$$

$$x_{c2} < x < a \tag{5.85}$$

$$\frac{\psi}{A} = \frac{1}{P^{1/2}} \exp\left(-\int_{a}^{x} P\,dx\right), \quad a < x \tag{5.86}$$

where A is a constant.

The coefficients of growing and decaying waves in (5.85) have been chosen in such a way that the connection formulae (5.71) and (5.74) are automatically satisfied at $x = x_{c2}$. The boundary condition remaining to be satisfied is the requirement for continuity of the fields and their derivatives at $x = a$. Denoting by $n(a^+)$ and $n(a^-)$ the values of refractive index at the discontinuity corresponding to approaching from $x > a$ and $x < a$, respectively, the field continuity equation yields:

$$\cos\left(\int_{x_{c1}}^{x_{c2}} Q\,dx\right) \exp\left(\int_{x_{c2}}^{a^-} P\,dx\right) + \frac{1}{2} \sin\left(\int_{x_{c1}}^{x_{c2}} Q\,dx\right) \exp\left(-\int_{x_{c2}}^{a^-} P\,dx\right) = \left(\frac{P(a^-)}{P(a^+)}\right)^{1/2}. \tag{5.87}$$

Noting the remarks concerning derivatives of first-order WKB solutions made in the previous subsection the condition for continuity of derivatives at $x = a$ yields:

$$\cos\left(\int_{x_{c1}}^{x_{c2}} Q\,dx\right) \exp\left(\int_{x_{c2}}^{a^-} P\,dx\right) - \frac{1}{2} \sin\left(\int_{x_{c1}}^{x_{c2}} Q\,dx\right) \exp\left(-\int_{x_{c2}}^{a^-} P\,dx\right) = -\left(\frac{P(a^+)}{P(a^-)}\right)^{1/2}. \tag{5.88}$$

Combining equations (5.87) and (5.88) yields the eigenvalue equation for 'buried' modes of order N (Janta and Čtyroký, 1978):

$$\int_{x_{c1}}^{x_{c2}} Q\,dx = N\pi + \tan^{-1}\left[2\left(\frac{P(a^-)+P(a^+)}{P(a^-)-P(a^+)}\right)\exp\left(2\int_{x_{c2}}^{a^-} P\,dx\right)\right] \quad (N = 0, 1, 2, \ldots). \quad (5.89)$$

Note that in the limit of the index discontinuity tending to zero, i.e. $n(a^-) \to n(a^+)$, equation (5.89) reduces to the form (5.79) derived earlier for the case of smoothly-varying index profiles.

The distinction between buried and unburied modes was noted first by Gedeon (1974) and Conwell (1975a). The eigenvalue equations (5.83) and (5.89) for buried and unburied modes, respectively, have been applied to calculations of mode spectra for exponential, Gaussian, and parabolic profiles with index discontinuities (Janta and Čtyroký, 1978). Such discontinuities occur for example at the interface between planar dielectric guides and the superstrate (frequently air). Comparison of the WKB results with those obtained by more accurate numerical computations revealed only small errors for most modes away from cut-off. The extension to anisotropic media has been given by Čada et al. (1979).

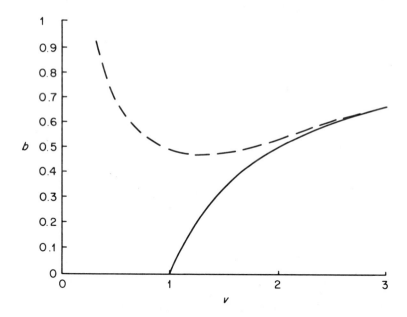

Figure 5.6 Results of the WKB approximation for the cladded-parabolic profile. Solid line, results from first-order theory, using derivatives of only the zero-order terms in the boundary condition; broken line, results using derivatives of both zero and first-order terms in the boundary condition. (From Adams (1978). Reproduced by permission of Chapman and Hall)

As noted earlier the derivation of equations (5.83) and (5.89) uses a matching condition of WKB solutions where only the derivative of the zero-order term is used. The result of including derivatives of both terms of the first-order WKB fields is to include terms containing the derivative of refractive index (dn/dx) evaluated as a limit on either side of the discontinuity. Such terms have been included for the case of linear profiles by Marcuse (1973b) and for cladded-parabolic profiles by Adams (1978), following the three-dimensional analogue presented by Olshansky (1977). However, the magnitude of the correction thus

152

obtained is somewhat uncertain in view of the neglect of higher-order terms in the WKB expansion. For the case of the cladded-parabolic profile, within the limits of the present first-order treatment, the result reduces to equation (5.79) which is indistinguishable from that found for extended parabolic media. Figure 5.6 shows plots of normalized propagation constant b versus normalized frequency v calculated for the two cases (i) including derivatives of both WKB terms, and (ii) including only the derivative of the zero-order term, for the lowest-order mode of the cladded-parabolic profile (Adams, 1978). Clearly neither case is able to describe adequately the mode solutions close to cut-off, and case (i) gives an exceedingly unlikely result in this region.

5.2.5 WKB tunnelling coefficient for leaky waves

Consider now the situation shown in Figure 5.7 which is in a sense the opposite to that of Figure 5.3, in that we now have a region of evanescent wave behaviour $k^2 n^2(x) < \beta^2$ ($x_{c1} < x < x_{c2}$) lying between two regions of oscillatory behaviour $k^2 n^2(x) > \beta^2$ ($x < x_{c1}$ and $x > x_{c2}$). In this case the overall solution may be described as a leaky or tunnelling wave, since the power of a guided wave on one side of the structure may penetrate through to the opposite side in exactly the same way that an electron may tunnel through a potential barrier in the quantum mechanical description. Indeed, the electromagnetic situation can be discussed in precisely the same terms as the corresponding quantum problem with an incident, reflected, and transmitted wave description. Furthermore, within the WKB framework we may calculate a tunnelling coefficient for arbitrary variation of the dielectric profile between the caustics x_{c1} and x_{c2} of Figure 5.7.

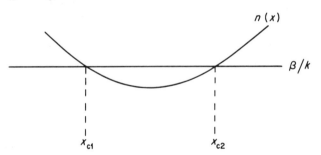

Figure 5.7 A region of evanescent field variation bounded by caustics x_{c1}, x_{c2} in a smoothly-varying index profile

Let us assume a wave incident on the structure of Figure 5.7 from the region $x < x_{c1}$; there is then also a reflected wave in the region $x < x_{c1}$ and a transmitted wave in the region $x > x_{c2}$. The transmitted wave may be written in the form (Bohm, 1951):

$$\frac{\psi}{A} = \frac{1}{Q^{1/2}} \exp\left[i\left(\int_{x_{c2}}^{x} Q\,dx - \frac{\pi}{4}\right)\right]$$

$$= \frac{1}{Q^{1/2}} \left\{ \cos\left[\int_{x_{c2}}^{x} Q\,dx - \frac{\pi}{4}\right] - i \cos\left[\int_{x_{c2}}^{x} Q\,dx + \frac{\pi}{4}\right] \right\} \quad (x > x_{c2}) \quad (5.90)$$

where the $\pi/4$ phase factor has been included merely for convenience in applying the connection formulae; A is a constant. Applying the connection formulae (5.71) and (5.74) at the caustic $x = x_{c2}$ leads to the expression for the field in the region $x_{c1} < x < x_{c2}$:

$$\frac{\psi(x)}{A} = \frac{1}{P^{1/2}} \left\{ \frac{1}{2} \exp\left(-\int_x^{x_{c2}} P\,dx\right) - i \exp\left(\int_x^{x_{c2}} P\,dx\right) \right\}$$

$$= \frac{1}{P^{1/2}} \left\{ \frac{1}{2} \exp\left(-\int_{x_{c1}}^{x_{c2}} P\,dx\right) \exp\left(\int_{x_{c1}}^{x} P\,dx\right) - i \exp\left(\int_{x_{c1}}^{x_{c2}} P\,dx\right) \exp\left(-\int_{x_{c1}}^{x} P\,dx\right) \right\}$$

$$(x_{c1} < x < x_{c2}). \tag{5.91}$$

Applying the connection formulae again at the caustic $x = x_{c1}$ gives the result for incident and reflected waves in the region $x < x_{c1}$:

$$\frac{\psi(x)}{A} = -\frac{1}{Q^{1/2}} \left\{ \frac{1}{2} \exp\left(-\int_{x_{c1}}^{x_{c2}} P\,dx\right) \sin\left[\int_x^{x_{c1}} Q\,dx - \frac{\pi}{4}\right] \right.$$

$$\left. + 2i \exp\left(\int_{x_{c1}}^{x_{c2}} P\,dx\right) \cos\left[\int_x^{x_{c1}} Q\,dx - \frac{\pi}{4}\right] \right\}$$

$$= \frac{-i}{Q^{1/2}} \left\{ \exp\left[i\left(\int_{x_{c1}}^{x} Q\,dx + \frac{\pi}{4}\right)\right] \left[\exp\left(\int_{x_{c1}}^{x_{c2}} P\,dx\right) + \frac{1}{4} \exp\left(-\int_{x_{c1}}^{x_{c2}} P\,dx\right) \right] \right.$$

$$\left. + \exp\left[-i\left(\int_{x_{c1}}^{x} Q\,dx + \frac{\pi}{4}\right)\right] \left[\exp\left(\int_{x_{c1}}^{x_{c2}} P\,dx\right) - \frac{1}{4} \exp\left(-\int_{x_{c1}}^{x_{c2}} P\,dx\right) \right] \right\}$$

$$x < x_{c1}. \tag{5.92}$$

Now, the tunnelling coefficient T is defined as the ratio of transmitted to incident intensities. From equations (5.90) and (5.92), it follows that

$$T = \left[\exp\left(\int_{x_{c1}}^{x_{c2}} P\,dx\right) + \frac{1}{4} \exp\left(-\int_{x_{c1}}^{x_{c2}} P\,dx\right) \right]^{-2}. \tag{5.93}$$

The second exponential on the r.h.s. of (5.93) is usually negligible in comparison with the first term provided the caustics at x_{c1} and x_{c2} are sufficiently far apart (which is also a necessary condition for the connection formulae to be applicable here). Hence for the situation shown in Figure 5.7, the tunnelling coefficient is given by

$$T \simeq \exp\left(-2 \int_{x_{c1}}^{x_{c2}} P\,dx\right). \tag{5.94}$$

Based on our previous experience of deriving eigenvalue equations within the WKB approximation in Sections 5.2.2–5.2.4, we may expect equation (5.94) for the tunnelling coefficient to be modified somewhat in the presence of index discontinuities. This is indeed the case and we shall take as an example the structure of Figure 5.8 where the region of evanescent behaviour $x_{c1} < x < x_{c2}$ is bounded by index discontinuities occurring at x_{c1} and x_{c2}. Then we can no longer apply the connection formula at these points and instead the familiar boundary conditions of matching the fields and their derivatives must be used. The general expressions for the fields in the three regions of Figure 5.8 are:

$$\psi(x) = \frac{1}{Q^{1/2}} \left\{ A \exp\left(i \int_x^{x_{c1}} Q\,dx\right) + B \exp\left(-i \int_x^{x_{c1}} Q\,dx\right) \right\}, \quad x < x_{c1} \tag{5.95}$$

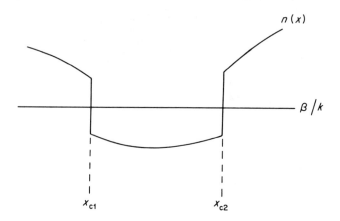

Figure 5.8 A region of evanescent field behaviour bounded by index discontinuities at x_{c1}, x_{c2}

$$\psi(x) = \frac{1}{P^{1/2}}\left\{C\exp\left(-\int_{x_{c1}}^{x} P\,dx\right) + D\exp\left(\int_{x_{c1}}^{x} P\,dx\right)\right\}, \quad x_{c1} < x < x_{c2} \quad (5.96)$$

$$\psi(x) = \frac{1}{Q^{1/2}}\exp\left(i\int_{x_{c2}}^{x} Q\,dx\right), \quad x > x_{c2}. \quad (5.97)$$

Matching the value of $\psi(x)$ and $d\psi(x)/dx$ at the boundary $x = x_{c1}$ leads to the relations:

$$A + B = (C + D)\left(\frac{Q_1}{P_1}\right)^{1/2} \quad (5.98)$$

$$A - B = i(C - D)\left(\frac{P_1}{Q_1}\right)^{1/2} \quad (5.99)$$

where we have used the shorthand notation P_1 for what was previously termed $P(x_{c1}^{-})$ and Q_1 for $Q(x_{c1}^{+})$ (see definitions in Section 5.2.3 with respect to a discontinuity at x_{c2}). Similarly, applying the boundary conditions at $x = x_{c2}$ leads to the relations:

$$Ce^{-\phi} + De^{\phi} = \left(\frac{P_2}{Q_2}\right)^{1/2} \quad (5.100)$$

$$-Ce^{-\phi} + De^{\phi} = i\left(\frac{Q_2}{P_2}\right)^{1/2} \quad (5.101)$$

where P_2, Q_2 are defined analogously to P_1, Q_1, and

$$\phi = \int_{x_{c1}}^{x_{c2}} P\,dx. \quad (5.102)$$

Eliminating C and D from equations (5.98–5.101) gives:

$$A = \frac{\cosh\phi\,(Q_1P_2 + P_1Q_2) + i\sinh\phi\,(P_1P_2 - Q_1Q_2)}{2(P_1P_2Q_1Q_2)^{1/2}} \quad (5.103)$$

$$B = \frac{\cosh\phi\,(Q_1P_2 - P_1Q_2) - i\sinh\phi\,(P_1P_2 + Q_1Q_2)}{2(P_1P_2Q_1Q_2)^{1/2}}. \quad (5.104)$$

Hence the tunnelling coefficient for the structure of Figure 5.8 is given by:

$$T = 1 - \left|\frac{B}{A}\right|^2$$

$$= \frac{4P_1P_2Q_1Q_2(1 - \tanh^2 \phi)}{(Q_1P_2 + P_1Q_2)^2 + \tanh^2 \phi (P_1P_2 - Q_1Q_2)^2}. \quad (5.105)$$

Note that in the case of $\phi \gg 1$, equation (5.105) for the tunnelling coefficient may be reduced to the form:

$$T \simeq \frac{16P_1P_2Q_1Q_2}{(P_1^2 + Q_1^2)(P_2^2 + Q_2^2)} \exp\left(-2 \int_{x_{c1}}^{x_{c2}} P \, dx\right). \quad (5.106)$$

Thus we see that the effect of index discontinuities at x_{c1} and x_{c2} is to modify the simple result (5.94) by a factor, usually of order unity, as given in equation (5.106). This factor is in fact the product of the appropriate generalized Fresnel transmission coefficients for each interface (Love and Winkler, 1977), cf. Section 2.1.

The tunnelling coefficient T may be used to calculate a power attenuation coefficient α by an extension of the ray theory arguments used in Section 2.2.4. If the waveguide has an associated periodic length z_p, defined as the axial displacement corresponding to successive turning points of a ray at which power is lost by the tunnelling mechanism, then the attenuation due to tunnelling is:

$$\alpha = \frac{T}{z_p}. \quad (5.107)$$

As an example of the use of this result we may consider again the W-guide discussed in Section 2.6.4. Using the notation of Section 2.6.4, we note the relationships:

$$P_1 = P_2 = \frac{t''}{a}; \quad Q_1 = \frac{u}{a}; \quad Q_2 = \frac{w''}{a};$$

$$z_p \simeq \frac{2kn_1}{u} a^2.$$

Hence equations (5.106) and (5.107) lead to

$$\alpha \simeq \frac{8t''^2 u^2 w''}{(u^2 + t''^2)(w''^2 + t''^2)kn_1 a^2} \exp\left[-2t''\left(\frac{b}{a} - 1\right)\right] \quad (5.108)$$

as the attenuation for a leaky wave in the W-guide. For the symmetric case $w'' = u$, application of the zero-order approximation (2.230) $u \simeq (N+1)\pi/2$ leads directly to the result (2.233) which was previously derived by an entirely different argument commencing with the eigenvalue equation.

5.3 Variational Methods

We give here a brief introduction to the very broad subject of variational techniques; more detailed accounts may be found in a number of standard texts (see, for example, Morse and Feshbach, 1953; Schechter, 1967). In particular we are concerned with the proof of a variational theorem for the propagation constant β viewed as the eigenvalue of the scalar wave equation. With the aid of this theorem it is then relatively simple to derive a general relationship between phase and group velocities for a mode in a waveguide of arbitrary refractive index distribution. Finally in this section we give an example of the solution of a stationary expression for β by the Rayleigh–Ritz method and apply the technique to specific index profiles.

5.3.1 The variational principle for eigenvalues

Consider the scalar wave equation in the general form given by equation (5.23), which we repeat here for convenience:

$$H\psi(x) = \beta^2 \psi(x) \qquad (5.109)$$

where $\psi(x)$ is related to the appropriate field component and H is given by (5.24) or (5.25) for TE or TM modes, respectively. Pre-multiplying (5.109) by $\psi(x)$ and integrating over the waveguide transverse dimension x yields:

$$\beta^2 = \frac{\int_{-\infty}^{\infty} \psi(x) H \psi(x) \, dx}{\int_{-\infty}^{\infty} \psi^2(x) \, dx}. \qquad (5.110)$$

Note that if the eigenfunctions $\psi(x)$ are correctly normalized according to equation (5.28), then the denominator of (5.110) will be unity. We now prove that (5.110) is a stationary expression for β^2 with respect to small variations of the function $\psi(x)$. Let $\psi(x)$ be incremented by a small, arbitrary function $\delta\psi(x)$ which is sufficiently small that only first-order terms need be considered. Equation (5.110) then yields for the corresponding change in eigenvalue $\delta\beta^2$:

$$\delta\beta^2 \simeq \frac{\int_{-\infty}^{\infty} [\psi(x) + \delta\psi(x)] H[\psi(x) + \delta\psi(x)] \, dx}{\int_{-\infty}^{\infty} [\psi^2(x) + 2\delta\psi(x)\psi(x)] \, dx} - \frac{\int_{-\infty}^{\infty} \psi(x) H \psi(x) \, dx}{\int_{-\infty}^{\infty} \psi^2(x) \, dx}$$

$$\simeq \frac{\int_{-\infty}^{\infty} [\delta\psi(x) H \psi(x) + \psi(x) H \delta\psi(x)] \, dx}{\int_{-\infty}^{\infty} \psi^2(x) \, dx} - \frac{\beta^2 2 \int_{-\infty}^{\infty} \delta\psi(x) \psi(x) \, dx}{\int_{-\infty}^{\infty} \psi^2(x) \, dx}. \qquad (5.111)$$

Note also that the operator H is self-adjoint (or symmetric), since

$$\int_{-\infty}^{\infty} \left[\delta\psi(x) \frac{\partial^2 \psi(x)}{\partial x^2} - \psi(x) \frac{\partial^2}{\partial x^2} (\delta\psi(x)) \right] dx = \left[\delta\psi(x) \frac{\partial \psi(x)}{\partial x} - \psi(x) \frac{\partial (\delta\psi(x))}{\partial x} \right]_{-\infty}^{\infty}$$

$$= 0,$$

provided $\psi(x)$ and $\partial\psi(x)/\partial x$ are chosen to satisfy the appropriate boundary conditions of vanishing as $x \to \pm\infty$. With the aid of this result equation (5.111) reduces to

$$\delta\beta^2 \simeq \frac{2\int_{-\infty}^{\infty} \delta\psi(x) [H - \beta^2] \psi(x) \, dx}{\int_{-\infty}^{\infty} \psi^2(x) \, dx} = 0 \qquad (5.112)$$

where the final step followed since $\psi(x)$ is a solution of the original differential equation. Hence equation (5.110) is a stationary expression for β^2 with respect to small variations of the function $\psi(x)$ chosen to satisfy the appropriate boundary conditions. In fact this represents an upper limit on the eigenvalue β^2 of the operator H. Explicitly, equation (5.110) may be rewritten, after using (5.24) and integrating the d^2/dx^2 term by parts, in the form (Matsuhara, 1973) for TE modes:

$$\beta^2 = \frac{\int_{-\infty}^{\infty}[\psi^2(x)k^2n^2(x)-(\partial\psi(x)/\partial x)^2]\,dx}{\int_{-\infty}^{\infty}\psi^2(x)\,dx}. \tag{5.113}$$

A similar expression for TM modes results from using (5.25) in (5.110):

$$\beta^2 = \frac{\int_{-\infty}^{\infty}\left\{\left[k^2n^2(x)+\frac{1}{2n^2(x)}\frac{d^2(n^2(x))}{dx^2}-\frac{3}{4n^4(x)}\left(\frac{d(n^2(x))}{dx}\right)^2\right]\psi^2(x)-\left(\frac{\partial\psi(x)}{\partial x}\right)^2\right\}dx}{\int_{-\infty}^{\infty}\psi^2(x)\,dx}. \tag{5.114}$$

The solution of equations (5.113) or (5.114) depends on the choice of trial functions $\psi(x)$ satisfying the appropriate boundary conditions. If $\psi(x)$ is chosen to be a function of a number of parameters then these parameters may be varied until the value of β^2 is a minimum. Clearly the process yields better values for β^2 if the trial function closely resembles the actual eigenfunction or if the range of dependent parameters is sufficiently large. A systematic approach for the choice of function is provided by the Ritz method where $\psi(x)$ is expanded as a series of functions $\phi_i(x)$ forming a complete orthonormal set:

$$\psi(x) = \sum_i a_i\phi_i(x). \tag{5.115}$$

The stationary expression for β^2 may then be differentiated with respect to each expansion coefficient a_i and the result equated to zero. This procedure yields a set of linear equations and the usual condition for a nontrivial solution gives a characteristic equation for the eigenvalue β^2. We shall return to a discussion of this approach in more detail below (in Section 5.3.4) in connection with a somewhat different stationary expression. However, it is worth noting here that the method as described above has been applied to graded-index guides of rectangular cross-section by a number of authors (beginning with Matsuhara, 1973) and these analyses will be discussed further in Chapter 6. In addition the method has been applied to graded-index fibres of circular cross-section (Geshiro et al., 1978a; Heyke and Kuhn, 1973) and these results will be mentioned in Chapter 8. The variational approach may be extended to include leaky modes by changing the path of integration in (5.113) or (5.114) by means of a complex coordinate system (Kawakami and Ogusu, 1978). A method for obtaining second-order perturbation corrections to the variational result has been presented by Lang (1979).

5.3.2 A general relationship between phase and group velocities

Since (5.110) is a stationary expression for the eigenvalue β^2, we may differentiate it with respect to k only where k occurs explicitly, i.e. ignoring the k-dependence of $\psi(x)$. Hence, using either of the forms (5.113) or (5.114) we obtain:

$$\beta\frac{d\beta}{dk} = k\frac{\int_{-\infty}^{\infty}n^2(x)\psi^2(x)\,dx}{\int_{-\infty}^{\infty}\psi^2(x)\,dx}. \tag{5.116}$$

Using the usual definitions of phase velocity $V_p = ck/\beta$ and group velocity $V_g = c\,dk/d\beta$, equation (5.116) may be re-written (Case, 1972):

$$\frac{c^2}{V_p V_g} = \frac{\int_{-\infty}^{\infty} n^2(x)\psi^2(x)\,dx}{\int_{-\infty}^{\infty} \psi^2(x)\,dx}. \tag{5.117}$$

Equation (5.117) gives a general relationship between the phase and group velocities for a guided mode in a two-dimensional graded-index waveguide in terms of the solutions $\psi(x)$ of the scalar wave equation. To re-interpret this result in terms of the axial power flow $S_z(x)$ at each position x of the guide cross-section, we must consider the definition of $\psi(x)$ for TE and TM modes separately.

For the case of TE modes, $\psi(x)$ represents the field component H_x (see equation (3.35)), and we may therefore calculate E_y from Maxwell's equations (see equation (2.78)) as:

$$E_y = -\frac{\omega\mu_0 H_x}{\beta} = -\frac{\omega\mu_0 \psi(x)}{\beta}. \tag{5.118}$$

Hence it follows that the axial power flow $S_z(x)$ for this case is given by the z-component of the Poynting vector as:

$$S_z(x) = -\frac{E_y H_x}{2} = \frac{\omega\mu_0}{2\beta}\psi^2(x). \tag{5.119}$$

Equation (5.117) may therefore be re-written for TE modes with the aid of (5.119) in terms of the power flow at position x:

$$\frac{c^2}{V_p V_g} = \frac{\int_{-\infty}^{\infty} n^2(x) S_z(x)\,dx}{\int_{-\infty}^{\infty} S_z(x)\,dx}. \tag{5.120}$$

For the case of TM modes, $\psi(x)$ is given from (3.37) in terms of the field component E_x as:

$$\psi(x) = n(x)E_x.$$

We may therefore calculate the component H_y from equation (2.83):

$$H_y = \frac{\omega n^2(x)\varepsilon_0}{\beta} E_x = \frac{\omega n(x)\varepsilon_0}{\beta}\psi(x). \tag{5.121}$$

The corresponding expression for axial power flow becomes:

$$S_z(x) = \frac{E_x H_y}{2} = \frac{\omega\varepsilon_0}{2\beta}\psi^2(x). \tag{5.122}$$

Hence equation (5.117) may be re-interpreted in terms of $S_z(x)$ for TM modes, when it reduces again to equation (5.120).

Equation (5.120) represents a general relationship for the product of phase and group velocities as a weighted average of the variation of refractive index distribution in a two-dimensional waveguide. The weighting is achieved by means of the corresponding axial power-flow densities. This relationship was originally derived by Brown (1966) from considerations of the electromagnetic equivalent of conservation of momentum. It has subsequently been re-derived by other methods (Kawakami, 1975; Nemoto and Makimoto, 1976; Haus and Kogelnik, 1976), and extend to vector modes of three-dimensional guides, including the case of anisotropic media (Nemoto and Makimoto, 1977). For dispersive media where $n(x)$ is a function of wave number k, the above argument is easily generalized to yield the result:

$$\frac{c^2}{V_p V_g} = \frac{\int_{-\infty}^{\infty} n(x)m(x)S_z(x)\,dx}{\int_{-\infty}^{\infty} S_z(x)\,dx} \quad (5.123)$$

where $m(x)$ is the group index, defined as:

$$m(x) = \frac{d(n(x)k)}{dk} = n(x) + k\frac{dn(x)}{dk}. \quad (5.124)$$

Special cases of the general relation (5.120) have already been encountered earlier in this book. For example, for the weakly-guiding symmetric three-layer slab waveguide of Section 2.3.5, equation (5.120) reduces to

$$\frac{c^2}{V_p V_g} = n_1^2 \frac{\int_{-a}^{a} S_z(x)\,dx}{\int_{-\infty}^{\infty} S_z(x)\,dx} + n_2^2 \frac{2\int_{a}^{\infty} S_z(x)\,dx}{\int_{-\infty}^{\infty} S_z(x)\,dx}$$

$$= n_1^2 \Gamma + n_2^2 (1 - \Gamma) \quad (5.125)$$

where Γ is the confinement factor defined in Section 2.3.5. Equation (5.125) is recognized as the result (2.133) originally derived by an entirely different method, namely differentiation of the eigenvalue equation and evaluation of Γ for the symmetric slab guide.

We note in passing that another result from Section 2.3.5 may be re-derived and generalized from the present approach by simply extending the argument to include media with complex permittivities. For the weakly-guiding case, the TM result (5.114) reduces to the same as that for TE modes (5.113). Taking $\beta = \beta_r + i\beta_i$, replacing $n^2(x)$ by $(n(x) + iK(x))^2$, and noting that $\psi(x)$ remains real, the imaginary part of (5.113) yields:

$$\beta_r \beta_i = \frac{\int_{-\infty}^{\infty} k^2 n(x) K(x) \psi^2(x)\,dx}{\int_{-\infty}^{\infty} \psi^2(x)\,dx}. \quad (5.126)$$

Re-writing $\beta_i = -\alpha_0/2$ and $K(x) = -\alpha(x)/2k$, and assuming the variation of $\alpha(x)$ with x greatly exceeds that of $n(x)$, so that $n(x)$ may be removed from the integral in (5.126), we find the approximate result:

$$\alpha_0 \simeq \frac{\int_{-\infty}^{\infty} \alpha(x) \psi^2(x)\,dx}{\int_{-\infty}^{\infty} \psi^2(x)\,dx}. \quad (5.127)$$

This equation represents a useful general relation for the modal attenuation coefficient α_0 in terms of the material loss coefficient $\alpha(x)$ for a weakly-guiding two-dimensional guide with relatively low losses ($\beta_r \simeq kn$). It reduces for the symmetric three-layer slab case to the result (2.128). An equivalent expression for the optical fibre case has been given by Gloge (1975).

Other useful inequalities may be derived from the weakly-guiding scalar result (5.113). Since $(\partial \psi(x)/\partial x)^2$ is positive definite, it follows that (Case, 1972):

$$\frac{c^2}{V_p^2} = \frac{\beta^2}{k^2} \leq \frac{\int_{-\infty}^{\infty} \psi^2(x) n^2(x)\,dx}{\int_{-\infty}^{\infty} \psi^2(x)}. \quad (5.128)$$

Combination of this result with (5.117) results in an inequality for the group velocity:

$$\frac{c^2}{V_g^2} \geq \frac{\int_{-\infty}^{\infty} \psi^2(x) n^2(x)\,dx}{\int_{-\infty}^{\infty} \psi^2(x)\,dx} \quad (5.129)$$

Equation (5.129) gives an upper limit for the group velocity in a two-dimensional graded-index waveguide in terms of the solution $\psi(x)$ of the scalar wave equation. Nemoto and

Makimoto (1976a) have extended these arguments somewhat by considering V_g as a function of phase velocity V_p for the three-layer slab guide and hence deriving an absolute lower limit for the value of V_g in terms of the refractive indices of the layers.

5.3.3 An alternative stationary expression

As an alternative approach to that outlined in Section 5.3.1, we give here a variational formulation derived first for three-dimensional circular waveguides by Okoshi and Okamoto (1974). Adapting this technique to the case of two-dimensional graded-index guides, the appropriate stationary expression for TE modes is given by

$$J(\psi) = \psi^2(a)\phi_A - \psi^2(-a)\phi_B - \int_{-a}^{a} \left(\frac{\partial \psi(x)}{\partial x}\right)^2 dx + \int_{-a}^{a} (k^2 n^2(x) - \beta^2)\psi^2(x)\,dx. \tag{5.130}$$

The quantities ϕ_A and ϕ_B appearing in (5.130) are related to the boundary conditions at $x = a$ and $x = -a$, respectively, and will be defined in more detail in due course. We now prove that $J(\psi)$ is a stationary expression with respect to small variations of the function $\psi(x)$, provided $\psi(x)$ satisfies the scalar wave equation and the appropriate boundary conditions. It should perhaps be emphasized at this point that we are restricting the present discussion to TE modes (or, equivalently, the weakly-guiding scalar wave approximation) and to refractive index distributions given by $n(x)$ for $-a \leqslant x \leqslant a$; outside this region it is assumed that explicit solutions of the scalar wave equation are available.

Consider small deviations of $\psi(x)$ of the form $\psi(x) = \psi_0(x) + \varepsilon \eta(x)$, where $\psi_0(x)$ is assumed to be the appropriate function which minimizes $J(\psi)$, ε is a real small quantity, and $\eta(x)$ is an arbitrary continuous function of x. Inserting this function in (5.130) we find (to first order in ε):

$$J(\psi) = \psi_0^2(a)\phi_A - \psi_0^2(-a)\phi_B + 2\varepsilon[\psi_0(a)\eta(a)\phi_A - \psi_0(-a)\eta(-a)\phi_B]$$

$$- \int_{-a}^{a} \left(\frac{\partial \psi_0}{\partial x}\right)^2 dx - 2\varepsilon \int_{-a}^{a} \left(\frac{\partial \psi_0}{\partial x}\right)\left(\frac{\partial \eta}{\partial x}\right) dx + \int_{-a}^{a} (k^2 n^2(x) - \beta^2)\psi_0^2(x)\,dx$$

$$+ 2\varepsilon \int_{-a}^{a} (k^2 n^2(x) - \beta^2)\psi_0(x)\eta(x)\,dx. \tag{5.131}$$

Varying the parameter ε in (5.131), it follows that, at $\varepsilon = 0$:

$$\frac{\partial J}{\partial \varepsilon} = 2[\psi_0(a)\eta(a)\phi_A - \psi_0(-a)\eta(-a)\phi_B] - 2\int_{-a}^{a} \frac{\partial \psi_0}{\partial x}\frac{\partial \eta}{\partial x}\,dx + 2\int_{-a}^{a} (k^2 n^2(x) - \beta^2)\psi_0 \eta\,dx. \tag{5.132}$$

The second term on the r.h.s. of (5.132) may be integrated by parts to give

$$\int_{-a}^{a} \frac{\partial \psi_0}{\partial x}\frac{\partial \eta}{\partial x}\,dx = \left[\frac{\partial \psi_0}{\partial x}\eta\right]_{-a}^{a} - \int_{-a}^{a} \frac{\partial^2 \psi_0}{\partial x^2}\eta\,dx.$$

Substituting this result in (5.132) and re-arranging the terms yields:

$$\frac{\partial J}{\partial \varepsilon} = 2\psi_0(a)\eta(a)\left[\phi_A - \frac{1}{\psi_0}\frac{\partial \psi_0}{\partial x}\right]_{x=a} + 2\psi_0(-a)\eta(-a)\left[\frac{1}{\psi_0}\frac{\partial \psi_0}{\partial x} - \phi_B\right]_{x=-a}$$

$$+ 2\int_{-a}^{a} \left[\frac{\partial^2 \psi_0}{\partial x^2} + (k^2 n^2(x) - \beta^2)\psi_0\right]\eta\,dx. \tag{5.133}$$

Since $\eta(x)$ is an arbitrary function of x, it follows from (5.133) that $[\partial J/\partial \varepsilon]_{\varepsilon=0} = 0$ provided that $\psi_0(x)$ is a solution of the scalar wave equation and that ϕ_A, ϕ_B are defined as:

$$\phi_A = \left[\frac{1}{\psi_0(x)} \frac{\partial \psi_0(x)}{\partial x}\right]_{x=a} \tag{5.134a}$$

$$\phi_B = \left[\frac{1}{\psi_0(x)} \frac{\partial \psi_0(x)}{\partial x}\right]_{x=-a}. \tag{5.134b}$$

Equations (5.134a) and (5.134b) clearly imply that ϕ_A, ϕ_B may be evaluated from the familiar boundary conditions of continuity of appropriate field components at $x = \pm a$. For the cases of interest here we may assume that the index distributions in the cladding regions $|x| \geq a$ are uniform and given by n_2 for $x \geq a$ and n_3 for $x \leq -a$. It follows that the corresponding solutions for $\psi_0(x)$ obey exponential laws, so that

$$\phi_A = -(\beta^2 - k^2 n_2^2)^{1/2} \tag{5.135a}$$
$$\phi_B = (\beta^2 - k^2 n_3^2)^{1/2}. \tag{5.135b}$$

Finally, we may cast the functional $J(\psi)$ into a more convenient form using $n^2(x) = n_1^2(1-f(x))$:

$$J(\psi) = [\psi^2(a)\phi_A - \psi^2(-a)\phi_B] + \frac{u^2}{a^2}\int_{-a}^{a}\psi^2(x)dx - \int_{-a}^{a}\left(\frac{\partial \psi(x)}{\partial x}\right)^2 dx$$

$$- k^2 n_1^2 \int_{-a}^{a} f(x)\psi^2(x)dx \tag{5.136}$$

where $u^2 = a^2(k^2 n_1^2 - \beta^2)$ as usual.

5.3.4 Solution by the Rayleigh–Ritz method

The stationary expression (5.136) may be solved by the Rayleigh–Ritz method with suitable choice of the trial function $\psi(x)$ as mentioned in Section 5.3.1. For simplicity we will restrict attention to the case of symmetric claddings $n_2 = n_3$ so that $\phi_A = -\phi_B = -w/a$, and deal separately with even and odd-order modes. For the case of even-order modes, a suitable orthonormal trial function is given by (Adams, 1978):

$$\psi(x) = \sum_{v=0}^{N} a_v \frac{1}{a^{1/2}} \cos\left(\frac{v\pi x}{a}\right) \tag{5.137}$$

where $N \to \infty$ for a rigorous solution, but the series may be terminated at a convenient value for numerical evaluation. With the aid of this choice, the first term on the r.h.s. of (5.136) becomes:

$$\psi^2(a)\phi_A - \psi^2(-a)\phi_B = -\frac{2w}{a^2}\sum_{\mu=0}^{N}\sum_{v=0}^{N}(-1)^{\mu+v}a_\mu a_v. \tag{5.138}$$

Similarly, the second term on the r.h.s. of (5.136) becomes:

$$\frac{u^2}{a^2}\int_{-a}^{a}\psi^2 dx = \frac{u^2}{a^2}\left[2a_0^2 + \sum_{\mu,v}{}' a_\mu a_v \delta_{\mu v}\right] \tag{5.139}$$

where the notation $\sum'_{\mu,v}$ implies that the case $\mu = v = 0$ is omitted from the sum; $\delta_{\mu v}$ is the

Kronecker delta. The third term in (5.136) becomes:

$$\int_{-a}^{a} \left(\frac{\partial \psi}{\partial x}\right)^2 dx = \sum_{\mu=0}^{N} \sum_{v=0}^{N} \frac{\mu^2 \pi^2}{a^2} a_\mu a_v \delta_{\mu v}. \qquad (5.140)$$

Hence equation (5.136) may be re-written using (5.137)–(5.140) as:

$$J(\psi) = -\frac{2w}{a^2} \sum_{\mu=0}^{N} \sum_{v=0}^{N} (-1)^{\mu+v} a_\mu a_v + \frac{u^2}{a^2} \left[2a_0^2 + \sum_{\mu,v}' a_\mu a_v \delta_{\mu v} \right]$$

$$- \sum_{\mu=0}^{N} \sum_{v=0}^{N} \frac{\mu^2 \pi^2}{a^2} a_\mu a_v \delta_{\mu v} - \frac{1}{a^2} \sum_{\mu=0}^{N} \sum_{v=0}^{N} a_\mu a_v C_{\mu v} \qquad (5.141)$$

where

$$C_{\mu v} = k^2 n_1^2 a \int_{-a}^{a} f(x) \cos\left(\frac{v \pi x}{a}\right) \cos\left(\frac{\mu \pi x}{a}\right) dx. \qquad (5.142)$$

$J(\psi)$ is now to be minimized in accordance with the Rayleigh–Ritz procedure by setting $\partial J/\partial a_\mu = 0$ for all μ. From (5.141) we have:

$$\frac{a^2}{2} \frac{\partial J}{\partial a_\mu} = -2w \sum_{v=0}^{N} (-1)^{\mu+v} a_v + \sum_{v=1}^{N} (u^2 - \mu^2 \pi^2) a_v \delta_{\mu v} - \sum_{v=0}^{N} a_v C_{\mu v} + 2u^2 a_0 \delta_{\mu 0}. \qquad (5.143)$$

Setting the l.h.s. of equation (5.143) equal to zero yields a system of linear homogeneous equations which can be written in the form:

$$\sum_{v=0}^{N} a_v S_{\mu v} = 0 \qquad (5.144)$$

where

$$S_{\mu v} = \begin{cases} -2w(-1)^{\mu+v} + (u^2 - \mu^2 \pi^2)\delta_{\mu v} - C_{\mu v} & (\mu \text{ and } v \text{ not both } 0) \\ -2w + 2u^2 - C_{00} & (\mu = v = 0). \end{cases} \qquad (5.145)$$

In order for a nontrivial solution of (5.144) to exist, the determinant of $S_{\mu v}$ must be zero:

$$\det(S_{\mu v}) = 0. \qquad (5.146)$$

This relation represents a characteristic equation for the eigenvalues of the scalar wave equation for a general graded-index two-dimensional guide. For given forms of the index profile as described by the function $f(x)$, $C_{\mu v}$ may be calculated from (5.142) and used in (5.145) to determine the matrix $S_{\mu v}$. For the sake of convenience, equation (5.145) may be re-written as (Okoshi and Okamoto, 1974):

$$S_{\mu v} = \Phi_{\mu v} + U_\mu \delta_{\mu v} - C_{\mu v} \qquad (5.147)$$

where

$$\left. \begin{array}{l} \Phi_{\mu v} = -2w(-1)^{\mu+v} \\ U_\mu = u^2 - \mu^2 \pi^2 \quad (\mu \neq 0) \\ U_0 = 2u^2 \end{array} \right\}. \qquad (5.148)$$

For the odd-order modes, an analogous formulation is possible, using as the trial function (Adams, 1978):

$$\psi(x) = \sum_{v=1}^{N} a_v \frac{1}{a^{1/2}} \sin\left[(v - \tfrac{1}{2}) \frac{\pi x}{a}\right]. \qquad (5.149)$$

Once again a characteristic equation of form (5.146) may be derived with the matrix $S_{\mu\nu}$ defined as in (5.147). In this case, however, the appropriate definitions of $\Phi_{\mu\nu}$ and U_μ are:

$$\left.\begin{array}{l}\Phi_{\mu\nu} = -2w(-1)^{\mu+\nu} \\ U_\mu = u^2 - (\mu - \tfrac{1}{2})^2 \pi^2\end{array}\right\}. \tag{5.150}$$

Note that all elements in this case commence with $\mu, \nu = 1$ rather than 0 as for the even-order modes.

As a general comment on the Ritz method adopted here, it should be noted that for each term of the series expansions (5.137) or (5.149), the appropriate boundary conditions (5.134) will not be satisfied, i.e. if the expansions are written as $\psi(x) = \sum_{\nu=1}^{N} a_\nu \chi_\nu(x)$, then we find

$$\left[\frac{1}{\chi_\nu(x)} \frac{\partial \chi_\nu(x)}{\partial x}\right]_{x=\pm a} = 0.$$

Hence when $\partial \psi(x)/\partial x$ is obtained from the expansion it will exhibit nonuniform convergence in the vicinity of $x = \pm a$ (Unger, 1977). However, this fact is not normally a significant disadvantage of the Ritz method provided sufficient terms are retained in the series expansion.

5.3.5 Results for specific index profiles

The formulation given in the previous subsection may be applied to the well-known three-layer symmetric slab problem as a test of the accuracy of the variational method. For the case of a uniform core, $f(x) = 0$ and hence $C_{\mu\nu} = 0$ for all μ, ν. Equations (5.146) and (5.147) therefore yield explicitly for this case (even-order modes):

$$\begin{vmatrix} \Phi_{00}+U_0 & \Phi_{01} & \Phi_{02} & \cdots & \Phi_{0N} \\ \Phi_{10} & \Phi_{11}+U_1 & \Phi_{12} & \cdots & \vdots \\ \Phi_{20} & \Phi_{21} & \Phi_{22}+U_2 & \cdots & \vdots \\ \vdots & \vdots & \vdots & & \vdots \\ \Phi_{N0} & \cdots & \cdots & \cdots & \Phi_{NN}+U_N \end{vmatrix} = 0. \tag{5.151}$$

This can be expanded to give

$$D - 2w \sum_{\mu=0}^{N} \sum_{\nu=0}^{N} (-1)^{\mu+\nu} Z_{\mu\nu} = 0 \tag{5.152}$$

where D is the determinant of the U's:

$$D = \prod_{\mu=0}^{N} U_\mu \tag{5.153}$$

and $Z_{\mu\nu}$ is the cofactor of the element $U_\mu \delta_{\mu\nu}$ of the matrix of D. Explicitly, $Z_{\mu\nu}$ can be written as

$$Z_{\mu\nu} = \begin{cases} \prod_{\mu=0,\mu\neq\nu}^{N} U_\mu & (\text{for } \mu \text{ and } \nu \text{ not both } 0) \\ 0 & (\text{for } \mu \neq \nu) \end{cases} \tag{5.154}$$

It follows that equation (5.152) may be written with the aid of (5.153) and (5.154) as:

$$\frac{1}{2w} = \sum_{\mu=0}^{N} \frac{1}{U_\mu}. \tag{5.155}$$

For an exact, closed-form solution of the problem, let $N \to \infty$, so that (5.155) yields with the aid of (5.148) (Abramowitz and Stegun, 1964):

$$\frac{1}{2w} = \frac{1}{2u^2} + \sum_{\mu=1}^{\infty} \frac{1}{u^2 - \mu^2 \pi^2} = \frac{\cot u}{2u}. \tag{5.156}$$

It is clear that equation (5.156) may be re-written as

$$\tan u = \frac{w}{u}$$

which is recognized from equation (2.147a) as the exact eigenvalue equation for even-order modes of a weakly-guiding symmetric slab.

For the odd-order modes, the analysis proceeds as outlined above, with the result from (5.150) (Gradshteyn and Ryzhik, 1965):

$$\frac{1}{2w} = \sum_{\mu=1}^{\infty} \frac{1}{u^2 - (\mu - \frac{1}{2})^2 \pi^2} = -\frac{\tan u}{2u}. \tag{5.157}$$

This is similarly recognized from Chapter 2 as the exact eigenvalue equation for odd-order modes of a weakly-guiding symmetric slab. Hence we conclude that by allowing $N \to \infty$ the exact results for the symmetric slab guide may be found from the variational approach. However, this case is an exception in that closed-form results may be found from the characteristic equation (5.146). In all other cases resort must be made to numerical solutions for the eigenvalues using suitably truncated series expansions for the trial functions. In order to test this numerical approach for the symmetric slab guide the value $N = 10$ was chosen for the expansions in (5.137) and (5.149) and results calculated for comparison with the exact solution as given by (5.156) and (5.157). The comparisons are shown in Figure 5.9 as plots of

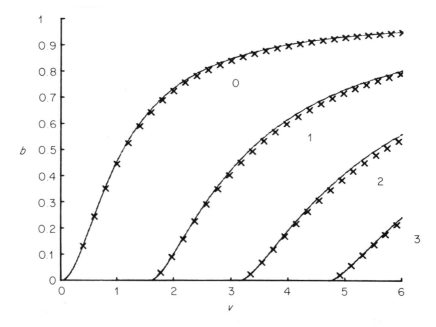

Figure 5.9 Results of the variational technique for the symmetric slab dielectric waveguide (first four modes). Solid lines, exact solution ($N \to \infty$); crosses, numerical solution with $N = 10$

normalized propagation constant b versus normalized frequency v for the first 4 modes. It is observed that the degree of agreement is probably quite satisfactory for $N = 10$; little improvement in the variational values was achieved by taking $N = 20$.

In order to illustrate the application of the variational method to a graded index profile where numerical solution of the characteristic equation is necessary, consider the cladded-parabolic profile discussed earlier. For this case it is convenient to consider the even- and odd-order modes together rather than separately as was the case in the previous subsection. Hence we choose as a new trial function:

$$\psi(x) = \sum_{v=0}^{N} a_v \frac{1}{a^{1/2}} \cos\left[\frac{v\pi}{2a}(x+a)\right]. \tag{5.158}$$

With this choice of ψ the analysis proceeds exactly as in Section 5.3.4 and leads to a characteristic equation again of the form (5.146) where $S_{\mu v}$ is defined as in (5.147). For the present case the appropriate definitions of $\Phi_{\mu v}$ and U_μ are (Adams, 1978):

$$\left.\begin{array}{l} \Phi_{\mu v} = -w[1+(-1)^{\mu+v}] \\ U_\mu = u^2 - \left(\dfrac{\mu\pi}{2}\right)^2 \quad (\mu \neq 0) \\ U_0 = 2u^2. \end{array}\right\}. \tag{5.159}$$

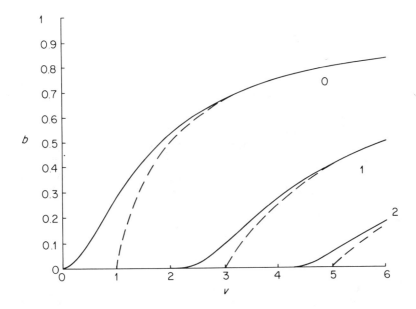

Figure 5.10 Results of the variational technique for the cladded-parabolic profile (first three modes). Solid lines, numerical variational solution with $N = 10$; broken lines, exact results for the infinitely extended parabolic profile. (From Adams (1978). Reproduced by permission of Chapman and Hall)

The definition of $C_{\mu v}$ becomes:

$$C_{\mu v} = k^2 n_1^2 a \int_{-a}^{a} f(x) \cos\left[\frac{v\pi}{2a}(x+a)\right] \cos\left[\frac{\mu\pi}{2a}(x+a)\right] dx \tag{5.160}$$

where the function $f(x)$ is given by:

$$f(x) = 2\Delta \left(\frac{x}{a}\right)^2 \qquad (5.161)$$

with the usual definition of relative index difference Δ. Using (5.161) in (5.160), the integral may be performed analytically with the result:

$$C_{\mu\nu} = \begin{cases} \dfrac{4v^2}{\pi^2}[1+(-1)^{\mu+\nu}]\left[\dfrac{1}{(\mu+\nu)^2}+\dfrac{1}{(\mu-\nu)^2}\right] & \text{for } \mu \neq \nu \\ v^2\left(\dfrac{2}{\pi^2\mu^2}+\dfrac{1}{3}\right) & \text{for } \mu = \nu \neq 0 \\ \dfrac{v^2 2}{3} & \text{for } \mu = \nu = 0 \end{cases} \qquad (5.162)$$

Computed solutions of the characteristic equation for the first few modes of the truncated parabolic profile are shown as b–v curves in Figure 5.10, together with results for the infinitely-extended parabolic profile. These results obtained by the present method with $N = 10$ are in fact as accurate as those by any available alternative numerical treatment. In the next section we will discuss other methods available for numerical computation of eigenvalues and give comparable results for the cladded-parabolic profile guide.

5.4 Other Numerical Methods

In a book of this nature it is not possible to give detailed accounts of all numerical techniques which may be applied to finding solutions to the wave equation in a graded-index waveguide subject to the appropriate boundary conditions. These range from the simple expedient of re-writing the wave equation as two coupled first-order differential equations and using a numerical integration technique, to more sophisticated methods of numerical analysis including, for example, finite element techniques (Yeh et al., 1975). However, in view of the limited space available for this discussion, we restrict attention here to four methods of a rather general nature which seem appropriate to the type of problems encountered. The choice of a specific method is very often dictated by the type of waveguide problem under consideration and, wherever possible, examples of problems where these methods may be applied are given.

5.4.1 Series solution

The wave equation may be solved by a series solution in cases where the dielectric profile can be expanded as a power series in the position coordinate x. The rate of convergence of the solution will of course depend on the details of the index profile. As an example we consider here the cladded-parabolic profile which has been discussed previously in this chapter; for simplicity we restrict attention to the weakly-guiding case of approximate TE/TM degeneracy. The solutions of the scalar wave equation are expanded as follows in the waveguide core region:

$$\psi(x) = \sum_{n=0}^{N} a_n \left(\frac{x}{a}\right)^n, \quad |x| \leq a. \qquad (5.163)$$

In the usual way, we may distinguish between even- and odd-order modes and retain only even- or odd-order terms, respectively, in the expansion (5.163). The appropriate field solutions for the cladding regions are given by

Even-order:
$$\psi(x) = \left(\sum_{n=0}^{N} a_{2n}\right) \exp\left[w\left(1 - \frac{|x|}{a}\right)\right], \quad |x| \geq a \tag{5.164}$$

Odd-order:
$$\psi(x) = \begin{cases} \left(\sum_{n=0}^{N} a_{2n+1}\right) \exp\left[w\left(1 - \frac{x}{a}\right)\right], & x \geq a \quad (5.165a) \\ \left(-\sum_{n=0}^{N} a_{2n+1}\right) \exp\left[w\left(\frac{x}{a} + 1\right)\right], & x \leq -a. \quad (5.165b) \end{cases}$$

The requirement for continuity of the field derivatives on the boundaries $x = \pm a$ yields the eigenvalue equations for the modes in the forms:

Even-order:
$$\sum_{n=0}^{N} 2n a_{2n} = -w \sum_{n=0}^{N} a_{2n} \tag{5.166a}$$

Odd-order:
$$\sum_{n=0}^{N} (2n+1) a_{2n+1} = -w \sum_{n=0}^{N} a_{2n+1}. \tag{5.166b}$$

The expansion coefficients a_n may be found by substituting the series solution (5.163) into the scalar wave equation and equating terms with the same power of x. This process yields (Adams, 1978):

$$a_n = \frac{v^2 a_{n-4} - u^2 a_{n-2}}{n(n-1)} \tag{5.167}$$

where the normalized parameters u, v are defined in the usual way. For even-order modes all the odd-order terms vanish and we may choose $a_0 = 1$ to facilitate the evaluation of the even-order terms. Similarly for the odd-order modes $a_1 = 1$ and $a_{2n} = 0$ (all n); the fields will of course not be correctly normalized with this convention, but this may be achieved by an appropriate multiplicative constant after sufficient of the other coefficients have been evaluated.

The numerical procedure for finding solutions by the series method described above is straightforward; it consists of evaluating the expansion coefficients from equation (5.167) and then using them in the appropriate eigenvalue equation (5.166), whence w (and hence u, b) may be found. Results for the cladded-parabolic guide found in this way are shown for the zero-order mode in Figure 5.11. Also shown for comparison are the corresponding results for step-index and extended-parabolic guides from Chapters 2 and 3, respectively. These results confirm those found for the cladded-parabolic profile by the variational technique of Section 5.3; they also illustrate the ranges of applicability of the two perturbation solutions of Section 5.1. The cut-off condition $w = 0$ has a real physical significance for this guide, as opposed to the situation for the extended-parabolic profile of Chapter 3. Furthermore it is anticipated that the series solution will remain accurate right down to cut-off, so that cut-off v-values may be evaluated with some confidence. For cases considered here, the convergence of the series solution was such that an adequate degree of accuracy was achieved in about 20–30 terms. The cut-off values for the first six modes of the cladded-parabolic, extended parabolic, and step-index waveguides are given in Table 5.1 (Adams, 1978).

A further quantity of interest for the present waveguide is the radiation confinement factor, Γ, introduced in Chapter 2 and defined as the ratio of power confined in the core to the total

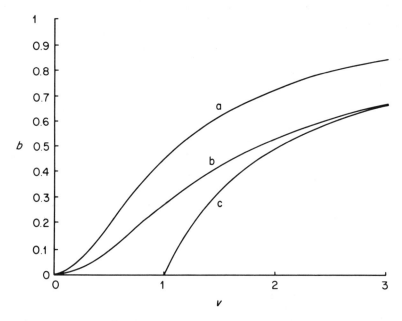

Figure 5.11 Results of the series solution for the cladded-parabolic profile (zero-order mode, weakly-guiding case): (a) step-index; (b) cladded-parabolic profile; (c) extended-parabolic profile. (From Adams (1978). Reproduced by permission of Chapman and Hall)

Table 5.1 Mode cut-off v-values for the first six modes of the step-index (Chapter 2), extended-parabolic (Chapter 3), and cladded-parabolic (series solution) waveguides

Mode Number	Step-index	Extended-parabolic	Cladded-Parabolic
0	0	1	0
1	1.571	3	2.263
2	3.142	5	4.287
3	4.712	7	6.298
4	6.283	9	8.304
5	7.854	11	10.308

power. This is also easily evaluated from the series solution, with the result:

$$\Gamma = \frac{1}{1 + \left[\left(\sum_{n=0}^{N} a_n \right)^2 \Big/ 2w \sum_{n=0}^{N} \sum_{m=0}^{M} \left(\frac{a_n a_m}{n+m+1} \right) \right]}. \tag{5.168}$$

A plot of Γ versus v from this expression is given in Figure 5.12, together with appropriate results for the step-index and extended-parabolic guides. As anticipated earlier, the extended-parabolic medium yields values of Γ markedly different from those of the cladded-parabolic guide over a substantial range of v-values.

The power series solution has been applied to a number of refractive index profiles by Kirchhoff (1973). His results indicate that profiles which grade into the cladding over a substantial distance are particularly unsuited for treatment by this method since very many

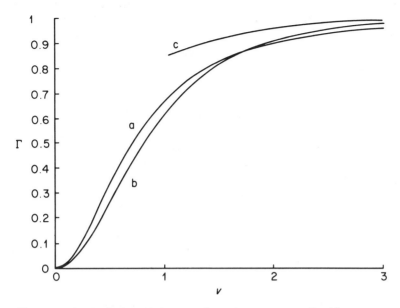

Figure 5.12 Radiation confinement factor Γ versus normalized frequency v for the cases: (a) step-index; (b) cladded-parabolic profile (series solution); (c) extended-parabolic profile. (From Adams (1978). Reproduced by permission of Chapman and Hall)

terms must be retained in the series solution for reasonable accuracy. The method has also been used for complex dielectric profiles (Hatz and Mohn, 1967; Shore and Adams, 1976) where it gives a reasonable approximation for low-order modes in injection laser structures where a linear variation of refractive index and gain profiles may sometimes occur.

5.4.2 Multilayer 'staircase' approximation

This technique, which is alternatively known as the stratification method, consists of replacing a known graded variation of refractive index by a multilayer structure where the index value and width of each step is chosen to yield a good approximation to the original profile. The procedure is illustrated schematically in Figure 5.13 where a symmetric profile is treated, one half of the core region being subdivided into N layers each of refractive index n_j ($j = 1-N$). Within each layer the scalar wave equation is to be solved and the solutions matched in the usual way at the interfaces x_1-x_N. In this way the fields of the multilayer structure may be calculated and the appropriate propagation constants found.

For the layers nearest the centre of the waveguide core ($x = 0$) the solutions are oscillatory and may be represented by a linear combination of sine and cosine functions in each layer:

$$\psi_j(x) = A_j \cos\left(\frac{u_j x}{a}\right) + B_j \sin\left(\frac{u_j x}{a}\right) \tag{5.169}$$

where A_j, B_j are unknown coefficients, a is the core half-width, and u_j is given by

$$u_j^2 = a^2(k^2 n_j^2 - \beta^2) \tag{5.170}$$

where β is the propagation eigenvalue to be determined. If the interface between layers j and

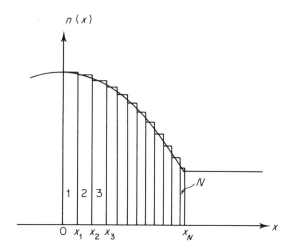

Figure 5.13 Schematic illustration of the stratification approximation for analysing graded-index waveguides

$j+1$ is denoted by x_j, then the continuity of ψ_j at this interface gives:

$$A_j \cos\left(\frac{u_j x_j}{a}\right) + B_j \sin\left(\frac{u_j x_j}{a}\right) = A_{j+1} \cos\left(\frac{u_{j+1} x_j}{a}\right) + B_{j+1} \sin\left(\frac{u_{j+1} x_j}{a}\right). \quad (5.171)$$

For TE modes the derivatives must also be continuous at x_j (an analogous equation results for TM modes in the usual way):

$$u_j\left[-A_j \sin\left(\frac{u_j x_j}{a}\right) + B_j \cos\left(\frac{u_j x_j}{a}\right)\right] = u_{j+1}\left[-A_{j+1} \sin\left(\frac{u_{j+1} x_j}{a}\right) + B_{j+1} \cos\left(\frac{u_{j+1} x_j}{a}\right)\right]. \quad (5.172)$$

Equations (5.171) and (5.172) may be conveniently combined in the matrix form:

$$\begin{pmatrix} \cos\left(\frac{u_j x_j}{a}\right) & \sin\left(\frac{u_j x_j}{a}\right) \\ -u_j \sin\left(\frac{u_j x_j}{a}\right) & u_j \cos\left(\frac{u_j x_j}{a}\right) \end{pmatrix} \begin{pmatrix} A_j \\ B_j \end{pmatrix}$$

$$= \begin{pmatrix} \cos\left(\frac{u_{j+1} x_j}{a}\right) & \sin\left(\frac{u_{j+1} x_j}{a}\right) \\ -u_{j+1} \sin\left(\frac{u_{j+1} x_j}{a}\right) & u_{j+1} \cos\left(\frac{u_{j+1} x_j}{a}\right) \end{pmatrix} \begin{pmatrix} A_{j+1} \\ B_{j+1} \end{pmatrix} \quad (5.173)$$

The solution of this equation for the $(j+1)$th vector in terms of the jth follows by premultiplying by the inverse of the matrix on the r.h.s. The result may be expressed in terms of a transfer matrix T_j:

$$\begin{pmatrix} A_{j+1} \\ B_{j+1} \end{pmatrix} = T_j \begin{pmatrix} A_j \\ B_j \end{pmatrix} \quad (5.174)$$

where the elements of T_j are (first subscript for rows, second subscript for columns):

$$T_{j11} = \cos\left(\frac{u_{j+1} x_j}{a}\right) \cos\left(\frac{u_j x_j}{a}\right) + \frac{u_j}{u_{j+1}} \sin\left(\frac{u_{j+1} x_j}{a}\right) \sin\left(\frac{u_j x_j}{a}\right) \quad (5.175a)$$

$$T_{j12} = \cos\left(\frac{u_{j+1}x_j}{a}\right)\sin\left(\frac{u_j x_j}{a}\right) - \frac{u_j}{u_{j+1}}\sin\left(\frac{u_{j+1}x_j}{a}\right)\cos\left(\frac{u_j x_j}{a}\right) \qquad (5.175b)$$

$$T_{j21} = \sin\left(\frac{u_{j+1}x_j}{a}\right)\cos\left(\frac{u_j x_j}{a}\right) - \frac{u_j}{u_{j+1}}\cos\left(\frac{u_{j+1}x_j}{a}\right)\sin\left(\frac{u_j x_j}{a}\right) \qquad (5.175c)$$

$$T_{j22} = \sin\left(\frac{u_{j+1}x_j}{a}\right)\sin\left(\frac{u_j x_j}{a}\right) + \frac{u_j}{u_{j+1}}\cos\left(\frac{u_{j+1}x_j}{a}\right)\cos\left(\frac{u_j x_j}{a}\right). \qquad (5.175d)$$

Thus for N layers there are $(N-1)$ matrix equations of the form given in (5.174). At the guide centre ($x = 0$) the requirement of even- or odd-order modes leads to:

$$\begin{pmatrix} A_1 \\ B_1 \end{pmatrix} = \begin{pmatrix} 0 \\ 1 \end{pmatrix} \quad \text{(odd-order modes)} \qquad (5.176a)$$

$$\begin{pmatrix} A_1 \\ B_1 \end{pmatrix} = \begin{pmatrix} 1 \\ 0 \end{pmatrix} \quad \text{(even-order modes)} \qquad (5.176b)$$

where the field normalization is such that the amplitudes are unity at $x = 0$.

For the outer layers the fields will be evanescent and the solution must be expressed as a linear superposition of growing and decaying exponentials. In this case the matrix equation (5.174) still holds, but the elements of T_j are now given by:

$$T_{j11} = \frac{e^{(w_j - w_{j+1})x_j/a}}{2}\left(1 + \frac{w_j}{w_{j+1}}\right) \qquad (5.177a)$$

$$T_{j12} = \frac{e^{-(w_j + w_{j+1})x_j/a}}{2}\left(1 - \frac{w_j}{w_{j+1}}\right) \qquad (5.177b)$$

$$T_{j21} = \frac{e^{(w_j + w_{j+1})x_j/a}}{2}\left(1 - \frac{w_j}{w_{j+1}}\right) \qquad (5.177c)$$

$$T_{j22} = \frac{e^{(w_{j+1} - w_j)x_j/a}}{2}\left(1 + \frac{w_j}{w_{j+1}}\right) \qquad (5.177d)$$

where

$$w_j^2 = a^2(\beta^2 - k^2 n_j^2). \qquad (5.178)$$

The existence of two forms of solution—oscillatory and evanescent—implies that for this approximation the caustic separating the two regions will coincide with one specific layer interface at $x_j = x_k$, say. For this interface equation (5.174) again holds true but with the appropriate matrix elements of T_k:

$$T_{k11} = \frac{e^{-w_{k+1}x_k/a}}{2}\left[\cos\left(\frac{u_k x_k}{a}\right) - \frac{u_k}{w_{k+1}}\sin\left(\frac{u_k x_k}{a}\right)\right] \qquad (5.179a)$$

$$T_{k12} = \frac{e^{-w_{k+1}x_k/a}}{2}\left[\sin\left(\frac{u_k x_k}{a}\right) + \frac{u_k}{w_{k+1}}\cos\left(\frac{u_k x_k}{a}\right)\right] \qquad (5.179b)$$

$$T_{k21} = \frac{e^{w_{k+1}x_k/a}}{2}\left[\cos\left(\frac{u_k x_k}{a}\right) + \frac{u_k}{w_{k+1}}\sin\left(\frac{u_k x_k}{a}\right)\right] \qquad (5.179c)$$

$$T_{k22} = \frac{e^{w_{k+1}x_k/a}}{2}\left[\sin\left(\frac{u_k x_k}{a}\right) - \frac{u_k}{w_{k+1}}\cos\left(\frac{u_k x_k}{a}\right)\right] \qquad (5.179d)$$

where we have assumed a simple monotonic decrease of refractive index ($n_k > n_{k+1}$); extension of the method to include more complicated forms of index profile is relatively simple and will not be discussed here.

At the core–cladding interface $x = x_N = a$ the usual continuity condition together with the requirement of a decaying field for $x \to \infty$ yield one further matrix equation:

$$\begin{pmatrix} e^{w_N} & e^{-w_N} \\ w_N e^{w_N} & -w_N e^{-w_N} \end{pmatrix} \begin{pmatrix} A_N \\ B_N \end{pmatrix} = c \begin{pmatrix} 1 \\ -w \end{pmatrix} \tag{5.180}$$

where C is a constant and w is defined in the usual way for cladding layers. If the matrix on the l.h.s. of equation (5.180) is denoted by T_N, and repeated use is made of (5.174), then the eigenvalue equation for the stratification method is found as:

$$(T_N T_{N-1} \ldots T_1) \begin{pmatrix} A_1 \\ B_1 \end{pmatrix} \equiv \left(\prod_{j=1}^{N} T_j \right) \begin{pmatrix} A_1 \\ B_1 \end{pmatrix} = C \begin{pmatrix} 1 \\ -w \end{pmatrix} \tag{5.181}$$

where the definitions of transfer matrices T_j for the various regions are as given above and A_1, B_1 are as in (5.176). Equation (5.181) must then be solved numerically for the propagation constant β, whence the field distributions may be found by repeated use of (5.174). The number N of layers required depends on the details of the specific index profile; values between five (Clarricoats and Chan, 1970) and eight (Suematsu and Furuya, 1972) occur in the literature. For the simple two-layer case ($N = 2$) equations (5.179)–(5.181) yield confirmation of the earlier result (2.225) for even-order TE modes of this structure ($kn_1 \geq \beta \geq kn_2$).

The staircase approximation as discussed above has been applied to planar waveguide structures (Brekhovskikh, 1960; Suematsu and Furuya, 1972), including the case of anisotropic media (Vassell, 1974; Yamanouchi et al., 1978). The method has found rather wider application in the analysis of graded-index optical fibres and we will therefore return to this technique in a later chapter. However, it should be noted at this point that recent numerical solutions (Arnold et al., 1977) have shown the stratification method to be less efficient than other forms of direct solution of the wave equation. Furthermore, Arnold (1977) has provided an explicit proof that this method is formally equivalent to solution by the Euler discretization method (Hildebrand, 1956), but involves substantially greater computation.

5.4.3 Evanescent field theory

Throughout this book we have emphasized the dual nature of ray and mode analyses of waveguide propagation, with the geometrical optics approach seen as an asymptotic representation of the full electromagnetic treatment. Thus we have compared results obtained by the two methods for a range of different waveguide structures. The ray approach may be followed to its logical conclusion by actually computing ray paths and transit times whilst invoking the concept of a phase resonance condition to identify specific ray congruences (Scheggi et al., 1975; Checcacci et al., 1975). However, when a guided mode is represented asymptotically as a ray congruence it is only possible to provide at best a very approximate result for the field distribution of the mode (Keller and Rubinow, 1960); a similar conclusion is reached for first-order WKB theory (see Section 5.2.). The reason for the breakdown of the ray approach in representing modal field distributions is the existence of the caustics where $\beta = kn(r)$ and the nature of the field distribution changes from oscillatory to evanescent. In the vicinity of the caustics the local plane-wave assumption becomes invalid and must be replaced by an improved representation, for example the linearized approximation leading to the WKB connection formulae (see above, Section 5.2).

An alternative approach to the local plane-wave treatment and one which avoids the problems associated with the caustics is furnished by evanescent field theory (for a review, see Felsen, 1976). In this method evanescent waves with complex phase are chosen as the local wave functions thus removing the caustics completely and ensuring continuous field distributions. In other words, the phase which was previously denoted by $S(x)$ in equation (5.56), where it was defined as real, is now permitted to take complex values. In order to take advantage of this assumption let us represent the solution of the scalar wave equation in the form (Choudhary and Felsen, 1978):

$$\psi(x, z) = A(x)e^{ikS(x)} \tag{5.182}$$

where

$$A(x) = \sum_{m=0}^{\infty} A_m(x)k^{-m} \tag{5.183a}$$

$$S(x) = n_1 z + iI(x) + z \sum_{m=1}^{\infty} p_m k^{-m}. \tag{5.183b}$$

In this approach the real amplitude $A(x)$ and the real part of the complex phase have been expanded in asymptotic series of inverse powers of k. The leading term is taken as $n_1 z$ where n_1 is the value of refractive index at $x = 0$; symmetric index profiles $n(x)$ are assumed. $I(x)$ is the imaginary part of the phase and represents the evanescent wave behaviour. The overall philosophy of this representation is similar to the treatment of Gaussian beams in terms of complex rays (Keller and Streifer, 1971; Deschamps, 1971).

The quantities p_m, A_m, and I are to be determined by substituting equations (5.182) and (5.183) into the scalar wave equation:

$$\left[\frac{d^2}{dx^2} + \frac{d^2}{dz^2} + k^2 n^2(x) \right] \psi(x, z) = 0. \tag{5.184}$$

Carrying out this substitution and equating to zero the coefficient of each power of k yields the results:

Terms in k^2:

$$\frac{dI(x)}{dx} = P(x) \tag{5.185}$$

where

$$P^2(x) = n_1^2 - n^2(x); \tag{5.186}$$

Terms in k:

$$\frac{d}{dx}(\ln A_0(x)) = -q(x) \tag{5.187}$$

where

$$q(x) = \frac{1}{2P(x)} \frac{dP(x)}{dx} + \frac{n_1 p_1}{P(x)}; \tag{5.188}$$

Terms in k^{1-m} ($m \geq 1$):

$$\frac{dA_m(x)}{dx} + A_m(x)q(x) = \frac{Q_m(x)}{P(x)} \tag{5.189}$$

where

$$Q_m(x) = \frac{1}{2} \frac{d^2 A_{m-1}(x)}{dx^2} - \sum_{j=2}^{m+1} A_{m-j+1}(x) \left[n_1 p_j + \frac{1}{2} \sum_{n=1}^{j-1} p_n p_{j-n} \right]. \tag{5.190}$$

The first few Q_m's are:

$$Q_1(x) = \frac{1}{2}\frac{d^2 A_0(x)}{dx^2} - \left(n_1 p_2 + \frac{p_1^2}{2}\right)A_0(x) \tag{5.191a}$$

$$Q_2(x) = \frac{1}{2}\frac{d^2 A_1(x)}{dx^2} - \left(n_1 p_2 + \frac{p_1^2}{2}\right)A_1(x) - (n_1 p_3 + p_1 p_2)A_0(x) \tag{5.191b}$$

$$Q_3(x) = \frac{1}{2}\frac{d^2 A_2(x)}{dx^2} - \left(n_1 p_2 + \frac{p_1^2}{2}\right)A_2(x) - (n_1 p_3 + p_1 p_2)A_1(x)$$

$$- \left(n_1 p_4 + p_1 p_3 + \frac{p_2^2}{2}\right)A_0(x). \tag{5.191c}$$

The solutions to equations (5.185)–(5.190) may be written as:

$$I(x) = \int_0^x P(x)\,dx \tag{5.192}$$

$$A_0(x) = \frac{1}{P^{1/2}(x)}\exp\left[-n_1 p_1 \int \frac{dx}{P(x)}\right] \tag{5.193}$$

$$A_m(x) = A_0(x)\int \frac{Q_m(dx)\,dx}{P(x)A_0(x)} \quad (m \geq 1) \tag{5.194}$$

where the normalization of $I(x)$ is such that $I(0) = 0$, whilst the normalization of the amplitude functions $A_m(x)$ is unspecified as yet. In practice the requirement that the functions $A_m(x)$ be analytic at $x = 0$ leads to a set of conditions on the coefficients p_m (Choudhary and Felsen, 1978) which enables the p_m to be evaluated uniquely for each mode.

Choudhary and Felsen (1978) have applied evanescent field theory as outlined above to a number of two-dimensional graded-index waveguides. For the case of an infinitely-extended parabolic medium, they have shown that the solution for the field distribution obtained by this method agrees term by term with the asymptotic expansion of the exact solution as given for example by equation (3.48) in Chapter 3. Similarly, for the symmetric Epstein layer profile discussed in Chapter 4 (see, for example, the form of equation (4.79)), the method yields term-by-term identification with the asymptotic expansion of the exact solution. Having thus confirmed the validity of this approach for these two cases where exact solutions are available, these authors have applied the method to a class of polynomial profiles for which exact results in terms of known functions are not known. In all the examples cited it was possible to integrate equations (5.192)–(5.194) explicitly and thus produce analytic results at least for the first few terms of the series in (5.183). The method therefore represents a novel and potentially useful approach to calculating modal fields in graded-index waveguides.

5.4.4 'Exact' solution for the cladded-parabolic profile

In this subsection we discuss an approach due to Hashimoto (1975, 1976) which is applicable in principle to cladded waveguides of any graded profile, but which is most conveniently illustrated by reference to the familiar cladded-parabolic profile. The method, which is comparable with the 'regularized WKB' theory of Ikuno (1978, 1979), yields results which are exact in the sense that they contain appropriate combinations of independent solutions of the appropriate wave equation chosen in such a way as to also satisfy the boundary conditions at the core–cladding interface. However, in evaluating the eigenvalue equation it is usually necessary to introduce some degree of approximation for the sake of computational convenience.

The method begins with the scalar wave equation for TE modes of a parabolic profile as discussed in Chapter 3 and given by (3.46):

$$\frac{d^2 X}{dx'^2} - 2x'\frac{dX}{dx'} + \left[(k^2 n_1^2 - \beta^2)\frac{w_0^2}{2} - 1\right]X = 0 \qquad (5.195)$$

where

$$x' = \frac{x\sqrt{2}}{w_0}$$

$$w_0 = a\sqrt{\frac{2}{v}}$$

and the conventional notation has been used. The field solution $\psi_N(x)$ for the Nth mode is given in terms of the soultions of (5.195) as:

$$\psi_N(x) = X(x)\exp\left(-\frac{x^2}{w_0^2}\right). \qquad (5.196)$$

Now since (5.195) is recognized as Hermite's differential equation, it follows that one set of solutions is the Hermite polynomials $H_N(x')$ as in Section 3.2, and the fields are given, in unnormalized form, by:

$$\psi_N(x) = H_N\left(\frac{\sqrt{2}x}{w_0}\right)\exp\left(-\frac{x^2}{w_0^2}\right) \qquad (5.197)$$

where, as in (3.47),

$$2N = (k^2 n_1^2 - \beta^2)\frac{w_0^2}{2} - 1 \qquad (N = 0, 1, 2, \ldots). \qquad (5.198)$$

We know also from WKB theory (Section 5.2) that the asymptotic form of the field solution may be represented as, for example,

$$\Psi_N(x) \simeq \frac{C_N}{(u^2 - v^2 y^2)^{1/4}} \cos\left(\int_0^{x/a} (u^2 - v^2 y^2)^{1/2} dy - \frac{N\pi}{2}\right) \qquad (5.199)$$

where C_N is a constant and conventional notation has been used. However, there exists also another solution of the wave equation for a parabolic profile, whose asymptotic representation is:

$$\Phi_N(x) \simeq \frac{C_N}{(u^2 - v^2 y^2)^{1/4}} \sin\left(\int_0^{x/a} (u^2 - v^2 y^2)^{1/2} dy - \frac{N\pi}{2}\right). \qquad (5.200)$$

This solution is not appropriate for the extended parabolic-index medium since it corresponds to the exponentially growing solution in the region beyond the caustic (see Section 5.2). However, for the case of a cladded waveguide, both types of solution $\Psi_N(x)$ and $\Phi_N(x)$ are needed and must be used together in a suitable combination. If the solution $\Psi_N(x)$ is known (as in this case from (5.197)), then $\Phi_N(x)$ may be uniquely determined by recognizing that Ψ_N and Φ_N satisfy the power conservation condition (from (5.199) and (5.200)):

$$\Psi_N(x)\Phi'_N(x) - \Psi'_N(x)\Phi_N(x) = \frac{C_N^2}{a}. \qquad (5.201)$$

The constant C_N required for equation (5.201) may be found by expanding (5.199) as a Taylor series about $x = 0$ (differentiating only the trigonometric part in accordance with the warning in Section 5.2.3):

$$\Psi_N(x) \simeq \frac{C_N}{u^{1/2}} \left[\cos\left(\frac{N\pi}{2}\right) + \frac{xu}{a} \sin\left(\frac{N\pi}{2}\right) \right].$$

It follows that at $x = 0$, C_N is given by:

$$C_N \simeq \begin{cases} u^{1/2} \Psi_N(0), & N \text{ even} \\ \dfrac{a}{u^{1/2}} \Psi'_N(0), & N \text{ odd}. \end{cases} \quad (5.202)$$

Applying this equation to the field solution (5.197), and taking account of (5.198), we find (Hashimoto, 1976):

$$C_N \simeq \begin{cases} (N-1)!! \, (2N+1)^{1/4} v^{1/4}, & N \text{ even} \\ \dfrac{N!! v^{1/4}}{(2N+1)^{1/4}}, & N \text{ odd} \end{cases} \quad (5.203)$$

where $(N-1)!! \equiv (N-1)(N-3)\ldots 1$ and $(-1)!! \equiv 1$.

In order to obtain the associated solution $\Phi_N(x)$, the Hermite function $h_N(x)$ of the second kind is used (Spain and Smith, 1970; Hashimoto, 1975, 1976). This function is a solution of Hermite's differential equation (5.195) which may be expressed in terms of parabolic cylinder functions and is related to the Hermite polynomial $H_N(x')$ as follows:

$$H_N(x')h'_N(x') - H'_N(x')h_N(x') = N! e^{x^2}. \quad (5.204)$$

It follows that in order to satisfy equation (5.201), the appropriate solution is given by

$$\Phi_N(x) = \frac{C_N^2}{N! v^{1/2}} h_N\left(\frac{\sqrt{2}x}{w_0}\right) \exp\left(-\frac{x^2}{w_0^2}\right). \quad (5.205)$$

This result will now be applied to the problem of the cladded-parabolic profile waveguide.

For the cladded profile we anticipate that the propagation constant β will be given by a similar equation to (5.198) with the difference that N may now take non-integer values, i.e.

$$k^2 n_1^2 - \beta^2 = \frac{2}{w_0^2}(2v+1) \quad (5.206)$$

where

$$v = N + \Delta v_N \quad (N = 0, 1, 2, \ldots). \quad (5.207)$$

The functions $\Psi_v(x)$ and $\Phi_v(x)$ for the independent solutions of the wave equation will likewise be analytic continuations of the functions $\Psi_N(x)$, $\Phi_N(x)$, respectively. Their asymptotic forms will be given by equations (5.199) and (5.200) with N replaced by v. It follows that the complete mode function for the cladded profile in the core region may be written as

$$\Psi_v(x)\cos\theta_v - \Phi_v(x)\sin\theta_v \simeq \frac{C_v}{(u^2 - v^2 y^2)^{1/4}} \cos\left(\int_0^{x/a} (u^2 - v^2 y^2)^{1/2}\, dy - \frac{v\pi}{2} + \theta_v\right). \quad (5.208)$$

Since it applies to the symmetric parabolic profile this solution is still required to be an even or odd function according to whether N is even or odd; it follows therefore that

$$-\frac{v\pi}{2} + \theta_v = -\frac{N\pi}{2}$$

or
$$\theta_v = \frac{\pi}{2}\Delta v_N. \tag{5.209}$$

All that remains to completely determine the solution is to ensure that it obeys the usual continuity conditions at the core–cladding interface. When these conditions are applied, assuming the usual exponential solutions in the cladding, the eigenvalue equation for the cladded-parabolic profile is obtained in the form:

$$\tan\left(\frac{\Delta v_N \pi}{2}\right) = \frac{a\Psi'_v + w\Psi_v}{a\Phi'_v + w\Phi_v}. \tag{5.210}$$

Since Δv_N is usually small compared with N, an explicit expression for the eigenvalue may be found by replacing the subscript v by N on the r.h.s. of (5.210). Substituting the appropriate expressions for Φ_N and Ψ_N from (5.205) and (5.197), this gives the result (Hashimoto, 1976):

$$\Delta v_N \simeq \left(\frac{2}{\pi}\right)\tan^{-1}\left\{\left(\frac{N!v^{1/2}}{C_N^2}\right)\frac{NH_{N-1}(v^{1/2}) + [(v-2N-1-2\Delta v_N)^{1/2} - v^{1/2}]H_N(v^{1/2})}{Nh_{N-1}(v^{1/2}) + [(v-2N-1-2\Delta v_N)^{1/2} - v^{1/2}]h_N(v^{1/2})}\right\} \tag{5.211}$$

where the results $H'_N = NH_{N-1}$, $h'_N = Nh_{N-1}$ have been used. Results for the eigenvalues of the cladded-parabolic profile may be obtained directly from equation (5.211) together with (5.203), (5.206), and (5.207). Numerical evaluation shows close agreement with the results of the series solution (Section 5.4.1) except close to cut-off of the lowest-order mode (Adams, 1978) where the approximation of replacing the subscript v by N on the r.h.s. of (5.210) fails. In general, however, the method described in the present subsection represents a very general technique for calculating eigenvalues and eigenfunctions of cladded graded-index waveguides.

Chapter 6
Guides of Rectangular Cross-section

The subject of waveguides with rectangular cross-section represents our first departure from the two-dimensional guides considered in previous chapters. As such it forms a natural extension of the theory discussed earlier in which all the respective components—conducting walls, dielectric slabs, multilayers, metal-clads, graded-index, etc.—have their analogous counterpart with an extra spatial dimension of wave confinement. Thus we consider first the relatively straightforward case of TE and TM mode propagation in rectangular conducting-wall waveguides. The treatment follows the standard microwave approach (see, for example, Collin, 1960) but has relevance also to optical frequencies in view of the recent interest in using metal guides for guidance of 10.6 μm CO_2 radiation (Garmire et al., 1980).

By contrast with metal-wall guides, the modes supported by rectangular dielectric guides are hybrid but essentially of the TEM type, i.e. the axial components of the electric and magnetic fields are very much less than the transverse components. An understanding of the properties of such guides is relevant to many integrated optics components, and this forms the subject of Section 6.2. Structures which are used in integrated optics to provide lateral confinement include the rib waveguide (Goell, 1973), the strip-loaded guide (Furuta et al., 1974), and the metal-clad optical strip line (Yamamoto et al., 1975). It is interesting to note that all these optical waveguide structures have their respective counterparts in the field of acoustic surface waveguides, and a recent review (Oliner, 1976) has served to emphasize the similarities between these subjects.

A third topic which is dealt with in this chapter is that of graded-index waveguides of rectangular cross-section. A typical example of such a structure is a channel waveguide fabricated by diffusion through a mask (Taylor, 1976). Apart from the problem of determining the two-dimensional dielectric profile produced by such a process, the question of calculating the modal propagation characteristics of this structure is extremely difficult. In Section 6.3 some approximate techniques for solving this latter problem are discussed. The increasing use of graded-index guides in integrated optics and new semiconductor laser structures make this topic an important one for further development.

6.1 Rectangular Conducting-wall Waveguides

Consider the structure shown in Figure 6.1, whose cross-section in the x–y plane is rectangular of dimensions $2a \times 2d$; the axis of propagation is in the z-direction. The modes supported by such a guide are either TE or TM; TEM modes cannot propagate in this structure since it is not possible to satisfy the boundary conditions for such modes. In the present section we are therefore concerned with finding the field distributions and eigenvalues for the TE and TM modes. As a starting-point, consider the wave equation (2.76) which, in rectangular Cartesian coordinates and suppressing the time- and z-dependencies,

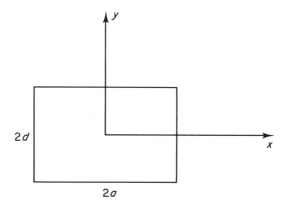

Figure 6.1 The rectangular waveguide with perfectly-conducting walls

may be written:

$$\frac{\partial^2 \psi}{\partial x^2} + \frac{\partial^2 \psi}{\partial y^2} + (k^2 n_1^2 - \beta^2)\psi = 0 \tag{6.1}$$

where $\psi(x, y)$ represents any field component, n_1 is the refractive index of the waveguide core, and the other symbols have their usual meaning.

Assuming separable solutions of the wave equation, in the form

$$\psi(x, y) = X(x)Y(y) \tag{6.2}$$

it follows from (6.1) that

$$\frac{X''}{X} + k_x^2 = 0 \tag{6.3}$$

$$\frac{Y''}{Y} + k_y^2 = 0 \tag{6.4}$$

where

$$k_x^2 + k_y^2 = k^2 n_1^2 - \beta^2 \equiv q^2. \tag{6.5}$$

From (6.5) we have $\beta = (k^2 n_1^2 - q^2)^{1/2}$, so that cut-off occurs for $q = kn_1$ and the cut-off wavelength λ_c may be defined as $\lambda_c = 2\pi/q$. If the guide wavelength λ_g is similarly defined as $\lambda_g = 2\pi/\beta$, then equation (6.5) may be re-written in an alternative form as

$$\frac{n_1^2}{\lambda^2} = \frac{1}{\lambda_g^2} + \frac{1}{\lambda_c^2} \tag{6.6}$$

which relates the free-space, guide, and cut-off wavelengths. The general solutions of equations (6.3) and (6.4) are

$$X = A \cos(k_x x) + B \sin(k_x x) \tag{6.7}$$
$$Y = C \cos(k_y y) + D \sin(k_y y) \tag{6.8}$$

where the values of A–D and k_x, k_y are to be determined from the appropriate boundary conditions.

6.1.1 TE modes

For this case $\mathbf{E} = (E_x, E_y, 0)e^{i\beta z - i\omega t}$, and Maxwell's equations yield:

$$-\beta E_y = \omega \mu_0 H_x \tag{6.9}$$

$$\beta E_x = \omega \mu_0 H_y \tag{6.10}$$

$$\frac{\partial E_y}{\partial x} - \frac{\partial E_x}{\partial y} = i\omega \mu_0 H_z \tag{6.11}$$

$$\frac{\partial H_z}{\partial y} - i\beta H_y = -i\omega \varepsilon E_x \tag{6.12}$$

$$i\beta H_x - \frac{\partial H_z}{dx} = -i\omega \varepsilon E_y \tag{6.13}$$

$$\frac{\partial H_y}{\partial x} - \frac{\partial H_x}{\partial y} = 0. \tag{6.14}$$

From these equations it is a simple matter to express all the field components in terms of H_z:

$$E_x = \frac{\omega \mu_0}{\beta} H_y = \frac{i\omega \mu_0}{q^2} \frac{\partial H_z}{\partial y} \tag{6.15}$$

$$H_x = -\frac{\beta E_y}{\omega \mu_0} = \frac{i\beta}{q^2} \frac{\partial H_z}{\partial x} \tag{6.16}$$

where the solution for H_z in the most general form is given by (6.2), (6.7), and (6.8) as:

$$H_z = [A\cos(k_x x) + B\sin(k_x x)][C\cos(k_y y) + D\sin(k_y y)]e^{i\beta z - i\omega t}. \tag{6.17}$$

The boundary conditions to be satisfied by the fields are that the normal component of the magnetic field and the tangential component of the electric field shall each be zero at the walls at $x = \pm a$ and $y = \pm d$. It follows that $E_x = 0 = H_y$ at $y = \pm d$ and $E_y = 0 = H_x$ at $x = \pm a$. Applying these boundary conditions to equations (6.15)–(6.17) yields the results for the field components (suppressing the factor $e^{i\beta z - i\omega t}$ throughout):

$$H_z = H_0 \cos\left(\frac{N'\pi(x+a)}{2a}\right)\cos\left(\frac{M'\pi(y+d)}{2d}\right) \tag{6.18}$$

$$E_x = \frac{\omega\mu_0}{\beta} H_y = -\frac{i\omega\mu_0 H_0 M'\pi}{2dq^2} \cos\left(\frac{N'\pi(x+a)}{2a}\right)\sin\left(\frac{M'\pi(y+d)}{2d}\right) \tag{6.19}$$

$$H_x = -\frac{\beta}{\omega\mu_0} E_y = -\frac{i\beta H_0 N'\pi}{2aq^2} \sin\left(\frac{N'\pi(x+a)}{2a}\right)\cos\left(\frac{M'\pi(y+d)}{2d}\right) \tag{6.20}$$

where H_0 is a constant and M', N' are integers.* Note, however, that the case $M' = N' = 0$ is not physically meaningful since this mode would not transport power in the z-direction. Hence the lowest order modes are TE_{01} and TE_{10}. Also, using these results in (6.5), the eigenvalue equation for the longitudinal propagation constant β is given by:

$$\beta^2 = k^2 n_1^2 - \left(\frac{N'\pi}{2a}\right)^2 - \left(\frac{M'\pi}{2d}\right)^2 \quad (M' = 0, 1, 2, \ldots; N' = 0, 1, 2, \ldots; M' \neq N' = 0). \tag{6.21}$$

---

* Note that M', N' count the number of field extrema in the y- and x-directions, respectively; these correspond to $M+1$, $N+1$ in other earlier notation, where M, N count the number of zeros in the field distribution.

6.1.2 TM modes

For this case $\mathbf{H} = (H_x, H_y, 0)e^{i\beta z - i\omega t}$, and Maxwell's equations yield:

$$\frac{\partial E_z}{\partial y} - i\beta E_y = i\omega\mu_0 H_x \tag{6.22}$$

$$i\beta E_x - \frac{\partial E_z}{\partial x} = i\omega\mu_0 H_y \tag{6.23}$$

$$\frac{\partial E_y}{\partial x} - \frac{\partial E_x}{\partial y} = 0 \tag{6.24}$$

$$-i\beta H_y = -i\omega\varepsilon E_x \tag{6.25}$$

$$i\beta H_x = -i\omega\varepsilon E_y \tag{6.26}$$

$$\frac{\partial H_y}{\partial x} - \frac{\partial H_x}{\partial y} = -i\omega\varepsilon E_z. \tag{6.27}$$

From these equations all the field components may be expressed in terms of E_z:

$$E_x = \frac{\beta}{\omega\varepsilon} H_y = \frac{i\beta}{q^2} \frac{\partial E_z}{\partial x} \tag{6.28}$$

$$H_x = -\frac{\omega\varepsilon}{\beta} E_y = -\frac{i\omega\varepsilon}{q^2} \frac{\partial E_z}{\partial y} \tag{6.29}$$

where the general solution for E_z again assumes the form of the r.h.s. of equation (6.17).
For the TM modes the boundary conditions are that $E_x = 0 = H_y$ at $y = \pm d$ and $E_y = 0 = H_x$ at $x = \pm a$, as for the TE modes. In this case these conditions imply $dE_z/dx = 0$ at $y = \pm d$ and $dE_z/dy = 0$ at $x = \pm a$; in addition, the requirement of zero tangential field at the walls also implies $E_z = 0$ at $x = \pm a$ and $y = \pm d$. Applying these conditions leads to the following expressions for the field components (again suppressing the time- and z-dependent factors):

$$E_z = E_0 \sin\left(\frac{N'\pi(x+a)}{2a}\right) \sin\left(\frac{M'\pi(y+d)}{2d}\right) \tag{6.30}$$

$$E_x = \frac{\beta}{\omega\varepsilon} H_y = \frac{i\beta N'\pi}{2aq^2} E_0 \cos\left(\frac{N'\pi(x+a)}{2a}\right) \sin\left(\frac{M'\pi(y+d)}{2d}\right) \tag{6.31}$$

$$H_x = -\frac{\omega\varepsilon}{\beta} E_y = -\frac{i\omega\varepsilon M'\pi E_0}{2dq^2} \sin\left(\frac{N'\pi(x+a)}{2a}\right) \cos\left(\frac{M'\pi(y+d)}{2d}\right) \tag{6.32}$$

where E_0 is a constant and $M' = 0, 1, 2, \ldots$, $N' = 0, 1, 2, \ldots$, $M' \neq N' = 0$. The propagation constant β is again given by equation (6.21).

6.1.3 Modal characteristics

The eigenvalue equation (6.21) may be expressed in terms of normalized variables (cf. chapter 1):

$$v^2 = a^2 k^2 n_1^2 \tag{6.33}$$

$$b = \beta^2/k^2 n_1^2 \tag{6.34}$$

when the result becomes

$$b = 1 - \left(\frac{\pi}{2v}\right)^2 \left[N'^2 + \left(\frac{M'a}{d}\right)^2\right] \quad (M' = 0, 1, 2, \ldots; N' = 0, 1, 2, \ldots; M' \neq N' = 0).$$
(6.35)

Plots of b versus v for the first few modes in the case $d/a = 2$ are given in Figure 6.2. In general,

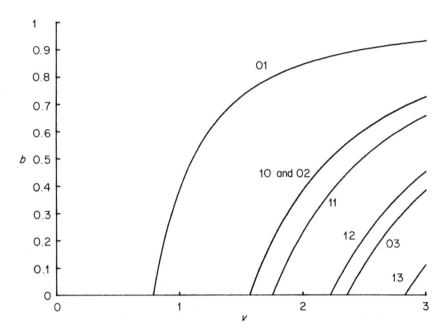

Figure 6.2 Normalized propagation constant b versus v for the rectangular guide with conducting walls and $d/a = 2$; labelling parameters give the values of mode indices (N', M')

the results are similar to those for the two-dimensional case (Figure 1.8); note the degeneracy of the (0, 2) and (1, 0) modes for the case $d/a = 2$. For $d > a$ the lowest-order mode is that with $N' = 0$, $M' = 1$; for $d < a$ the lowest-order mode is that with $N' = 1$, $M' = 0$.

From (6.21) it follows that the phase velocity V_p is given by:

$$V_p = \frac{\omega}{\beta} = \frac{ck}{\beta} \tag{6.36}$$

and that the group velocity V_g is given by:

$$V_g = \frac{d\omega}{d\beta} = c\frac{dk}{d\beta} = \frac{c\beta}{kn_1^2}. \tag{6.37}$$

As a consequence, the characteristic relationship

$$V_p V_g = \frac{c^2}{n_1^2} \tag{6.38}$$

is found to be satisfied for the rectangular waveguide with conducting walls.

6.2 Rectangular Dielectric Waveguides

Figure 6.3 shows the cross-section of a rectangular dielectric waveguide of dimensions $2a \times 2d$, whose core (refractive index n_1) is surrounded on four sides by materials of indices n_2–n_5. An exact analytical solution for such a structure is not possible; instead recourse must be made to various forms of approximation or numerical analysis (Schlosser and Unger, 1966; Goell, 1969). One such approximation is to consider the case of a structure whose modes are guided sufficiently strongly that there is little field penetration into the four cladding regions. It is then reasonable to assume that there is even less penetration into the four corner regions (shown shaded on Figure 6.3), so that fields in these regions may be ignored (Marcatili, 1969). This approximation will clearly be at its most accurate for modes well above cut-off. Under these circumstances it yields results which are sufficiently accurate for most purposes and give good agreement with computer calculations (Goell, 1969). If the further assumption is made that the cladding indices n_2–n_5 are only slightly smaller than the

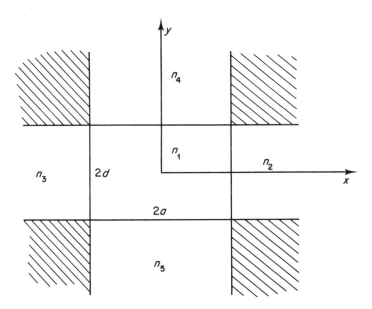

Figure 6.3 The rectangular dielectric waveguide

core index n_1 then the modal eigenvalue problem for the guide of Figure 6.3 becomes similar to that for two slab waveguides at right angles. This approach is discussed in Section 6.2.1.

A significant improvement in the accuracy of calculated results is obtained by a further approximation—the so-called effective index method (Knox and Toulios, 1970) described in Section 6.2.2. According to this method the equivalent slab guide in one direction is solved first and the results used to generate an effective refractive index for the core region. This effective index is then used in the solution of the other slab waveguide problem in the direction at right-angles to the first. In this way the solutions of the two slab guides are coupled, which was not the case in the former approach. The results of the effective index method give better agreement with numerical results in general and are valid over a larger range, even for modes reasonably close to cut-off. The method has also been applied to a wide range of optical guiding structures which may be described by the generic name of 'slab-coupled guides' (Marcatili, 1974). The properties of some of these structures form the subject of Section 6.2.3. Hollow waveguides of rectangular cross-section are dealt with in Section 6.2.4.

6.2.1 Approximate modal analysis

Let us seek separable solutions of the form (6.2) for the wave equation in each of the regions 1–5 of Figure 6.3. Without any assumptions about the polarization of the fields, Maxwell's equations yield:

$$i\omega\mu_0 H_x = \frac{\partial E_z}{\partial y} - i\beta E_y \tag{6.39}$$

$$i\omega\mu_0 H_y = i\beta E_x - \frac{\partial E_z}{\partial x} \tag{6.40}$$

$$i\omega\mu_0 H_z = \frac{\partial E_y}{\partial x} - \frac{\partial E_x}{\partial y} \tag{6.41}$$

$$-i\omega\varepsilon_j E_x = \frac{\partial H_z}{\partial y} - i\beta H_y \tag{6.42}$$

$$-i\omega\varepsilon_j E_y = i\beta H_x - \frac{\partial H_z}{\partial x} \tag{6.43}$$

$$-i\omega\varepsilon_j E_y = \frac{\partial H_y}{\partial x} - \frac{\partial H_x}{\partial y} \tag{6.44}$$

where the subscript j may refer to each of the regions 1–5 and the refractive index n_j, and permittivities ε_j are related by $\varepsilon_j = n_j^2 \varepsilon_0$. From these equations it is possible to write all the field components in terms of the longitudinal fields E_z and H_z:

$$E_x = \frac{i}{h_j^2}\left(\omega\mu_0 \frac{\partial H_z}{\partial y} + \beta \frac{\partial E_z}{\partial x}\right) \tag{6.45}$$

$$E_y = \frac{i}{h_j^2}\left(\beta \frac{\partial E_z}{\partial y} - \omega\mu_0 \frac{\partial H_z}{\partial x}\right) \tag{6.46}$$

$$H_x = \frac{i}{h_j^2}\left(\beta \frac{\partial H_z}{\partial x} - \omega\varepsilon_j \frac{\partial E_z}{\partial y}\right) \tag{6.47}$$

$$H_y = \frac{i}{h_j^2}\left(\omega\varepsilon_j \frac{\partial E_z}{\partial x} + \beta \frac{\partial H_z}{\partial y}\right) \tag{6.48}$$

where $h_j^2 = k^2 n_j^2 - \beta^2$ and the longitudinal field components E_z, H_z satisfy the scalar wave equation (6.1). Owing to our neglect of the shaded regions of Figure 6.3, the boundary conditions now involve only the matching of the tangential field components along the four sides of the core region, i.e. at $x = \pm a$, $y = \pm d$.

The modes supported by this waveguide are of two types. The first type, denoted E^x_{NM} (Marcatili, 1969), is polarized principally in the x-direction; the other type, denoted E^y_{NM}, is polarized principally in the y-direction. In each case the subscripts N, M give the number of field zeros in the x- and y-directions. By analogy with the familiar slab guides we anticipate sinusoidal solutions for the fields within the core and evanescent fields in the four cladding regions. Within region 1, therefore, the general longitudinal field components will take the form of equation (6.17). Let us consider first the case of E^x_{NM} modes; suppressing the time- and

z-dependence, a general solution for E_z is (Marcuse, 1974):

$$E_z = A \cos[k_x(x+\xi)] \cos[k_y(y+\eta)] \quad (6.49)$$

where A is a constant and ξ, η are general phase parameters. Since there is a degree of freedom in the choice of phase and amplitude coefficients, we may choose:

$$H_x = 0. \quad (6.50)$$

It follows from (6.47) that H_z is then given by:

$$H_z = -\frac{A\omega\varepsilon_1 k_y}{\beta k_x} \sin[k_x(x+\xi)] \sin[k_y(y+\eta)]. \quad (6.51)$$

From equations (6.45)–(6.48) the remaining field components are:

$$E_x = -\frac{iA(n_1^2 k^2 - k_x^2)}{\beta k_x} \sin[k_x(x+\xi)] \cos[k_y(y+\eta)] \quad (6.52)$$

$$E_y = \frac{iA k_y}{\beta} \cos[k_x(x+\xi)] \sin[k_y(y+\eta)] \quad (6.53)$$

$$H_y = \frac{iA\omega\varepsilon_1}{k_x} \sin[k_x(x+\xi)] \cos[k_y(y+\eta)] \quad (6.54)$$

where

$$h_1^2 = k^2 n_1^2 - \beta^2 = k_x^2 + k_y^2. \quad (6.55)$$

Now if the refractive indices n_2–n_5 are only a little smaller than n_1 (cf. the 'weakly-guiding' approximation of Chapter 2), then $\beta \simeq kn_1$, and equation (6.55) implies

$$k_x, k_y \ll \beta. \quad (6.56)$$

These inequalities when applied to (6.53) imply $E_y \ll E_z$, whilst applying them to (6.52) implies $E_z \ll E_x$. Hence it becomes clear that under these conditions E_y may be neglected as being of second order in (k_y/β), so that the E_{NM}^x modes are predominantly polarized in the x-direction. Similar considerations apply to the field components in regions 2–5. In these regions the component E_z is given by (Marcuse, 1974):

Region 2: $\quad E_z = A \cos[k_x(a+\xi)] \cos[k_y(y+\eta)] \exp[-\gamma_2(x-a)] \quad (6.57)$

Region 3: $\quad E_z = A \cos[k_x(\xi-a)] \cos[k_y(y+\eta)] \exp[\gamma_3(x+a)] \quad (6.58)$

Region 4: $\quad E_z = \frac{A\varepsilon_1}{\varepsilon_4} \cos[k_x(x+\xi)] \cos[k_y(d+\eta)] \exp[-\gamma_4(y-d)] \quad (6.59)$

Region 5: $\quad E_z = \frac{A\varepsilon_1}{\varepsilon_5} \cos[k_x(x+\xi)] \cos[k_y(\eta-d)] \exp[\gamma_5(y+d)]. \quad (6.60)$

The other field components are given by equations (6.45)–(6.48) and the condition $H_x = 0$; the decay parameters γ_j are given by

$$h_j^2 = k^2 n_j^2 - \beta^2 = k_y^2 - \gamma_j^2 \quad (j = 2, 3) \quad (6.61)$$

$$h_j^2 = k^2 n_j^2 - \beta^2 = k_x^2 - \gamma_j^2 \quad (j = 4, 5). \quad (6.62)$$

In equations (6.57)–(6.60) the coefficients were chosen so that E_z is continuous at $x = \pm a$; similarly the strong field component E_x is continuous at $y = \pm d$. It is not possible, even within this limited approximation, to match all the required field components on the boundaries and E_z is *not* continuous at $y = \pm d$ except in the limit $\varepsilon_1 = \varepsilon_4 = \varepsilon_5$. The eigenvalue equations are obtained by matching the remaining tangential field components on the appropriate boundaries. Matching H_z (and hence H_y) at $x = \pm a$ (for a more complete discussion, see Marcuse (1974)), we obtain:

$$\tan(2k_x a) = \frac{n_1^2 k_x (\gamma_2 n_3^2 + \gamma_3 n_2^2)}{(n_2^2 n_3^2 k_x^2 - n_1^4 \gamma_2 \gamma_3)}. \tag{6.63}$$

Similarly, continuity of H_z at $y = \pm d$ yields:

$$\tan(2k_y d) = \frac{k_y(\gamma_4 + \gamma_5)}{k_y^2 - \gamma_4 \gamma_5}. \tag{6.64}$$

From (6.55), (6.61), and (6.62), it follows that:

$$\gamma_j^2 = (n_1^2 - n_j^2)k^2 - k_x^2 \quad (j = 2, 3) \tag{6.65}$$

$$\gamma_j^2 = (n_1^2 - n_j^2)k^2 - k_y^2 \quad (j = 4, 5). \tag{6.66}$$

Equation (6.63) is clearly recognizable as the eigenvalue equation for TM modes of a dielectric slab in the x-direction (cf. equation (2.87)). This is intuitively obvious when we recall that $E_y \simeq 0$ and $E_x \gg E_z$, so that the mode is polarized principally with the electric field normal to the interface at $x = \pm a$. Similarly, Equation (6.64) corresponds to the case of TE modes of a dielectric slab in the y-direction (cf. equation (2.82)). The eigenvalue problem is now reduced to the solution of two transcendental equations (6.63) and 6.64 for k_x, k_y, respectively, whence the propagation constant β is given from (6.55) as:

$$\beta^2 = k^2 n_1^2 - k_x^2 - k_y^2. \tag{6.67}$$

The E_{NM}^y modes can be derived in strict analogy to the E_{NM}^x field distributions merely by changing E to H and μ_0 to $-\varepsilon$, and vice versa. In the case of the E_{NM}^y modes, the only nonzero components (to first-order) are E_z, H_z, E_y, and H_x. The eigenvalue equations are given by:

$$\tan(2k_x a) = \frac{k_x(\gamma_2 + \gamma_3)}{k_x^2 - \gamma_2 \gamma_3} \tag{6.68}$$

$$\tan(2k_y b) = \frac{n_1^2 k_y (\gamma_4 n_5^2 + \gamma_5 n_4^2)}{(n_4^2 n_5^2 k_y^2 - n_1^4 \gamma_4 \gamma_5)} \tag{6.69}$$

where the γ's are again given by (6.65) and (6.66), and β is defined by (6.67).

For the rectangular dielectric waveguide considered here, several methods of numerical solution are also available (Schlosser, 1964; Schlosser and Unger, 1966; Goell, 1969; Pregla, 1974; Ogusu, 1977) which may be used to test the accuracy of the approximate analysis presented above. In particular, detailed comparisons have been published (Marcatili, 1969) with the results of Goell (1969). Goell's method consists of expanding the electromagnetic fields, both inside and outside the core region, in terms of circular harmonics; the expansion coefficients are then determined by matching the fields at selected points on the core–cladding boundary. Plots of the resulting field distributions and eigenvalues may be found in Goell's paper. When compared with the results of the above approximate analysis it is found that the agreement of propagation constant is very good for each mode far above cut-off. Close to cut-off, however, the approximation fails to predict the correct behaviour; the results become progressively closer in agreement for higher mode numbers and for structures where the aspect ratio d/a departs further from unity.

The approximate analysis presented above has been extended to include rectangular guides fabricated in anisotropic media by Steinberg and Giallorenzi (1977). It has also been applied to the analysis of stripe-geometry injection lasers, allowing for complex dielectric permittivities, by Cross and Adams (1972) and Voges (1974). Further comparisons of predictions from the approximate analysis have also been made with the results of a variational analysis (Pregla, 1974), the finite element technique (Yeh et al., 1975), the method of generalized telegraphist's equations (Ogusu, 1977), and perturbation theory (Eyges, 1978).

For the special case of dielectric symmetry $n_2 = n_3 = n_4 = n_5$, the results of this subsection may be cast into normalized form. For the case of weak guidance when $(n_1 - n_2) \ll n_1$ the E_{NM}^x and E_{NM}^y modes become degenerate and the pairs of equations (6.63), (6.68) and (6.64), (6.69) become respectively identical. If we define:

$$v_x^2 = a^2 k^2 (n_1^2 - n_2^2) \tag{6.70}$$

$$v_y^2 = d^2 k^2 (n_1^2 - n_2^2) \equiv v_x^2 (d/a)^2 \tag{6.71}$$

$$u_x^2 = a^2 k_x^2 \tag{6.72}$$

$$u_y^2 = d^2 k_y^2 \tag{6.73}$$

$$w_x^2 = a^2 \gamma_2^2 = a^2 \gamma_3^2 \equiv v_x^2 - u_x^2 \tag{6.74}$$

$$w_y^2 = d^2 \gamma_4^2 = d^2 \gamma_5^2 \equiv v_y^2 - u_y^2 \tag{6.75}$$

then the eigenvalue equations are given by:

$$\tan 2u_x = \frac{2 u_x w_x}{u_x^2 - w_x^2} \tag{6.76}$$

$$\tan 2u_y = \frac{2 u_y w_y}{u_y^2 - w_y^2}. \tag{6.77}$$

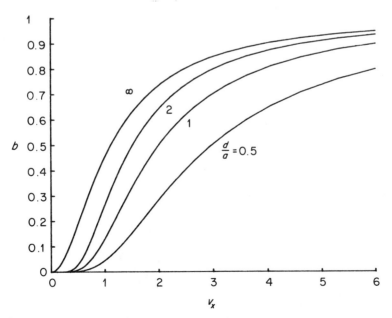

Figure 6.4 b versus v_x for the lowest-order mode of the weakly-guiding rectangular dielectric waveguide, calculated by the method of Section 6.2.1; labelling parameter gives the value of aspect ratio d/a

Finally, the normalized propagation constant $b = (\beta^2 - k^2 n_2^2)/(k^2 n_1^2 - k^2 n_2^2)$ is given from (6.67) as:

$$b = 1 - \frac{[u_x^2 + u_y^2 (a/d)^2]}{v_x^2} \tag{6.78}$$

and these normalized results for each mode are then only functions of v_x and the parameter d/a. Some typical plots of b versus v_x calculated by this approximation for the lowest-order (0, 0) mode of the weakly-guiding symmetric rectangular waveguide are shown in Figure 6.4 for three values of the aspect ratio d/a. For comparison the slab-guide result ($d/a \to \infty$) is also shown on the figure. More detailed plots of this type together with comparisons of the present approximation with Goell's computer solutions may be found in Marcatili's 1969 paper.

6.2.2 The effective index method

In order to obtain an improvement in the accuracy of the results determined by the approximate method of Section 6.2.1, the effective index method may be used (Knox and Toulios, 1970). This approach takes as its starting point the equivalence of the approximate rectangular waveguide problem of Section 6.2.1 to the problem of two slab guides (one in the x-direction and the other in the y-direction), as illustrated in Figure 6.5. Hence we consider first the guiding action in the x-direction as shown in Figure 6.5(a); for the case of E_{NM}^x modes the eigenvalue equation of the rectangular guide corresponds to that for TM modes of this slab, and is given by equation (6.63) with γ_2, γ_3 defined as in (6.65). The solution of this equation, k_x, may be used to define an effective index, n_{eff}, for the other slab guide shown in Figure 6.5(b), as follows:

$$n_{\text{eff}}^2 = n_1^2 - \left(\frac{k_x}{k}\right)^2. \tag{6.79}$$

The eigenvalue equation for the guide of Figure 6.5(b) is then given, as in (6.64), by the appropriate TE mode equation:

$$\tan(2k_y d) = \frac{k_y (\gamma_{4e}^2 + \gamma_{5e}^2)}{k_y^2 - \gamma_{4e} \gamma_{5e}} \tag{6.80}$$

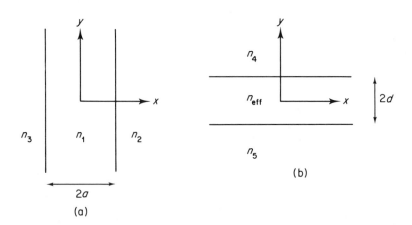

Figure 6.5 Schematic illustration of the effective index method as applied to the rectangular dielectric guide: (a) guidance in the x-direction; (b) guidance in the y-direction

where
$$\gamma_{je}^2 = (n_{\text{eff}}^2 - n_j^2)k^2 - k_y^2 \quad (j = 4, 5). \tag{6.81}$$

Finally, the solution for k_y is used to find the propagation constant, β:
$$\beta^2 = k^2 n_{\text{eff}}^2 - k_y^2. \tag{6.82}$$

Note that equations (6.79) and (6.82) are equivalent to (6.67) in the earlier approximation.

For the case of E_{NM}^y modes, an entirely analogous procedure is possible, beginning with (6.68) for the TE modes of the first slab, and using the resultant k_x to define n_{eff} as in (6.79). The second guide then obeys the eigenvalue equation

$$\tan(2k_y d) = \frac{n_{\text{eff}}^2 k_y (\gamma_{4e} n_5^2 + \gamma_{5e} n_4^2)}{(n_4^2 n_5^2 k_y^2 - n_{\text{eff}}^4 \gamma_{4e} \gamma_{5e})} \tag{6.83}$$

where γ_{4e}, γ_{5e} are defined as in (6.81). Similarly, β is given as for the E_{NM}^x modes by an equation like (6.82).

In general, the results calculated by the effective index approach are in closer agreement with those found by numerical solutions than was the case for the method of Section 6.2.1. This is especially noticeable for the region close to cut-off, particularly for low-order modes (Knox and Toulios, 1970; Hocker and Burns, 1977). For the case of dielectric symmetry $n_2 = n_3 = n_4 = n_5$ and weak-guidance $(n_1 - n_2) \ll n_1$, the results may be expressed in normalized form, using the definitions of (6.70), (6.72), and (6.74). The eigenvalue equation for u_x is given, as in Section 6.2.1, by:

$$\tan 2u_x = \frac{2u_x w_x}{u_x^2 - w_x^2}. \tag{6.84}$$

If we now make the definitions

$$v_{ye}^2 = d^2 k^2 (n_{\text{eff}}^2 - n_2^2) = \left(\frac{d}{a}\right)^2 w_x^2 \tag{6.85}$$

$$u_y^2 = d^2 k_y^2 \tag{6.86}$$

$$w_{ye}^2 = d^2 \gamma_{4e}^2 = d^2 \gamma_{5e}^2 \equiv v_{ye}^2 - u_y^2 \tag{6.87}$$

then the eigenvalue equation for u_y is again of the form

$$\tan 2u_y = \frac{2u_y w_{ye}}{u_y^2 - w_{ye}^2}. \tag{6.88}$$

Similarly, the result for normalized propagation constant b is given in terms of the solutions of (6.84) and (6.88) by equation (6.78).

Figure 6.6 shows plots of b versus v_x for the lowest-order mode of the weakly-guiding rectangular dielectric waveguide calculated by the effective index approach as outlined above. To facilitate comparison with the results of the approximate method described in Section 6.2.1, the values of aspect ratio chosen in Figure 6.6 are the same as those for Figure 6.4. Note the different nature of the curves close to cut-off where the present approximation yields closer agreement with the results of Goell's numerical analysis. Further comparison of the results of the effective index method with those of Marcatili's approximation and Goell's computer analysis may be found elsewhere in the literature (Knox and Toulios, 1970; Hocker and Burns, 1977; Unger, 1977).

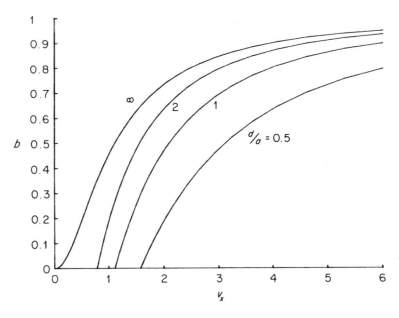

Figure 6.6 b versus v_x for the lowest-order mode of the weakly-guiding rectangular dielectric waveguide, calculated by the effective index method; labelling parameter gives the value of aspect ratio d/a

6.2.3 Slab-coupled guides

In the previous two subsections we were concerned with methods of solution for the idealized rectangular dielectric guide whose cross-section is shown in Figure 6.3. In the present subsection, however, we will deal with rather more specific examples of the rectangular guide, and in particular with guides which are of interest for applications in integrated optics. Consider first the four possible waveguide cross-sections illustrated in Figure 6.7; all of these guides can be fabricated by the conventional techniques of masked etching, sputtering, diffusion, or ion implantation, although in some cases there may be gradual variations of refractive index rather than the abrupt transitions shown (see Section 6.3). Figure 6.7(a) shows the raised strip guide consisting, in the simplest case, of a strip of high-index (n_1) material upon a substrate (index n_2), surmounted by a superstrate (index n_0) which may be air ($n_0 = 1$). In the case of Figure 6.7(b) the strip of material to which radiation is largely confined is actually embedded in the substrate. From a theoretical point of view the only difference between (a) and (b) lies in the likely values of refractive indices n_1, n_2, n_0; in the cases most commonly encountered for waveguides in integrated optics, n_1 and n_2 are likely to be somewhat similar (frequently there is the weakly-guiding situation $(n_1 - n_2) \ll n_1$), whilst n_0 is most probably rather smaller (as in the case of air when $n_0 = 1$). In each case the properties of the waveguides may be analysed using the approaches given in Sections 6.2.1 and 6.2.2.

A practical difficulty associated with the structures of Figure 6.7(a) and (b) is the requirement for accurate control of the edge walls of the guides; in order to avoid excessively high scattering losses from these guides, the walls must be extremely smooth. This requirement is relaxed somewhat for the structures shown in Figure 6.7(c) and (d)—the strip-loaded (Furuta et al., 1974) and rib (Goell, 1973) waveguides respectively. These guides may not be treated directly by the methods discussed above in relation to Figure 6.3, and they form examples of the class of slab-coupled guides to be discussed here. We will apply the effective-

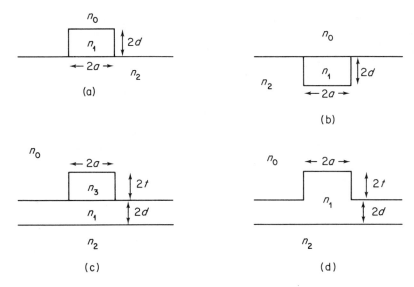

Figure 6.7 Cross-sections of some slab-coupled waveguides: (a) raised strip guide; (b) embedded strip guide; (c) strip-loaded guide; (d) rib guide

index approach to the strip-loaded and rib waveguides in order to understand their operation in more detail.

Consider first the strip-loaded guide (or optical stripline) shown in Figures 6.7(c) and 6.8(a); applying the effective index method to this structure in the spirit of Section 6.2.2 (see also McLevige et al., 1975), we begin with the equivalent guides in the y-direction shown in

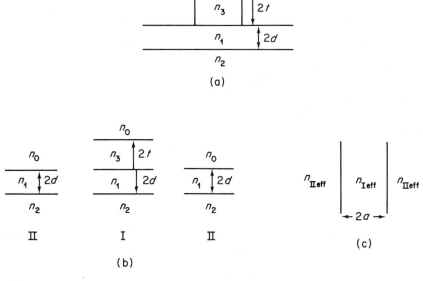

Figure 6.8 The strip-loaded guide treated by the effective index method: (a) the strip-loaded guide; (b) three equivalent guides in the y-direction; (c) resultant effective guide in the x-direction

Figure 6.8(b). Corresponding to the strip region there is a four-layer waveguide (I) in the y-direction, whose solution may be formulated in terms of an effective index n_{Ieff}, defined analogously to that in equation (6.79). In the regions outside the strip there are three-layer waveguides (II), whose solutions may be written similarly in terms of an effective index n_{IIeff}. Finally in Figure 6.8(c) is shown the effective symmetric slab waveguide formed by the x-direction guiding structure, whose core has index n_{Ieff} and claddings have index n_{IIeff}.

For the case $n_1 > n_3$ (low-index strip), the four-layer guide I of Figure 6.8(b) falls into the category already considered in Section 2.6.1. Making the appropriate changes of notation and restricting attention to the case of an evanescent wave in the strip layer (n_3), i.e. $kn_1 \geqslant \beta \geqslant kn_3$ the eigenvalue equation (2.204) may be written for TE modes:

$$2dh_1 = M\pi + \tan^{-1}\left(\frac{h_2}{h_1}\right) + \tan^{-1}\left\{\frac{h_3''}{h_1}\tanh\left[\tanh^{-1}\left(\frac{h_0}{h_3''}\right) + 2th_3''\right]\right\} \quad (M = 0, 1, 2, \ldots) \tag{6.89}$$

where

$$\left.\begin{array}{l} h_1^2 = k^2 n_1^2 - \beta^2 \\ h_2^2 = \beta^2 - k^2 n_2^2 \\ h_3''^2 = \beta^2 - k^2 n_3^2 \\ h_0^2 = \beta^2 - k^2 n_0^2 \end{array}\right\}. \tag{6.90}$$

Expanding the argument of the second arctangent function in equation (6.89) enables the equation to be written in the alternative form (Ramaswamy, 1974a):

$$2dh_1 = M\pi + \tan^{-1}\left(\frac{h_2}{h_1}\right) + \tan^{-1}\left(\frac{h_3''}{h_1}\frac{1 - \eta e^{-4h_3''t}}{1 + \eta e^{-4h_3''t}}\right) \quad (M = 0, 1, 2, \ldots) \tag{6.91}$$

where

$$\eta = \frac{h_3'' - h_0}{h_3'' + h_0}. \tag{6.92}$$

The effective index n_{Ieff} is obtained by numerical solution of equation (6.91) for the propagation constant β, and setting $n_{\text{Ieff}} = \beta/k$. The corresponding value n_{IIeff} for guide II of Figure 6.8(b) is found by solving (6.91) for the limiting case $t \to 0$, when the equation reduces to that for the asymmetric 3-layer slab considered in Section 2.3.

For a numerical example, let us take the case where the strip layer possesses the same refractive index as the substrate, i.e. $n_3 = n_2$, $h_3'' = h_2$. Then equation (6.91) may be re-written in terms of normalized variables as:

$$2u_0 = M\pi + \tan^{-1}\left(\frac{w_0}{u_0}\right) + \tan^{-1}\left(\frac{w_0}{u_0}\frac{1 - \eta e^{-4w_0 t/d}}{1 + \eta e^{-4w_0 t/d}}\right) \quad (M = 0, 1, 2, \ldots) \tag{6.93}$$

where

$$\left.\begin{array}{l} u_0^2 = d^2 h_1^2 = d^2(k^2 n_1^2 - \beta^2) \\ w_0^2 = d^2 h_2^2 = d^2(\beta^2 - k^2 n_2^2) \\ \eta = \dfrac{w_0 - w_0'}{w_0 + w_0'} \\ w_0'^2 = d^2 h_0^2 = d^2(\beta^2 - k^2 n_0^2) \end{array}\right\}. \tag{6.94}$$

For consistency with earlier work we may define a normalized frequency v_0 and asymmetry parameter c_0 such that:

$$\left.\begin{array}{l} v_0^2 = k^2 d^2 (n_1^2 - n_2^2) \\ c_0^2 = \dfrac{n_1^2 - n_0^2}{n_1^2 - n_2^2} \\ w_0^2 = v_0^2 - u_0^2 \\ w_0'^2 = v_0^2 c_0^2 - u_0^2 \end{array}\right\} \quad (6.95)$$

The solutions of equation (6.93) may then be expressed as functions of v_0 in terms of only two parameters, t/d and c_0, for each mode. Note that in the limiting case of an infinitely thick cover strip ($t \to \infty$), equation (6.93) reduces to

$$2u_0 = M\pi + 2\tan^{-1}\left(\frac{w_0}{u_0}\right) \quad (M = 0, 1, 2, \ldots) \quad (6.96)$$

which is the appropriate eigenvalue equation for a symmetric 3-layer slab. Similarly, for the limit $t \to 0$, we find the corresponding result for an asymmetric 3-layer slab:

$$2u_0 = M\pi + \tan^{-1}\left(\frac{w_0}{u_0}\right) + \tan^{-1}\left(\frac{w_0'}{u_0}\right) \quad (M = 0, 1, 2, \ldots) \quad (6.97)$$

Numerical results for the solutions of equation (6.93) for the lowest order mode are illustrated in Figure 6.9 in the form of plots of b_I versus v_0, where b_I is defined as w_0^2/v_0^2. The results were calculated for the case $c_0^2 = 4$ and $t/d = 0, 1, 10, 100$; the curve for $t/d = 0$ yields the appropriate values of b_II for the guide II of Figure 6.8(b). Clearly, at each value of v_0, the curve for b_II lies below those for b_I so that a guiding action is produced in the region beneath the strip. The strength of this guiding action may be expressed in terms of an effective normalized frequency, v_eff, defined (from Figure 6.8(c)) as:

$$v_\text{eff}^2 = a^2 k^2 (n_\text{Ieff}^2 - n_\text{IIeff}^2). \quad (6.98)$$

This quantity is easily expressed in terms of the propagation constants b_I, b_II as (Kogelnik, 1975):

$$v_\text{eff}^2 = a^2 k^2 (n_1^2 - n_2^2)(b_\text{I} - b_\text{II}) \equiv v^2 (b_\text{I} - b_\text{II}) \quad (6.99)$$

where the r.h.s. of (6.99) has been expressed in terms of v, as conventionally defined. Hence values of b_I and b_II may be read from the curves of Figure 6.9 and used to define v_eff; the corresponding result for the symmetric slab guide of Figure 6.8(c) may then be taken immediately from Figure 2.8 where the normalized solutions of this waveguide are presented graphically.

Apart from the effective index approach used here, other treatments of the strip-loaded guide have also appeared in the literature. Marcatili (1974) has analysed the general class of slab-coupled guides by an 'equivalent guide' method where each of the component guides I and II of Figure 6.8(b) are replaced by an equivalent slab guide whose width is chosen appropriately. The results of this method have been compared with those of the effective index method and also with those from a variational technique by Ohtaka et al. (1976). The numerical results from the variational technique tend to give closer agreement with the effective index results than with those from Marcatili's method, thus giving some degree of confidence in the accuracy of the analytic approximation presented above. The variational results have in turn been confirmed by application of a finite element technique (Yeh et al., 1979). An alternative numerical treatment has been given by Butler et al. (1976), who assumed

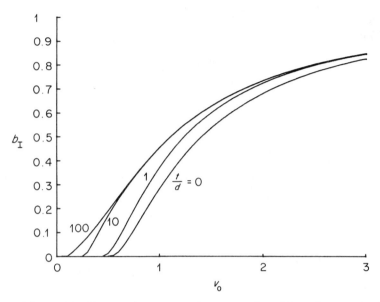

Figure 6.9 Propagation constant b_1 versus v_0 found from the numerical solution of equation (6.93) for the lowest-order mode of the four-layer slab described in the text. Parameters are $c_0^2 = 4$ (defined as in (6.95)) and $t/d = 0, 1, 10, 100$

a continuum of plane waves in the guiding layer and substrate (layers of index n_1 and n_2, respectively, in Figure 6.7(c)); transverse confinement then occurs via spatial interference effects. Numerical results (using a finite number of plane waves) gave good agreement with experimentally-observed field profiles in the lateral direction on GaAs optical striplines.

The effective index method has been extended to deal with the case of complex permittivities in the strip-loaded guide and applied to channelled-substrate-planar (CSP) laser structures by Aiki et al. (1978). In these structures the lasing mode is confined laterally to a region overlying a channelled substrate; this confinement is principally due to excess absorption loss outside the channel region, and the effective index method provides a useful aid to designing appropriate dimensions of guides for stable single-mode operation (Kuroda et al., 1978).

In the discussion given above of the strip-loaded guide, we confined attention to the case of a low-index strip, i.e. $n_3 < n_1$ in Figure 6.8(a). The opposite case of a high-index loading strip has been considered by Uchida (1976). This situation leads to oscillatory solutions in regions 1 and 3 of guide I in Figure 6.8(b), so that the appropriate 4-layer eigenvalue equation is now given by (2.202). Making the appropriate substitutions, the TE mode eigenvalue equation becomes:

$$2dh_1 = M\pi + \tan^{-1}\left(\frac{h_2}{h_1}\right) + \tan^{-1}\left\{\frac{h_3}{h_1}\tan\left[\tan^{-1}\left(\frac{h_0}{h_3}\right) - 2th_3\right]\right\} \quad (M = 0, 1, 2, \ldots) \tag{6.100}$$

where

$$\left.\begin{array}{l} h_1^2 = k^2 n_1^2 - \beta^2 \\ h_3^2 = k^2 n_3^2 - \beta^2 \\ h_0^2 = \beta^2 - k^2 n_0^2 \\ h_2^2 = \beta^2 - k^2 n_2^2 \end{array}\right\}. \tag{6.101}$$

Solutions of equation (6.100) may be obtained numerically for given values of the indices and guide thicknesses (Uchida, 1976; Uchida et al., 1976); they show a remarkable dependence of propagation constant β on the loading strip thickness t for some values of the other parameters. This effect may be understood qualitatively by noting that the loading strip with $n_3 > n_1$ itself may form a waveguide capable of confining a mode with an evanescent behaviour in layer 1. The cut-off condition for the zero-order mode of this elementary waveguide is given by

$$\tan^{-1}\left(\frac{h_0}{h_3}\right) = 2th_3 \qquad (6.102)$$

which corresponds to zero argument of the second $\tan^{-1}$ function on the r.h.s. of equation (6.100). Hence we may expect large changes in the variation of β with t for values close to those satisfying equation (6.102). This effect provides a means of controlling the field confinement and power distribution in the strip-loaded guide by accurate control of the strip thickness t. In particular if the parameters can be arranged such that the optical field is principally concentrated close to the interface of the strip and the guiding layer, then it has been suggested that this would improve the modulation efficiency of a planar modulator whose structure includes a loading film (Uchida et al., 1976). Another application of a high-index strip-loaded guide has been the development of the strip buried heterostructure (SBH) laser (Tsang et al., 1978; Tsang and Logan, 1979). In this application the regions 0, 3, 1, 2 of Figure 6.8(a) are respectively p-$Al_{0.3}Ga_{0.7}As$, p-GaAs, n-$Al_{0.1}Ga_{0.9}As$, and n-$Al_{0.3}Ga_{0.7}As$. The strip serves to control the effective index change along the junction plane and hence to encourage stable fundamental mode operation of the laser.

Turning to the rib waveguide illustrated in Figure 6.7(d), it is clear from the discussion of the strip-loaded guide that this structure may be treated as a special case of the latter with $n_3 = n_1$. The effective index treatment then consists of treating the rib-region as a simple three-layer slab and proceeding otherwise as for the case of the strip-loaded guide. Rib waveguides may be fabricated as asymmetric (Reinhart et al., 1974; Shelton et al., 1979) or symmetric (Lee et al., 1975) structures, i.e. with $n_0 \neq n_2$ or $n_0 = n_2$, respectively. A particular case of interest is the symmetric structure designed for single mode operation. Such a structure has been used for a stripe laser (Lee et al., 1975) and for the single material optical fibre (Kaiser et al., 1973; Kaiser and Astle, 1974); the cross-section of the single-material fibre is shown in Figure 6.10. It consists of a rib waveguide enclosed in a capillary which serves the purpose merely of supporting and protecting the guiding structure, i.e. optical confinement is provided by the rib. It is worth investigating, at least approximately, the nature of the

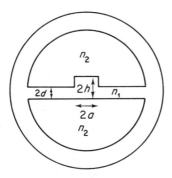

Figure 6.10 Cross-section of the single-material optical fibre

conditions for single-mode operation of such a structure. In the case of the single-material fibre, the material is usually silica, of refractive index $n_1 \simeq 1.5$, whilst the outer material is air, i.e. $n_2 = n_0 = 1$. It follows that the fields are usually strongly confined to the guiding structure in the y-direction. Hence as a zero-order approximation to the eigenvalue problem we may use the condition $v \gg 1$ so that in region I of the equivalent effective guide,

$$u_1 \simeq \frac{(M+1)\pi}{2} \quad (M = 0, 1, 2, \ldots) \tag{6.103}$$

where $u_1{}^2 = h^2 k^2 (n_1{}^2 k^2 - \beta^2)$, where $2h$ is the rib height (see Figure 6.10). A similar approximation holds in region II of the equivalent effective guide. Hence, from equation (6.99) the effective normalized frequency is given for zero-order modes ($M = 0$) by:

$$v_{\text{eff}}{}^2 = a^2 \left(\frac{\pi}{2}\right)^2 \left[\frac{1}{d^2} - \frac{1}{h^2}\right] \tag{6.104}$$

where $2a$ is the rib width and $2d$ the thickness of the supporting slab. For a single transverse mode we require $v_{\text{eff}} \leq \pi/2$ (see equation (2.115)) so that from equation (6.104) the condition for monomode operation becomes (Kaiser et al., 1973):

$$\frac{1}{a^2} \geq \frac{1}{d^2} - \frac{1}{h^2}. \tag{6.105}$$

A further result which follows is that the numerical aperture, NA, of the single-mode single material fibre, defined as the sine of the maximum acceptance angle (see Chapter 7), is given by (Kaiser et al., 1973):

$$\text{NA} = (n_{\text{Ieff}}{}^2 - n_{\text{IIeff}}{}^2)^{1/2} = \left(\frac{\pi}{2k}\right)\left(\frac{1}{d^2} - \frac{1}{h^2}\right)^{1/2}$$

$$\simeq \frac{\pi}{2ka} = \frac{\lambda}{4a} \tag{6.106}$$

where the equality in equation (6.105) has been used to obtain the maximum equivalent numerical aperture in (6.106). For example, for a square-core fibre with $2a = 2h = 7\ \mu\text{m}$, equation (6.105) says that the support width $2d \geq 4.95\ \mu\text{m}$; at $\lambda = 0.633\ \mu\text{m}$ the effective numerical aperture of such a fibre is given by (6.106) as 0.045.

More detailed analyses of the single-material fibre and rib waveguides than the simple theory given above have also appeared in the literature. Marcatili (1974) has applied an equivalent slab model to these structures to obtain improved estimates for the single-mode condition, the effective numerical aperture, field distributions, and other properties. Ohtaka et al. (1976) have used a variational method of analysis to calculate field distributions and propagation constants and have compared the latter with Marcatili's results and those of the effective index method; as for the case of the strip-loaded guide, the numerical results tend to confirm those found from the effective index approach. Further numerical comparisons have been performed by Bird (1977), using a finite element analysis; these results indicate some departure from those of Marcatili (1974) in the region of cut-off, but with increasing agreement as the optical fields become more closely confined to the guide. Rütze (1977) has presented an alternative method of numerical analysis for the rib guide by enclosing the structure with conducting walls; the fields are then expanded in terms of the modes of the rectangular dielectric loaded waveguide. Further numerical results have been obtained for the rib guide by Marcuse (1974a) using a point-matching technique, by Voges (1976) from a variational approach, and in the case of trapezoidal cross-section guides by Yasuura et al.

(1980) and Miyamoto (1980) by a mode-matching method, by Pelosi et al. (1978) who used the finite element method, and by Gallagher (1979) using the effective index approach.

It is important to realize that even in the region of single-mode operation considered above for the single-material fibre, the mode can still exist in two different polarizations, i.e. there are in fact two fundamental modes denoted by E_{00}^x and E_{00}^y corresponding respectively to polarization principally in the x and y directions. The existence of these two modes is of relevance to the possible use of a single-material fibre for transmitting information over long distances, since the different modal transit times may lead to substantial pulse spreading. A measure of this pulse spreading is given by the difference $\Delta \tau$ between transit times of the two modes E_{00}^x and E_{00}^y. This has been calculated by Rütze (1977) and Petermann (1976) and shown to vary as $(kh)^{-3}$. In other words, the modal delay differences varies as the inverse third power of the rib height and can thus be made small by using sufficiently high ribs, subject of course to the single-mode condition (6.105). A further complication arising from the presence of the two polarizations of the lowest order mode is the possibility of leakage effects. In the above analysis we have always assumed only one mode polarization, TE or TM, present in the component guides of the rib structure. However, on reflection on the rib interfaces each such mode can also excite a reflected and transmitted curve of the other polarization. This TE-TM coupling can lead to leakage of power from the rib, and the magnitude of this effect has been calculated by Peng and Oliner (1977); the loss increases with increase of rib height and decreases as the rib width increases (see also Ogusu et al., 1979).

Apart from the strip-loaded guide, the rib guide, and the single-material optical fibre, there is one other waveguide which may conveniently be discussed under the classification of slab-coupled guiding structures. This is the metal-clad optical strip line whose structure is illustrated in Figure 6.11. It consists of a guiding layer of index n_1 above a substrate of index n_2, and partially clad by a metallic layer of complex index n_3. The strip left uncovered by the metal has the effect of confining radiation to the area beneath the strip since the effective mode index is larger there than in regions beneath the metal cladding. This guide is clearly amenable therefore to analysis by the effective index method and corresponds to the strip-loaded structure of Figure 6.8 in the special case of an infinitely thick strip layer ($t \to \infty$). The complex refractive index of the metal cladding implies an absorption loss for each propagating mode (see Section 2.5 for a discussion of metal-clad guides). Numerical results for propagation constants, losses, and far-field patterns have been calculated using this method by Yamamoto et al. (1975). The results were compared with experimental measurements on a structure composed of SiO_2 ($n_2 = 1.48$), Al_2O_3 ($n_1 = 1.68$), Al ($n_3 = 1.2 - i7$), and air ($n_0 = 1$). Good agreement was found between measured and predicted far-field patterns and likewise between experimental and theoretical values for coupling angle for E_{00}^y and E_{00}^x modes. From the theory the propagation loss is expected to decrease rapidly with

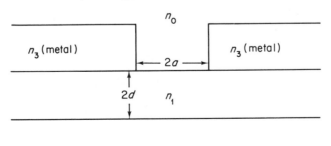

Figure 6.11 Cross-section of the metal-clad optical strip line

increasing strip width ($2a$), and experimentally the losses were found to be below 0.5 dB/cm at He–Ne wavelength for $2a = 50\,\mu$m.

Before leaving this subsection it is worth noting that the methods of analysis discussed above may easily be extended to the corresponding case of multilayer guides (in both x- and y-directions). For example, the effects of a thin overlying film on guiding structures for use in integrated optics have been analysed by Vincent and Lit (1976). Similarly, detailed studies of new semiconductor laser structures, including the effects of all layers present, have been made with the aid of these simple techniques (Rozzi et al., 1977, 1978; Delaney and Butler, 1979).

6.2.4 Hollow guides

We have already mentioned in Section 2.5 the uses of planar hollow waveguides for waveguide lasers and for guidance of CO_2 laser radiation. In the present subsection we will extend some of the previous work to include rectangular structures by considering the waveguide cross-section shown in Figure 6.12. In this structure the core region is of dimensions $2a \times 2d$ and has the permittivity ε_0 of free space. In order to make the analysis reasonably general, the layers on either side of the core in the x-direction are assumed to have dielectric permittivity ε_2 and those in the y-direction to have permittivity ε_3. We thus allow the confining layers to be dielectrics or metals or a pair of each type of material. We will analyse this structure using Marcatili's (1969) approximation that the radiation penetrating the shaded regions of Figure 6.12 may be ignored. Assuming, in addition, that $ka \gg 1$ and $kd \gg 1$, we ensure that the modes are sufficiently far away from cut-off that Marcatili's approximation is valid and that the modes are approximately of the types E^x_{NM} or E^y_{NM} (see Section 6.2.1).

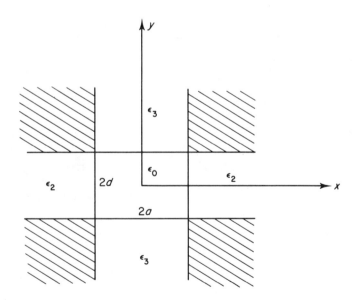

Figure 6.12 Cross-section of the hollow rectangular waveguide

With the definitions and assumptions given above, the fields of the rectangular hollow waveguide may be written in the forms given in Section 6.2.1, with appropriate allowance being made for the symmetry of the present structure and the possibility of complex permittivities ε_2, ε_3. It follows that the eigenvalue equations for E^x_{NM} modes are given from (6.63)–6.67) as:

$$\tan(2k_x a) = \frac{2\varepsilon_0 \varepsilon_2 k_x \gamma_2}{\varepsilon_2^2 k_x^2 - \varepsilon_0^2 \gamma_2^2} \quad (6.107)$$

$$\tan(2k_y d) = \frac{2k_y \gamma_3}{k_y^2 - \gamma_3^2} \quad (6.108)$$

where

$$\gamma_2^2 = (1 - v_2^2)k^2 - k_x^2 \quad (6.109)$$
$$\gamma_3^2 = (1 - v_3^2)k^2 - k_y^2 \quad (6.110)$$
$$v_j^2 = \frac{\varepsilon_j}{\varepsilon_0} \quad (j = 2, 3) \quad (6.111)$$

and the propagation constant β is finally related to k_x, k_y as:

$$\beta^2 = k^2 - k_x^2 - k_y^2. \quad (6.112)$$

With the assumptions that $ka \gg 1$, $kd \gg 1$, it follows that the modes are far above cut-off and we may use the same zero-order approximate solutions for these equations as for the planar case in Section 2.5.3, i.e.

$$\left. \begin{array}{l} k_x a \simeq \left(\dfrac{N+1}{2}\right)\pi \\[6pt] k_y d \simeq \left(\dfrac{M+1}{2}\right)\pi \end{array} \right\}. \quad (6.113)$$

It follows that the first-order approximate solutions to equations (6.107) and (6.108) are given from (2.184) as:

$$ak_x \simeq \frac{(N+1)\pi}{2}\left[1 - \frac{v_2^2}{ak(1-v_2^2)^{1/2}}\right] \quad (6.114)$$

$$dk_y \simeq \frac{(M+1)\pi}{2}\left[1 - \frac{1}{dk(1-v_3^2)^{1/2}}\right]. \quad (6.115)$$

Hence from (6.112) with $\beta = \beta_r + i\beta_i$, confining attention to the low loss case $|\beta_r| \gg |\beta_i|$,

$$\beta_r \simeq k\left\{1 - \frac{\pi^2}{2}\left(\frac{N+1}{2ak}\right)^2\left[1 - \frac{2}{ak}\text{Re}\left(\frac{v_2^2}{(1-v_2^2)^{1/2}}\right)\right]\right.$$
$$\left. - \frac{\pi^2}{2}\left(\frac{M+1}{2dk}\right)^2\left[1 - \frac{2}{dk}\text{Re}\left(\frac{1}{(1-v_3^2)^{1/2}}\right)\right]\right\} \quad (6.116)$$

$$\beta_i \simeq \left(\frac{N+1}{2}\right)^2 \frac{\pi^2}{a^3 k^2}\text{Im}\left(\frac{v_2^2}{(1-v_2^2)^{1/2}}\right) + \left(\frac{M+1}{2}\right)^2 \frac{\pi^2}{d^3 k^2}\text{Im}\left(\frac{1}{(1-v_3^2)^{1/2}}\right). \quad (6.117)$$

Equations (6.116) and (6.117) are approximate relations for the real and imaginary parts of propagation constant for the E^x_{NM} mode of the rectangular hollow waveguide, for the case far above cut-off ak, $dk \gg 1$. Similar equations may analogously be derived for the E^y_{NM} modes under the same conditions. Field components for these modes may be derived from the equations given in Section 6.2.1 suitably modified for the hollow guide. The approximate results are conveniently collected together for ease of reference in Table 6.1 (Krammer, 1976; Laakmann and Steier, 1976). When the mode indices N, M are even (odd) the modes are of the familiar symmetric (antisymmetric) forms. Note that the imaginary part of the propagation constant (β_i) consists in each case of a sum of the β_i-values for the TE and TM modes in the corresponding hollow slab guides of width $2a$, $2d$ (cf. Section 2.5). Similarly the real part of the propagation constant (β_r) contains terms relating to the modes of these slab guides. In the limit of a square, symmetrically-clad guide, i.e. $a = d$, $\varepsilon_2 = \varepsilon_3$, the E^x_{NM} and E^y_{NM} modes are degenerate, corresponding to the analogous case of TE/TM mode degeneracy in hollow slab waveguides.

From the above analysis of the rectangular guide we may state the conditions on dielectric constants for the approximate solutions of Table 6.1 to hold. These may be simply derived

Table 6.1 Approximate core field components and eigenvalues for the hollow rectangular waveguide. (Note: (i) field components expressed as $\simeq 0$ are strictly zero only to order λ/a; (ii) C_{NM} is normalization constant)

E^x_{NM} modes

$$E_x = -H_y \left(\frac{\mu_0}{\varepsilon_0}\right)^{1/2} = C_{NM} \begin{cases} \cos\left(\frac{(N+1)\pi x}{2a}\right)\cos\left(\frac{(M+1)\pi y}{2d}\right); & M, N \text{ even} \\ \sin\left(\frac{(N+1)\pi x}{2a}\right)\sin\left(\frac{(M+1)\pi y}{2d}\right); & M, N \text{ odd} \end{cases}$$

$H_x = 0$; $E_y \simeq 0$; $E_z \simeq 0$; $H_z \simeq 0$

$$\beta_r \simeq k\left\{1 - \frac{\pi^2}{2}\left(\frac{N+1}{2ak}\right)^2\left[1 - \frac{2}{ak}\text{Im}\left(\frac{v_2^2}{(v_2^2-1)^{1/2}}\right)\right] - \frac{\pi^2}{2}\left(\frac{M+1}{2dk}\right)^2 \times \left[1 - \frac{2}{dk}\text{Im}\left(\frac{1}{(v_3^2-1)^{1/2}}\right)\right]\right\}$$

$$\beta_i \simeq \left(\frac{N+1}{2}\right)^2 \frac{\pi^2}{a^3k^2}\text{Re}\left(\frac{v_2^2}{(v_2^2-1)^{1/2}}\right) + \left(\frac{M+1}{2}\right)^2 \frac{\pi^2}{d^3k^2}\text{Re}\left(\frac{1}{(v_3^2-1)^{1/2}}\right)$$

E^y_{NM} modes

$$E_y = H_x \left(\frac{\mu_0}{\varepsilon_0}\right)^{1/2} = C_{NM} \begin{cases} \cos\left(\frac{(N+1)\pi x}{2a}\right)\cos\left(\frac{(M+1)\pi y}{2d}\right); & M, N \text{ even} \\ \sin\left(\frac{(N+1)\pi x}{2a}\right)\sin\left(\frac{(M+1)\pi y}{2d}\right); & M, N \text{ odd} \end{cases}$$

$E_x = 0$; $H_y \simeq 0$; $H_z \simeq 0$; $E_z \simeq 0$

$$\beta_r \simeq k\left\{1 - \frac{\pi^2}{2}\left(\frac{N+1}{2ak}\right)^2\left[1 - \frac{2}{ak}\text{Im}\left(\frac{1}{(v_2^2-1)^{1/2}}\right)\right] - \frac{\pi^2}{2}\left(\frac{M+1}{2dk}\right)^2 \times \left[1 - \frac{2}{dk}\text{Im}\left(\frac{v_3}{(v_3^2-1)^{1/2}}\right)\right]\right\}$$

$$\beta_i \simeq \left(\frac{N+1}{2}\right)^2 \frac{\pi^2}{a^3k^2}\text{Re}\left(\frac{1}{(v_2^2-1)^{1/2}}\right) + \left(\frac{M+1}{2}\right)^2 \frac{\pi^2}{d^3k^2}\text{Re}\left(\frac{v_3}{(v_3^2-1)^{1/2}}\right)$$

from inspection of the r.h.s. of equations (6.107) and (6.108) for the E^x_{NM} modes as follows:

$$\left. \begin{array}{r} |v_3^2 - 1|^{1/2} \gg \dfrac{(M+1)\lambda}{4d} \\ \\ \dfrac{|v_2^2 - 1|^{1/2}}{v_2} \gg \dfrac{(N+1)\lambda}{4a} \end{array} \right\}. \quad (6.118)$$

For the E^y_{NM} modes, the corresponding conditions required for the approximate solutions given in the table are (Laakmann and Steier, 1976):

$$\left. \begin{array}{r} |v_2^2 - 1|^{1/2} \gg \dfrac{(N+1)\lambda}{4a} \\ \\ \dfrac{|v_3^2 - 1|^{1/2}}{v_3} \gg \dfrac{(M+1)\lambda}{4d} \end{array} \right\}. \quad (6.119)$$

Subject to these conditions, the approximate formulae of Table 6.1 may be applied to guides whose walls are composed of metals or dielectrics. In particular, the formulae have been used to estimate losses in rectangular hollow guides designed to transmit CO_2 laser radiation (Garmire, 1976). Such guides have already been discussed in some detail in Section 2.5 where the planar version was analysed. Another major area for application of these results is for the waveguide laser of rectangular cross-section (Abrams and Bridges, 1973). They have been applied by Laakmann and Steier (1976) to estimate modal attenuation for CO_2 waveguide lasers with walls of SiO_2, BeO, and Al_2O_3. For a guide with two walls of BeO and two walls of aluminium, the lowest-loss mode is polarized perpendicular to the metal walls and has a loss of approximately 0.016 dB/m. The coupling between a hollow rectangular waveguide and external spherical mirrors in such a laser has been calculated by Avrillier and Verdonck (1977); optimal mirror positions and radii may be found by such computations. The extension of the treatment outlined above to the case of a hollow rectangular waveguide with anisotropic surrounding media has been given by Chaudhuri and Paul (1978).

6.3 Three-dimensional Graded-index Waveguides

A convenient fabrication method for three-dimensional optical waveguides consists of diffusion of impurities into a substrate through a mask. The result is a guide whose refractive index is graded in both x and y directions normal to the direction of propagation. The analysis of such structures forms the subject of Section 6.3.1 where we consider first the somewhat idealized situation of a graded-index guide in the y-direction directly beneath the mask used to define the guiding region, sandwiched between layers of homogeneous material. The more general case of two-dimensional diffusion and hence index-grading is considered subsequently. We will emphasize again the convenience and power of the effective-index approach for treating these structures; in addition some details will be given of a variational method which may also be applied when more detailed results are required, especially close to modal cut-off.

In Section 6.3.2 we will be concerned with another form of three-dimensional guide suitable for integrated optics applications—that formed by adding a loading strip on top of a diffused planar guide in order to provide transverse confinement. Finally, in Section 6.3.3 we return to the subject of stripe-geometry semiconductor lasers to see how the effective permittivity approach may be applied to analysis of these devices.

6.3.1 Diffused channel waveguides

Consider first the case of one-dimensional diffusion which produces a structure like that shown in Figure 6.13(a). Here a region of width $2a$ is assumed to have a graded variation of refractive index characterized by some depth $2d$ as shown in Figure 6.13(b); on either side of this region, uniform material of refractive index n_2 is assumed. Such a structure has been analysed using Marcatili's (1969) method by Hammer (1976); here, however, we will apply the effective index method (Hocker and Burns, 1977). To take a specific example for the purposes of illustration we will consider an exponential index profile in the channel region, although other likely profiles may include Gaussian and complementary error functions; these have also been analysed in the literature (Hocker and Burns, 1975). Hence in the region $-a \leqslant x \leqslant a$ we assume the form for $n(y)$ given by analogy with equation (4.1):

$$n^2(y) = \begin{cases} n_1^2[1 - 2\Delta(1 - e^{y/d})], & y \leqslant 0 \\ n_0^2, & y > 0 \end{cases} \qquad (6.120)$$

where there is strong asymmetry, i.e. $n_0^2 \ll n_1^2$. For this guide, the normalized frequency is defined in the usual way as

$$v_0 = dkn_1(2\Delta)^{1/2} \qquad (6.121)$$

and the effective mode index b_1 follows by the geometric optics theory from (4.7):

$$(1 - b_1)^{1/2} - b_1^{1/2} \cos^{-1}(b_1^{1/2}) = \frac{\pi}{4v_0}(2M + \tfrac{3}{2}) \quad (M = 0, 1, 2, \ldots). \qquad (6.122)$$

To apply the effective index method in this case we assume the equivalent planar slab guide in the x-direction shown in Figure 6.13(c) consisting of a layer, width $2a$, index n_{eff}, between

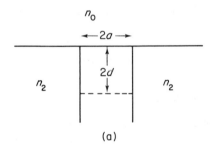

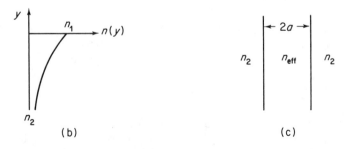

Figure 6.13 (a) Diffused channel waveguide with no sideways diffusion; (b) index profile in the channel region; (c) equivalent planar effective guide in the x-direction

two uniform layers of index n_2 ($= n_1(1-2\Delta)^{1/2}$). The effective index n_{eff} is defined as

$$n_{\text{eff}}^2 = b_1 n_1^2 + (1-b_1)n_2^2 \tag{6.123}$$

and the corresponding modes of the equivalent x-direction guide are given in the usual way by the results from Chapter 2. In particular, the guide is characterized by an effective normalized frequency v_{eff} and a propagation constant b, defined as:

$$v_{\text{eff}}^2 = a^2 k^2 (n_{\text{eff}}^2 - n_2^2) = a^2 k^2 (n_1^2 - n_2^2) b_1 \tag{6.124}$$

$$b = \frac{(\beta/k)^2 - n_2^2}{n_1^2 - n_2^2} \tag{6.125}$$

where β is the longitudinal propagation constant. Hence, to treat a specific case: the value of b_1 is obtained from the numerical solution of (6.122) or, equivalently, from the normalized plots of Figure 4.2. This is then used to define v_{eff} in equation (6.124), whence b may be found from solution of the appropriate eigenvalue equation (2.114) or from the plots of Figure 2.8.

For the mode cut-offs, we apply the condition $b = 0$ in (6.125) and find (from equation (2.115)):

$$v_{\text{eff}} = \frac{N\pi}{2} \quad (N = 0, 1, 2, \ldots). \tag{6.126}$$

Hence (6.124), (6.126), and (6.122) may be combined to yield an equation for the cut-off of the mode (N, M):

$$v_0 = \frac{\pi(2M + \frac{3}{2})/4}{[1 - (N\pi/2v)^2]^{1/2} - (N\pi/2v)\cos^{-1}(N\pi/2v)} \tag{6.127}$$

where

$$v = ak(n_1^2 - n_2^2)^{1/2} = \frac{a}{d} v_0. \tag{6.128}$$

The results of equation (6.127) are expressed graphically in Figure 6.14 as cut-off lines for each mode on a plot of v_0 against v. Note that in using this approach we have assumed the weak-guidance situation, $2\Delta \ll 1$, so that E^x_{NM} and E^y_{NM} modes are approximately degenerate. Hence, although the above analysis applies strictly only for TE modes in the equivalent slab, it remains approximately valid for both polarizations.

Turning to the case of two-dimensional diffusion through a mask, the first question concerns the distribution of diffusant in the substrate. If the diffusant concentration is denoted by $C(x, y)$, then this is given by the solution of the diffusion equation (Crank, 1956):

$$D\left(\frac{\partial^2 C}{\partial x^2} + \frac{\partial^2 C}{\partial y^2}\right) = \frac{\partial C}{\partial t} \tag{6.129}$$

where D is the diffusion constant and the substrate occupies the region $y \leq 0$ as in Figure 6.13(a). The boundary conditions are that the region $-a \leq x \leq a, y = 0$ possesses a constant concentration C_0 and that at time $t = 0$, $C = 0$ everywhere in the substrate. The solutions of equation (6.129) may then be found numerically in general (see, for example, Taylor, 1976; Kennedy and Murley, 1966). However, if instead we consider an instantaneous-source diffusion process in which the diffusion source is removed after $t = 0$, then an analytic solution to equation (6.129) may be obtained by integrating the solution for a line source over the strip $-a \leq x \leq a$. The result gives (Kennedy and Murley, 1966; Hocker and Burns, 1977):

$$C(x, y) = \frac{C_0}{2d\sqrt{\pi}} e^{-y^2/d^2} \left[\text{erf}\left(\frac{a+x}{d}\right) + \text{erf}\left(\frac{a-x}{d}\right)\right] \tag{6.130}$$

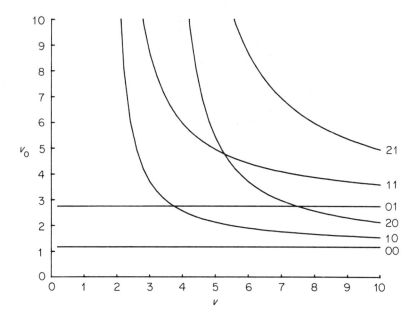

Figure 6.14 Cut-off lines for the modes of the diffused channel guide of Figure 6.13; labelling parameters give the values of mode indices (N, M)

where erf is the error function (Abramowitz and Stegun, 1964) and d is the diffusion length given by

$$d = 2(Dt)^{1/2}. \tag{6.131}$$

Equation (6.130) may be normalized to the value at $x = 0$, $y = 0$, when we find:

$$\frac{C(x, y)}{C(0, 0)} = \frac{e^{-y^2/d^2}}{2 \, \text{erf}(a/d)} \left[\text{erf}\left(\frac{a+x}{d}\right) + \text{erf}\left(\frac{a-x}{d}\right) \right]. \tag{6.132}$$

Some representative contours for $C(x, y)/C(0, 0)$ are given on a plot of y/d versus x/d in Figure 6.15. The result (6.132) should be a good approximation for diffusion from a finite strip source for times t much larger than that taken for all the source material to diffuse into the substrate. If the dielectric permittivity is assumed to be a linear function of diffusant concentration, then we may write:

$$n^2(x, y) = n_2^2 + (n_1^2 - n_2^2) C(x, y)/C(0, 0) \tag{6.133}$$

where n_2 is the bulk index of the substrate and n_1 is the surface index at $x = 0$, $y = 0$. Hence the curves of Figure 6.15 also give the contour lines of the refractive index distribution. Curves of this form have been shown to describe adequately the case of Ti-diffused LiNbO$_3$ guides (Fukuma et al., 1978).

One simplifying feature of the form of index distribution given by equations (6.132) and (6.133) is that it is separable in x and y, i.e. we may re-write (6.132) in the form

$$\frac{C(x, y)}{C(0, 0)} = f\left(\frac{y}{d}\right) g\left(\frac{x}{a}\right) \tag{6.134}$$

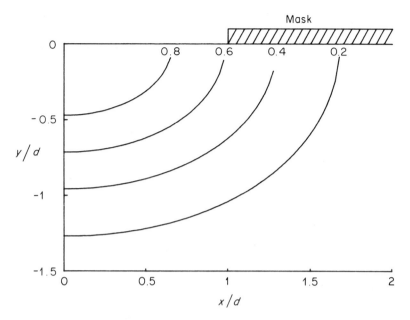

Figure 6.15 Contour lines of equal diffusant concentration from equation (6.132) in the case $a/d = 1$. Labelling parameter gives the value of $C(x, y)/C(0, 0)$

so that (6.133) becomes

$$n^2(x, y) = n_2^2 + (n_1^2 - n_2^2) f\left(\frac{y}{d}\right) g\left(\frac{x}{a}\right). \quad (6.135)$$

It follows that for index distributions of this kind the effective index method may be used (Hocker and Burns, 1976), solving first for the y-direction and then defining an equivalent effective guide for the x-direction in the usual way. Hence we characterize the guide in the y-direction at each x-value by a normalized frequency $v_0(x)$:

$$v_0^2(x) = d^2 k^2 (n_1^2 - n_2^2) g(x/a). \quad (6.136)$$

Similarly the modes of this guide are characterized at each x by

$$b_0(x) = \frac{(\beta(x)/k)^2 - n_2^2}{(n_1^2 - n_2^2) g(x/a)}. \quad (6.137)$$

It follows that the equivalent guide in the x-direction is characterized by an effective index distribution:

$$n_{\text{eff}}^2(x) = n_2^2 + (n_1^2 - n_2^2) g(x/a) b_0(x). \quad (6.138)$$

For a given distribution which is separable in x and y in the sense of (6.134) the effective profile $n_{\text{eff}}^2(x)$ may be built up point by point from solutions of the appropriate y-direction guide characterized by $b_0(x)$ and $v_0(x)$ in (6.137) and (6.136). After this procedure, the equivalent x-direction guide defined by $n_{\text{eff}}^2(x)$ in (6.138) must be solved (e.g. by ray or WKB methods) to give the modal propagation constant. Numerical results for the distribution of (6.132) have been given by Hocker and Burns (1977) using WKB calculations for the eigenvalue equations of the constituent graded-index waveguides. The results were expressed in terms of the parameters (normalized frequency and propagation constant) of the equivalent guide with no sideways diffusion, i.e. a simple Gaussian distribution in the y-direction.

An alternative method of analysis for guides fabricated by two-dimensional diffusion is furnished by the variational technique. As already discussed in Chapter 5, a variational formula for propagation constant β is given by equation (5.113). Making the appropriate modifications for three-dimensional guides, the result becomes:

$$\beta^2 = \frac{\int_{-\infty}^{\infty}\int_{-\infty}^{\infty}[k^2n^2(x,y)\psi^2 - (\partial\psi/\partial x)^2 - (\partial\psi/\partial y)^2]dx\,dy}{\int_{-\infty}^{\infty}\int_{-\infty}^{\infty}\psi^2\,dx\,dy} \qquad (6.139)$$

where $\psi(x, y)$ is a suitable trial function and the integrals are carried out over the entire x–y plane. If the expression (6.135) for $n^2(x, y)$ is used in (6.139) and the result expressed in terms of the normalized propagation constant b as conventionally defined, then:

$$b = \frac{\int_{-\infty}^{\infty}\int_{-\infty}^{\infty}\left\{f\left(\frac{y}{d}\right)g\left(\frac{x}{a}\right)\psi^2 - \frac{1}{k^2(n_1^2 - n_2^2)}\left[\left(\frac{\partial\psi}{\partial x}\right)^2 + \left(\frac{\partial\psi}{\partial y}\right)^2\right]\right\}dx\,dy}{\int_{-\infty}^{\infty}\int_{-\infty}^{\infty}\psi^2\,dx\,dy}. \qquad (6.140)$$

Using series of parabolic cylinder functions in the x- and y-directions for the trial function ψ, this equation may be solved by the Rayleigh–Ritz method (Matsuhara, 1973). Numerical results for a calculated distribution of diffusant $g(y/d)f(x/a)$ have been computed by this approach by Taylor (1976). A total of 21 terms in the trial function were required to ensure convergence; these involved parabolic cylinder functions of the 7 lowest orders in the y-direction and 3 appropriate orders (depending on the mode symmetry) in the x-direction.

The problem of two-dimensional diffused guides has also been treated numerically by the finite-element method (Yeh et al., 1979). Graded-index optical channel waveguides have been analysed using a combination of the effective-index method and WKB theory by Suhara et al. (1979).

6.3.2 Strip-loaded diffused guides

As in the case of the homogeneous core guides discussed in Section 6.2, for graded-index waveguides there is again the possibility of achieving transverse confinement by the use of a loading strip. The situation is illustrated in Figure 6.16(a), where the structure is indicated, and in Figure 6.16(b), where the index profile $n(y)$ is shown schematically. The strip is assumed to have index n_3 and to be of thickness $2t$ and width $2a$; the profile $n(y)$ is again characterized by same distance $2d$ in the y-direction in order to conform with earlier notation. Once again the effective index method may be applied to this structure by solving the y-direction guides in regions I and II as indicated on Figure 6.16(a). The solutions are then used to define effective indices n_{Ieff} and n_{IIeff} for these regions and finally the equivalent x-direction guide shown in Figure 6.16(c) is solved. As in previous cases the y-direction guides may be characterized by a normalized frequency $v_0 = dkn_1(2\Delta)^{1/2}$ as in (6.95) and (6.121). The effective indices n_{Ieff} and n_{IIeff} resulting from the solution of the y-direction guides are defined in the usual way:

$$n_{j\text{eff}}^2 = n_1^2 b_j + n_2^2(1 - b_j) \quad (j = \text{I or II}). \qquad (6.141)$$

Similarly, the result (6.99) still holds for the effective normalized frequency v_{eff} of the equivalent x-direction guide shown in Figure 6.16(c):

$$v_{\text{eff}}^2 = v^2(b_\text{I} - b_\text{II}). \qquad (6.142)$$

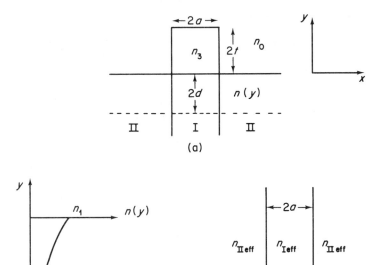

Figure 6.16 (a) Strip-loaded diffused waveguide; (b) index profile in the y-direction ($y \leq 0$); (c) equivalent planar effective guide in the x-direction

The final solution is found for the calculated value of v_{eff} from the solution of the usual three-layer symmetric slab as discussed in Section 2.3.

Numerical results and a discussion of the effective index approach to strip-loaded diffused guides have been given by Hocker (1976) for index profiles $n(y)$ given by exponential, Gaussian, and complementary error function forms. For the exponential profile, the treatment has been extended by Findakly and Chen (1978) to include the case when region II is covered by a metal film, i.e. the loading strip is formed by opening a slot in a metal overlay and depositing a dielectric film. The metal overlay causes n_{IIeff} to decrease from its value without the overlay, whilst the dielectric film results in an increased n_{Ieff}, thus leading to tighter transverse confinement of guided waves than in the simpler structure discussed above. The variational technique described in Section 6.3.1 has been applied to the analysis of strip-loaded diffused guides by Geshiro et al. (1978). Numerical results were presented by these authors for the cases of exponential, Gaussian, and complementary error function profiles. This method has the advantage over the effective-index approach of giving accurate results for any aspect ratio a/d and for modes close to cut-off; in addition field distributions are also deduced in the process of solving the eigenvalue problem.

Further effective-index calculations for strip-loaded guides with an exponential dielectric profile have been given by Noda et al. (1978). In addition these authors have experimentally demonstrated guidance of the lowest-order mode in such a guide fabricated by sputtering a loading strip of chalcogenide glass on top of a Ti-diffused $LiNbO_3$ planar waveguide.

For the case of metal overlays on either side of a dielectric strip Oliner and Peng (1978) have pointed out that coupling of the TM modes to the plasmon mode in the metal-coated regions (see Section 2.5) will tend to increase the E^y mode losses as calculated by Findakly and Chen (1978).

6.3.3 Stripe-geometry heterostructure lasers

The basic structure of the heterojunction laser has already been discussed in Section 2.3.6. Furthermore, in Sections 2.4.4 and 3.3.3 reference has been made to the variation of complex dielectric permittivity within the active layer of the laser as a result of a stripe-geometry contact. It remains therefore to relate the confinement mechanisms in the two transverse directions for this interesting guiding structure and to explore the modification to the earlier theories resulting from this interaction. The model to be used will involve a real three-layer slab waveguide for the heterostructure layers, together with a simple parabolic variation of the complex dielectric permittivity within the central active layer. The structure is illustrated schematically in Figure 6.17; once again the effective index method forms a useful and

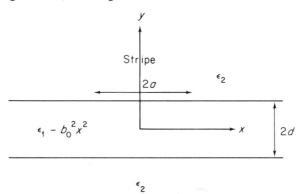

Figure 6.17 Stripe-geometry laser model

powerful tool with which to analyse such a structure.

The dielectric distribution to be analysed is defined by (Paoli, 1977):

$$\varepsilon(x,y) = \begin{cases} \varepsilon_1 - b_0^2 x^2, & |y| \leq d \\ \varepsilon_2, & |y| > d \end{cases} \tag{6.143}$$

where x is the transverse direction along the junction plane and y is normal to this plane (See Figure 6.17). The constants ε_1 and b_0 are defined as in (3.72) and (3.73):

$$\varepsilon_1 = n_1^2 + \frac{in_1 g_1}{k} \tag{6.144}$$

$$b_0^2 = \frac{2\Delta n_1^2}{a^2} + \frac{in_1 \delta g_1}{ka^2} \tag{6.145}$$

where $2a$ is the stripe width, n_1 and g_1 are the refractive index and gain, respectively, at the stripe centre, and Δn_1 and δg_1 are the changes in these quantities between stripe centre and edge. In applying the effective index method to this structure we must allow for the facts (a) that the stripe width, $2a$, is usually much greater than the active layer thickness, $2d$, and (b) that the guidance in the y-direction is much stronger (relative index difference as high as 10 per cent) than that in the x-direction (relative index difference of order 10^{-3}, gain differences of order $10–100\,\mathrm{cm}^{-1}$). Hence we may first solve for each position x the corresponding three-layer slab problem in the y-direction, and then use the result to define an effective permittivity for an equivalent graded-index guide in the x-direction (Kirkby and Thompson, 1976). The effective permittivity is defined as:

$$\varepsilon_{\mathrm{eff}}(x) = B(x)(\varepsilon_1 - b_0^2 x^2) + (1 - B(x))\varepsilon_2 \tag{6.146}$$

where the x-dependence of the normalized propagation constant B has been explicitly included.

The value of B is given as the solution of the complex symmetric slab problem characterized in this case by normalized frequency V_0:

$$V_0^2 = d^2 k^2 (\varepsilon_1 - \varepsilon_2 - b_0^2 x^2) \tag{6.147}$$

where ε_2 is defined in terms of index n_2 and loss α_2 as:

$$\varepsilon_2 = n_2^2 - \frac{i n_2 \alpha_2}{k}. \tag{6.148}$$

However, the rather simpler real slab problem is characterized by normalized frequency $v_0 = dk(n_1^2 - n_2^2)^{1/2}$ and has normalized propagation constant b. Thus we may define a more conventional effective dielectric constant $\varepsilon'_{\text{eff}}$ for the real slab:

$$\varepsilon'_{\text{eff}} = b n_1^2 + (1-b) n_2^2. \tag{6.149}$$

Now recall that the imaginary parts of the dielectric permittivities are always much less than the real parts for stripe-geometry lasers (typically n_1^2 is of order 10 and $n_1 g_1/k$, $n_2 \alpha_2/k$ of order 10^{-2}); furthermore the variation of $B(x)$ is rather small. Hence we may write $B(x) = b + \delta b(x)$ and $V_0 = v_0 + \delta v_0(x)$ and perform a perturbation analysis to estimate the contribution of $\delta b(x)$ (Streifer and Kapon, 1979; Buus, 1979; Thompson, 1980). From the definitions (6.144), (6.146), (6.148), and (6.149), it follows that:

$$\varepsilon_{\text{eff}}(x) - \varepsilon'_{\text{eff}} \simeq \delta b(x)(n_1^2 - n_2^2) - i\frac{n_2 \alpha_2}{k} + b\left(i\frac{n_1 g_1}{k} + i\frac{n_2 \alpha_2}{k} - b_0^2 x^2\right). \tag{6.150}$$

If we now write

$$\delta b(x) = \frac{db}{dv_0} \delta v_0(x)$$

where v_0 and $\delta v_0(x)$ are as defined above, then we find:

$$\delta b(x) \simeq \frac{db}{dv_0} \frac{v_0}{2(n_1^2 - n_2^2)} \left(i\frac{n_1 g_1}{k} + i\frac{n_2 \alpha_2}{k} - b_0^2 x^2\right). \tag{6.151}$$

Using the result (6.151) in (6.150) yields:

$$\varepsilon_{\text{eff}}(x) - \varepsilon'_{\text{eff}} \simeq \left(b + \frac{v_0}{2}\frac{db}{dv_0}\right)\left(i\frac{n_1 g_1}{k} + i\frac{n_2 \alpha_2}{k} - b_0^2 x^2\right) - i\frac{n_2 \alpha_2}{k}$$

$$= \Gamma\left(i\frac{n_1 g_1}{k} - b_0^2 x^2\right) - (1-\Gamma)i\frac{n_2 \alpha_2}{k} \tag{6.152}$$

where the final result on the r.h.s. of (6.152) follows from the definition (2.125) and the relation (2.122) for the symmetric slab waveguide (Buus and Adams, 1979). Equation (6.152) tells us that the appropriate weighting factors for small perturbations in dielectric constant are given by the power fraction in the appropriate layer, i.e. Γ for the core and $(1-\Gamma)$ for the cladding (Streifer and Kapon, 1979; Buus, 1979). This confirms the *ad hoc* use of the filling factor Γ by Paoli (1977) and others (Streifer et al., 1978, 1979; Butler and Sommers, 1978; Tsang, 1978; Petermann, 1978; Asbeck et al., 1979), and replaces earlier estimates of alternative weighting factors (Buus, 1978, 1979a; Suhara et al., 1979).

Using the result (6.152) the scalar wave equation for guidance in the x-direction becomes:

$$\frac{1}{\psi}\frac{d^2\psi}{dx^2} + k^2\left[bn_1^2 + (1-b)n_2^2 + \Gamma\left(i\frac{n_1 g_1}{k} - b_0^2 x^2\right) - (1-\Gamma)i\frac{n_2 \alpha_2}{k}\right] - \beta^2 = 0 \tag{6.153}$$

which yields the familiar Hermite–Gaussian modes (see Chapter 3):

$$\psi_N(x) = H_N(\Gamma^{1/4}(\pm kb_0)^{1/2}x)\exp\left(\mp\frac{kb_0\Gamma^{1/2}}{2}x^2\right) \quad (6.154)$$

where the upper and lower signs correspond again to positive and negative values of $\text{Re}(b_0)$. For the case of the zero-order mode along the junction plane, equation (6.154) assumes a simple Gaussian form and may be written in terms of a spot size w_0 and a phase-front radius of curvature R:

$$\psi_0(x) = \exp\left[\frac{ikn_1 x^2}{2R} - \left(\frac{x}{w_0}\right)^2\right]. \quad (6.155)$$

Using the definition of b_0^2 given in (6.145), together with (6.154) and (6.155), yields the results:

$$\Gamma\text{Re}(v^2) \equiv \Gamma a^2 k^2 2\Delta n_1^2 = 4\left(\frac{a}{w_0}\right)^4 - \left(\frac{a^2 kn_1}{R}\right)^2 \quad (6.156a)$$

$$\Gamma\text{Im}(v^2) \equiv \Gamma a^2 kn_1 \delta g_1 = -4\left(\frac{a}{w_0}\right)^2\left(\frac{a^2 kn_1}{R}\right). \quad (6.156b)$$

Equations (6.156a) and (6.156b) are the analogous results to (3.102) and (3.103) now that allowance has been made for the interaction of wave-guiding effects in the two transverse directions via the effective index method. They relate the physically observable parameters w_0 and R to the unknown quantities Δn_1 and δg_1. The form of these equations suggests a simple universal chart relating these quantities with axes corresponding to the observables $(a/w_0)^2$ and $(a^2 kn_1/R)$; such a chart is given in Figure 6.18. Each contour on the figure represents a constant value of $\Gamma\text{Re}(v^2)$ or $\Gamma\text{Im}(v^2)$, as given by equations (6.156a) and (6.156b). A number of methods of measuring w_0 and R are available based on observations of near and far fields (Cook and Nash, 1975), spontaneous emission intensity (Paoli, 1977), or apparent beam waists and their positions (Kirkby et al., 1977). The plots of Figure 6.18 should therefore enable the relative contributions of real guidance, $\text{Re}(v^2)$, and virtual guidance, $\text{Im}(v^2)$, to be determined, provided that the appropriate values of Γ is used. The parameter Γ may be found either by direct calculation in cases where the dielectric discontinuity and guide thickness, $2d$, are known with sufficient accuracy, or from experimental measurements of the confinement factor.

With the aid of the Gaussian solution (6.155), the wave equation (6.153) yields the eigenvalue $\beta = \beta_r + i\beta_i$ where $\beta_r \gg \beta_i$ as follows:

$$\beta_r^2 \simeq k^2\left[n_1^2 b + n_2^2(1-b)\right] - \frac{2}{w_0^2} \quad (6.157)$$

$$2\beta_i \simeq \Gamma\left[g_1 - \delta g_1\left(\frac{w_0}{2a}\right)^2\right] - (1-\Gamma)\alpha_2. \quad (6.158)$$

The quantity in equation (6.158) may be interpreted as the nett modal gain per unit length; it is the appropriate parameter to use in calculations of laser threshold and efficiency (cf. equation (2.164)). Measurements of this parameter as a function of drive current may also be used to determine the strength δg_1 of the gain-induced optical confinement in the active layer (Paoli, 1977).

The analysis in the present subsection has concentrated on a parabolic variation of permittivity in the active layer, as defined by equation (6.143). However, similar equations occur when, instead of a parabolic dielectric distribution, we consider a parabolic variation of

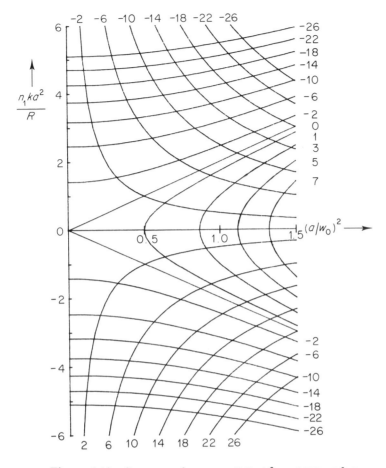

Figure 6.18 Contours of constant $\Gamma \operatorname{Re}(v^2)$ and $\Gamma \operatorname{Im}(v^2)$ from equation (6.156); labelling parameters give the values of $\Gamma \operatorname{Im}(v^2)$ (at top and bottom) and $\Gamma \operatorname{Re}(v^2)$ (at right)

active layer *thickness*. This case has been analysed along similar lines by Suematsu and Yamada (1973) and by Kirkby and Thompson (1976) who dealt with a specific laser structure involving such a variation of active layer thickness, viz. the channelled substrate buried heterostructure. A similar analysis has also been applied to a thin-film beam splitter by Vincent and Lit (1977); in this case the variation of guide thickness occurs as a result of the fabrication process, e.g. sputtering with a shadowing effect produced by an optical fibre placed on the substrate surface.

The dielectric distribution considered in this subsection, as defined in (6.143), is not the only one which has been applied to analyses of stripe-geometry heterostructure lasers. In addition to the parabolic variation with x, Streifer *et al.* (1978, 1979) have also considered squared-tangent and Epstein-layer variations of gain and refractive index in the active layer. The symmetric Epstein model (sech^2) has been used by Asbeck *et al.* (1978) and Osinski and Eliseev (1979), whilst Petermann (1978) employed a power-law dependence of dielectric constant on x. All these analyses considered abrupt steps in dielectric permittivity at the hetero-interfaces between the three layers of Figure 6.17. However, Zachos and Ripper (1969) used parabolic models for both x- and y-directions in stripe-geometry homojunction lasers, and Nunes *et al.* (1979) found good agreement with experiments on double heterostructure devices by using a parabolic variation in the y-direction with a sech^2 model for the x-dependence. In these cases the wave equation is separable and it is not necessary to use the effective index method. The accuracy of the effective index method has been investigated by Butler and Delaney (1978) who compared its results with those from a rigorous numerical method. In this approach the solution of the wave equation in the active layer was written as a linear combination of Hermite–Gaussian functions and that for the passive layer was expressed as a linear combination of plane waves. The spot-size calculated from the effective index method was found to differ by as much as 30 per cent from the result of this numerical analysis. Other numerical methods of analysis for stripe-geometry lasers have been reported by Chinone (1977), Buus (1978, 1979, 1979a), Asbeck *et al.* (1979), and Lang (1979).

Chapter 7
Circular Waveguides and Step-index Fibres

The first theoretical study of wave propagation on circular cylindrical dielectric structures was made by Hondros and Debye (1910). Most of the theoretical and experimental work which followed was concerned with propagation at radio and microwave frequencies, and it was not until 1966 that the proposal to use circular glass fibres as optical transmission lines was made (Kao and Hockham, 1966). For the types of glass available to form fibres, the refractive index difference is sufficiently small that the weakly-guiding condition ($n_1 - n_2$) $\ll n_1$ is usually met, and as a result the theoretical analysis of optical fibres is greatly simplified (Snyder, 1969; Gloge, 1971). The achievement of low-loss glass fibres (< 20 dB/km) in 1970 (Kapron et al., 1970) has resulted in a spectacular growth of interest in the subject of optical communications. In this chapter we will explore some of the fundamental properties of uniform-core circular waveguides with especial reference to their use in optical communications systems.

In keeping with the spirit of treatments of other waveguides in this book, the chapter commences with a discussion of propagation in hollow pipes with conducting walls. The results of geometrical optics are compared with those from a more rigorous electromagnetic approach. For the dielectric rod the electromagnetic mode treatment is given in the general case and later simplified to reveal the linearly-polarized modes of the weakly-guiding limit. The principal causes of pulse-spreading in single-mode and multimode fibres are discussed and the pertinent properties of real fibres are described. The effects of small departures from perfect circularity of the core are also considered. The chapter concludes with analyses of more complicated structures than the simple cladded-core fibre, viz. hollow and metallic waveguides and multilayer fibres.

7.1 Propagation in Hollow Circular Pipes

As in the cases of planar and rectangular waveguides we commence the discussion of circular guides with a brief introduction to the theory of structures with perfectly-conducting walls. The simplicity of this treatment as compared with that for dielectric guides serves as a useful introduction to the main characteristics of the solutions and has some relevance in addition to the use of hollow cylindrical pipes for guidance of CO_2 laser radiation (Garmire et al., 1976). As in the rectangular structures of Chapter 6, the modes of circular waveguides fall into two classes corresponding to TE and TM polarization. Before proceeding, however, to the full modal treatment, we begin here with a derivation of the eigenvalue equation from a geometric optics viewpoint based on a method developed for optical fibres by Love and Snyder (1976). Consideration of the wave equation in circular cylindrical coordinates then leads to identification of the ray angles in terms of the modal parameters via the local plane-wave decomposition of the wave vector. The resulting eigenvalues (Keller and Rubinow,

1960) are then compared with those derived from the full electromagnetic treatment of TE and TM modes.

7.1.1 Geometric optics derivation of the eigenvalue equation

Consider a circular cylinder, whose axis is the z-axis, with radius a and refractive index n_1 in the core, surrounded by perfectly-conducting walls. The path of a skew ray in such a structure is illustrated in Figure 7.1; the ray propagates by repeated reflection from the walls of the

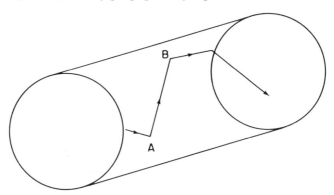

Figure 7.1 Path of a skew ray propagating by repeated reflection from the conducting walls of a circular guide

guide. Our objective here is to find the permitted angles at which such a ray may propagate in order to form a bound ray of the structure. As is by now familiar from earlier chapters, the ray treatment will involve calculations of the phase changes suffered along a segment of the ray path and the derivation of a condition on the phase changes in order that the repetition of such path segments makes up the entire ray path in the guide. In more detail the method here (following Love and Snyder, 1976) involves calculating the phase change along the path segment AB of Figure 7.1 in two different ways:

(i) By adding the phase change due to the geometrical path length to that occurring on reflection at the wall and that resulting from the path touching an inner caustic; Figure 7.2

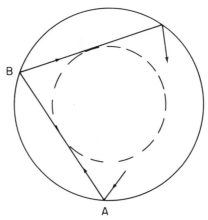

Figure 7.2 Projection of the skew ray path of Figure 7.1 upon a cross-section of the guide; the broken line shows the caustic which is touched tangentially by each ray path segment

shows the ray path viewed from the end of the waveguide and illustrates the existence of an inner circular caustic which is touched by each ray path segment.

(ii) By determining the geometrical transformations of rotation $\Delta\phi$ and translation Δz of the ray path that moves A to B; the rotation is also illustrated on Figure 7.2. The corresponding phase change consists of two parts:
  (a) the translation Δz multiplied by the axial component of wave vector and
  (b) the rotation $\Delta\phi$ multiplied by the trangential component of wave vector and by the core radius a. The total phase change is the sum of these components (a) and (b).

The eigenvalue equation is then obtained by requiring that the total phase changes as calculated by methods (i) and (ii) be the same within a multiple of 2π.

Figure 7.3 shows the segment AB of the ray path together with the relevant angles; AC is the axial translation denoted by Δz in the discussion above. The angle between AB and the normal at the core–cladding interface is denoted by θ_1 (to correspond with the notation of Chapter 2), that between AB and the axial direction z is denoted by γ, and that between AB and the tangential direction at the interface is denoted by α. The angle between the ray projection BC and the tangential direction is denoted by ϕ. It follows from Figure 7.3 that

$$\sin\phi = \frac{\cos\theta_1}{\sin\gamma}; \quad \cos\phi = \frac{\cos\alpha}{\sin\gamma} \qquad (7.1)$$

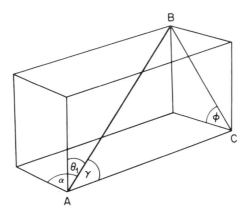

Figure 7.3 Segment AB of a skew ray path, together with its projection BC on a perpendicular cross-section of the guide

Figure 7.4 shows the ray projection BC on the circular cross-section of the guide, radius a. The angular rotation $\Delta\phi$ is seen to be given by 2ϕ; in addition we have that

$$BC = 2a\sin\phi. \qquad (7.2)$$

Hence, using (7.1) and (7.2), it follows that:

$$AC = 2a\sin\phi\cot\gamma \equiv \frac{2a\cos\theta_1\cos\gamma}{\sin^2\gamma} \qquad (7.3)$$

$$AB = \frac{2a\sin\phi}{\sin\gamma} \equiv \frac{2a\cos\theta_1}{\sin^2\gamma}. \qquad (7.4)$$

We are now in a position to calculate the phase changes by methods (i) and (ii) as described

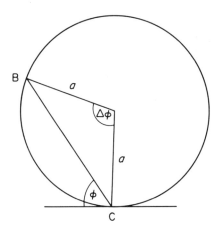

Figure 7.4 The projection BC of ray segment AB upon the guide cross-section; $\Delta\phi$ is the angular rotation required to move B to C

above. Let us denote the phase shift on reflection at the wall by δ; this will be equal (from Chapter 1) to either π or 2π corresponding to polarization with the electric field vector either perpendicular or parallel to the plane of incidence, respectively. For the phase change occurring when the ray path touches a caustic, we know from Section 3.1.3 that this is given by $\pi/2$, independent of the wave polarization. Adding on the phase change due to the geometrical path length AB, we find:

Total phase change in going from A to B $= kn_1 \,(\mathrm{AB}) - \delta - \dfrac{\pi}{2}$

$$= \frac{2akn_1 \cos\theta_1}{\sin^2\gamma} - \delta - \frac{\pi}{2} \quad (7.5)$$

where equation (7.4) has been used. For calculation of this phase change by method (ii) above we need to evaluate first the change due to the angular rotation:

$$kn_1 a \cos\alpha \, \Delta\phi = kn_1 a \cos\alpha \sin^{-1}\left(\frac{2\cos\alpha \cos\theta_1}{\sin^2\gamma}\right) \quad (7.6)$$

and also the axial shift due to translation:

$$kn_1 \cos\gamma \, \Delta z = 2kn_1 a \cos\theta_1 \cot^2\gamma \quad (7.7)$$

where equations (7.1) and (7.3) have been used. We now demand that the total phase change made up of (7.7) and (7.6) combined be equal to the expression in (7.5), within a multiple of 2π:

$$2akn_1 \cos\theta_1 \cot^2\gamma + 2akn_1 \cos\alpha \cos^{-1}\left(\frac{\cos\alpha}{\sin\gamma}\right) + 2N\pi$$
$$= \frac{2akn_1 \cos\theta_1}{\sin^2\gamma} - \delta - \frac{\pi}{2} \quad (N = 0, 1, 2, \ldots) \quad (7.8)$$

Rearranging this equation, we obtain the final result for the eigenvalue equation derived by a ray optics argument:

$$akn_1 \cos\theta_1 - akn_1 \cos\alpha \cos^{-1}\left(\frac{\cos\alpha}{\sin\gamma}\right) = \left(N + \frac{1}{4}\right)\pi + \frac{\delta}{2} \quad (N = 0, 1, 2, \ldots) \quad (7.9)$$

where δ is to be chosen according to the polarization as either π or 2π. In order to identify the angles appearing in (7.9) in terms of the corresponding modal parameters we now turn to a consideration of the wave equation in circular cylindrical coordinates.

7.1.2 Maxwell's equations in circular cylindrical coordinates

In cylindrical polar coordinates (r, θ, z) Maxwell's equations (3.27) and (3.28) for homogeneous, isotropic media can be written, suppressing the time- and z-dependence in the usual way, as:

$$\frac{1}{r}\frac{\partial H_z}{\partial \theta} - i\beta H_\theta = -in_1^2 \varepsilon_0 \omega E_r \tag{7.10}$$

$$i\beta H_r - \frac{\partial H_z}{\partial r} = -in_1^2 \varepsilon_0 \omega E_\theta \tag{7.11}$$

$$\frac{1}{r}\frac{\partial}{\partial r}(r H_\theta) - \frac{1}{r}\frac{\partial H_r}{\partial \theta} = -in_1^2 \varepsilon_0 \omega E_z \tag{7.12}$$

$$\frac{1}{r}\frac{\partial E_z}{\partial \theta} - i\beta E_\theta = i\omega\mu_0 H_r \tag{7.13}$$

$$i\beta E_r - \frac{\partial E_z}{\partial r} = i\omega\mu_0 H_\theta \tag{7.14}$$

$$\frac{1}{r}\frac{\partial}{\partial r}(rE_\theta) - \frac{1}{r}\frac{\partial E_r}{\partial \theta} = i\omega\mu_0 H_z \tag{7.15}$$

where n_1 is the refractive index and $\mathbf{E} = (E_r, E_\theta, E_z)$, $\mathbf{H} = (H_r, H_\theta, H_z)$. For the special cases of TE and TM waves we have seen in the previous chapter on rectangular waveguides that all field components may be written in terms of the axial components E_z, H_z which are then given by the solutions of the wave equation (2.76). It is therefore plausible in the present instance to look again for solutions of the wave equation which may be written in cylindrical coordinates as:

$$\frac{\partial^2 \psi}{\partial r^2} + \frac{1}{r}\frac{\partial \psi}{\partial r} + \frac{1}{r^2}\frac{\partial^2 \psi}{\partial \theta^2} + (n_1^2 k^2 - \beta^2)\psi = 0 \tag{7.16}$$

where $\psi(r, \theta)$ is either E_z or H_z. Since we are dealing with circularly-symmetric structures, we look for solutions periodic in the angular coordinate θ, i.e. solutions of the form:

$$\psi(r, \theta) = R(r)e^{iv\theta} \quad (v = 0, 1, 2, \ldots). \tag{7.17}$$

Substituting (7.17) in (7.16) we arrive at the scalar wave equation for circular waveguides:

$$\frac{\partial^2 R}{\partial r^2} + \frac{1}{r}\frac{\partial R}{\partial r} + \left(n_1^2 k^2 - \beta^2 - \frac{v^2}{r^2}\right)R = 0. \tag{7.18}$$

With the azimuthal periodicity of (7.17) assumed for each field component, the fields may be expressed from (7.10)–(7.15) in terms of E_z and H_z:

$$E_r = -\frac{a^2}{u^2}\left(\frac{v\omega\mu_0}{r}H_z - i\beta\frac{\partial E_z}{\partial r}\right) \tag{7.19}$$

$$E_\theta = -\frac{a^2}{u^2}\left(i\omega\mu_0\frac{\partial H_z}{\partial r} + \frac{\beta v}{r}E_z\right) \tag{7.20}$$

$$H_r = \frac{a^2}{u^2}\left(\frac{vn_1^2\varepsilon_0\omega E_z}{r} + i\beta\frac{\partial H_z}{\partial r}\right) \tag{7.21}$$

$$H_\theta = \frac{a^2}{u^2}\left(in_1^2\varepsilon_0\omega\frac{\partial E_z}{\partial r} - \frac{v\beta}{r}H_z\right) \tag{7.22}$$

where the usual definitions $u^2 = a^2(k^2n_1^2 - \beta^2)$, $k^2 = \omega^2\mu_0\varepsilon_0$ have been used. In general, the axial components E_z and H_z may be found from solutions of the wave equation (7.18). This equation is in fact the well-known differential equation for Bessel functions; there are two independent solutions, the choice of which must be made with due regard to the boundary conditions. For an oscillatory solution in the region $r \leqslant a$ with no singularity at $r = 0$, we are constrained to choose the Bessel function $J_v(ur/a)$ (Abramowitz and Stegun, 1964). In determining the eigenvalues β, v we must take account of the polarization, and this will be postponed until the next subsection.

For the present, we are concerned with finding the relationships between the mode eigenvalues (β, v) and the ray angles (γ, α) of Section 7.1.1. Since the axial variation of the mode is of the form $\exp(i\beta z)$ and in the ray picture the ray direction is at angle γ to the z-axis (Figure 7.3), it follows that

$$\beta = kn_1\cos\gamma. \tag{7.23}$$

Similarly, since the ray direction is at angle α to the tangential direction (Figure 7.3), it follows that the azimuthal variation may be written as $\exp[i(kn_1\cos\alpha)(\theta a)]$. Comparing this expression with the azimuthal periodicity from the mode picture, as expressed in (7.17), we find:

$$v = akn_1\cos\alpha. \tag{7.24}$$

Using the relationship between θ_1, α, and γ as defined on Figure 7.3, we have:

$$\left(k^2n_1^2 - \beta^2 - \frac{v^2}{a^2}\right)^{1/2} = kn_1\cos\theta_1. \tag{7.25}$$

The relationships (7.23)–(7.25) are summarized in a more physical way by the local plane wave picture of Figure 7.5, which shows the decomposition of the wave vector $n_1\mathbf{k}$ into its Cartesian components at the point $r = a$. Note that this interpretation may also be obtained by a WKB analysis of the wave equation (7.18) when specialized to the point $r = a$ (Gloge and Marcatili, 1973).

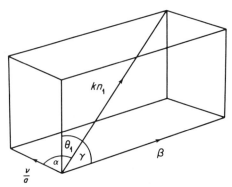

Figure 7.5 Local plane wave decomposition of the wave vector $\mathbf{k}n_1$ into its radial, tangential, and axial components at $r = a$

With the aid of the relations (7.23)–(7.25), the ray optics eigenvalue equation (7.9) derived previously may now be expressed in terms of the modal parameters u, v:

$$(u^2 - v^2)^{1/2} - v\cos^{-1}\left(\frac{v}{u}\right) = \left(N + \frac{1}{4}\right)\pi + \frac{\delta}{2} \quad (v, N = 0, 1, 2, \ldots). \tag{7.26}$$

Equation (7.26) is now in a suitable form for comparison with the results of the modal analysis to be given below.

7.1.3 TE and TM modes

We may solve Maxwell's equations (7.19)–(7.22) for the two classes of modes: TE (i.e. $E_z = 0$) and TM ($H_z = 0$). As discussed in the previous subsection, an appropriate choice of solution for the wave equation (7.18) is the Bessel function $J_v(ur/a)$. The boundary conditions to be satisfied are that the tangential components of electric field, E_θ and E_z, and the normal magnetic field component, H_r, should all vanish on the wall $r = a$. It follows that the TE mode solutions are

$$E_r = -\frac{a^2 v\omega\mu_0}{u^2 r} H_0 J_v\left(\frac{ur}{a}\right) \tag{7.27}$$

$$E_\theta = -\frac{ia\omega\mu_0 H_0}{u} J_v'\left(\frac{ur}{a}\right) \tag{7.28}$$

$$E_z = 0 \tag{7.29}$$

$$H_r = \frac{ia\beta}{u} H_0 J_v'\left(\frac{ur}{a}\right) \tag{7.30}$$

$$H_\theta = -\frac{a^2 v\beta}{u^2 r} H_0 J_v\left(\frac{ur}{a}\right) \tag{7.31}$$

$$H_z = H_0 J_v\left(\frac{ur}{a}\right) \tag{7.32}$$

where H_0 is a constant. The corresponding TE eigenvalue equation is given by:

$$J_v'(u) = 0. \tag{7.33}$$

For TM modes, it follows from equations (7.19)–(7.22) that the field components are:

$$E_r = \frac{ia\beta}{u} E_0 J_v'\left(\frac{ur}{a}\right) \tag{7.34}$$

$$E_\theta = -\frac{a^2 \beta v}{u^2 r} E_0 J_v\left(\frac{ur}{a}\right) \tag{7.35}$$

$$E_z = E_0 J_v\left(\frac{ur}{a}\right) \tag{7.36}$$

$$H_r = \frac{a^2 v n_1^2 \varepsilon_0 \omega}{u^2 r} E_0 J_v\left(\frac{ur}{a}\right) \tag{7.37}$$

$$H_\theta = \frac{ian_1^2 \varepsilon_0 \omega}{u} E_0 J_v'\left(\frac{ur}{a}\right) \tag{7.38}$$

$$H_z = 0 \tag{7.39}$$

where E_0 is a constant, and the corresponding TM eigenvalues are the solutions of:

$$J_\nu(u) = 0. \tag{7.40}$$

We may compare the modal eigenvalues as given by equations (7.33) and (7.40) with those found from the geometrical optics derivation given earlier. Hence, choosing $\delta = 2\pi$ in equation (7.26), we find:

$$(u^2 - v^2)^{1/2} - v\cos^{-1}\left(\frac{v}{u}\right) = \left(N + \frac{5}{4}\right)\pi \quad (v, N = 0, 1, 2, \ldots). \tag{7.41}$$

Similarly, for comparison with the TM modal solution (7.40), we choose $\delta = \pi$ in (7.26) and thus we have:

$$(u^2 - v^2)^{1/2} - v\cos^{-1}\left(\frac{v}{u}\right) = \left(N + \frac{3}{4}\right)\pi \quad (v, N = 0, 1, 2, \ldots). \tag{7.42}$$

The eigenvalues (u) as given by equations (7.33) and (7.41) for TE polarization, and (7.40) and (7.42) for TM polarization, are compared in Table 7.1 (Keller and Rubinow, 1960). In general, the level of agreement improves for higher values of the mode numbers v and N.

Table 7.1 Comparison of eigenvalues found from modal solutions with those from the ray optics approximation

		TE		TM	
v	N	equation (7.33)	equation (7.41)	equation (7.40)	equation (7.42)
0	0	3.832	3.927	2.405	2.356
	1	7.016	7.069	5.520	5.498
	2	10.173	10.210	8.654	8.639
1	0	1.841	2.115	3.832	3.795
	1	5.331	5.405	7.016	6.997
	2	8.536	8.581	10.173	10.161
2	0	3.054	3.300	5.136	5.101
	1	6.706	6.771	8.417	8.401
	2	9.969	10.010	11.620	11.609

The physical interpretation of the azimuthal mode number v is already clear from its definition in terms of azimuthal periodicity as in (7.17). On the other hand, the meaning of N has not been discussed as yet; the radial field distributions plotted in Figures 7.6 and 7.7 will remedy this situation. Figure 7.6 shows the variation of H_z for a few low-order TE modes, as given by equations (7.32) and (7.33). It is clear that the integer N specifies a radial mode number whose value is given by one less than the number of zeros of the field distribution. Similarly, Figure 7.7 plots E_z versus r for some TM modes as given by equations (7.35) and (7.40) and again illustrates the definition of N (provided we count only field zeros other than that at $r = 0$).

Further insight may be obtained by calculating the radial distribution of intensity $I(r)$ rather than field components $R(r)$. Hence we consider the axial component of the time-averaged Poynting vector for TE modes:

$$I(r) = \tfrac{1}{2}\mathrm{Re}(\mathbf{E} \wedge \mathbf{H}^*)_z = \tfrac{1}{2}(E_r H_\theta^* - H_r E_\theta^*) = \frac{a^2\beta\omega\mu_0}{2u^2}H_0^2\left[J_\nu'^2\left(\frac{ur}{a}\right) + \left(\frac{av}{ur}\right)^2 J_\nu^2\left(\frac{ur}{a}\right)\right]$$

$$= \frac{a^2\beta\omega\mu_0}{4u^2}H_0^2\left[J_{\nu-1}^2\left(\frac{ur}{a}\right) + J_{\nu+1}^2\left(\frac{ur}{a}\right)\right] \tag{7.43}$$

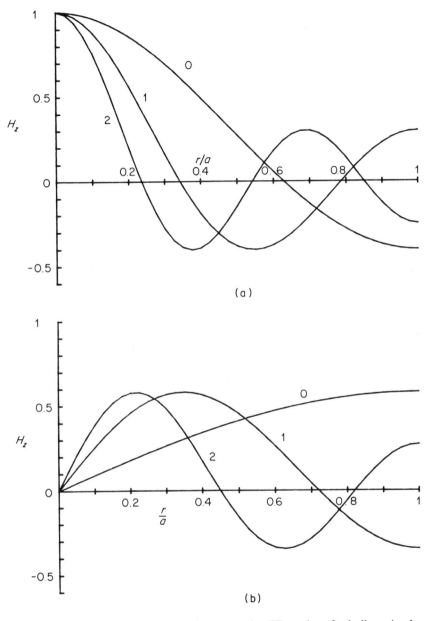

Figure 7.6 Radial variation of H_z for low-order TE modes of a hollow circular pipe: (a) $v = 0$; (b) $v = 1$. Labelling parameter gives the value of the radial mode number N

where equations (7.27), (7.28), (7.30), (7.31) and the usual Bessel function relationships (Abramowitz and Stegun, 1964) have been used. Similarly, for TM modes we find from equations (7.34), (7.35), (7.37), and (7.38):

$$I(r) = \frac{a^2 \beta \omega n_1^2 \varepsilon_0 E_0^2}{4u^2} \left[J_{v-1}^2\left(\frac{ur}{a}\right) + J_{v+1}^2\left(\frac{ur}{a}\right) \right]. \qquad (7.44)$$

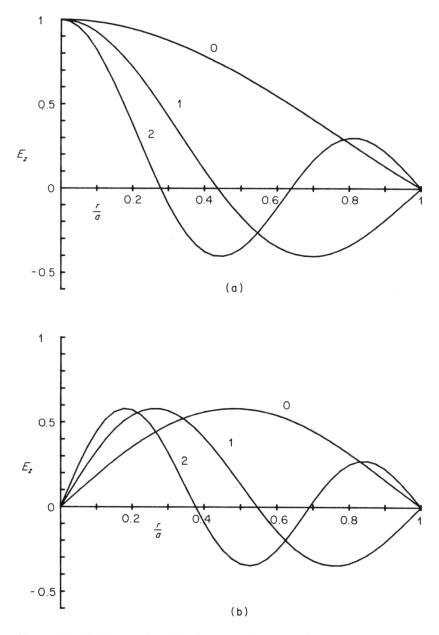

Figure 7.7 Radial variation of E_z for low-order TM modes of a hollow circular pipe: (a) $v = 0$; (b) $v = 1$. Labelling parameter gives the value of the radial mode number N

It follows from (7.43) and (7.44), together with the eigenvalue equations (7.33) and (7.40), respectively, that the radial intensity distributions of the TE_{vN} and TM_{vN} modes are circularly symmetric with $N+1$ maxima in the radial direction, not counting the central maximum which occurs for $v = 1$. An exception is the TE_{1N} mode which possesses only N subsidiary maxima. Further discussion of these modes, their polarization, and other features are given in the book by Kapany and Burke (1972). As we would expect from earlier discussions in the present work, the relation $V_p V_g = c^2/n_1^2$ is again satisfied by these modes, as may be easily verified from the eigenvalue equations. Similarly, the eigenvalues themselves may be represented on normalized b–v plots as in earlier sections, but we will omit the result here.

7.2 Step-index Optical Fibres

The theory of the circular dielectric waveguide is now well established and we therefore begin this section with a brief review of the relevant results. The main application of the theory from our point of view is for optical fibres with a uniform core made from a glass of refractive index n_1 embedded in a relatively large cladding glass of index $n_2 (< n_1)$. The glasses used in practice ensure that the weakly-guiding condition $(n_1 - n_2) \ll n_1$ is always applicable and, as a consequence, considerable simplification ensues in the theoretical analysis. In particular it is then possible to find linearly-polarized (LP) modes of the structure which form useful approximations to the true modes (Snyder, 1969; Gloge, 1971). This section therefore includes a discussion of the properties of these modes together with a comparison of the LP modal results with those from the ray-optics picture of the step-index fibre.

The dispersive properties of optical fibres also form an important topic; pulse-spreading in a multimode fibre stems from three principal causes: (i) material dispersion, (ii) transit time differences between modes, and (iii) waveguide dispersion. Item (i) occurs simply as a consequence of the wavelength dependence of the refractive indices n_1, n_2 of the core and cladding glasses and is present in all fibres. Item (ii) is a serious limitation on the bandwidth of operation of a multimode step-index fibre, but should be absent in a single-mode fibre. Item (iii) occurs as a consequence of the wavelength dependence of the waveguide properties of the fibre. Note that the material and waveguide dispersion effects are relevant only when the source used for a fibre communications system is of finite bandwidth; for a truly monochromatic source these effects would vanish.

Whilst it is possible in principle to avoid intermodal delay differences by operating in the single-mode regime, in fact the lowest-order mode occurs in two orthogonal polarizations. Whilst these polarizations lead to the same eigenvalue equation and dispersion properties for a perfectly circular fibre, for a noncircular guide this degeneracy is removed and there can be a difference in propagation times of the two polarizations. Since small departures from circularity often occur in practical fibres by the nature of the fabrication process, it is therefore appropriate to include within this section some discussion of the properties of elliptical dielectric waveguides of small eccentricity.

7.2.1 Circular dielectric rod

Consider the waveguide structure of Figure 7.8 consisting of a homogeneous core of radius a, index n_1, surrounded by an infinite uniform cladding of index n_2. The axis of the rod is the z-axis of cylindrical polar coordinates so that we may again apply Maxwell's equations in the forms (7.10)–(7.15) for the core region, with a similar set of equations for the cladding where n_1 is replaced by n_2. Similarly, the transverse field components may all be expressed in terms of the longitudinal components E_z, H_z as in equations (7.19)–(7.22) which are solutions of the wave equation (7.18), with appropriate changes of n_1 to n_2 and u^2 to $-w^2$ for the cladding. Using the usual definitions of $u^2 = a^2(k^2 n_1^2 - \beta^2)$, $w^2 = a^2(\beta^2 - k^2 n_2^2)$, the wave equations become:

Core:
$$\frac{\partial^2 R}{\partial r^2} + \frac{1}{r}\frac{\partial R}{\partial r} + \left(\frac{u^2}{a^2} - \frac{v^2}{r^2}\right) R = 0 \quad (r \leqslant a) \tag{7.45}$$

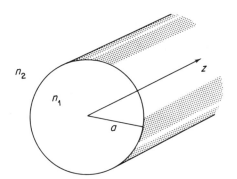

Figure 7.8 The circular dielectric rod waveguide

Cladding: $$\frac{\partial^2 R}{\partial r^2} + \frac{1}{r}\frac{\partial R}{\partial r} - \left(\frac{w^2}{a^2} + \frac{v^2}{r^2}\right)R = 0 \qquad (r > a). \tag{7.46}$$

For the core equation (7.45) the solution $J_v(ur/a)$ is appropriate, as in Section 7.1, since it remains finite at $r = 0$. For the cladding region we require a solution to (7.46) which approaches zero asymptotically for large distances from the core; such a solution is provided by the modified Bessel function of the second kind $K_v(wr/a)$, also called the modified Hankel function (Abramowitz and Stegun, 1964).

In the metallic guides of Section 7.1 we found two sets of solutions—the TE and TM modes. For the dielectric rods presently under consideration, however, only the cylindrically symmetric TE_{0N}, TM_{0N} modes are of this type; all others are hybrid, i.e. they have both electric and magnetic axial field components. Hence we assume the general solutions to (7.45) and (7.46) in the forms (omitting time-, z-, and θ-dependence):

$$E_z = \begin{cases} \dfrac{AJ_v(ur/a)}{J_v(u)}, & r \leqslant a \qquad (7.47a) \\[6pt] \dfrac{AK_v(wr/a)}{K_v(w)}, & r > a \qquad (7.47b) \end{cases}$$

$$H_z = \begin{cases} \dfrac{BJ_v(ur/a)}{J_v(u)}, & r \leqslant a \qquad (7.48a) \\[6pt] \dfrac{BK_v(wr/a)}{K_v(w)}, & r > a \qquad (7.48b) \end{cases}$$

where A and B are constants to be determined. The remaining field components are then given from (7.19)–(7.22) as:

$$E_r = \begin{cases} -\dfrac{a^2}{u^2}\left[\dfrac{v\omega\mu_0 BJ_v(ur/a)}{rJ_v(u)} - \dfrac{i\beta AuJ_v'(ur/a)}{aJ_v(u)}\right], & r \leqslant a \qquad (7.49a) \\[10pt] \dfrac{a^2}{w^2}\left[\dfrac{v\omega\mu_0 BK_v(wr/a)}{rK_v(w)} - \dfrac{i\beta AwK_v'(wr/a)}{aK_v(w)}\right], & r > a \qquad (7.49b) \end{cases}$$

$$E_\theta = \begin{cases} -\dfrac{a^2}{u^2}\left[\dfrac{i\omega\mu_0 BuJ_v'(ur/a)}{aJ_v(u)} + \dfrac{\beta vAJ_v(ur/a)}{rJ_v(u)}\right], & r \leqslant a \qquad (7.50a) \\[10pt] \dfrac{a^2}{w^2}\left[\dfrac{i\omega\mu_0 BwK_v'(wr/a)}{aK_v(w)} + \dfrac{\beta vAK_v(wr/a)}{rK_v(w)}\right], & r > a \qquad (7.50b) \end{cases}$$

$$H_r = \begin{cases} \dfrac{a^2}{u^2}\left[\dfrac{vn_1{}^2\varepsilon_0\omega A J_v(ur/a)}{rJ_v(u)} + \dfrac{i\beta Bu J_v'(ur/a)}{aJ_v(u)}\right], & r \leq a \quad (7.51\text{a}) \\[2ex] -\dfrac{a^2}{w^2}\left[\dfrac{vn_2{}^2\varepsilon_0\omega A K_v(wr/a)}{rK_v(w)} + \dfrac{i\beta Bw K_v'(wr/a)}{aK_v(w)}\right], & r > a \quad (7.51\text{b}) \end{cases}$$

$$H_\theta = \begin{cases} \dfrac{a^2}{u^2}\left[\dfrac{in_1{}^2\varepsilon_0\omega A u J_v'(ur/a)}{aJ_v(u)} - \dfrac{v\beta B J_v(ur/a)}{rJ_v(u)}\right], & r \leq a \quad (7.52\text{a}) \\[2ex] -\dfrac{a^2}{w^2}\left[\dfrac{in_2{}^2\varepsilon_0\omega A w K_v'(wr/a)}{aK_v(w)} - \dfrac{v\beta B K_v(wr/a)}{rK_v(w)}\right], & r > a. \quad (7.52\text{b}) \end{cases}$$

It remains to satisfy the boundary conditions at $r = a$ and hence to evaluate the ratio of the constants A, B and find the eigenvalue equation. At the core–cladding interface $r = a$ it is required that the tangential components of the electric and magnetic fields be continuous. From the choice of axial components in (7.47) and (7.48) we see that E_z and H_z are automatically continuous at $r = a$. It therefore remains only to equate at $r = a$ the corresponding expressions for E_θ in (7.50a) and (7.50b), and similarly for H_θ in (7.52a) and (7.52b). Carrying out this procedure, we find from (7.50)

$$\frac{A}{B} = -\frac{i\omega\mu_0}{\beta v}\left(\frac{u^2 w^2}{v^2}\right)\left[\frac{J_v'(u)}{uJ_v(u)} + \frac{K_v'(w)}{wK_v(w)}\right] \quad (7.53)$$

and from (7.52):

$$\frac{A}{B} = -i\frac{v\beta}{\omega\varepsilon_0}\left(\frac{v^2}{u^2 w^2}\right)\left[n_1{}^2\frac{J_v'(u)}{uJ_v(u)} + n_2{}^2\frac{K_v'(w)}{wK_v(w)}\right]^{-1}. \quad (7.54)$$

On equating the right-hand sides of (7.53) and (7.54) we obtain the well-known eigenvalue equation for modes of a dielectric rod:

$$\left[\frac{J_v'(u)}{uJ_v(u)} + \frac{K_v'(w)}{wK_v(w)}\right]\left[\frac{J_v'(u)}{uJ_v(w)} + (1-2\Delta)\frac{K_v'(w)}{wK_v(w)}\right] = \left(\frac{v\beta}{kn_1}\right)^2\left(\frac{v}{uw}\right)^4 \quad (7.55)$$

where we have used the conventional definition $n_2{}^2 = n_1{}^2(1-2\Delta)$ as in earlier chapters.

Consider now the special case $v = 0$ when the r.h.s. of (7.55) vanishes. From equation (7.53) we see that in order to keep A finite we must choose $B = 0$, i.e. $H_z = 0$ and we have the special case of TM modes. For nonzero A in this case, the expression in square brackets on the r.h.s. of (7.53) cannot be zero and hence from (7.55) it follows that the eigenvalue equation for TM modes is:

$$\frac{J_1(u)}{uJ_0(u)} + (1-2\Delta)\frac{K_1(w)}{wK_0(w)} = 0 \quad (7.56)$$

where we have used the relations $J_0'(u) = -J_1(u)$, $K_0'(w) = -K_1(w)$ (Abramowitz and Stegun, 1964). Similarly, when $v = 0$ in (7.54) we must choose $A = 0$, so that this case corresponds to TE modes; it follows from (7.55) that the TE eigenvalue equation is:

$$\frac{J_1(u)}{uJ_0(u)} + \frac{K_1(w)}{wK_0(w)} = 0, \quad (7.57)$$

Thus for the special case $v = 0$ we find that the dielectric rod waveguide supports modes which may be designated as TM_{0M} and TE_{0M} where M is a radial mode number.

In the general case $v \geq 1$ we are concerned with eigenvalues which are solutions of the rather complicated transcendental equation (7.55). However, considerable simplification can be achieved when we recall that for most circular dielectric guides of interest at optical wavelengths the relative difference Δ between core and cladding refractive indices is of order 1 per cent or less. Hence we may usually apply the weakly-guiding approximation $\Delta \ll 1$. Writing $(\beta/kn_1)^2 = 1 - 2\Delta u^2/v^2$ on the r.h.s. of (7.55) and taking the limit as $\Delta \to 0$, we see that both sides of the equation become perfect squares and we have the result:

$$\frac{J_v'(u)}{uJ_v(u)} + \frac{K_v'(w)}{wK_v(w)} = \pm \frac{vv^2}{u^2w^2}. \tag{7.58}$$

This result may also be simplified somewhat by using the Bessel and Hankel function relationships (Abramowitz and Stegun, 1964):

$$J_v'(u) = \mp \frac{vJ_v(u)}{u} \pm J_{v\mp 1}(u) \tag{7.59}$$

$$K_v'(w) = \mp \frac{vK_v(w)}{w} - K_{v\mp 1}(w) \tag{7.60}$$

Taking the + sign in (7.58) with the lower signs in (7.59) and (7.60) yields:

$$\frac{J_{v+1}(u)}{uJ_v(u)} + \frac{K_{v+1}(w)}{wK_v(w)} = 0 \quad (v > 0) \tag{7.61}$$

whilst the opposite combination gives:

$$\frac{J_{v-1}(u)}{uJ_v(u)} - \frac{K_{v-1}(w)}{wK_v(w)} = 0 \quad (v > 0). \tag{7.62}$$

The modes corresponding to equation (7.61) are termed EH modes and those corresponding to (7.62) are called HE modes following the nomenclature of Snitzer (1961). The reason for this terminology is related to the ratio of axial components of electric and magnetic field, $E_z : H_z$. From equations (7.53), (7.54) and (7.61), (7.62) this ratio is given as:

EH modes: $$\frac{A}{B} = -\frac{i}{n_1}\left(\frac{\mu_0}{\varepsilon_0}\right)^{1/2} \frac{v}{|v|} \tag{7.63}$$

HE modes: $$\frac{A}{B} = +\frac{i}{n_1}\left(\frac{\mu_0}{\varepsilon_0}\right)^{1/2} \frac{v}{|v|}. \tag{7.64}$$

Whilst these simple relations hold true only for the limit $\Delta = 0$, a similar distinction between HE and EH modes can be made for general Δ by considering the ratio $E_z : H_z$ far from cut-off (Snitzer, 1961). The reason for including the term $v/|v|$ in (7.63) and (7.64) is related to our original choice of the azimuthal dependence as $\exp(iv\theta)$ in equation (7.17). Since we could equally well have chosen a variation as $\exp(-iv\theta)$ there is a two-fold degeneracy of modes of azimuthal order $v (\neq 0)$. Since we wish to take advantage of this in deriving the LP modes in Section 7.2.2, it is therefore necessary to distinguish carefully between v's occurring as a result of the azimuthal variation and those occurring as the order of a Bessel function; the latter are invariant with change of sign and are therefore denoted by $|v|$ (Marcuse, 1972).

The mode cut-offs also form a distinction between the various classes of modes on the circular dielectric rod. The cut-off condition $w = 0$ ($u = v \equiv v_c$) may be investigated with the aid of the following approximations for the Hankel functions of small argument:

$$K_0(w) \simeq \ln\left(\frac{2}{\gamma w}\right) \quad \text{where } \gamma = 1.781 \text{ (Euler's constant)} \tag{7.65}$$

$$K_\nu(w) \simeq \frac{(\nu-1)!}{2}\left(\frac{2}{w}\right)^\nu \quad (\nu > 0). \tag{7.66}$$

It follows that $K_1(w)/wK_0(w) \to \infty$ as $w \to 0$ and hence from (7.56) and (7.57) the cut-off condition for TE_{0M} and TM_{0M} modes is given by

TE_{0M}, TM_{0M} modes: $\qquad\qquad J_0(v_c) = 0. \tag{7.67}$

Similarly, for EH modes in the limit $\Delta = 0$ we may use the result from (7.66) that $K_{\nu+1}(w)/wK_\nu(w) \to \infty$ as $w \to 0$ ($\nu > 0$), and hence from equation (7.61) the cut-off condition is given by:

$EH_{\nu N}$ modes: $\qquad\qquad J_\nu(v_c) = 0 \quad (\nu > 0, v_c > 0). \tag{7.68}$

The case $v_c = 0$ is explicitly excluded from (7.68) since for this case the term $J_{\nu+1}(u)/uJ_\nu(u)$ in (7.61) can never become infinite and hence cannot fulfil the cut-off condition. We may prove this by using the result for *small u*:

$$J_\nu(u) \simeq \frac{1}{\nu!}\left(\frac{u}{2}\right)^\nu \quad (\nu > 0). \tag{7.69}$$

It follows that in the limit of small u,

$$\frac{J_{\nu+1}(u)}{uJ_\nu(u)} \to \frac{1}{2(\nu+1)};$$

hence we must exclude the case $v_c = 0$ from (7.68).

For HE mode cut-offs in the limit $\Delta = 0$ we must examine two separate cases, $\nu = 1$ and $\nu > 1$. For $\nu = 1$, we use (7.65) and (7.66) to find that $K_0(w)/wK_1(w) \to \infty$ as $w \to 0$; hence from the HE eigenvalue equation (7.62) for $\nu = 1$ it follows that the cut-off condition is:

HE_{1M} modes: $\qquad\qquad J_1(v_c) = 0. \tag{7.70}$

The case $v_c = 0$ is not excluded from (7.70) as was the case for EH modes in (7.68). This is because the term $J_0(u)/uJ_1(u) \to \infty$ as $u \to 0$ in equation (7.62) as a consequence of the small argument approximation $J_0(u) \simeq 1$. Hence the HE_{11} mode is the fundamental mode of the circular-dielectric rod waveguide since it propagates for all frequencies v. It is therefore this mode which is of interest for use in optical communications systems with single-mode fibres. Note that this mode has $M = 1$ rather than 0, since it is common practice to use M to count field extrema for the dielectric rod rather than field zeros as in Section 7.1 ($M = N + 1$).

HE mode cut-offs for $\nu > 1$ are given from (7.65) and (7.66) by the condition

$$\frac{K_{\nu-1}(w)}{wK_\nu(w)} \to \frac{1}{2(\nu-1)}$$

for $w \to 0$. It follows from (7.62) that the cut-off condition may be written:

$$\frac{J_{\nu-1}(v_c)}{v_c J_\nu(v_c)} = \frac{1}{2(\nu-1)}. \tag{7.71}$$

If we use the Bessel function identity (Abramowitz and Stegun, 1964):

$$J_\nu(v_c) + J_{\nu-2}(v_c) = \frac{2(\nu-1)}{v_c}J_{\nu-1}(v_c) \tag{7.72}$$

then (7.71) reduces to the simpler form (Biernson and Kingsley, 1965):

$HE_{\nu M}$ modes: $\qquad\qquad J_{\nu-2}(v_c) = 0 \quad (\nu > 1, v_c > 0) \tag{7.73}$

where the case $v_c = 0$ is excluded for similar reasons to those for (7.68).

The cut-off conditions (7.68), (7.70), and (7.73) have been derived here for hybrid modes in the limit $\Delta = 0$. For nonzero Δ it may be shown that (7.68) and (7.70) still hold but that the r.h.s. of (7.73) is replaced by a term of order Δ (Snitzer, 1961; Biernson and Kingsley, 1965; Marcuse, 1972). For the general case of nonzero Δ, to find hybrid mode eigenvalues it is of course necessary to solve equation (7.55). Normalized plots of u versus v for the first twelve modes for the cases of zero and nonzero Δ have been given by Biernson and Kingsley (1965). Other than in optical fibres, a further application of the theory of the dielectric rod waveguide occurs in the mechanism of colour vision in the human eye. The outer segments of the retinal cones are cylindrical waveguides of diameter about one micron. Waveguide mode patterns corresponding to the first 12 modes may be observed in these photoreceptors; calculated plots of the intensity distributions for these modes using appropriate parameters have been given by Biernson and Kingsley (1965).

7.2.2 Weakly-guiding optical fibres

Let us now confine attention to the weakly-guiding situation $\Delta \ll 1$ and investigate the nature of the hybrid modes subject to the approximate EH and HE eigenvalue equations (7.61) and (7.62). We wish to show that the modes can be re-expressed in terms of linearly-polarized (LP) 'pseudo-modes' (Snyder, 1969; Gloge, 1971) with greatly simplified properties. To that end we use the relations (7.63) and (7.64) in order to simplify the field components (7.47)–(7.52). Using the convention of upper sign applying to EH modes and lower sign to HE modes, we find:

$$E_z = \begin{cases} \dfrac{A J_\nu(ur/a)}{J_\nu(u)}, & r \leqslant a \\[2mm] \dfrac{A K_\nu(wr/a)}{K_\nu(w)}, & r > a \end{cases} \quad (7.74)$$

$$H_z = \begin{cases} \pm i A n \left(\dfrac{\varepsilon_0}{\mu_0}\right)^{1/2} \dfrac{\nu}{|\nu|} \dfrac{J_\nu(ur/a)}{J_\nu(u)}, & r \leqslant a \\[2mm] \pm i A n \left(\dfrac{\varepsilon_0}{\mu_0}\right)^{1/2} \dfrac{\nu}{|\nu|} \dfrac{K_\nu(wr/a)}{K_\nu(w)}, & r > a \end{cases} \quad (7.75)$$

$$E_r = \begin{cases} \mp i A \dfrac{akn}{u} \dfrac{J_{\nu\pm1}(ur/a)}{J_\nu(u)}, & r \leqslant a \\[2mm] i A \dfrac{akn}{w} \dfrac{K_{\nu\pm1}(wr/a)}{K_\nu(w)}, & r > a \end{cases} \quad (7.76)$$

$$E_\theta = \begin{cases} -A \dfrac{akn}{u} \dfrac{\nu}{|\nu|} \dfrac{J_{\nu\pm1}(ur/a)}{J_\nu(u)}, & r \leqslant a \\[2mm] \pm A \dfrac{akn}{w} \dfrac{\nu}{|\nu|} \dfrac{K_{\nu\pm1}(wr/a)}{K_\nu(w)}, & r > a \end{cases} \quad (7.77)$$

$$H_r = \begin{cases} A\dfrac{akn^2}{u}\left(\dfrac{\varepsilon_0}{\mu_0}\right)^{1/2}\dfrac{v}{|v|}\dfrac{J_{v\pm 1}(ur/a)}{J_v(u)}, & r \leqslant a \\ \mp A\dfrac{akn^2}{w}\left(\dfrac{\varepsilon_0}{\mu_0}\right)^{1/2}\dfrac{v}{|v|}\dfrac{K_{v\pm 1}(wr/a)}{K_v(w)}, & r > a \end{cases} \quad (7.78)$$

$$H_\theta = \begin{cases} \mp iA\dfrac{akn^2}{u}\left(\dfrac{\varepsilon_0}{\mu_0}\right)^{1/2}\dfrac{J_{v\pm 1}(ur/a)}{J_v(u)}, & r \leqslant a \\ iA\dfrac{akn^2}{w}\left(\dfrac{\varepsilon_0}{\mu_0}\right)^{1/2}\dfrac{K_{v\pm 1}(wr/a)}{K_v(w)}, & r > a. \end{cases} \quad (7.79)$$

Equations (7.74)–(7.79) were obtained with the aid of the Bessel function relations (7.59) and (7.60); the symbol $n = n_1 \simeq n_2$ is used throughout to emphasize that these are approximations which hold strictly only in the limit $\Delta \to 0$.

Recalling that the θ-dependence of the fields has been suppressed in (7.74)–(7.79) and that we have the freedom to choose $\exp(iv\theta)$ or $\exp(-iv\theta)$ since the terms $v/|v|$ are included in these equations, we may construct new modes by adding and subtracting pairs of solutions. In this way we may find trigonometric functions to describe the azimuthal dependence, rather than complex exponentials. At the same time we may re-normalize the fields for future use by multiplying each component by the factor $uJ_v(u)/(2J_{v\pm 1}(u))$ where again the upper and lower signs apply to EH and HE modes, respectively. By virtue of the eigenvalue equations (7.61) and (7.62), this is equivalent to multiplying by the factor $\mp wK_v(w)/(2K_{v\pm 1}(w))$. Hence, using this latter factor for the fields with $r > a$ and the former factor for $r \leqslant a$, and explicitly including the θ-dependence, we find the new field components as:

$$E_z = \begin{cases} \dfrac{AuJ_v(ur/a)}{J_{v\pm 1}(u)}\begin{Bmatrix}\cos v\theta \\ i\sin v\theta\end{Bmatrix}, & r \leqslant a \\ \mp \dfrac{AwK_v(wr/a)}{K_{v\pm 1}(w)}\begin{Bmatrix}\cos v\theta \\ i\sin v\theta\end{Bmatrix}, & r > a \end{cases} \quad (7.80)$$

$$H_z = \begin{cases} \pm An\left(\dfrac{\varepsilon_0}{\mu_0}\right)^{1/2}\dfrac{uJ_v(ur/a)}{J_{v\pm 1}(u)}\begin{Bmatrix}-\sin v\theta \\ i\cos v\theta\end{Bmatrix}, & r \leqslant a \\ -An\left(\dfrac{\varepsilon_0}{\mu_0}\right)^{1/2}\dfrac{wK_v(wr/a)}{K_{v\pm 1}(w)}\begin{Bmatrix}-\sin v\theta \\ i\cos v\theta\end{Bmatrix}, & r > a \end{cases} \quad (7.81)$$

$$E_r = \begin{cases} \mp Aakn\dfrac{J_{v\pm 1}(ur/a)}{J_{v\pm 1}(u)}\begin{Bmatrix}i\cos v\theta \\ -\sin v\theta\end{Bmatrix}, & r \leqslant a \\ \mp Aakn\dfrac{K_{v\pm 1}(wr/a)}{K_{v\pm 1}(w)}\begin{Bmatrix}i\cos v\theta \\ -\sin v\theta\end{Bmatrix}, & r > a \end{cases} \quad (7.82)$$

$$E_\theta = \begin{cases} -Aakn\dfrac{J_{v\pm 1}(ur/a)}{J_{v\pm 1}(u)}\begin{Bmatrix}i\sin v\theta \\ \cos v\theta\end{Bmatrix}, & r \leqslant a \\ -Aakn\dfrac{K_{v\pm 1}(wr/a)}{K_{v\pm 1}(w)}\begin{Bmatrix}i\sin v\theta \\ \cos v\theta\end{Bmatrix}, & r > a \end{cases} \quad (7.83)$$

$$H_r = \begin{cases} Aakn^2 \left(\frac{\varepsilon_0}{\mu_0}\right)^{1/2} \frac{J_{v\pm 1}(ur/a)}{J_{v\pm 1}(u)} \begin{Bmatrix} i\sin v\theta \\ \cos v\theta \end{Bmatrix}, & r \leq a \\ Aakn^2 \left(\frac{\varepsilon_0}{\mu_0}\right)^{1/2} \frac{K_{v\pm 1}(wr/a)}{K_{v\pm 1}(w)} \begin{Bmatrix} i\sin v\theta \\ \cos v\theta \end{Bmatrix}, & r > a \end{cases} \quad (7.84)$$

$$H_\theta = \begin{cases} \mp Aakn^2 \left(\frac{\varepsilon_0}{\mu_0}\right)^{1/2} \frac{J_{v\pm 1}(ur/a)}{J_{v\pm 1}(u)} \begin{Bmatrix} i\cos v\theta \\ -\sin v\theta \end{Bmatrix}, & r \leq a \\ \mp Aakn^2 \left(\frac{\varepsilon_0}{\mu_0}\right)^{1/2} \frac{K_{v\pm 1}(wr/a)}{K_{v\pm 1}(w)} \begin{Bmatrix} i\cos v\theta \\ -\sin v\theta \end{Bmatrix}, & r > a \end{cases} \quad (7.85)$$

where each choice of trigonometric functions represents an alternative multiplicative factor related to the mode polarization.

Since we are seeking LP modes we now transform the transverse field components from polar to Cartesian coordinates with the usual transformation:

$$\begin{pmatrix} E_x \\ E_y \end{pmatrix} = \begin{pmatrix} \cos\theta & -\sin\theta \\ \sin\theta & \cos\theta \end{pmatrix} \begin{pmatrix} E_r \\ E_\theta \end{pmatrix} \quad (7.86)$$

with a similar equation for H_x, H_y in terms of H_r, H_θ. The result is:

$$E_x = \begin{cases} \mp Aakn \frac{J_{v\pm 1}(ur/a)}{J_{v\pm 1}(u)} \begin{Bmatrix} i\cos[(v\pm 1)\theta] \\ -\sin[(v\pm 1)\theta] \end{Bmatrix}, & r \leq a \\ \mp Aakn \frac{K_{v\pm 1}(wr/a)}{K_{v\pm 1}(w)} \begin{Bmatrix} i\cos[(v\pm 1)\theta] \\ -\sin[(v\pm 1)\theta] \end{Bmatrix}, & r > a \end{cases} \quad (7.87)$$

$$E_y = \begin{cases} -Aakn \frac{J_{v\pm 1}(ur/a)}{J_{v\pm 1}(u)} \begin{Bmatrix} i\sin[(v\pm 1)\theta] \\ \cos[(v\pm 1)\theta] \end{Bmatrix}, & r \leq a \\ -Aakn \frac{K_{v\pm 1}(wr/a)}{K_{v\pm 1}(w)} \begin{Bmatrix} i\sin[(v\pm 1)\theta] \\ \cos[(v\pm 1)\theta] \end{Bmatrix}, & r > a \end{cases} \quad (7.88)$$

$$H_x = \begin{cases} Aakn^2 \left(\frac{\varepsilon_0}{\mu_0}\right)^{1/2} \frac{J_{v\pm 1}(ur/a)}{J_{v\pm 1}(u)} \begin{Bmatrix} i\sin[(v\pm 1)\theta] \\ \cos[(v\pm 1)\theta] \end{Bmatrix}, & r \leq a \\ Aakn^2 \left(\frac{\varepsilon_0}{\mu_0}\right)^{1/2} \frac{K_{v\pm 1}(wr/a)}{K_{v\pm 1}(w)} \begin{Bmatrix} i\sin[(v\pm 1)\theta] \\ \cos[(v\pm 1)\theta] \end{Bmatrix}, & r > a \end{cases} \quad (7.89)$$

$$H_y = \begin{cases} \mp Aakn^2 \left(\frac{\varepsilon_0}{\mu_0}\right)^{1/2} \frac{J_{v\pm 1}(ur/a)}{J_{v\pm 1}(u)} \begin{Bmatrix} i\cos[(v\pm 1)\theta] \\ -\sin[(v\pm 1)\theta] \end{Bmatrix}, & r \leq a \\ \mp Aakn^2 \left(\frac{\varepsilon_0}{\mu_0}\right)^{1/2} \frac{K_{v\pm 1}(wr/a)}{K_{v\pm 1}(w)} \begin{Bmatrix} i\cos[(v\pm 1)\theta] \\ -\sin[(v\pm 1)\theta] \end{Bmatrix}, & r > a. \end{cases} \quad (7.90)$$

Equations (7.80), (7.81), and (7.87)–(7.90) are to be interpreted with upper and lower signs corresponding to EH and HE modes, respectively, whilst the bracketed pairs of trigonometric functions refer to the polarization of each mode; the choice of upper or lower circular function is to be made consistently throughout, but is of course independent of the sign convention for the modes. Once again the reader is reminded that these results hold only for the limit $\Delta \to 0$ and that in general the more complicated field components (7.47)–(7.52) must be used. However, as an approximation to the modes of the weakly-guiding fibre the above equations are extremely useful, as will become apparent below.

There is clearly a degeneracy of modes inherent in equations (7.87)–(7.90) resulting from the choice of signs for EH and HE modes. To make this apparent we need only set $l = v+1$ for EH modes and $l = v-1$ for HE modes. The eigenvalue equations (7.61) and (7.62) may then be written as:

$$\frac{J_l(u)}{uJ_{l\pm 1}(u)} = \pm \frac{K_l(w)}{wK_{l\pm 1}(w)}. \tag{7.91}$$

This is the weakly-guiding step-index fibre eigenvalue equation which holds for all l. Note that the case $l = 0$ corresponds to $v = 1$ so that the fundamental HE_{11} mode is referred to as LP_{01} with this convention. Also, the TE and TM eigenvalue equations (7.57) and (7.56) in the limit $\Delta \to 0$ can be recovered from (7.91) for $l = 1$ with the choice of the lower signs. Analytic approximations for the solution of (7.91) have been developed by Snyder (1969), Gloge (1971), Rudolph and Neumann (1976), Marcuse (1978), Miyaji and Nishida (1979), and Snyder and Sammut (1979a); Sammut (1979) has compared the relative accuracy of these approximations for the HE_{11} mode.

To see the effects of this mode degeneracy of the EH and HE modes of appropriate azimuthal orders, we form the following combinations:

$$(E_x)_{HE_{l+1}} + (E_x)_{EH_{l-1}} = 0 \tag{7.92}$$

$$(E_y)_{HE_{l+1}} + (E_y)_{EH_{l-1}} = \begin{cases} -A2akn \dfrac{J_l(ur/a)}{J_l(u)} \begin{Bmatrix} i\sin l\theta \\ \cos l\theta \end{Bmatrix}, & r \leqslant a \\[1em] -A2akn \dfrac{K_l(wr/a)}{K_l(w)} \begin{Bmatrix} i\sin l\theta \\ \cos l\theta \end{Bmatrix}, & r > a \end{cases} \tag{7.93}$$

$$(E_z)_{HE_{l+1}} + (E_z)_{EH_{l-1}} = \begin{cases} Au\left[\dfrac{J_{l+1}(ur/a)}{J_l(u)}\begin{Bmatrix}\cos[(l+1)\theta]\\ i\sin[(l+1)\theta]\end{Bmatrix}\right. \\[1em] \left.+\dfrac{J_{l-1}(ur/a)}{J_l(u)}\begin{Bmatrix}\cos[(l-1)\theta]\\ i\sin[(l-1)\theta]\end{Bmatrix}\right], \quad r \leqslant a \\[1em] Aw\left[\dfrac{K_{l+1}(wr/a)}{K_l(w)}\begin{Bmatrix}\cos[(l+1)\theta]\\ i\sin[(l+1)\theta]\end{Bmatrix}\right. \\[1em] \left.-\dfrac{K_{l-1}(wr/a)}{K_l(w)}\begin{Bmatrix}\cos[(l-1)\theta]\\ i\sin[(l-1)\theta]\end{Bmatrix}\right], \quad r > a \end{cases} \tag{7.94}$$

$$(H_x)_{HE_{l+1}} + (H_x)_{EH_{l-1}} = \begin{cases} A2akn^2 \left(\dfrac{\varepsilon_0}{\mu_0}\right)^{1/2} \dfrac{J_l(ur/a)}{J_l(u)} \begin{Bmatrix} i\sin l\theta \\ \cos l\theta \end{Bmatrix}, & r \leqslant a \\[1em] A2akn^2 \left(\dfrac{\varepsilon_0}{\mu_0}\right)^{1/2} \dfrac{K_l(wr/a)}{K_l(w)} \begin{Bmatrix} i\sin l\theta \\ \cos l\theta \end{Bmatrix}, & r > a \end{cases} \tag{7.95}$$

$$(H_y)_{HE_{l+1}} + (H_y)_{EH_{l-1}} = 0 \tag{7.96}$$

$$(H_z)_{HE_{l+1}} + (H_z)_{EH_{l-1}} = \begin{cases} Aun\left(\dfrac{\varepsilon_0}{\mu_0}\right)^{1/2}\left[\dfrac{J_{l+1}(ur/a)}{J_l(u)}\begin{Bmatrix}\sin[(l+1)\theta]\\-i\cos[(l+1)\theta]\end{Bmatrix}\right.\\\left.-\dfrac{J_{l-1}(ur/a)}{J_l(u)}\begin{Bmatrix}\sin[(l-1)\theta]\\-i\cos[(l-1)\theta]\end{Bmatrix}\right],\quad r\leqslant a\\ Awn\left(\dfrac{\varepsilon_0}{\mu_0}\right)^{1/2}\left[\dfrac{K_{l+1}(wr/a)}{K_l(w)}\begin{Bmatrix}\sin[(l+1)\theta]\\-i\cos[(l+1)\theta]\end{Bmatrix}\right.\\\left.+\dfrac{K_{l-1}(wr/a)}{K_l(w)}\begin{Bmatrix}\sin[(l-1)\theta]\\-i\cos[(l-1)\theta]\end{Bmatrix}\right],\quad r>a \end{cases}$$ (7.97)

From these equations it is clear that the transverse fields are linearly polarized. Furthermore, inspection of the coefficients of the axial fields shows that they are each of order Δ times the magnitudes of the nonvanishing transverse field components. Hence these approximate LP modes are almost TEM for very small Δ. A further set of solutions may be constructed from (7.87)–(7.90) by forming the differences ($HE_{l+1} - EH_{l-1}$) rather than the sums as in (7.92)–(7.97). The result is a set of LP field components rotated by 90° from the first set, i.e. with $E_y = H_x = 0$. Consequently there is always a two-fold degeneracy of polarization of the LP modes; in addition there is a further two-fold degeneracy arising from the choice of trigonometric functions ($l \neq 0$). Hence all LP modes with $l \neq 0$ are 4-fold degenerate, and all those with $l = 0$ are 2-fold degenerate. Table 7.2 illustrates these degeneracies by tabulating the corresponding HE, EH, TE, and TM modes for each LP mode. The cut-off condition $w = 0$ ($u = v = v_c$) is used to order the modes; from equation (7.91) this yields

LP_{lM} modes: $\qquad\qquad\qquad J_{l-1}(v_c) = 0 \qquad\qquad\qquad$ (7.98)

which includes the case $l = 0$ (for which $J_{-1}(v_c) = -J_1(v_c)$) which is counted so as to include $J_1(0) = 0$ as the first root; the values of v_c for the first 11 LP modes are given in the table. From the table it is also clear that each LP_{lM} mode for $l \neq 1$ is equivalent to the $HE_{l+1,M}$ and $EH_{l-1,M}$ modes, whilst each LP_{1M} mode is equivalent to HE_{2M}, TE_{0M}, and TM_{0M}. The final column gives the total number of modes propagating at the cut-off of the next higher LP mode, including this degeneracy as well as the two-fold polarization degeneracy referred to above.

Table 7.2 Degeneracies of the LP modes (including polarization degeneracy)

Cut-off condition	v_c	LP mode	Equivalent modes	Total number of modes propagating for $v = v_c$ of next LP mode
$J_1 = 0$	0	LP_{01}	HE_{11}	2
$J_0 = 0$	2.405	LP_{11}	HE_{21}, TE_{01}, TM_{01}	6
$J_1 = 0$	3.832	LP_{02}, LP_{21}	HE_{12}, EH_{11}, HE_{31}	12
$J_2 = 0$	5.136	LP_{31}	EH_{21}, HE_{41}	16
$J_0 = 0$	5.520	LP_{12}	HE_{22}, TE_{02}, TM_{02}	20
$J_3 = 0$	6.380	LP_{41}	EH_{31}, HE_{51}	24
$J_1 = 0$	7.016	LP_{03}, LP_{22}	HE_{13}, EH_{12}, HE_{32}	30
$J_4 = 0$	7.588	LP_{51}	EH_{41}, HE_{61}	34
$J_2 = 0$	8.417	LP_{32}	EH_{22}, HE_{42}	38
$J_0 = 0$	8.654	LP_{13}	HE_{23}, TE_{03}, TM_{03}	42
$J_5 = 0$	8.771	LP_{61}	EH_{51}, HE_{71}	46

The degeneracy of the LP modes which we have discussed is of course only exact for the case $\Delta = 0$. In practice even for a small Δ, the $HE_{l+1,M}$ and $EH_{l-1,M}$ modes have slightly different propagation constants β_1 and β_2. Hence as the modes propagate the superposition of their fields—the LP_{lM} 'pseudo-mode'—actually will change its field distribution. Only after a distance of a beat wavelength, i.e. $2\pi/(\beta_1 - \beta_2)$, will the two constituent modes be exactly in phase again so that a linearly-polarized distribution is achieved. In spite of this disadvantage, the field descriptions and eigenvalues resulting from the LP mode treatment of the weakly-guiding fibre have provided useful results for fibre design. It is worth noting that all the results (7.91)–(7.97) can be equally well derived by directly assuming linearly-polarized modes (Gloge, 1971) without going through the detailed derivation from the exact modes. It has also been pointed out (Arnaud, 1974a) that these results are also achieved by a simple scalar analysis. Techniques for generating the more accurate hybrid modes from the approximate LP solutions have been discussed by Kurtz (1975), Snyder and Young (1978), and Tjaden (1978).

7.2.3 Properties of LP modes

Eigenvalues for LP modes of the weakly-guiding step-index fibre may be found from the solutions of equation (7.91). Following Gloge (1971), it is now customary to present these results as plots of b versus v where b is the normalized propagation constant (w^2/v^2). These results may be compared with those from the geometric-optics approximation for the eigenvalue equation. A derivation of this latter equation for the hollow circular pipe was given in Section 7.1.1; the general result (7.26) may be applied to the weakly-guiding fibre by suitable choice of the phase shift δ. Results for δ for two possible polarizations are given in Chapter 2, equations (2.32) and (2.34). For the weakly-guiding situation these equations both reduce to the form:

$$\delta = 2 \tan^{-1} \left[\frac{(n_1^2 \sin^2 \theta_1 - n_2^2)^{1/2}}{n_1 \cos \theta_1} \right]. \tag{7.99}$$

With the aid of the relation (7.25) between ray angle θ_1 and mode constants β, v (or, equivalently, l), we may re-cast (7.99) into the form:

$$\delta = 2 \cos^{-1} \left[\frac{(u^2 - l^2)^{1/2}}{v} \right]. \tag{7.100}$$

Using this result in (7.26), we find the ray-optics approximation for the eigenvalue equation as (Hartog and Adams, 1977; Love and Snyder, 1976):

$$(u^2 - l^2)^{1/2} - l \cos^{-1} \left(\frac{l}{u} \right) = \cos^{-1} \left[\frac{(u^2 - l^2)^{1/2}}{v} \right] + (M - \tfrac{3}{4})\pi$$
$$(l = 0, 1, 2, \ldots; M = 1, 2, 3, \ldots). \tag{7.101}$$

Note that equation (7.101) uses the mode numbers $l = 0, 1, 2, \ldots$ and $M = 1, 2, 3, \ldots$ so that it is in a form for direct comparison with the solution of (7.91) using the integer M as radial mode number for the LP modes. Note also that the asymptotic forms of equations (7.91) and (7.101) for large v are identical.

In order to determine the accuracy of the approximation represented by (7.101) we have plotted results obtained from equations (7.91) and (7.101) in Figures 7.9 and 7.10 for $v = 0$ and 1, respectively (Hartog and Adams, 1977). The figures show that (7.101) provides a very good approximation to the weakly-guiding mode eigenvalues even for lower-order modes (except for the fundamental mode close to cut-off). The r.m.s. difference between each pair of curves (calculated from equations (7.91) and (7.101)) is less than 10^{-2} for all the cases of Figures 7.9 and 7.10.

Let us consider now the dispersive properties of the weakly-guiding fibre. To maintain generality we will include here the effects of material dispersion as well as those resulting from the waveguide, i.e. we allow the refractive indices n_1 and n_2 to be functions of wavelength λ. The corresponding group indices m_1, m_2 are defined as in equation (2.54), viz:

$$m_j = n_j - \lambda \frac{dn_j}{d\lambda} \equiv n_j + k \frac{dn_j}{dk} \quad (j = 1, 2). \tag{7.102}$$

In terms of these definitions and the usual interpretation of normalized frequency $v = ak(n_1^2 - n_2^2)^{1/2}$, we find by differentiation:

$$\frac{dv}{d\lambda} = -\frac{v}{\lambda}\left(\frac{m_1 n_1 - m_2 n_2}{n_1^2 - n_2^2}\right). \tag{7.103}$$

In order to involve the waveguide dispersion in our discussion, let us define an effective index $n = \beta/k$ which includes waveguide effects, so that in terms of b we have:

$$n^2 = bn_1^2 + (1-b)n_2^2. \tag{7.104}$$

The corresponding effective group index $m = d\beta/dk$ is given in terms of an equation exactly analogous to (7.102) merely by omitting subscripts in that equation. Hence, differentiating

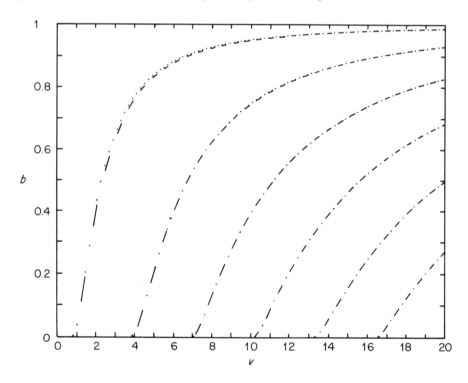

Figure 7.9 Comparison between eigenvalues for $l = 0$ obtained by (a) the geometric optics approximation (7.101) (dotted lines), and (b) the weakly-guiding mode theory (7.91) (broken lines). (From Hartog and Adams (1977). Reproduced by permission of Chapman and Hall)

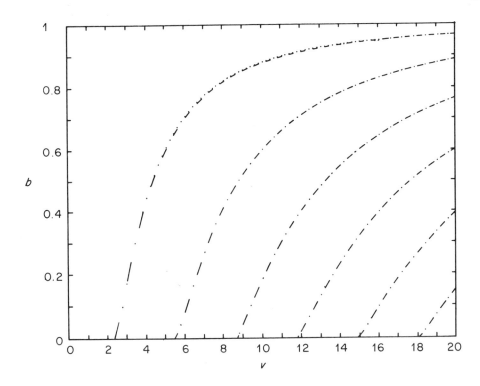

Figure 7.10 As Figure 7.9, except that here $l = 1$. (From Hartog and Adams (1977). Reproduced by permission of Chapman and Hall)

(7.104) with respect to λ and using (7.102) and (7.103), we derive the following general relationship between effective phase and group indices (Gloge, 1971a):

$$\frac{\beta}{k}\frac{d\beta}{dk} \equiv nm = n_2 m_2 + \frac{1}{2}\left[b + \frac{d(bv)}{dv}\right](n_1 m_1 - n_2 m_2). \quad (7.105)$$

It is important to realize the value of equation (7.105) in clarifying the relative components of material and waveguide dispersion; whilst n_1, n_2, m_1, m_2 are all material properties, only the term in square brackets on the r.h.s. contains the waveguide effects. This makes it possible to generalize earlier numerical calculations of dispersive effects in fibres (Laybourn, 1968; Gambling and Laybourn, 1970; Kapron and Keck, 1971; Dyott and Stern, 1971) by separating the relative effects of material and waveguide dispersion. Equation (7.105) is also related as we shall see below to the general phase/group velocity connection, equation (5.123), derived earlier for dispersive planar guides.

In order to discuss dispersive effects further, we must derive an expression for the term $d(bv)/dv$ for the weakly-guiding fibre. Differentiating the eigenvalue equation (7.91) on both sides with respect to v, and using (7.59), (7.60), and other standard Bessel function relations, we find (Gloge, 1971):

$$\frac{du}{dv} = \frac{u}{v}[1 - \kappa_l(w)] \quad (7.106)$$

where

$$\kappa_l(w) = \frac{K_l^2(w)}{K_{l-1}(w)K_{l+1}(w)}. \quad (7.107)$$

Hence it follows that

$$\frac{d(vb)}{dv} = \frac{d}{dv}\left(v - \frac{u^2}{v}\right) = 1 - \frac{u^2}{v^2}[1 - 2\kappa_l(w)]. \tag{7.108}$$

Once again it is of interest to compare this LP mode result with the corresponding one from ray theory. Differentiating the approximate eigenvalue equation (7.101) with respect to v, we obtain:

$$\frac{du}{dv} = \frac{u}{v}\left[(w^2 + l^2)^{1/2} + \frac{u^2}{u^2 - l^2}\right]^{-1}. \tag{7.109}$$

Comparing this result with (7.106), we see that the term of interest, $d(vb)/dv$, may be expressed again in the form of equation (7.108) with the difference that the expression (7.107) for $\kappa_l(w)$ is replaced in the ray theory by the approximation:

$$\kappa_l(w) \simeq 1 - \left[(w^2 + l^2)^{1/2} + \frac{u^2}{u^2 - l^2}\right]^{-1}. \tag{7.110}$$

This result may also be derived (White and Pask, 1977) by a simple extension of the theory of Section 2.2 where it was shown that for a planar dielectric guide the group velocity may be found by including the Goos–Haenchen shift in a modified geometric optics treatment.

The results of the ray and mode theories of dispersion are illustrated in Figure 7.11 in terms of plots of $d(vb)/dv$ versus v for $l = 0$ from equation (7.108). The ray and mode results for $\kappa_l(w)$, equations (7.110) and (7.107) respectively, are seen to give a generally high level of agreement. The figure also illustrates the nature of these curves for $l = 0$, commencing with the value zero (corresponding to the velocity of a plane wave in the cladding) at cut-off, and tending asymptotically to unity for large v (corresponding to the velocity of a plane wave in the core material). In between these two extremes, each curve goes through a maximum. We

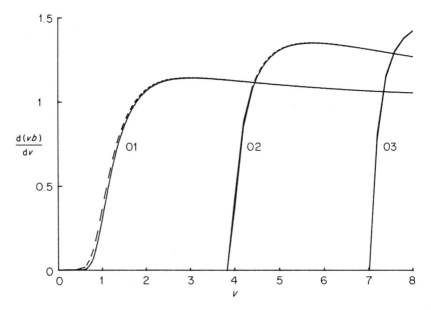

Figure 7.11 $d(vb)/dv$ versus v for $l = 0$ modes of the weakly-guiding step-index fibre, as calculated from (7.108). Solid line, LP mode result (7.107); broken line, ray theory result (7.110)

may understand the reason for this behaviour as follows: as v increases from cut-off more of the mode power propagates in the core and hence the group velocity decreases (equivalent to the diminishing importance of the Goos–Haenchen shift in the ray picture). Eventually, however, we reach a point where the group velocity can be simply thought of in terms of the familiar zig-zag path. As v increases further, the angle of the ray to the axis decreases and hence the power travels with a greater velocity until it eventually approaches the speed of a plane wave in the core material.

For general values of l, at cut-off $d(vb)/dv = 2\kappa_l(0)$. This means that the curves of $d(vb)/dv$ versus v are only zero at cut-off for $l = 0$ and 1 within the limits of the LP mode result (7.107). For other values of l, (7.107) shows that $d(vb)/dv$ at cut-off is given by $2[1-(1/l)]$ (Gloge, 1971). General plots of $d(vb)/dv$ versus v for many LP modes have been calculated by Gloge (1971) and these results are reproduced in many books, including those of Marcuse (1974), Unger (1977), and Midwinter (1979). For a multimode step-index fibre the spread in group velocities resulting from the different values of $d(vb)/dv$ for the various modes is a major cause of pulse-spreading and this subject will be discussed further in the next subsection. Comparisons of impulse responses of multimode step-index fibres calculated by ray and mode theories have been given by Sharma et al. (1980).

Returning to the properties of LP modes in general, it is appropriate at this point to consider the power confinement factor, Γ, defined as the ratio of power confined in the core to total power. The Pointing vector in the axial direction is calculated from the cross product of the transverse fields given in equations (7.92)–(7.97). Integration over the fibre cross-section yields the following expressions for power in the core (P_{core}) and in the cladding (P_{clad}):

$$P_{\text{core}} = \frac{A^2}{2}(2akn)^2 n\left(\frac{\varepsilon_0}{\mu_0}\right)^{1/2} \int_0^a \int_0^{2\pi} r \frac{J_l^2(ur/a)}{J_l^2(u)} \left\{\begin{array}{l}\sin^2 l\theta \\ \cos^2 l\theta\end{array}\right\} d\theta dr \quad (7.111)$$

$$P_{\text{clad}} = \frac{A^2}{2}(2akn)^2 n\left(\frac{\varepsilon_0}{\mu_0}\right)^{1/2} \int_a^\infty \int_0^{2\pi} r \frac{K_l^2(wr/a)}{K_l^2(w)} \left\{\begin{array}{l}\sin^2 l\theta \\ \cos^2 l\theta\end{array}\right\} d\theta dr. \quad (7.112)$$

Fortunately, the integrals in (7.111) and (7.112) are tabulated (Snyder, 1969) and may be written in terms of $\kappa_l(w)$ as defined in (7.107). The results are:

$$P_{\text{core}} = \pi A^2 (a^2 kn)^2 n \left(\frac{\varepsilon_0}{\mu_0}\right)^{1/2} \left[1 + \frac{w^2}{u^2 \kappa_l(w)}\right] \quad (7.113)$$

$$P_{\text{clad}} = \pi A^2 (a^2 kn)^2 n \left(\frac{\varepsilon_0}{\mu_0}\right)^{1/2} \left[\frac{1}{\kappa_l(w)} - 1\right]. \quad (7.114)$$

Equations (7.113) and (7.114) are correct for $l \neq 0$; for $l = 0$ the r.h.s of each must be multiplied by 2. It follows that the power confinement factor (for all l) is given by

$$\Gamma = \frac{P_{\text{core}}}{P_{\text{core}} + P_{\text{clad}}} = 1 - \frac{u^2}{v^2}(1 - \kappa_l(w)). \quad (7.115)$$

An interesting relationship between Γ, b, and $d(vb)/dv$ may be derived by comparing (7.115), (7.108), and the definition of $b = w^2/v^2$; the result is

$$\Gamma = \tfrac{1}{2}\left[b + \frac{d(bv)}{dv}\right]. \quad (7.116)$$

This relation is also true for planar dielectric guides, as may be seen from inspection of equations (2.122) and (2.125) of Chapter 2. With the aid of (7.116), equation (7.105) may be re-

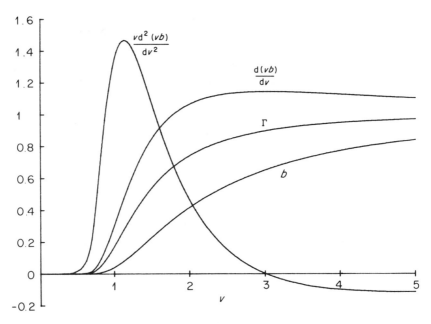

Figure 7.12 $\Gamma, b, d(vb)/dv$, and $vd^2(vb)/dv^2$ versus v for the LP_{01} mode of the weakly-guiding fibre

written as (Hartog, 1979):

$$nm = n_1 m_1 \Gamma + n_2 m_2 (1-\Gamma) \qquad (7.117)$$

and is now easily recognized as the general phase/group velocity relation (5.123) when applied to uniform-core waveguides. The result (7.116) is further illustrated in Figure 7.12 where plots of Γ, b, and $d(bv)/dv$ versus v are given for the lowest order LP mode of the weakly-guiding fibre. As expected, the mode power is concentrated in the core far away from cut-off. Plots of Γ versus v for higher-order modes have been given by Gloge (1971) and show that for low-order modes ($l = 0, 1$) as cut-off is approached the power withdraws into the cladding, whilst for $l \geqslant 2$ the modes maintain a fixed ratio of $(l-1)$ between the power in the core and cladding at cut-off.

In discussing dispersion effects thus far we have assumed the only causes of pulse-spreading to come from differences in transit times between the modes of a multimode fibre. A further cause of pulse-spreading arises from the use of sources with a finite spectral spread, since signals impressed on a fibre at different wavelengths will have different group velocities. In general, the transit time $\tau(\lambda)$ of a mode at wavelength λ may be related to that at the mean source wavelength λ_s by expanding τ as a Taylor series about λ_s:

$$\tau(\lambda) = \tau(\lambda_s) + (\lambda - \lambda_s)\frac{d\tau}{d\lambda}\bigg|_{\lambda_s} + \frac{(\lambda - \lambda_s)^2}{2}\frac{d^2\tau}{d\lambda^2}\bigg|_{\lambda_s} + \ldots \qquad (7.118)$$

For practical sources used in optical communications systems, such as LED's or semiconductor lasers, the spectral widths involved vary between about 2 nm for a typical GaAs laser to 30–40 nm for an LED at 0.9 μm or 90–100 nm for an LED whose wavelength is near 1.3 μm. Hence the dominant term in the expansion (7.118) will be that containing the first derivative $d\tau/d\lambda$. We may evaluate this derivative by noting that $\tau = Lm/c$ where L is the fibre length, c the velocity of light in vacuo, and m the group index for the mode which is given by equation

(7.117). Hence we need to evaluate the total dispersion coefficient (Gloge, 1971a)

$$\frac{1}{c}\frac{dm}{d\lambda} = -\frac{\lambda}{c}\frac{d^2n}{d\lambda^2}. \tag{7.119}$$

Note first that the derivative of a group and phase index product assumes the simple form:

$$\frac{d(nm)}{d\lambda} = \frac{m}{\lambda}(n-m) - n\lambda\frac{d^2n}{d\lambda^2}. \tag{7.120}$$

Also, from (7.103) and (7.116) we may find the derivative of the confinement factor as:

$$\frac{d\Gamma}{d\lambda} = -\frac{1}{2\lambda}\left(\frac{m_1n_1 - m_2n_2}{n_1^2 - n_2^2}\right)\left[2(\Gamma - b) + v\frac{d^2(bv)}{dv^2}\right]. \tag{7.121}$$

With the aid of (7.120) and (7.121), equation (7.117) may be differentiated, without any approximations, to yield the result:

$$-\lambda\frac{d^2n}{d\lambda^2} = -\frac{(m_1n_1 - m_2n_2)^2}{2\lambda n(n_1^2 - n_2^2)}\left[2(\Gamma - b) + v\frac{d^2(bv)}{dv^2}\right] + \frac{m^2}{\lambda n}$$
$$-\frac{\Gamma}{n}\left[n_1\lambda\frac{d^2n_1}{d\lambda^2} + \frac{m_1^2}{\lambda}\right] - \frac{(1-\Gamma)}{n}\left[n_2\lambda\frac{d^2n_2}{d\lambda^2} + \frac{m_2^2}{\lambda}\right]. \tag{7.122}$$

Equation (7.122) may be simplified and expressed in terms of components arising from waveguide dispersion, material dispersion, and profile dispersion with the aid of suitable approximations. We postpone further discussion of these effects until the next subsection. For the present it is sufficient to note that the only waveguide term occurring in (7.122) which has not been discussed so far is $vd^2(bv)/dv^2$. This quantity may be evaluated by differentiating equation (7.108); the result is (Chang, 1979):

$$\frac{vd^2(bv)}{dv^2} = \frac{2u^2\kappa_l(w)}{v^2w^2}\left\{[3w^2 - 2\kappa_l(w)(w^2 - u^2)]\right.$$
$$\left. + w(w^2 + u^2\kappa_l(w))(\kappa_l(w) - 1)\left[\frac{K_{l-1}(w) + K_{l+1}(w)}{K_l(w)}\right]\right\} \tag{7.123}$$

where $\kappa_l(w)$ is defined as in (7.107). A plot of $vd^2(bv)/dv^2$ versus v from (7.123) is given in Figure 7.12. Commencing at a value of 0 at cut-off the function reaches a maximum near $v = 1.2$, goes through zero at $v = 3$ and then a minimum in the region of $v = 5$, before tending asymptotically to 0 for large v. We shall see later that this parameter is the dominant waveguide dispersion term leading to pulse-broadening in single-mode fibres.

7.2.4. Relevant properties of practical fibres

A significant measure of the progress made in fibre fabrication technology over the last decade has been the dramatic reduction of optical loss from several thousands of decibels per kilometre to the current best figure of 0.2 dB/km (Miya et al., 1979). This progress is illustrated by a plot of loss versus calendar year in Figure 7.13 (after Li, 1978). The lowest-loss fibres are made usually by the method of modified chemical vapour deposition (MCVD) which was first developed in 1974 (Black et al., 1974; French et al., 1974; Payne and Gambling, 1974). In this process $SiCl_4$ gas together with oxygen is passed through a silica substrate tube;

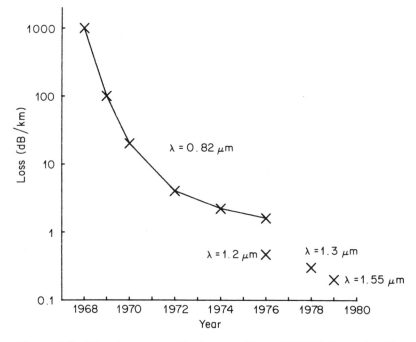

Figure 7.13 Fibre loss versus calendar year. (From Li (1978). Reproduced by permission of the IEEE)

at sufficiently high temperatures a heterogeneous reaction occurs and layers of SiO_2 are deposited on the inside surface of the tube. Other gases, e.g. $GeCl_4$, BCl_3, or $POCl_3$, can be used to deposit dopant oxides (GeO_2, B_2O_3, P_2O_5) in order to vary the refractive index of the deposited layers. The method lends itself readily to the production of graded-index fibres (whose properties are discussed in Chapter 8) since the refractive index may be varied in a controlled manner over many layers. After deposition is completed the tube is collapsed at a higher temperature to form a solid rod which is termed the preform. This contains a similar refractive index distribution to that which will be present in the resulting fibre. The fibre is drawn from the preform by heating the latter to about 1900 °C and applying tension by winding the fibre onto a drum at a precisely controlled rate; several kilometres of fibre may be drawn from the same preform. For further details and descriptions of other fibre fabrication techniques the reader is referred to the book by Midwinter (1979).

A plot of measured attenuation versus wavelength for a fibre fabricated by the MCVD process is shown in Figure 7.14. The loss on the low-wavelength side of the plot is dominated by Rayleigh scattering due to density and compositional fluctuations in the material which are frozen-in when the fibre cools. The increase of attenuation towards the long-wavelength side is determined by the absorption of OH ions and the tail of the intrinsic vibrational absorption of the glass constituents which occurs in the region 7–11 μm. The peaks near 1.4 μm and 0.95 μm are respectively the first and second harmonics of the fundamental OH absorption which occurs at about 2.8 μm. The peak at about 1.25 μm is a combination tone resulting from the second harmonic OH absorption and that due to the fundamental bond-stretching vibration of SiO_2 at 12.5 μm (Keck et al., 1973).

The fundamental loss mechanisms of absorption and scattering which were discussed above may occur in both the core and cladding regions of the fibre. If the loss in the core is denoted by α_1 and that in the cladding by α_2, then the overall modal loss α will be given by an

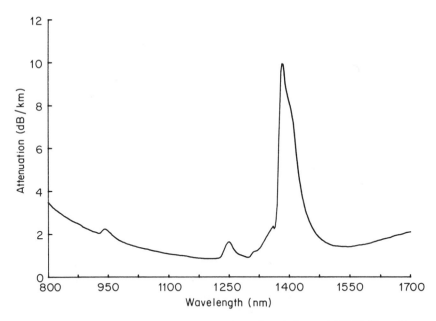

Figure 7.14 Measured attenuation versus wavelength for an MCVD fibre. (From W. A. Gambling, *The Radio and Electronic Engineer*, **49**, 182–186, April 1979. Reproduced by permission of the IERE)

equation of the form (2.128) which has already been discussed for slab waveguides (Snyder, 1972):

$$\alpha = \alpha_1 \Gamma + \alpha_2 (1 - \Gamma) \quad (7.124)$$

where Γ is the power confinement factor for the mode of interest, as given in (7.115). The form of the $\Gamma-v$ curves already discussed in the preceding subsection is sufficient to show that for a multimode fibre the loss will be dominated by the attenuation in the core for all but a very few modes which are close to cut-off. However, for a single-mode fibre a substantial amount of power may propagate in the cladding and hence the value of the loss coefficient α_2 is also a relevant design parameter. In order to minimize α_2 it is important in a single-mode fibre to ensure that the cladding has minimum impurity content. It is therefore common practice to deposit cladding as well as core layers on the inside of the substrate tube in the MCVD process in order to ensure that a carefully-controlled material composition is achieved for the cladding. Perhaps the principal problem encountered in preparing low-loss fibres is the presence of OH absorption bands probably arising from the hydroxyl content in the substrate SiO_2 tube. The use of deposited cladding layers can substantially reduce the OH absorption bands, particularly that corresponding to the second harmonic at 0.95 μm (Payne and Gambling, 1974a). Clearly the thickness of the deposited cladding is also of relevance, since if this is too thin some optical power may penetrate to the substrate and result in loss. In order to understand this effect in more detail and thus provide estimates of cladding thickness required, we defer the discussion to the relevant section on multilayer dielectric rod waveguides below.

In addition to the fundamental loss mechanisms and those arising from the design of the cladding structure as outlined above, there are various other sources of loss which may be introduced during fibre fabrication, cabling, and deployment in the field. These may be collected into four main categories (Gloge, 1975; Marcuse, 1976a):

(i) radiation due to fibre curvature
(ii) microbending loss
(iii) leaky mode loss
(iv) losses due to mode conversion and differential mode attenuation.

With regard to item (i), it is clear that in traversing a bend in a waveguide the field distribution of a guided mode must attempt to maintain phase fronts on planes normal to the local fibre axis. Hence some fraction of the field must apparently exceed the plane wave velocity in the cladding medium; this of course cannot occur in practice and hence the energy associated with this part of the field will be radiated (Marcatili and Miller, 1969). The theoretical treatment of calculating radiation loss from bent waveguides is an important topic in its own right and, one which will not be addressed here; the book by Lewin et al., (1977) provides a comprehensive account of the theory together with a survey of the relevant literature.

Item (ii) on the above list refers to radiation loss due to minute but numerous bends in fibres which may occur during the coating and cabling process and which will be statistically distributed along the length of any fibre. Analyses of attenuation due to microbends in fibres have been presented by (amongst others) Gloge (1975a), Olshansky (1976), Petermann (1976a), and Rousseau and Arnaud (1977, 1978).

Leaky modes are modes below cut-off which may yet propagate for significant distances with varying degrees of attenuation. They correspond to a subset of the skew rays and their ability to accept and carry power over long distances comes about from the circular nature of the fibre geometry. Their properties will be discussed in some detail in Chapter 8 and we therefore postpone calculation of their attenuation coefficients until later.

We have already commented above on the different attenuation suffered by different groups of modes in a multimode fibre as a consequence of their evanescent fields penetrating the cladding structure. Whilst such a loss (differential mode attenuation) is only expected to be significant for guided modes close to cut-off, it is important to realize that power exchange between the modes of an imperfect waveguide may occur and thus power may be coupled into modes of high attenuation. The form of power coupling to which we refer is that due to statistical fluctuations of fibre properties along the length of the waveguide, a subject which has been extensively investigated by Marcuse (1974). Power may also be coupled in this manner directly to the radiation modes and thus be lost from the fibre (Marcuse, 1976a, b).

Clearly the question of fibre attenuation and its causes is a most important one and doubtless efforts will continue to be made to reduce the losses of cabled and installed fibres still further. However, the values already achieved are sufficiently low to enable a wide range of potential applications to be fulfilled by present-day fibres (with the possible exception of long-distance, repeaterless submarine cables). The brief discussion of loss mechanisms given above will suffice to indicate the range of theoretical problems raised by the study of fibre attenuation. A further group of problems is raised by the question of dispersion in multimode and single-mode fibres. With the advent of ultra-low loss waveguides the likely lengths of cable between repeater stations become of the order of tens of kilometres; hence the requirement for fibres with a high bandwidth–length product becomes rather stringent. We shall investigate here the effects of pulse-spreading in optical fibre waveguides of uniform core and circular cross-section.

For a multimode step-index fibre, the dominant cause of pulse-spreading is likely to be the range of transit times associated with the various modes. From Section 7.2.3 we know that the spread of values for the dispersion term $d(vb)/dv$ is of the order of unity. From equation

(7.105), in the absence of material dispersion ($m_1 = n_1, m_2 = n_2$), it follows that the spread of group delays per unit length of fibre is of order $(n_1 - n_2)/c$, where c is the velocity of light in vacuo. For a 1 per cent index difference this means that the difference in modal delay times of the fastest and slowest modes can be as high as 50 ns/km. The severe limitation on bandwidth implied by this figure can be avoided either by the use of single-mode fibres or by the adoption of graded-index multimode fibres where the core refractive index is graded in order to equalize intermodal delay differences. This latter alternative has the advantage of preserving the larger physical dimensions associated with multimode fibres (as compared with monomode) and thus allowing greater tolerance in jointing and connecting to other optical components. Theoretical analysis of the properties of graded-index fibres forms the subject of Chapter 8.

A second important contribution to pulse-spreading, and one which applies to both multimode and single-mode fibres, arises from material dispersion and the use of sources with nonzero spectral width. In order to examine the effects of material dispersion in isolation for multimode fibres, we may use equation (7.122). Since almost all the modes of a multimode fibre of typical v-value (20–50) are well-confined and well above cut-off, it is a reasonable approximation to use $\Gamma \simeq b \simeq 1, n \simeq n_1$. The r.h.s. of equation (7.122) then reduces to the single expression $(-d^2 n_1/d\lambda^2)$. It follows that the material dispersion coefficient for the fibre is dominated by that for the core (Di Domenico, 1972; Laybourn and Gambling, 1973; Timmermann, 1974):

$$\frac{1}{c}\frac{dm}{d\lambda} \simeq -\frac{\lambda}{c}\frac{d^2 n_1}{d\lambda^2} \equiv -M_1. \tag{7.125}$$

In order to evaluate the magnitude of the material dispersion coefficient M_1, we may make use of refractive index data for fibre materials (Mallitson, 1965; Fleming, 1976; Kobayashi et al., 1978; Fleming, 1978) which is conventionally expressed in terms of a three-term Sellmeier equation:

$$n_1^2 - 1 = \sum_{i=1}^{3} \frac{A_i \lambda^2}{(\lambda^2 - \lambda_i^2)} \tag{7.126}$$

where the parameters A_i and λ_i ($i = 1-3$) are determined by an r.m.s. fit to experimental data (Sutton and Stavroudis, 1961). Differentiating (7.126) and using the definition of group index of the core material as in (7.102) yields:

$$m_1 n_1 - n_1^2 = \lambda^2 \sum_{i=1}^{3} \frac{A_i \lambda_i^2}{(\lambda^2 - \lambda_i^2)^2}. \tag{7.127}$$

A further differentiation yields the material dispersion coefficient M_1 as:

$$M_1 = \frac{\lambda}{c n_1}\left[\frac{m_1(n_1 - m_1)}{\lambda^2} + \lambda^2 4 \sum_{i=1}^{3} \frac{A_i \lambda_i^2}{(\lambda^2 - \lambda_i^2)^3}\right]. \tag{7.128}$$

Table 7.3 compiles values of the coefficients A_i and resonant wavelengths λ_i for various glass compositions employed in optical fibres. Figure 7.15 gives plots of refractive index n_1, group index m_1, and material dispersion coefficient M_1 for SiO_2 as calculated from equations (7.126)–(7.128) and Table 7.3. From the figure we see that at the GaAs emission wavelength of 0.85 µm the value of M_1 is about 80 ps/km/nm, thus implying that for a source spectral width of 40 nm the pulse-spreading due to material dispersion alone will amount to about 3 ns in a kilometre length. However, at longer wavelengths M_1 decreases and passes through zero at $\lambda = 1.27$ µm thus implying greatly reduced pulse-spreading at these longer wavelengths

Table 7.3. Values of the parameters A_i, λ_i ($i = 1–3$) in the Sellmeier expansion (7.126),

Composition	A_1	A_2	A_3
SiO_2	0.6961663	0.4079426	0.8974794
13.5 $^m/_0$ GeO_2, 86.5 $^m/_0$ SiO_2	0.73454395	0.42710828	0.82103399
7.0 $^m/_0$ GeO_2, 93.0 $^m/_0$ SiO_2	0.68698290	0.44479505	0.79073512
4.1 $^m/_0$ GeO_2, 95.9 $^m/_0$ SiO_2	0.68671749	0.43481505	0.89656582
9.1 $^m/_0$ GeO_2, 7.7 $^m/_0$ B_2O_3, 83.2 $^m/_0$ SiO_2	0.72393884	0.41129541	0.79292034
4.03 $^m/_0$ GeO_2, 9.7 $^m/_0$ B_2O_3, 86.27 $^m/_0$ SiO_2	0.70420420	0.41289413	0.95238253
0.1 $^m/_0$ GeO_2, 5.4 $^m/_0$ B_2O_3, 94.5 $^m/_0$ SiO_2	0.69681388	0.40865177	0.89374039
13.5 $^m/_0$ B_2O_3, 86.5 $^m/_0$ SiO_2	0.70724622	0.39412616	0.63301929
13.5 $^m/_0$ B_2O_3, 86.5 $^m/_0$ SiO_2 (chilled)	0.67626834	0.42213113	0.58339770
3.1 $^m/_0$ GeO_2, 96.9 $^m/_0$ SiO_2	0.7028554	0.4146307	0.8974540
3.5 $^m/_0$ GeO_2, 96.5 $^m/_0$ SiO_2	0.7042038	0.4160032	0.9074049
5.8 $^m/_0$ GeO_2, 94.2 $^m/_0$ SiO_2	0.7088876	0.4206803	0.8956551
7.9 $^m/_0$ GeO_2, 92.1 $^m/_0$ SiO_2	0.7136824	0.4254807	0.8964226
3.0 $^m/_0$ B_2O_3, 97.0 $^m/_0$ SiO_2	0.6935408	0.4052977	0.9111432
3.5 $^m/_0$ B_2O_3, 96.5 $^m/_0$ SiO_2	0.6929642	0.4047468	0.9154064
3.3 $^m/_0$ GeO_2, 9.2 $^m/_0$ B_2O_3, 87.5 $^m/_0$ SiO_2	0.6958807	0.4076588	0.9401093
2.2 $^m/_0$ GeO_2, 3.3 $^m/_0$ B_2O_3, 94.5 $^m/_0$ SiO_2	0.6993390	0.4111269	0.9035275
Quenched SiO_2	0.696750	0.408218	0.890815
13.5 $^m/_0$ GeO_2, 86.5 $^m/_0$ SiO_2	0.711040	0.451885	0.704048
9.1 $^m/_0$ P_2O_5, 90.9 $^m/_0$ SiO_2	0.695790	0.452497	0.712513
13.3 $^m/_0$ B_2O_3, 86.7 $^m/_0$ SiO_2	0.690618	0.401996	0.898817
1.0 $^m/_0$ F, 99.0 $^m/_0$ SiO_2	0.691116	0.399166	0.890423
16.9 $^m/_0$ Na_2O, 32.5 $^m/_0$ B_2O_3, 50.6 $^m/_0$ SiO_2	0.796468	0.497614	0.358924

(Payne and Gambling, 1975). For wavelengths at and near the zero of M_1, higher-order material dispersion terms arising from the expansion of equation (7.118) are clearly important (Kapron, 1977; Unger, 1977a; Miyagi and Nishida, 1979, b, c; Marcuse, 1980a). However, in going to this higher level of approximation it is important also to consider (a) the effects of differing material dispersion coefficients for core and cladding glasses (Adams et al., 1978) and (b) modal effects arising from other terms in equation (7.122) (Gambling et al., 1979).

The wavelength corresponding to zero material dispersion is a function of glass composition, and measurements in the $GeO_2 - SiO_2$ system have shown variations over the range 1.27–1.39 μm. It is also possible to determine material dispersion values directly for fibres, rather than from measurements on bulk glasses, by measuring the total transit time of a short pulse as a function of wavelength. Such measurements have been reported by Luther-Davies et al. (1975), Miyashita et al. (1977), Payne and Hartog (1977), Cohen and Chinlon Lin (1978), Lin et al. (1978), and Horiguchi et al. (1979); a comparison of the available data for the wavelength of zero material dispersion for germania-doped silica has been given by Adams et al. (1978a). The range of wavelengths involved coincides with a low-loss 'window' in the fibre attenuation plot (see, for example, Figure 7.14) and this provides a second reason for the current interest in fibre systems operating at these wavelengths. Semiconductor sources for the 1.3 μm range have been developed based on the quaternary $In_xGa_{1-x}As_yP_{1-y}$ (see, for example, Kressel and Butler, 1977). Lasers fabricated in this material on InP substrates are as determined from refractive index measurements on bulk glass samples

$\lambda_1 (\mu m)$	$\lambda_2 (\mu m)$	$\lambda_3 (\mu m)$	Reference
0.0684043	0.1162414	9.896161	Mallitson (1965)
0.86976930	0.11195191	10.846540	Fleming (1976)
0.78087582	0.11551840	10.436628	,,
0.072675189	0.11514351	10.002398	,,
0.085826532	0.10705260	9.3772959	,,
0.067974973	0.12147738	9.6436219	,,
0.070555513	0.11765660	9.8754801	,,
0.080478054	0.10925792	7.8908063	,,
0.076053015	0.11329618	7.8486094	,,
0.0727723	0.1143085	9.896161	Kobayashi et al. (1978)
0.0514415	0.1291600	9.896156	,,
0.0609053	0.1254514	9.896162	,,
0.0617167	0.1270814	9.896161	,,
0.0717021	0.1256396	9.896154	,,
0.0604843	0.1239609	9.896152	,,
0.0665654	0.1211422	9.896140	,,
0.0617482	0.1242404	9.896158	,,
0.069066	0.115662	9.900559	Fleming (1978)
0.064270	0.129408	9.425478	,,
0.061568	0.119921	8.656641	,,
0.061900	0.123662	9.098960	,,
0.068227	0.116460	9.993707	,,
0.094359	0.093386	5.999652	,,

capable of c.w. room temperature operation (Hsieh et al., 1976) and long operating lives (Shen et al., 1977) whilst LED's in the same material can produce high power outputs into optical fibres (Goodfellow et al., 1979).

In the foregoing discussion of material dispersion it has been assumed that all optical power was confined to the fibre core. However, for a single-mode fibre this is not the case and it is then necessary in general to consider material dispersion for both core and cladding materials, as well as waveguide dispersion effects. Equation (7.122) can be used to find the total dispersion in step-index monomode fibres, although it may also be simplified and expressed as a sum of components arising from waveguide and material terms. As a first step in this direction, note that a composite material dispersion term M_{cmd} may be expressed as

$$M_{cmd} = \frac{\lambda}{c}\left[\frac{n_1 \Gamma}{n}\frac{d^2 n_1}{d\lambda^2} + \frac{n_2}{n}(1-\Gamma)\frac{d^2 n_2}{d\lambda^2}\right]$$

$$\simeq \frac{\lambda}{c}\left[\Gamma\frac{d^2 n_1}{d\lambda^2} + (1-\Gamma)\frac{d^2 n_2}{d\lambda^2}\right]. \quad (7.129)$$

This expression is a weighted average of the material dispersion coefficients of core and cladding with the weights given by the proportion of optical power carried in each region.

In order to simplify (7.122) further and reveal the dominant waveguide dispersion term, it is convenient to eliminate the cladding index n_2 by introducing the relative index difference Δ.

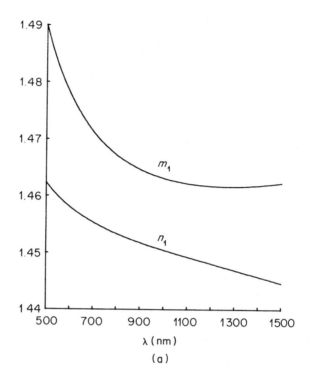

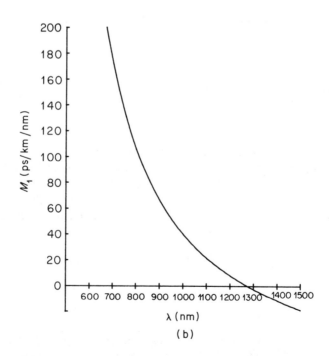

Figure 7.15 (a) Refractive index n_1 and group index m_1 plotted versus wavelength for fused SiO_2. (b) Material dispersion parameter M_1 versus wavelength for SiO_2

With this approach the coefficient of the first term of (7.122) yields:

$$\frac{(m_1n_1 - m_2n_2)^2}{2\lambda n(n_1^2 - n_2^2)} = \frac{m_1^2\Delta}{\lambda n} + \frac{\lambda n_1^2\Delta'^2}{4n\Delta} - \frac{n_1 m_1 \Delta'}{n}. \qquad (7.130)$$

Now terms in Δ' ($\equiv d\Delta/d\lambda$) result from the wavelength dependence of the relative index difference and are usually termed 'profile dispersion' terms for reasons which will be discussed in Chapter 8. Hence, leaving these terms aside for the present, it is clear that the dominant purely waveguide dispersion term M_{wd} in (7.122) may be expressed as:

$$M_{wd} = \frac{m_1^2 \Delta}{c \lambda n} v \frac{d^2(bv)}{dv^2}$$

$$\simeq \frac{n_1 \Delta}{c\lambda} \left(\frac{m_1}{n_1}\right)^2 v \frac{d^2(bv)}{dv^2}. \qquad (7.131)$$

Consider next the profile dispersion terms containing Δ' in equation (7.122) via the substitution (7.130). Collecting these terms we find the composite profile dispersion coefficient M_{cpd} as:

$$M_{cpd} = \frac{n_1^2 \Delta'}{cn} \left(\frac{\lambda \Delta'}{4\Delta} - \frac{m_1}{n_1}\right) \left[2(\Gamma - b) + v\frac{d^2(bv)}{dv^2}\right]$$

$$\simeq \frac{n_1 \Delta'}{c} \left(\frac{\lambda \Delta'}{4\Delta} - \frac{m_1}{n_1}\right) \left[2(\Gamma - b) + v\frac{d^2(bv)}{dv^2}\right]. \qquad (7.132)$$

Finally, equation (7.122) may be written in terms of all its component parts as:

$$-\frac{\lambda}{c}\frac{d^2 n}{d\lambda^2} = -M_{wd} - M_{cmd} - M_{cpd} - M_r \qquad (7.133)$$

where M_{wd}, M_{cmd}, M_{cpd} are given in equations (7.131), (7.129), (7.132), and the remainder term M_r is composed of cross-products which are not easily categorized:

$$M_r = -\frac{1}{\lambda cn}\left[m_1^2 2\Delta(\Gamma - b) - m^2 + m_1^2 \Gamma + m_2^2(1 - \Gamma)\right]. \qquad (7.134)$$

It may be shown that M_r is small for most materials and wavelength ranges of interest and hence this term may usually be neglected in (7.133).

From the practical point of view a significant property of the total dispersion (7.133) is that the waveguide and composite material terms M_{wd} and M_{cmd} can have opposite signs over a range of wavelength. This occurs since the material dispersion coefficient changes sign at a wavelength near 1.3 µm (see Figure 7.15(b)); hence for wavelengths longer than this, the effect of the waveguide dispersion M_{wd} may be used to compensate to some extent for the material effects. Hence it is possible to shift the zero of total dispersion to longer wavelengths than 1.3 µm by suitable choice of the waveguide parameters. This effect has been known to be possible in principle for some time (Kapron and Keck, 1971; Smith and Snitzer, 1973, 1974; Jürgensen, 1974, 1975); however, it is only recently that sufficient material data has become available for quantitative estimates to be made (South, 1979; White and Nelson, 1979; Gambling et al. 1979, 1979a; Tsuchiya and Imoto, 1979; Jeunhomme, 1979; Marcuse, 1979; Sammut, 1979; Jürgensen, 1979; Chang, 1979, 1979a). In the above analysis we have followed the spirit (although not the detail) of Gambling et al. (1979, 1979a) in separating the various contributions to total chromatic dispersion in monomode fibres. The advantage of this formalism is that it permits experimental data for profile dispersion (Δ') and material dispersion ($d^2n_2/d\lambda^2$) to be used provided that the index of one material (either core or cladding) is known accurately as a function of wavelength. In the case discussed above it is the core index data (n_1, m_1) which are assumed known; for the notation of Gambling *et al.* (1979, 1979a) it is the cladding index data.

As an example of the application of the above theory to the calculation of total chromatic dispersion for a step-index monomode fibre, consider the case when the core contains 13.5 $^m/_0$ GeO$_2$, 86.5 $^m/_0$ SiO$_2$ and the cladding is pure SiO$_2$. The refractive index data may then be taken from lines 19 (Fleming, 1978) and 1 (Mallitson, 1965) of Table 7.3. Figure 7.16 shows computed results for this case (with core diameter 4 μm) as a function of wavelength, using

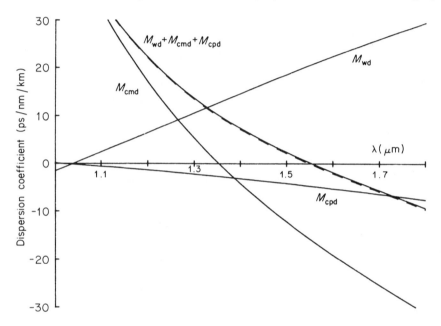

Figure 7.16 Dispersion coefficients as calculated from equations (7.129), (7.131), (7.132) for a step-index fibre whose core contains 13.5$^m/_0$ GeO$_2$ (line 19, Table 7.3), cladded with pure SiO$_2$ (line 1, Table 7.3) and core diameter 4 μm. Broken line gives the total dispersion as calculated from Equation (7.122) for the same parameters

equation (7.129) for composite material dispersion M_{cmd} and equation (7.131) for waveguide dispersion M_{wd}. For wavelengths above about 1.35 μm, these two components have opposite signs and may thus compensate each other in the total dispersion coefficient. The composite profile dispersion M_{cpd} calculated from equation (7.132) is also shown on Figure 7.16 and this clearly contributes a smaller term with less wavelength dependence than the other two; however, its effect can be significant when the terms M_{wd} and M_{cmd} are nearly equal. The sum of these three contributions ($M_{wd} + M_{cmd} + M_{cpd}$) is also plotted on the figure, as is the total dispersion coefficient as given by (7.122) (broken line). The close agreement of these latter two curves confirms our earlier assertion that the remainder term M_r of (7.134) can safely be neglected for most cases of interest. A significant feature of the total dispersion curves is the shift of the wavelength corresponding to zero total dispersion; in the case of the parameters used for Figure 7.16 this wavelength occurs near 1.55 μm as compared with 1.35 μm for the composite material dispersion alone. Smaller core diameters would shift the wavelength of zero dispersion to higher values and larger cores would result in a lower value. This degree of flexibility in the design of a monomode fibre for high bandwidth operation is especially important when considered in conjunction with the attenuation versus wavelength plots of fibre materials. In particular the recently reported ultra-low loss of 0.2 dB/km (Miya et al., 1979) occurs at 1.55 μm. Hence it is now possible by use of relatively large index-difference, small core fibres to 'tune' the wavelength of zero dispersion to the value required for minimum attenuation.

Experimental measurements of total chromatic dispersion in single-mode fibres have been reported by a number of authors (Miyashita et al., 1977; Cohen and Chinlon Lin, 1977; Daikoku and Sugimura, 1978; Cohen et al., 1979). The shift of the zero dispersion wavelength discussed here has been observed by Cohen et al. (1979) in a monomode fibre of core diameter 4.8 μm with 13m/o GeO$_2$ peak dopant concentration; for this fibre the measured zero dispersion wavelength was 1.54 μm. Hence the overall trend of waveguide/material dispersion compensation has been confirmed, viz. small core and relatively large index difference ($\Delta n = 0.018$ for this fibre). However, more quantitative comparison with the theory presented here is not justified since the CVD fabrication process for monomode fibres usually produces substantial grading of the index profile. In this respect it is worth noting that the calculations of Gambling et al. (1979, 1979a) have been applied to a range of graded profiles and the results indicate that somewhat larger cores can be used than for the step-index case, with all other parameters the same, and still achieve the desired optimum wavelength. This trend explains the difference between experimental core diameter of 4.8 μm referred to above, as compared with the theoretical 4 μm core for similar sets of parameters in Figure 7.16. Calculations of total chromatic dispersion based on measurements of monomode fibre index profiles have been reported by Sugimura et al. (1980) and Cohen et al. (1980); in both cases good agreement was obtained with experimental determination of dispersion as a function of wavelength. The numerical methods of calculation used by these authors will be discussed in Chapter 8.

To conclude this subsection we leave the topic of chromatic dispersion and return to a simple question concerning multimode step-index fibres: for a given normalized frequency v, what is the number of modes supported by such a fibre? In order to answer this question we may use the simple ray-optics eigenvalue equation (7.101) relating radial mode number M and azimuthal mode number l. For guided modes we know that $u \leqslant v$ and thus the total number of such modes can be found by adding up all values of l, M yielding $u \leqslant v$ in (7.101). The mode density may be visualized as a uniform array of points on a plot of l/v versus M/v (Gloge, 1975). For the usual cases of interest in a multimode fibre $v \gg 1$ and hence the total mode number M_0 can be found simply by integrating (7.101) over l and putting $u = v$:

$$M_0 = \int_0^v \left\{ \frac{3}{4} + \frac{(v^2 - l^2)^{1/2}}{\pi} - \frac{l}{\pi}\cos^{-1}\left(\frac{l}{v}\right) - \frac{1}{\pi}\cos^{-1}\left[\frac{(v^2 - l^2)^{1/2}}{v}\right] \right\} dl$$

$$= \frac{3v}{4} + \frac{v^2}{4} - \frac{v^2}{8} - \frac{v}{2} + \frac{v}{\pi}$$

$$\simeq \frac{v^2}{8} \quad (v \gg 1). \tag{7.135}$$

However, we should also recall that each (l, M) mode is four-fold degenerate (except for the $l = 0$ modes which are doubly degenerate) and hence a further factor of 4 is required in (7.135) to give the total number of modes as (Gloge, 1971):

$$M_0 \simeq \frac{v^2}{2}. \tag{7.136}$$

Equation (7.136) is a good approximation for the number of modes on a step-index fibre with

$v \gg 1$. For comparison the number of $l = 0$ modes (corresponding to meridional rays) is given direct from (7.101), for $v \gg 1$ and allowing for twofold degeneracy, as $2v/\pi$.

7.2.5 Effects of small departures from circularity

In the CVD fabrication process described briefly in the previous subsection it is quite normal for departures from circularity to occur, especially at the stage where the silica tube is collapsed to form the preform. This part of the process calls for considerable skill if the preform is to remain circular. Hence it is of interest to enquire what effects may be expected from the resultant noncircularity in the fibre. In practice, the degree of circularity achieved is remarkably good in many cases, and we may therefore confine the investigation here to small departures from circularity and, for convenience, to the case of elliptical fibres of small eccentricity. Therefore we introduce elliptical coordinates ξ, η (Abramowitz and Stegun, 1964) which are related to the usual Cartesian cooridinates x, y as follows:

$$\left.\begin{array}{l} x = q \cosh \xi \cos \eta \\ y = q \sinh \xi \sin \eta \end{array}\right\} \tag{7.137}$$

where q is the semi-focal distance of the family of confocal ellipses defined by ξ = constant. We take one such ellipse, say $\xi = \xi_0$, as the core–cladding interface of a fibre with core refractive index n_1 and cladding index n_2. The major and minor axes of this ellipse are then given by $2a$ and $2b$, respectively, where

$$a = q \cosh \xi_0, \quad b = q \sinh \xi_0. \tag{7.138}$$

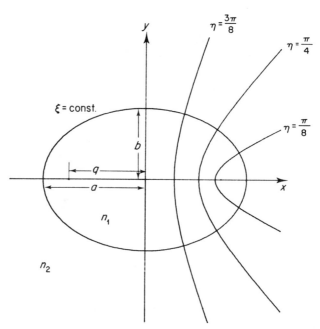

Figure 7.17 Cross-section of the elliptical step-index fibre, core refractive index n_1, cladding index n_2. The line ξ = constant indicates the core–cladding boundary whilst the hyperbolae are given by different values of η. Semi-major and semi-minor axes are of length a and b respectively, whilst the semi-focal distance is denoted by q

Figure 7.17 illustrates the fibre cross-section in the x–y plane and also gives plots of the hyperbolae defined by η = constant. The eccentricity e of the core–cladding boundary is given by

$$e^2 = 1 - \left(\frac{b}{a}\right)^2 = \operatorname{sech}^2 \xi_0. \qquad (7.139)$$

With the definitions (7.137) the scalar wave equation for a field component $\psi(\xi, \eta)$, assuming the usual z- and time-dependence, becomes

$$\frac{\partial^2 \psi}{\partial \xi^2} + \frac{\partial^2 \psi}{\partial \eta^2} + \frac{q^2}{2}(\cosh 2\xi - \cos 2\eta)(n_j^2 k^2 - \beta^2)\psi = 0 \quad (j = 1, 2). \qquad (7.140)$$

Assuming separable solutions $\psi(\xi, \eta) = f(\xi)g(\eta)$ and a separation constant h yields

$$\frac{d^2 f}{d\xi^2} - \left[h - \frac{q^2 \cosh 2\xi}{2}(n_j^2 k^2 - \beta^2)\right] f = 0 \qquad (7.141\text{a})$$

$$\frac{d^2 g}{d\eta^2} + \left[h - \frac{q^2 \cos 2\eta}{2}(n_j^2 k^2 - \beta^2)\right] g = 0 \qquad (7.141\text{b})$$

where $j = 1$ for $\xi \leqslant \xi_0$ and $j = 2$ for $\xi > \xi_0$.

Equation (7.141b) is Mathieu's equation and (7.141a) is Mathieu's modified equation (Abramowitz and Stegun, 1964; McLachlan, 1947); the solutions are Mathieu functions and modified Mathieu functions, respectively. However, before proceeding with the formal solution let us look first at the general nature of the functions f and g. The behaviour of f and g depends on the magnitude of the separation constant h. In the case where $2h > q^2(n_1^2 k^2 - \beta^2)$, then put

$$\frac{2h}{q^2(n_1^2 k^2 - \beta^2)} = \cosh 2\xi_1 \qquad (7.142)$$

whence equation (7.141) yields inside the core ($\xi \leqslant \xi_0$):

$$\frac{d^2 f}{d\xi^2} + \frac{q^2(n_1^2 k^2 - \beta^2)}{2}(\cosh 2\xi - \cosh 2\xi_1)f = 0 \qquad (7.143\text{a})$$

$$\frac{d^2 g}{d\eta^2} + \frac{q^2(n_1^2 k^2 - \beta^2)}{2}(\cosh 2\xi_1 - \cos 2\eta)g = 0. \qquad (7.143\text{b})$$

In this case the factor multiplying g in (7.143b) is always positive definite and hence there is a periodic solution for g corresponding to a guided mode. The inner caustic for this mode will be given in the usual way by the zero of the function multiplying f in (7.143a), i.e. at $\xi = \xi_1$. Thus the oscillatory part of the field distribution is contained between the confocal ellipses defined by $\xi = \xi_1$ and $\xi = \xi_0$ (the core–cladding boundary). In the corresponding ray picture this corresponds to a ray suffering total internal reflection at the core–cladding interface and meeting the inner caustic tangentially in the usual way (cf. Figure 7.2 for the circular fibre). Such a ray, sometimes termed a 'whispering' ray, is shown projected on the fibre cross-section in Figure 7.18 (Keller and Rubinow, 1960; Bykov and Vaĭnshteĭn, 1965; Checcacci et al., 1979; Love et al., 1979). Clearly, as the eccentricity e tends to zero the ray path of Figure 7.18 tends towards the circular case shown in Figure 7.2.

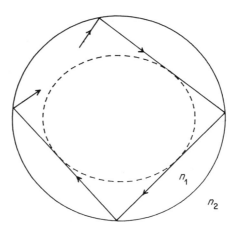

Figure 7.18 Projection of the ray path of a 'whispering' ray in a step-index elliptical fibre; the broken line shows the caustic which is the ellipse defined by $\xi = \xi_1$ as in equation (7.142)

For the opposite case where $2h < q^2(n_1^2 k^2 - \beta^2)$, define

$$\frac{2h}{q^2(n_1^2 k^2 - \beta^2)} = \cos 2\eta_1 \tag{7.144}$$

whence equation (7.141) yields inside the core ($\xi \leq \xi_0$):

$$\frac{d^2 f}{d\xi^2} + \frac{q^2(n_1^2 k^2 - \beta^2)}{2}(\cosh 2\xi - \cos 2\eta_1)f = 0 \tag{7.145a}$$

$$\frac{d^2 g}{d\eta^2} + \frac{q^2(n_1^2 k^2 - \beta^2)}{2}(\cos 2\eta_1 - \cos 2\eta)g = 0. \tag{7.145b}$$

In this case the factor multiplying f in (7.145a) is always positive definite, whilst that multiplying g in (7.145b) is only positive for $\eta_1 \leq |\eta| \leq \pi/2$. This means that the oscillatory part of the field distribution is bound by the core–cladding boundary $\xi = \xi_0$ and the confocal hyperbolic caustics given by $\eta = \eta_1$. On the ray picture this corresponds to a so-called 'bouncing' ray which undergoes total internal reflection at the core–cladding interface and touches the hyperbolic caustics tangentially, as shown in cross-section in Figure 7.19 (Keller and Rubinow, 1960; Bykov and Vaĭnshteĭn, 1965; Checcacci et al., 1979; Love et al., 1979). For rays or modes of this type, as the eccentricity of the core–cladding boundary approaches zero the caustics become progressively closer to a single straight line running through the core centre. In the limit the ray becomes a meridional ray ($e = 0$) of the circular fibre.

Based on the general nature of the solutions discussed above it would be possible to determine approximate eigenvalue equations by the usual methods of geometric optics, or equivalently by applying the WKB approximation. However, since the solutions of (7.141) are known in terms of Mathieu functions it is preferred here to proceed with the exact solution and simplify the results later by taking account of the conditions (i) of weak guidance $(n_1 - n_2) \ll n_1, n_2$, and (ii) of small eccentricity $e \ll 1$. Complete solutions for the elliptical dielectric rod problem have been formulated by Lyubimov et al. (1961), Yeh (1962), and Piefke (1964); in addition, approximate results for the case $e \ll 1$ have been developed by Schlosser (1965, 1972), and for the case $(n_1 - n_2) \ll n_1, n_2$ by Yeh (1976). In the following we

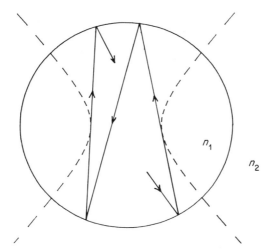

Figure 7.19 Projection of the ray path of a 'bouncing' ray in a step-index elliptical fibre; the broken line shows the caustic which is the hyperbola defined by $\eta = \eta_1$ as in equation (7.144)

begin with the full theory of Yeh (1962) and then apply both simplifying assumptions in order to produce approximate results for a few quantities of interest (Adams *et al.*, 1979; Love *et al.*, 1979a).

The general solutions of the wave equations (7.141a), (7.141b) may be divided into even and odd functions:

$$\psi(\xi,\eta) = \begin{cases} \mathrm{Ce}_v(\xi,\gamma_1^2)\mathrm{ce}_v(\eta,\gamma_1^2) & (\xi \leqslant \xi_0, \text{ even}) \quad (7.146a) \\ \mathrm{Se}_v(\xi,\gamma_1^2)\mathrm{se}_v(\eta,\gamma_1^2) & (\xi \leqslant \xi_0, \text{ odd}) \quad (7.146b) \end{cases}$$

$$\psi(\xi,\eta) = \begin{cases} \mathrm{Fek}_v(\xi,-\gamma_2^2)\mathrm{ce}_v(\eta,-\gamma_2^2) & (\xi \geqslant \xi_0, \text{ even}) \quad (7.147a) \\ \mathrm{Gek}_v(\xi,-\gamma_2^2)\mathrm{se}_v(\eta,-\gamma_2^2) & (\xi \geqslant \xi_0, \text{ odd}) \quad (7.147b) \end{cases}$$

where the notation for the Mathieu functions is that of McLachlan (1947), and

$$\gamma_1^2 = \frac{(n_1^2 k^2 - \beta^2)q^2}{4} \quad (7.148a)$$

$$\gamma_2^2 = \frac{(\beta^2 - n_2^2 k^2)q^2}{4}. \quad (7.148b)$$

As for the general case of the circular dielectric rod, the modes here are hybrid with both components E_z, H_z present to satisfy the boundary conditions (Yeh, 1962). A further problem, however, peculiar to the elliptical waveguide, is that the angular Mathieu functions are not only functions of η, but also of the refractive indices n_1, n_2. Hence the functions $\mathrm{ce}_v(\eta,\gamma_1^2)$ and $\mathrm{ce}_v(\eta,-\gamma_2^2)$ are not the same in general, and it is necessary to represent the solutions as series of these functions summed over the subscript v; similar considerations apply to the functions $\mathrm{se}_v(\eta,\gamma_1^2)$ and $\mathrm{se}_v(\eta,-\gamma_2^2)$. The subscript v takes integer values related to the azimuthal periodicity of the field distributions. Therefore, even though a series of functions is involved, we expect the strongest contribution to come from the value of v corresponding to the modal azimuthal periodicity. In the weakly-guiding slightly-elliptical case it is in fact a very good approximation to include only this value of v in the series (Yeh,

1962). The general forms of the functions ce_v, se_v may be written as (McLachlan, 1947):

$$\left.\begin{aligned} ce_{2n}(\eta, \gamma^2) &= \sum_{r=0}^{\infty} A_{2r}^{(2n)} \cos(2r\eta) \\ ce_{2n+1}(\eta, \gamma^2) &= \sum_{r=0}^{\infty} A_{2r+1}^{(2n+1)} \cos((2r+1)\eta) \\ se_{2n+1}(\eta, \gamma^2) &= \sum_{r=0}^{\infty} B_{2r+1}^{(2n+1)} \sin((2r+1)\eta) \\ se_{2n+2}(\eta, \gamma^2) &= \sum_{r=0}^{\infty} B_{2r+2}^{(2n+2)} \sin((2r+2)\eta) \end{aligned}\right\} \quad (7.149)$$

where the A's and B's are functions of γ^2. The separation coefficient h in (7.141) may also be written in terms of v and γ^2. Using these expansions and retaining terms only up to order q^2 (small eccentricity) it is possible to show (by using orthogonality relations) that the only terms of real significance for the weakly-guiding, slightly elliptical case are those involving cos ($v\eta$) for ce_v and sin ($v\eta$) for se_v. Readers interested in the rigorous treatment for the more general case are referred to Appendix A of the book by Kapany and Burke (1972) where Yeh gives in full the details of his 1962 paper.

As a result of the asymmetry of the elliptical waveguide it is possible to have two orientations for the field configuration, denoted by even or odd according to which combinations of Mathieu functions are used from (7.146), (7.147). The axial magnetic and electric fields of an even wave are represented by even and odd Mathieu functions respectively; similarly the axial magnetic and electric fields of an odd wave are represented by odd and even Mathieu functions respectively. As mentioned previously the fields are hybrid and thus the notations $_eHE_{vM}$ and $_eEH_{vM}$ are used for even modes, and $_oHE_{vM}$, $_oEH_{vM}$ for odd modes.

Based on the above comments we write a simplified form for the axial field components of the $_eHE_{vM}$ wave as:

$$H_z = \begin{cases} A \, Ce_v(\xi, \gamma_1^2) \cos v\eta, & \xi \leq \xi_0 \quad (7.150a) \\ L \, Fek_v(\xi, -\gamma_2^2) \cos v\eta, & \xi \geq \xi_0 \quad (7.150b) \end{cases}$$

$$E_z = \begin{cases} B \, Se_v(\xi, \gamma_1^2) \sin v\eta, & \xi \leq \xi_0 \quad (7.151a) \\ P \, Gek_v(\xi, -\gamma_2^2) \sin v\eta, & \xi \geq \xi_0 \quad (7.151b) \end{cases}$$

where A, B, L, P are constants to be evaluated.

Writing Maxwell's equations in terms of elliptical coordinates (Morse and Feshbach, 1953) we may express all other field components in terms of H_z, E_z. Since we are only interested in imposing boundary conditions on the tangential electric and magnetic fields at the interface $\xi = \xi_0$ we restrict attention to E_η, H_η which are given by (assuming the usual z- and time-dependence):

$$E_\eta = \frac{i}{(k^2 n_j^2 - \beta^2) q (\cosh^2 \xi - \cos^2 \eta)^{1/2}} \left[\beta \frac{\partial E_z}{\partial \eta} - \omega \mu_0 \frac{\partial H_z}{\partial \xi} \right] \quad (7.152)$$

$$H_\eta = \frac{i}{(k^2 n_j^2 - \beta^2) q (\cosh^2 \xi - \cos^2 \eta)^{1/2}} \left[\beta \frac{\partial H_z}{\partial \eta} + \omega n_j^2 \varepsilon_0 \frac{\partial E_z}{\partial \xi} \right] \quad (7.153)$$

where $j = 1$ for $\xi \leq \xi_0$ and $j = 2$ for $\xi \geq \xi_0$. Using (7.150)–(7.153) and matching the field components at $\xi = \xi_0$, we obtain the relations:

$$A \, Ce_v(\xi_0, \gamma_1^2) = L \, Fek_v(\xi_0, -\gamma_2^2) \quad (7.154)$$

$$B\text{Se}_v(\xi_0, \gamma_1^2) = P\text{Gek}_v(\xi_0, -\gamma_2^2) \tag{7.155}$$

$$-\frac{[A\beta v\text{Ce}_v(\xi_0, \gamma_1^2) + \omega\varepsilon_0 n_1^2 B\text{Se}'_v(\xi_0, \gamma_1^2)]}{(k^2 n_1^2 - \beta^2)}$$

$$= \frac{[-L\beta v\text{Fek}_v(\xi_0, -\gamma_2^2) + \omega\varepsilon_0 n_2^2 P\text{Gek}'_v(\xi_0, -\gamma_2^2)]}{(k^2 n_2^2 - \beta^2)} \tag{7.156}$$

$$\frac{[B\beta v\text{Se}_v(\xi_0, \gamma_1^2) - \omega\mu_0 A\text{Ce}'_v(\xi_0, \gamma_1^2)]}{(k^2 n_1^2 - \beta^2)}$$

$$= \frac{[P\beta v\text{Gek}_v(\xi_0, -\gamma_2^2) - \omega\mu_0 L\text{Fek}'_v(\xi_0, -\gamma_2^2)]}{(k^2 n_2^2 - \beta^2)} \tag{7.157}$$

where the notations $\text{Se}'_v(\xi_0, \gamma_1^2)$, etc. have been used as abbreviations for

$$\left.\frac{d}{d\xi}(\text{Se}_v(\xi, \gamma_1^2))\right|_{\xi=\xi_0},$$

etc.

Eliminating the coefficients A, B, L, P from equations (7.154)–(7.157) we find the eigenvalue equation for $_e\text{HE}_{vM}$ modes on the weakly-guiding elliptical dielectric rod of small eccentricity in the form (Adams et al., 1979; Cozens and Dyott, 1979):

$$\left[\frac{\text{Ce}'_v(\xi_0, \gamma_1^2)}{u^2 \text{Ce}_v(\xi_0, \gamma_1^2)} + \frac{\text{Fek}'_v(\xi_0, -\gamma_2^2)}{w^2 \text{Fek}_v(\xi_0, -\gamma_2^2)}\right]\left[\frac{\text{Se}'_v(\xi_0, \gamma_1^2)}{u^2 \text{Se}_v(\xi_0, \gamma_1^2)} + \frac{(1-2\Delta)\text{Gek}'_v(\xi_0, -\gamma_2^2)}{w^2 \text{Gek}_v(\xi_0, -\gamma_2^2)}\right]$$

$$= \left(\frac{v\beta}{kn_1}\right)^2 \left(\frac{v}{uw}\right)^4 \tag{7.158}$$

where the conventional definitions of u, v, w, and Δ have been used. This equation is analogous to (7.55) for the circular dielectric rod; the analogy may be made more precise by re-defining the derivatives of the Mathieu functions, as pointed out by Cozens and Dyott (1979).

For the $_o\text{HE}_{vM}$ modes on the dielectric rod, we may apply a very similar argument to that given above, with the result that the corresponding eigenvalue equation becomes:

$$\left[\frac{\text{Se}'_v(\xi_0, \gamma_1^2)}{u^2 \text{Se}_v(\xi_0, \gamma_1^2)} + \frac{\text{Gek}'_v(\xi_0, -\gamma_2^2)}{w^2 \text{Gek}_v(\xi_0, -\gamma_2^2)}\right]\left[\frac{\text{Ce}'_v(\xi_0, \gamma_1^2)}{u^2 \text{Ce}_v(\xi_0, \gamma_1^2)} + \frac{(1-2\Delta)\text{Fek}'_v(\xi_0, -\gamma_2^2)}{w^2 \text{Fek}_v(\xi_0, -\gamma_2^2)}\right]$$

$$= \left(\frac{v\beta}{kn_1}\right)^2 \left(\frac{v}{uw}\right)^4. \tag{7.159}$$

Once again the similarity to equation (7.55) is obvious and it is easily shown that in the limit $e \to 0$ both (7.158) and (7.159) reduce to (7.55).

Numerical solutions of equations (7.158) and (7.159) may be computed by using expansions of the Mathieu functions in terms of series of Bessel functions (McLachlan, 1947). For small eccentricities these series may be terminated after terms of order e^2 which makes the computation a relatively simple task. More general solutions have been calculated by Yeh (1962) and Lyubimov et al. (1961). For the multimode weakly-guiding, slightly elliptical case the dispersion characteristics are not of great interest since the spread of group velocities is similar to that for the corresponding circular fibre. Once again interest is confined to the cases of graded index profiles or of single-mode operation. Mode cut-offs may also be found from (7.158) and (7.159) in the usual limit $\beta = kn_2$. The only modes with zero cut-off are the $_e\text{HE}_{11}$

and $_o\text{HE}_{11}$ modes which are therefore the fundamental modes, just as for the circular fibre the HE_{11} mode is the fundamental.

By analogy with circular fibres, the first higher-order mode is expected to be associated with $v = 0$ in (7.158) and (7.159). Since the function Se_0 does not exist, there are only TE even modes and TM odd modes with characteristic equations (Cozens and Dyott, 1979).

$$\frac{\text{Ce}_0'(\xi_0, \gamma_1^2)}{u^2 \text{Ce}_0(\xi_0, \gamma_1^2)} + \frac{\text{Fek}_0'(\xi_0, -\gamma_2^2)}{w^2 \text{Fek}_0(\xi_0, -\gamma_2^2)} = 0 \qquad (7.160)$$

$$\frac{\text{Ce}_0'(\xi_0, \gamma_1^2)}{u^2 \text{Ce}_0(\xi_0, \gamma_1^2)} + \frac{(1 - 2\Delta)\text{Fek}_0'(\xi_0, -\gamma_2^2)}{w^2 \text{Fek}_0(\xi_0, -\gamma_2^2)} = 0, \qquad (7.161)$$

As $w \to 0$, $(\text{Fek}_0'/\text{Fek}_0) \to 1$ and hence at cut-off for both $_e\text{TE}_0$ and $_o\text{TM}_0$, we must have the condition:

$$\text{Ce}_0(\xi_0, \gamma_1^2) = 0. \qquad (7.162)$$

Note that equations (7.160)–(7.162) hold for weakly-guiding elliptical waveguides with small eccentricity. The effects of the latter limitation have been discussed by Citerne (1980) and Rengarajan and Lewis (1980). A theory which is valid for arbitrary eccentricity in dielectric rod and tube waveguides has been developed by Lewis and Deshpande (1979) and Rengarajan and Lewis (1980a); these authors have presented numerical results for higher-mode cut-offs (Rengarajan and Lewis, 1980). These results indicate that equation (7.162) is only valid for a/b ratios very close to unity.

Returning to the fundamental $_e\text{HE}_{11}$ and $_o\text{HE}_{11}$ modes in the 'single-mode' region $v < v_c$ as given by (7.162) for small eccentricity, it is clear that we now have two fundamental modes with different propagation characteristics, phase and group velocities. The polarization degeneracy of the lowest-order HE_{11} mode on the circular dielectric rod has been broken by the ellipticity of the waveguide and this has important implications both for pulse dispersion and polarization-dependent propagation. The resultant difference $\delta\beta$ between propagation constants of the two orthogonally-polarized states of the fundamental mode now gives rise to a birefringence effect so that the fibre exhibits a linear retardation (Ramaswamy et al., 1978) whose magnitude depends on the fibre length. This leads to a polarization state which is, in general, elliptical but which varies periodically along the fibre with characteristic length L_B, where

$$L_B = \frac{2\pi}{\delta\beta}. \qquad (7.163)$$

For elliptical guides of arbitrary eccentricity, numerical solutions of the $_e\text{HE}_{11}$ and $_o\text{HE}_{11}$ mode eigenvalue equations have been given by Dyott et al. (1979). Here, however, we confine attention to small eccentricity to develop an approximate result for $\delta\beta$ based on equations (7.158) and (7.159). The method follows that of Adams et al. (1979) although a minor algebraic error has been subsequently corrected by Love et al. (1979a).

Using the standard expressions for the Mathieu functions in terms of Bessel functions (McLachlan, 1947) and retaining terms to order Δ and e^2, we obtain for the required ratios of Mathieu functions:

$$\frac{\text{Ce}_1'(\xi_0, \gamma_1^2)}{\text{Ce}_1(\xi_0, \gamma_1^2)} \simeq u(1 - e^2)^{1/2} \left[\frac{J_1'(u_e) + (u^2 e^2/32) J_3'(u_e)}{J_1(u_e) + (u^2 e^2/32) J_3(u_e)} \right] \qquad (7.164)$$

$$\frac{\text{Se}_1'(\xi_0, \gamma_1^2)}{\text{Se}_1(\xi_0, \gamma_1^2)} \simeq \frac{e^2}{(1 - e^2)^{1/2}} + u(1 - e^2)^{1/2} \left[\frac{J_1'(u_e) + (3u^2 e^2/32) J_3'(u)}{J_1(u_e) + (3u^2 e^2/32) J_3'(u_e)} \right] \qquad (7.165)$$

$$\frac{\mathrm{Fek}_1'(\xi_0, -\gamma_2^2)}{\mathrm{Fek}_1(\xi_0, -\gamma_2^2)} \simeq w(1-e^2)^{1/2}\left[\frac{K_1'(w_e)+(w^2e^2/32)K_3'(w_e)}{K_1(w_e)+(w^2e^2/32)K_3(w_e)}\right] \quad (7.166)$$

$$\frac{\mathrm{Gek}_1'(\xi_0, -\gamma_2^2)}{\mathrm{Gek}_1(\xi_0, -\gamma_2^2)} \simeq \frac{e^2}{(1-e^2)^{1/2}} + w(1-e^2)^{1/2}\left[\frac{K_1'(w_e)+(3we^2/32)K_3'(w_e)}{K_1(w_e)+(3w^2e^2/32)K_3(w_e)}\right] \quad (7.167)$$

where the definitions of u_e, w_e occurring as arguments of Bessel functions are (Yeh, 1976):

$$\left.\begin{array}{l} u_e = 2\gamma_1 \cosh \xi_0 \\ w_e = 2\gamma_2 \cosh \xi_0 \end{array}\right\}. \quad (7.168)$$

Note that the definitions reduce to the customary ones for u, w as $e \to 0$. Inserting the results (7.164)–(7.167) into the eigenvalue equations for $_e\mathrm{HE}_{11}$ and $_o\mathrm{HE}_{11}$ from (7.158) and (7.159), we obtain simplified equations of the form:

$$F(\beta_e) + e^2 2\Delta G_e(\beta_e) = 0 \quad (7.169\mathrm{a})$$

$$F(\beta_o) + e^2 2\Delta G_o(\beta_o) = 0 \quad (7.169\mathrm{b})$$

where, omitting the arguments of the Bessel functions for brevity,

$$F = \left(\frac{J_0}{uJ_1} - \frac{K_0}{wK_1}\right)\frac{2v^2u^2}{w^2} \quad (7.170)$$

$$G_e = \frac{v^2u^2}{32w}\left[\frac{K_3'K_1 - K_1'K_3}{K_1^2}\right]$$
$$+ \frac{u^2}{w}\left(\frac{K_0}{K_1}+\frac{1}{w}\right)\left[\frac{2v^2}{w^2} + \frac{3u^3}{32}\left(\frac{J_3'J_1 - J_1'J_3}{J_1^2}\right) + \frac{3u^2w}{32}\left(\frac{K_3'K_1 - K_3K_1'}{K_1^2}\right)\right] \quad (7.171)$$

$$G_o = \frac{3v^2u^2}{32w}\left[\frac{K_3'K_1 - K_1'K_3}{K_1^2}\right]$$
$$+ \frac{u^2}{w}\left(\frac{K_0}{K_1}+\frac{1}{w}\right)\left[\frac{v^2}{w^2} + \frac{u^3}{32}\left(\frac{J_3'J_1 - J_1'J_3}{J_1^2}\right) + \frac{u^2w}{32}\left(\frac{K_3'K_1 - K_3K_1'}{K_1^2}\right) + \frac{u^2v^2}{w^4}\right]. \quad (7.172)$$

A simple result for $\delta\beta = \beta_e - \beta_o$ may be found by writing equations (7.169a) and (7.169b) as Taylor series expansions around the weakly-guiding circular-fibre eigenvalue equation (7.91) for the HE_{11} mode ($F = 0$). The result is:

$$\delta\beta = \beta_e - \beta_o = e^2 2\Delta \left[\frac{G_e(\beta) - G_o(\beta)}{\partial F/\partial \beta}\right] \quad (7.173)$$

where the arguments of functions on the r.h.s. of (7.173) are those for the circular case. Using (7.170)–(7.172) in (7.173) together with the usual Bessel function relationships, we obtain finally:

$$\delta\beta = \frac{e^2(2\Delta)^{3/2}}{a}\left(\frac{u^2w^2}{8v^3}\right)\left[1 + \left(\frac{u^2-w^2}{u^2}\right)\left(\frac{J_0(u)}{J_1(u)}\right)^2 + \frac{w^2}{u}\left(\frac{J_0(u)}{J_1(u)}\right)^3\right]. \quad (7.174)$$

This result is plotted in the normalized form $a\delta\beta/e^2(2\Delta)^{3/2}$ as a function of v in Figure 7.20. Equation (7.174) has also been derived from perturbation theories by Tsuchiya and Sakai (1974), Tjaden (1978a), and Love et al. (1979a). An interesting feature of the result of Figure

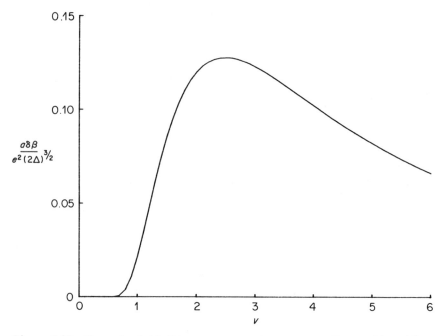

Figure 7.20 Normalized birefringence parameter versus v for weakly-guiding slightly-elliptical fibres, as calculated from equation (7.174). (From Love et al. (1979). Reproduced by permission of the IEE)

7.20 is that the maximum value of $\delta\beta$ is seen to occur for $v \simeq 2.5$ which is close to the cut-off of the first higher-order mode in slightly elliptical fibres and hence the preferred operating point for a single-mode fibre. Numerical results for $\delta\beta$ in graded-index fibres have been presented by Sammut (1980).

The practical implications of birefringence due to core ellipticity are important for two groups of fibre devices. As an example of the first group we may cite the Faraday-effect current transducer (Smith, 1978) where polarized light is used in a coil of fibre encircling a current-carrying conductor in order to measure the current by means of the rotation of polarization. For this application it is clearly necessary to minimize birefringence in the fibre and ensure a beat length L_B in (7.163) long enough ($\simeq 50$ m) to prevent significant polarization variation in the length of fibre used. The implications of (7.174) for the fabrication of low-retardance fibres may be found by examination of the normalization parameter. We see that a greater ellipticity can be tolerated in fibres having large cores and small values of index difference Δ. However, the circularity required to produce a fibre with a periodic length L_B of 50 m (i.e. $\delta\beta = 0.13$ rad/m) remains exceptionally high. A fibre designed to operate at 0.633 μm and $v = 2.4$, having a core diameter of 6 μm, requires the difference between major and minor axes of the core to be only 0.7 per cent. In practice good core circularity is not the only requirement for low-birefringence fibres since the presence of an asymmetrical residual stress distribution within the fibre results in stress birefringence (Ramaswamy et al., 1978a) and similarly contributes to the observed retardation. A fibre having an exceptionally circular core and reduced stress levels is therefore necessary (Schneider et al., 1978). The best results achieved to date using this approach have resulted in a fibre with a beat length L_B of 140 m (Norman et al., 1979).

The second group of devices requiring a fibre that can maintain a fixed state of polarization over an extended length includes a range of applications in fibre interferometers, e.g. the fibre gyro (Vali and Shorthill, 1977), and integrated optics (Steinberg and Giallorenzi, 1976). In this case the requirement is for a fibre with a defined output polarization, a problem which is further complicated by the fact that power may be transferred from one polarization to the other by bending of the fibre, and that the stress-birefringence effect is temperature-dependent. The appropriate solution here is to induce deliberately a large degree of birefringence in the fibre by fabricating a highly elliptical core (Ramaswamy et al., 1978a) or introducing an asymmetrical stress distribution (Ramaswamy et al., 1978b). By this means a preferred polarization direction is established into which all power may be launched. If the beat length L_B is short compared with the period of the spatial undulations of the fibre axis to be found in practice ($\simeq 1$ mm), transfer of power to the orthogonal mode will be precluded and the output will remain linearly polarized in the direction of the preferred axis. From equation (7.174) it is clear that large values of $\delta\beta$ may be achieved for large index differences and large eccentricities, a trend which is maintained when accurate numerical solutions applicable for arbitrary values of e are employed (Dyott et al., 1979). Fabrication of fibres with $\Delta n_1 = 0.065$ and $a/b = 2.5$ has resulted in a beat length L_B of 0.75 mm (Dyott et al., 1979), an achievement which is attributed solely to the elliptical geometry. On the other hand, the technique of introducing an asymmetrical stress distribution by differential thermal expansion in a fibre prepared from a circular preform by the exposed-cladding technique has resulted in $L_B \simeq 2$ mm at $0.633 \,\mu$m (Kaminow et al., 1979).

The presence of two nondegenerate modes of orthogonal polarization in the slightly-elliptical fibre leads also to a difference in modal transit times and hence to a source of intermodal pulse-spreading even in a nominally 'single-mode' fibre. The magnitude of this effect may be calculated by differentiating $\delta\beta$ from equation (7.174) in the usual way, to obtain the time delay difference per unit length, $\delta\tau$:

$$\delta\tau = \frac{1}{c}\frac{d(\delta\beta)}{dk} = \frac{n_1}{c}e^2(2\Delta)^2\Phi \qquad (7.175)$$

where we have neglected the wavelength-dependence of n_1 and Δ, and Φ is given by (Tjaden, 1978a):

$$\Phi = \frac{d}{dv}\left\{\frac{u^2w^2}{8v^3}\left[1+(u^2-w^2)H^2+u^2w^2H^3\right]\right\}$$

$$= \frac{u^2w^2}{8v^4}\left\{1-2(u^2-w^2)H+(5u^2-5w^2-3u^2w^2)H^2+(13u^2w^2-2v^4)H^3\right.$$

$$\left.+[2v^4-12u^2w^2-3u^2w^2(u^2-w^2)]H^4+6u^2w^2(u^2-w^2)H^5+3u^4w^4H^6\right\} \qquad (7.176)$$

and

$$H = \frac{J_0(u)}{uJ_1(u)} = \frac{K_0(w)}{wK_1(w)}. \qquad (7.177)$$

In evaluating the derivatives in (7.176) the weakly-guiding eigenvalue equation (7.91) has been used in the form (7.177), together with the usual relationships between u, v, w.

A plot of Φ from (7.176) versus v is given in Figure 7.21. A typical fibre having $v = 2.4$, $\Delta = 4 \times 10^{-3}$ and a difference between major and minor axes of 5 per cent would give a pulse separation of about 1 ps/km, a very small value. Note, however, that if the stress-induced birefringence is such as to produce a periodic length L_B of 10 cm at 1.3 μm, the pulse separation $\delta\tau = \lambda/cL_B$ (in the absence of mode coupling) is expected to be 43 ps/km, a dispersion which may well be significant on long-distance routes. An experimental

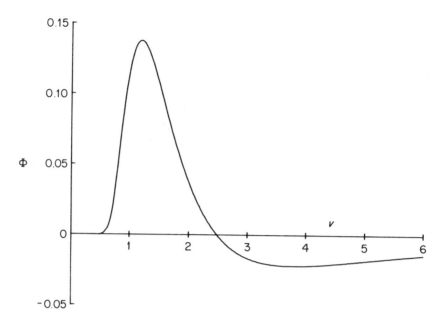

Figure 7.21 Normalized group delay difference Φ between orthogonally-polarized modes, as claculated from equation (7.176)

measurement for such a fibre has been reported with the result of 30 ps/km at $\lambda = 0.69\,\mu$m (Rashleigh and Ulrich, 1978). On the other hand, a reduction in birefringence to give a periodic length of 10 m would render the effect negligible.

7.3 Step-index Guides of More Complicated Structure

The present section is concerned with the three-dimensional circular analogues of the structures discussed for two dimensions in Sections 2.5 and 2.6. In principle, of course, detailed classification of all the possible structures could proceed along the lines of the discussion given in Chapter 2 where slab waveguides were distinguished by the relative magnitudes of the real and imaginary parts of the complex permittivity in core and cladding layers. However, for the circular structures there is only practical interest at present in a subset of the full range of possibilities; in particular, circular guides in which gain and loss play a significant role in the guiding action do not seem to have been investigated in the literature (see comments by Lim and Lit, 1978). Therefore we confine attention here to the cases of hollow and metallic circular waveguides which are of relevance, respectively, to waveguide lasers (Degnan, 1976) and as guides for CO_2 laser radiation (Garmire et al., 1976). Further structures of interest are the multilayer circular cylindrical guides which will also be analysed in this section. The applications here include the 'buffer layer' commonly used in fabricating single-mode fibres (Tasker et al., 1978), the W-fibre specifically designed to take advantage of certain features of propagation in multilayer guides (Kawakami and Nishida, 1974a), and the use of absorbing jacket materials in order to decrease crosstalk between adjacent fibres (Cherin and Murphy, 1975).

7.3.1 Hollow and metallic waveguides

Consider the circular waveguide whose cross-section is illustrated in Figure 7.22, with core of radius a, refractive index n_1, and cladding of dielectric constant ε_2. This represents a

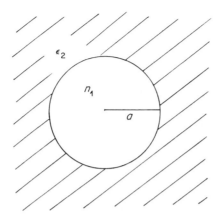

Figure 7.22 Cross-section of general hollow or metallic circular guide

generalization of the dielectric rod shown in Figure 7.8 where the cladding was assumed to have real refractive index n_2. However, the analysis of Section 7.2.1 still holds for this situation provided we recognize that the quantities Δ, β, u, v, w can now all take complex values as a consequence of the use of complex permittivity ε_2. In particular the eigenvalue equation (7.55) remains valid with the proviso that the Bessel functions may now have complex arguments. We will analyse this structure subject to the restraints (Marcatili and Schmeltzer, 1964) of low-loss modes with wavelength much less than the core radius a:

$$ka \gg 1, \quad \left|\frac{\beta}{n_1 k} - 1\right| \ll 1. \tag{7.178}$$

Using these conditions in (7.55) yields $K_v'(w)/K_v(w) \simeq 1$ for large w, and hence we find the zero-order real approximation $u^{(0)}$ for the parameter u as follows:

$$J_{v \pm 1}(u^{(0)}) = 0. \tag{7.179}$$

In this equation the choice of $+$ or $-$ signs corresponds to the EH or HE modes, respectively (Degnan, 1973), according to the definitions of the modal notation given in Section 7.2.1. For the TE and TM modes of this guide, $v = 0$ and the simpler eigenvalue equations (7.57) and (7.56), respectively, yield (using 7.178):

TE: $$\frac{J_1(u)}{u J_0(u)} \simeq \frac{-i}{ak(\varepsilon_2 - n_1^2)^{1/2}} \tag{7.180}$$

TM: $$\frac{J_1(u)}{u J_0(u)} \simeq \frac{-i\varepsilon_2}{ak n_1^2 (\varepsilon_2 - n_1^2)^{1/2}}. \tag{7.181}$$

Returning to the EH and HE modes ($v \neq 0$), in order to find a first-order approximation for the eigenvalue, we write equation (7.55) in the form (using (7.178) and the usual Bessel function relations):

$$\left(\pm v \pm \frac{u J_{v \mp 1}(u)}{J_v} + \frac{u^2}{w}\right)\left(\pm v \pm \frac{u J_{v \mp 1}(u)}{J_v} + \frac{\varepsilon_2 u^2}{n_1^2 w}\right) \simeq v^2. \tag{7.182}$$

It follows, neglecting terms of higher order than u^2/w, that (Marcatili and Schmeltzer, 1964;

Degnan, 1973):

EH/HE:
$$\frac{J_{v\pm 1}(u)}{uJ_v(u)} \simeq \mp \frac{i(\varepsilon_2 + n_1^2)}{2akn_1^2(\varepsilon_2 - n_1^2)^{1/2}}. \tag{7.183}$$

The first-order approximate eigenvalues may now be obtained by noting that the r.h.s. of equations (7.180), (7.181), and (7.182) are all close to zero by virtue of (7.178). Using a straightforward perturbation technique and retaining only the first term of the perturbation, it follows that the first-order approximation $u^{(1)}$ is given by:

TE:
$$u^{(1)} \simeq u^{(0)}\left[1 - \frac{i}{ak(\varepsilon_2 - n_1^2)^{1/2}}\right] \tag{7.184a}$$

TM:
$$u^{(1)} \simeq u^{(0)}\left[1 - \frac{i\varepsilon_2}{akn_1^2(\varepsilon_2 - n_1^2)^{1/2}}\right] \tag{7.184b}$$

EH/HE:
$$u^{(1)} \simeq u^{(0)}\left[1 - \frac{i(\varepsilon_2 + n_1^2)}{2akn_1^2(\varepsilon_2 - n_1^2)^{1/2}}\right]. \tag{7.184c}$$

Note that we follow here the EH/HE notation of Degnan (1973), in accordance with Section 7.2.

With the aid of the restrictions (7.178), these results may be used to find approximations for the propagation constant β:

$$\beta = \left[k^2 n_1^2 - \left(\frac{u^{(1)}}{a^2}\right)^2\right]^{1/2} \simeq kn_1\left[1 - \frac{1}{2}\left(\frac{u^{(1)}}{akn_1}\right)^2\right]. \tag{7.185}$$

The real part β_r of the propagation constant, and the power attenuation coefficient α, are related to β as:

$$\beta = \beta_r - \frac{i\alpha}{2}. \tag{7.186}$$

With the aid of (7.184)–(7.186), we obtain for β_r and α:

$$\beta_r \simeq kn_1\left[1 - \left(\frac{u^{(0)}}{akn_1}\right)^2\left(\frac{1}{2} + \frac{\text{Im}(\mu)}{ak}\right)\right] \tag{7.187}$$

$$\alpha \simeq 2\left(\frac{u^{(0)}}{ak}\right)^2 \frac{\text{Re}(\mu)}{an_1} \tag{7.188}$$

where

$$\mu = \begin{cases} \dfrac{1}{(\varepsilon_2 - n_1^2)^{1/2}}, & \text{for TE modes} \\[2mm] \dfrac{\varepsilon_2}{n_1^2(\varepsilon_2 - n_1^2)^{1/2}}, & \text{for TM modes} \\[2mm] \dfrac{(\varepsilon_2 + n_1^2)}{2n_1^2(\varepsilon_2 - n_1^2)^{1/2}}, & \text{for EH/HE modes.} \end{cases} \tag{7.189}$$

With the aid of these results we may now discuss propagation in hollow and metallic waveguides of circular cross-section. Let us begin with the hollow circular waveguide where ε_2 is real and may be expressed as the square of the refractive index n_2, such that $n_2 > n_1$. It follows that the mode losses are given from (7.188), (7.189) as:

$$\alpha \simeq \begin{cases} \dfrac{2}{an_1}\left(\dfrac{u^{(0)}}{ak}\right)^2 \dfrac{1}{(n_2^2-n_1^2)^{1/2}}, & \text{for TE modes} \quad (7.190a) \\ \dfrac{2}{an_1}\left(\dfrac{u^{(0)}}{ak}\right)^2 \dfrac{n_2^2}{n_1^2(n_2^2-n_1^2)^{1/2}}, & \text{for TM modes} \quad (7.190b) \\ \dfrac{2}{an_1}\left(\dfrac{u^{(0)}}{ak}\right)^2 \dfrac{(n_1^2+n_2^2)}{2n_1^2(n_2^2-n_1^2)^{1/2}}, & \text{for EH/HE modes.} \quad (7.190c) \end{cases}$$

It is worth noting that these results may also be derived from ray theory as well as from the modal eigenvalue equation as above. For the TE case, the polarization of a ray at the core/cladding interface is perpendicular to the plane of incidence and the reflectivity $R_\perp$ is given by equation (2.12) as:

$$R_\perp = \frac{\sin^2(\theta_2-\theta_1)}{\sin^2(\theta_2+\theta_1)} \quad (7.191)$$

where θ_1, θ_2 are angles of incidence and refraction, respectively. For the TM case, the polarization of a ray at the core/cladding interface lies in the plane of incidence and the reflectivity $R_\parallel$ is given by equation (2.20) as:

$$R_\parallel = \frac{\tan^2(\theta_1-\theta_2)}{\tan^2(\theta_1+\theta_2)}. \quad (7.192)$$

Now applying the conditions expressed by (7.178) to the ray picture gives

$$\cos\theta_1 = \frac{u}{akn_1} \ll 1 \quad (7.193a)$$

$$\sin\theta_1 = \frac{\beta}{kn_1} \simeq 1 \quad (7.193b)$$

$$\sin\theta_2 = \frac{n_1}{n_2}\sin\theta_1 \simeq \frac{n_1}{n_2} \quad (7.193c)$$

$$\cos\theta_2 \simeq \left[1-\left(\frac{n_1}{n_2}\right)^2\right]^{1/2} \quad (7.193d)$$

where equation (7.25) has been used with $v = 0$ for TE/TM modes. Using the relations (7.193) in (7.191) and (7.192) yields:

$$1 - R_\perp^2 \simeq \frac{8u^{(0)}}{ak(n_2^2-n_1^2)^{1/2}} \quad (7.194)$$

$$1 - R_\parallel^2 \simeq \left(\frac{n_2}{n_1}\right)^2 \frac{8u^{(0)}}{ak(n_2^2-n_1^2)^{1/2}} \quad (7.195)$$

where $u^{(0)}$ for the ray approach is given strictly by equation (7.41) with $v = 0$, as

$$u^{(0)} = (M+\tfrac{1}{4})\pi \quad (M = 1, 2, 3, \ldots) \quad (7.196)$$

which is already known (from Section 7.1) to give reasonable approximations for the roots of (7.179) with $v = 0$.

Equations (7.194) and (7.195) have been expressed in these forms to facilitate the calculation of attenuation coefficient α from the ray result (cf. Section 2.2 and equation (2.60)):

$$\alpha = \frac{1-R^2}{4a \tan \theta_1} \quad (7.197)$$

where we have taken account of equal reflections at each interface and noted that the Goos–Haenchen shift is zero for partial reflection. Using (7.194) and (7.195) in (7.197), with the aid of (7.193a) and (7.193b), we obtain:

$$\alpha = \begin{cases} \dfrac{2}{an_1}\left(\dfrac{u^{(0)}}{ak}\right)^2 \dfrac{1}{(n_2^2-n_1^2)^{1/2}}, & \text{for polarization perpendicular to the plane of incidence} \\[2ex] \dfrac{2}{an_1}\left(\dfrac{u^{(0)}}{ak}\right)^2 \dfrac{n_2^2}{n_1^2(n_2^2-n_1^2)^{1/2}}, & \text{for polarization parallel to the plane of incidence.} \end{cases} \quad (7.198)$$

Clearly, equation (7.198) reproduces (7.190a) and (7.190b) for TE and TM modes, respectively. EH and HE modes may be regarded as composed of plane waves of both types— those with polarization parallel and perpendicular to the plane of incidence at the core–cladding interface. Hence it is reasonable to expect for these cases an average value of the two expressions for α given in (7.198). That this is indeed the case is confirmed by (7.190c), which was derived from the eigenvalues for EH/HE modes.

It is of interest to identify lowest-loss modes and means of achieving minimum attenuation in hollow waveguides. From (7.190b) it follows that TM mode loss is minimized for $n_2 = n_1 2^{1/2}$; from (7.190c), EH/HE losses are minimized for $n_2 = n_1 3^{1/2}$. By comparing (7.190a) and (7.190c) we find that the lowest-loss mode is TE_{01} if $n_2 > 2.02 n_1$ and HE_{11} if $n_2 < 2.02 n_1$. It follows that for hollow quartz waveguides the lowest-loss mode is HE_{11}. For example, with $n_2 = 1.5$, $n_1 = 1$, $\lambda = 0.633\,\mu m$, $a = 1\,mm$, the loss of this mode is given by (7.190c) as $1.7 \times 10^{-4}\,m^{-1}$ or, equivalently, $0.74\,dB/km$. The waveguide lasers discussed briefly in Section 2.5 frequently use the configuration of a hollow quartz tube to contain the active material, following the suggestion of Marcatili and Schmeltzer (1964). Note that the loss expressions (7.190) and (7.198) retain the a^{-3} dependence found for two-dimensional hollow waveguides as in equation (2.197), and thus the same argument concerning optimization of waveguide laser dimensions is carried over into the circular geometry. Experimental loss measurements at $\lambda = 10\,\mu m$ (Hall et al., 1977) have confirmed the a^{-3} dependence predicted by theory.

Turning now to the metal-walled waveguide, we allow ε_2 to be complex in the results (7.187)–(7.189). These equations may be simplified by making an assumption similar to (2.62) which expresses the fact that the extinction coefficient K_2 is usually large ($\varepsilon_2 = (n_2 + iK_2)^2$). Hence for most combinations of dielectric cores and metallic claddings at optical frequencies:

$$n_1^2 \ll |n_2^2 - K_2^2|. \quad (7.199)$$

With the aid of this assumption, equations (7.188), (7.189) give:

$$\alpha \simeq \begin{cases} \dfrac{2}{a}\left(\dfrac{u^{(0)}}{akn_1}\right)^2 \dfrac{n_1 n_2}{(n_2^2+K_2^2)}, & \text{for TE modes} & (7.200a) \\[2ex] \dfrac{2}{a}\left(\dfrac{u^{(0)}}{akn_1}\right)^2 \dfrac{n_2}{n_1}, & \text{for TM modes} & (7.200b) \\[2ex] \dfrac{2}{a}\left(\dfrac{u^{(0)}}{akn_1}\right)^2 \dfrac{n_2}{n_1}, & \text{for EH/HE modes.} & (7.200c) \end{cases}$$

In order to confirm the results of (7.200) we may again use ray theory; the results for reflectivity at core–cladding boundaries are readily available from equations (2.63) and (2.64) where the assumption (7.199) was also involved. For polarization perpendicular and parallel to the plane of incidence these equations give, respectively:

$$R_\perp \simeq \frac{(n_1 \cos\theta_1 - n_2)^2 + K_2^2}{(n_1 \cos\theta_1 + n_2)^2 + K_2^2} \tag{7.201}$$

$$R_\parallel \simeq \frac{(n_2 \cos\theta_1 - n_1)^2 + K_2^2 \cos^2\theta_1}{(n_2 \cos\theta_1 + n_1)^2 + K_2^2 \cos^2\theta_1}. \tag{7.202}$$

Using the condition (7.193a) on these relations yields the simplified approximations (Eaglesfield, 1962):

$$1 - R_\perp^2 \simeq \frac{8 n_2 u^{(0)}}{ak(n_2^2 + K_2^2)} \tag{7.203}$$

$$1 - R_\parallel^2 \simeq \frac{8 n_2 u^{(0)}}{akn_1^2} \tag{7.204}$$

where $u^{(0)}$ is given by (7.196) for the ray approximation. Insertion of equations (7.203) and (7.204) in the loss formula (7.197), together with (7.193) for $\tan\theta_1$, leads directly to confirmation of equations (7.200a) and (7.200b) for TE and TM modes, respectively. Once again we may interpret the result (7.200c) for EH/HE modes in ray terms by recognizing that these modes are composed of plane waves of both polarizations; hence (7.200c) is an average value of the results (7.200a) and (7.200b) for polarizations perpendicular and parallel to the plane of incidence, respectively.

In view of the restriction (7.199), it is clear from (7.200) that the TE modes of the metal-clad guide always have considerably lower loss than the corresponding TM or EH/HE modes. It follows that the lowest-loss mode is TE_{01}. For a copper guide with air as the core material, using the data of Table 2.2 ($n_2 = 12.6$, $K_2 = -64.3$), equation (7.200a) yields (Garmire et al., 1976)

$$\alpha_{01}^{TE} \simeq 0.0021 \left(\frac{\lambda^2}{a^3}\right).$$

For example, for a guide with $a = 200\,\mu m$, $\lambda = 10.6\,\mu m$, the loss is only $0.03\,m^{-1}$, so that about 97 per cent of the TE_{01} mode intensity will be transmitted in a metre. By contrast, the loss of the corresponding TM mode is typically a hundred times greater than the TE mode loss, for most metallic claddings. Whilst these results indicate the promise of circular metallic waveguides for low-loss optical and i-r transmission, the difficulty of efficiently exciting the TE_{01} mode with commercially available lasers has meant that more attention has been given to metal guides of rectangular cross-section (see Section 6.2.4). For these latter structures a principal advantage is that when the guide is bent the optical power propagates as a 'whispering-gallery' mode with relatively low loss. As a further step towards the efficient transmission of light in hollow guides, a helical tube has been proposed and demonstrated with flexible copper tubing (Marhic et al., 1978, 1978a; Casperson and Garfield, 1979). In this case the light is made to follow the outermost helix and is thus confined in both transverse directions whilst propagating with relatively low loss (Marhic, 1979). A helical structure of this type provides a convenient flexible guide for CO_2 laser radiation, which should find many applications in engineering and medicine.

To conclude this subsection we list the approximate field components for hollow waveguides, using the condition (7.178) and neglecting terms of order (λ/a) and above. For TE modes $v = 0$ and we choose $A = 0$ in equation (7.47), and hence (7.49)–(7.52) yield the approximate fields ($r \leqslant a$) omitting time- and z-dependence, and re-normalizing (Degnan, 1976):

$$
\left.\begin{aligned}
E_r &= 0 \\
E_\theta &= J_1\left(\frac{ur}{a}\right) \\
E_z &= 0 \\
H_r &= -n_1\left(\frac{\varepsilon_0}{\mu_0}\right)^{1/2} J_1\left(\frac{ur}{a}\right) \\
H_\theta &= 0 \\
H_z &= O\left(\frac{\lambda}{a}\right)
\end{aligned}\right\}. \tag{7.205}
$$

Similarly, for TM modes $B = 0$, $v = 0$, and equations (7.47)–(7.52) give:

$$
\left.\begin{aligned}
E_r &= J_1\left(\frac{ur}{a}\right) \\
E_\theta &= 0 \\
E_z &= O\left(\frac{\lambda}{a}\right) \\
H_r &= 0 \\
H_\theta &= n_1\left(\frac{\varepsilon_0}{\mu_0}\right)^{1/2} J_1\left(\frac{ur}{a}\right) \\
H_z &= 0
\end{aligned}\right\}. \tag{7.206}
$$

For EH/HE modes, equations (7.53) and (7.54) with $\lambda \ll a$ yield:

$$
\frac{A}{B} \simeq \mp \frac{i}{n_1}\left(\frac{\mu_0}{\varepsilon_0}\right) \tag{7.207}
$$

where the upper and lower signs refer, respectively, to EH and HE modes. It follows that the corresponding fields are given (neglecting time-, z-, and θ-dependences) approximately by:

$$
\left.\begin{aligned}
E_r &= J_{v\pm 1}\left(\frac{ur}{a}\right) \\
E_\theta &= \mp i J_{v\pm 1}\left(\frac{ur}{a}\right) \\
E_z &= O\left(\frac{\lambda}{a}\right) \\
H_r &= \pm i n_1\left(\frac{\varepsilon_0}{\mu_0}\right)^{1/2} J_{v\pm 1}\left(\frac{ur}{a}\right) \\
H_\theta &= n_1\left(\frac{\varepsilon_0}{\mu_0}\right)^{1/2} J_{v\pm 1}\left(\frac{ur}{a}\right) \\
H_z &= O\left(\frac{\lambda}{a}\right)
\end{aligned}\right\}. \tag{7.208}
$$

The simplified field components given in (7.205), (7.206), (7.208) form the basis for theories of waveguide laser resonators (Degnan, 1976). Higher-order approximations, including the field behaviour in the cladding regions ($r \geq a$), may also be derived in the usual way (Marcatili and Schmeltzer, 1964; Degnan, 1973).

7.3.2 Multilayer guides

Consider now the multilayer circular dielectric waveguide whose cross-section is shown in Figure 7.23. The core, of refractive index n_1 and radius a, is surrounded by a cladding of index n_2 and radius d, which in turn is embedded in an outer cladding of index n_3 and which is assumed to be of infinite extent. We assume $n_1 \geq n_3 > n_2$ and weak-guidance, i.e. $(n_1 - n_2) \ll n_1$ and $|n_3 - n_2| \ll n_1$. Whilst the modal solutions for this structure could be derived by a procedure similar to that used in Section 7.2, it is simpler to recognize that for the

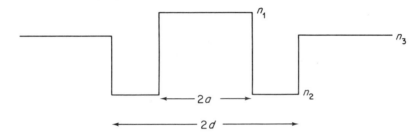

Figure 7.23 Cross-section of a general three-layer circular dielectric waveguide

weakly-guiding situation the modes will be approximately linearly-polarized and to seek the appropriate LP solutions directly. Following this line of thought, it is natural to assume solutions with Cartesian field components $(0, E_y, E_z)$, $(H_x, 0, H_z)$ where $|E_z|, |H_z| \ll |E_y|, |H_x|$ (Gloge, 1971). Omitting the usual time- and z-dependence, the transverse field components in the three layers of the guide may be expressed as (Unger, 1977):

$$E_y = \begin{cases} -\dfrac{1}{n_1}\left(\dfrac{\mu_0}{\varepsilon_0}\right)^{1/2} H_x = A J_l\!\left(\dfrac{ur}{a}\right)\cos l\theta, & r \leq a \quad (7.209\text{a}) \\[6pt] -\dfrac{1}{n_2}\left(\dfrac{\mu_0}{\varepsilon_0}\right)^{1/2} H_x = \left[B I_l\!\left(\dfrac{t''r}{a}\right) + C K_l\!\left(\dfrac{t''r}{a}\right)\right]\cos l\theta, & a \leq r \leq d \quad (7.209\text{b}) \\[6pt] -\dfrac{1}{n_3}\left(\dfrac{\mu_0}{\varepsilon_0}\right)^{1/2} H_x = D K_l\!\left(\dfrac{wr}{a}\right)\cos l\theta, & d \leq r \quad (7.209\text{c}) \end{cases}$$

where A, B, C, D are constants to be evaluated, J_l is a Bessel function, and I_l, K_l are modified Bessel functions. In (7.209) only the case of a cosine function of the azimuthal coordinate θ has been used; there is the usual degeneracy of the LP modes with regard to the alternative sine function for all except the $l = 0$ mode. The normalized transverse propagation constants used in (7.209) are defined in the same way as in Section 2.6:

$$u^2 = a^2(k^2 n_1^2 - \beta^2) \qquad (7.210\text{a})$$
$$t''^2 = a^2(\beta^2 - k^2 n_2^2) \qquad (7.210\text{b})$$
$$w^2 = a^2(\beta^2 - k^2 n_3^2). \qquad (7.210\text{c})$$

The axial field components may be found from the transverse components via the use of Maxwell's equations, with the result:

$$E_z \simeq \frac{-i}{kn_j^2}\left(\frac{\mu_0}{\varepsilon_0}\right)^{1/2}\frac{\partial H_x}{\partial y} \quad (j=1,2,3) \tag{7.211}$$

$$H_z \simeq \frac{-i}{k}\left(\frac{\varepsilon_0}{\mu_0}\right)^{1/2}\frac{\partial E_y}{\partial x}. \tag{7.212}$$

In order to calculate H_z from (7.212), we make use of the relation:

$$\frac{\partial E_y}{\partial x} = \cos\theta \frac{\partial E_y}{\partial r} - \frac{\sin\theta}{r}\frac{\partial E_y}{\partial \theta}. \tag{7.213}$$

From equation (7.209) and the usual Bessel function relationships, we obtain:

$$H_z = \begin{cases} -\dfrac{Aiu}{2ak}\left(\dfrac{\varepsilon_0}{\mu_0}\right)^{1/2}\left[J_{l-1}\left(\dfrac{ur}{a}\right)\cos(l-1)\theta - J_{l+1}\left(\dfrac{ur}{a}\right)\cos(l+1)\theta\right], & r \leqslant a \\ & \quad (7.214\text{a}) \\ -\dfrac{it''}{2ak}\left(\dfrac{\varepsilon_0}{\mu_0}\right)^{1/2}\Bigg\{\left[BI_{l-1}\left(\dfrac{t''r}{a}\right)-CK_{l-1}\left(\dfrac{t''r}{a}\right)\right]\cos(l-1)\theta \\ \quad +\left[BI_{l+1}\left(\dfrac{t''r}{a}\right)-CK_{l+1}\left(\dfrac{t''r}{a}\right)\right]\cos(l+1)\theta\Bigg\}, & a\leqslant r \leqslant d \quad (7.214\text{b}) \\ \dfrac{iDw}{2ak}\left(\dfrac{\varepsilon_0}{\mu_0}\right)^{1/2}\left[K_{l-1}\left(\dfrac{wr}{a}\right)\cos(l-1)\theta + K_{l+1}\left(\dfrac{wr}{a}\right)\cos(l+1)\theta\right] & d \leqslant r. \\ & \quad (7.214\text{c}) \end{cases}$$

In order to calculate E_z from (7.211), we make use of the relation:

$$E_z \simeq \frac{i}{kn_1}\frac{\partial E_y}{\partial y} = \sin\theta\frac{\partial E_y}{\partial r} + \frac{\cos\theta}{r}\frac{\partial E_y}{\partial \theta}. \tag{7.215}$$

It follows from equation (7.209) and the usual Bessel function relationships that E_z is given by:

$$E_z = \begin{cases} -\dfrac{Aiu}{2akn_1}\left[J_{l-1}\left(\dfrac{ur}{a}\right)\sin(l-1)\theta + J_{l+1}\left(\dfrac{ur}{a}\right)\sin(l+1)\theta\right], & r \leqslant a \quad (7.216\text{a}) \\ -\dfrac{it''}{2akn_1}\Bigg\{\left[BI_{l-1}\left(\dfrac{t''r}{a}\right)-CK_{l-1}\left(\dfrac{t''r}{a}\right)\right]\sin(l-1)\theta \\ \quad -\left[BI_{l+1}\left(\dfrac{t''r}{a}\right)-CK_{l+1}\left(\dfrac{t''r}{a}\right)\right]\sin(l+1)\theta\Bigg\}, & a\leqslant r \leqslant d \quad (7.216\text{b}) \\ \dfrac{iDw}{2akn_1}\left[K_{l-1}\left(\dfrac{wr}{a}\right)\sin(l-1)\theta - K_{l+1}\left(\dfrac{wr}{a}\right)\sin(l+1)\theta\right], & d \leqslant r. \\ & \quad (7.216\text{c}) \end{cases}$$

In order to satisfy the boundary conditions at $r=a$ and $r=d$, it is necessary to evaluate the azimuthal field components E_θ, H_θ; these are given in terms of the transverse components E_y, H_x as:

$$E_\theta = E_y \cos\theta \tag{7.217}$$

$$H_\theta = -H_x \sin\theta. \tag{7.218}$$

Using these relations with (7.209) and noting that for the case of weak-guidance $n_1 \simeq n_2 \simeq n_3$, we obtain two equations by matching E_θ and H_θ at $r = a$ and $r = d$:

$$AJ_l(u) = BI_l(t'') + CK_l(t'') \tag{7.219}$$

$$DK_l\left(\frac{wd}{a}\right) = BI_l\left(\frac{t''d}{a}\right) + CK_l\left(\frac{t''d}{a}\right). \tag{7.220}$$

Similarly, matching the values of H_z and E_z, respectively, at $r = a$ and $r = d$, we obtain from (7.214) and (7.216):

$$\frac{Au}{t''}J_{l\pm 1}(u) = \mp[BI_{l\pm 1}(t'') - CK_{l\pm 1}(t'')] \tag{7.221}$$

$$\frac{Dw}{t''}K_{l\pm 1}\left(\frac{wd}{a}\right) = -BI_{l\pm 1}\left(\frac{t''d}{a}\right) + CK_{l\pm 1}\left(\frac{t''d}{a}\right). \tag{7.222}$$

The constants A and D may easily be eliminated from equations (7.219)–(7.222), whereupon one is left with two simultaneous equations for the remaining constants B and C. Invoking the condition for a nontrivial solution yields the requirement for a determinant to vanish and hence the eigenvalue equation for the guide. The result may be expressed as (Kawakami and Nishida, 1975; Arnold, 1977a; Unger, 1977):

$$X_1(t'', u) X_2(t'', w) = \frac{I_{l\pm 1}(t'') K_{l\pm 1}\left(\frac{t''d}{a}\right)}{I_{l\pm 1}\left(\frac{t''d}{a}\right) K_{l\pm 1}(t'')} Y_1(t'', u) Y_2(t'', w) \tag{7.223}$$

where

$$X_1(t'', u) = \frac{K_l(t'')}{t'' K_{l\pm 1}(t'')} \mp \frac{J_l(u)}{u J_{l\pm 1}(u)} \tag{7.224}$$

$$X_2(t'', w) = \frac{I_l(t''d/a)}{t'' I_{l\pm 1}(t''d/a)} + \frac{K_l(wd/a)}{w K_{l\pm 1}(wd/a)} \tag{7.225}$$

$$Y_1(t'', u) = \frac{I_l(t'')}{t'' I_{l\pm 1}(t'')} \pm \frac{J_l(u)}{u J_{l\pm 1}(u)} \tag{7.226}$$

$$Y_2(t'', w) = \frac{K_l(t''d/a)}{t'' K_{l\pm 1}(t''d/a)} - \frac{K_l(wd/a)}{w K_{l\pm 1}(wd/a)}. \tag{7.227}$$

Numerical solutions of equations (7.223)–(7.227) for a number of different cases have been published by Kawakami and Nishida (1974a, 1974b), Belanov et al. (1973, 1976), Kuhn (1974, 1975, 1975a), and Roberts (1970, 1975). In the following, however, we confine ourselves to discussing the relevant physical properties of the solutions without the necessity for lengthy numerical computation. A first-order approximation to the solutions of (7.223)–(7.227) may be obtained by considering the outer medium as a perturbation on the modes of the fibre formed by core and infinite cladding. This approach is illustrated in Figure 7.24(a) where the reference fibre SC_1, obtained by setting $n_3 = n_2$, is shown. The eigenvalue equation of this reference fibre is well known from Section 7.2; in the present notation, putting $n_3 = n_2$ in equation (7.223) yields $X_1(t'', u) = 0$ which is identical with equation (7.91). It follows that the zeroth order approximation for the mode cut-offs is given by

$$w = 0 \tag{7.228a}$$

$$X_1(t_0'', u) = 0. \tag{7.228b}$$

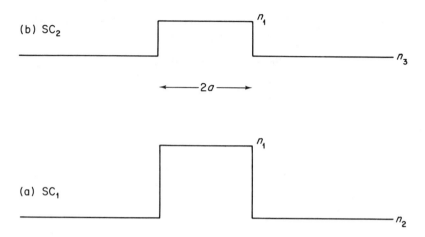

Figure 7.24 Reference fibre index distributions (a) SC_1 and (b) SC_2 for use with the 3-layer structure of Figure 7.23

The normalized frequency v for this structure may be defined as

$$v = ak(n_1^2 - n_3^2)^{1/2}. \tag{7.229}$$

Hence, using (7.228a) together with the definitions of (7.210), the cut-off frequency v_0 is given by

$$v_0 = u = t_0'' \left(\frac{n_1^2 - n_3^2}{n_3^2 - n_2^2} \right)^{1/2}. \tag{7.230}$$

The solutions of equation (7.228b) are represented graphically in Figure 7.25 as a plot of u versus $ak(n_1^2 - n_2^2)^{1/2}$ for the two lowest order modes LP_{01} and LP_{11}. The lines $t'' = 0$ and $w = 0$ are also marked and the intersections of the latter line with the mode curves give the mode cut-offs. We see that the figure is thus divided into three distinct regions corresponding to guided, leaky, and refracted modes. These three classes correspond respectively to three ranges of propagation constant β:

guided modes $n_3 \leq \beta/k \leq n_1$
leaky modes $n_2 \leq \beta/k \leq n_3$
refracted modes $\beta/k \leq n_2$.

Of these classes, the guided modes propagate without loss in the absence of absorption in the guide layers, the leaky modes possess a range of attenuation coefficients, whilst the refracted modes are characterized by very large losses. Note that for the case $n_1 \geq n_2 \geq n_3$ there are no leaky modes, whilst for the case $n_3 \geq n_1 \geq n_2$ there are no guided modes. The structure chosen for analysis here (Figure 7.23) is the so-called 'W-fibre' for which all three mode types are present.

The approximate cut-offs found by the above procedure and illustrated in Figure 7.25 have no dependence on the cladding radius d. In order to obtain the true mode cut-offs, which of course are in general functions of d, it is necessary to examine the cut-off condition found by inserting $w = 0$ in equations (7.223)–(7.227). This yields:

$$X_1(t_c'', v_c) = -\frac{I_{l\pm1}(t_c'')K_{l\pm1}(t_c''d/a)}{I_{l\pm1}(t_c''d/a)K_{l\pm1}(t_c'')} Y_1(t_c'', v_c) \tag{7.231}$$

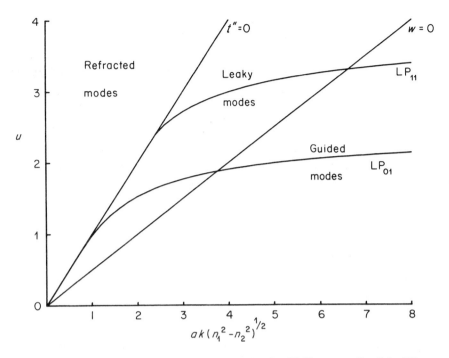

Figure 7.25 The solutions of equation (7.228) for W-fibre cut-offs of the LP_{11} and LP_{01} modes on a plot of u versus $ak(n_1^2 - n_2^2)^{1/2} \equiv v_0 c$, where c is defined in (7.233)

where

$$t_c''^2 = v_c^2(c^2 - 1) \tag{7.232}$$

$$c^2 = \frac{n_1^2 - n_2^2}{n_1^2 - n_3^2}. \tag{7.233}$$

Numerical results for the two lowest-order modes ($l = 1$ and $l = 0$) have been presented by Unger (1977) and Sammut (1978). A second-order approximation for the cut-offs may be obtained by treating (7.231) as a perturbation of the first-order equation (7.228). Writing $v_c = v_0 + \Delta v$ and $t_c'' = t_0'' + \Delta t''$ where t_0'' is given by (7.230), we find

$$\frac{\Delta t''}{t_0''} = \frac{\Delta v}{v_0} \tag{7.234}$$

and hence the l.h.s. of equation (7.231) may be written as

$$X_1(t_c'', v_c) = \Delta t'' \frac{\partial X_1}{\partial t''} + \Delta v \frac{\partial X_1}{\partial v} = \Delta v \left(\frac{t''}{v_0} \frac{\partial X_1}{\partial t''} + \frac{\partial X_1}{\partial v} \right) \tag{7.235}$$

where the derivatives are evaluated at (t_0'', v_0) as given by the solutions of (7.228) and (7.230). When these derivatives are substituted into (7.231) the second-order approximate cut-offs are obtained (Kawakami and Nishida, 1975):

$$v_c = v_0 - v_0 \frac{I_{l \pm 1}(t_0'') K_{l \pm 1}(t_0'' d/a)}{I_{l \pm 1}(t_0'' d/a) K_{l \pm 1}(t_0'')} \frac{Y_1(t_0'', v_0)}{\left[(K_l(t_0'')/K_{l \pm 1}(t_0''))^2 + (J_l(v_0)/J_{l \pm 1}(v_0))^2 \right]}. \tag{7.236}$$

In this equation all the functions appearing in the complicated expression on the r.h.s. are positive and hence the effect is to shift the cut-off v's to lower values than those given by the first approximation v_0. As the cladding-to-core ratio d/a increases, the value of v_c tends asymptotically to v_0 for each given value of c^2 according to (7.233). The particular cases of interest for single-mode fibres are the cut-offs of the HE_{11} and LP_{11} modes. For the LP_{11} mode, the cut-offs increase monotonically for increasing d/a, starting from the cut-off ($v = 2.405$) for the reference fibre SC_2 shown in Figure 7.24(b); the analogous two-dimensional situation was illustrated in Chapter 2, Figure 2.33. For the HE_{11} mode, in contrast, the introduction of the inner cladding does not necessarily lead to an increase of the cut-off above the SC_2 value of zero. It requires a certain combination of the diameter ratio d/a and the index ratio c^2 in order to produce a nonzero HE_{11} cut-off. An expression for this combination can be obtained by taking the case $l = 0$ and $v_c = 0$ in equation (7.231). The result is (Sammut, 1978):

$$\left(\frac{d}{a}\right)^2 \geq \frac{c^2}{c^2 - 1}. \tag{7.237}$$

Sets of d/a and c^2 which satisfy (7.237) will yield HE_{11} mode cut-offs greater than zero. A plot of d/a versus c^2 from equation (7.237) is shown in Figure 7.26. A further discussion of mode cut-offs in 3-layer guides is given by Safaai-Jazi and Yip (1978).

As already discussed in Section 2.6, the W-guide offers the possibility of an extended range of single-mode operation. In order to fully utilize this possibility it is necessary to ensure that the LP_{11} mode suffers sufficient loss to make its presence negligible, since, as shown in the discussion of Figure 7.25, this mode still propagates below cut-off as a leaky mode. The loss may be estimated by solving the eigenvalue equation (7.223) approximately by assuming that

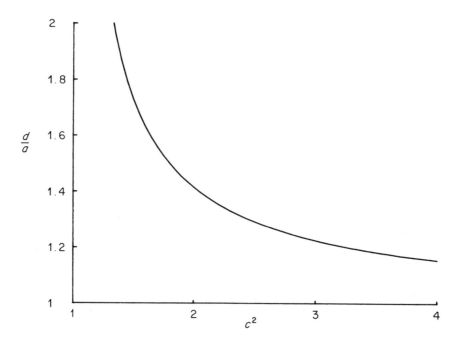

Figure 7.26 The ratio of outer and inner cladding radii, d/a, versus refractive index ratio c^2 (equation 7.233)) for the W-fibre from equation (7.237). Guides with parameters below the line have HE_{11} mode cut-offs of zero frequency

the outer cladding is a perturbation on the reference fibre SC_1. Clearly this approach is valid provided the field at the interface $r = d$ is not too large (Kawakami and Nishida, 1975). This implies large values of t'' so that the asymptotic forms of the modified Bessel functions in (7.223) may be used and it is easy to show that the r.h.s. is of order $\exp[-2(d/a-1)w] \ll 1$ for these conditions. For a perturbation solution, we begin with the values of t'', u obtained from the step-index eigenvalue equation $X_1(t'', u) = 0$, and let $t'' \to t'' + \Delta t''$, $u \to u + \Delta u$. The eigenvalue equation (7.223) may then be written approximately as:

$$\Delta t'' \frac{\partial X_1}{\partial t''} + \Delta u \frac{\partial X_1}{\partial u} = \frac{I_{l\pm 1}(t'') K_{l\pm 1}(t''d/a)}{I_{l\pm 1}(t''d/a) K_{l\pm 1}(t'')} \frac{Y_1(t'', u) Y_2(t'', w)}{X_2(t'', w)}. \tag{7.238}$$

From the definitions (7.210) we have the result:

$$u^2 + t''^2 = a^2 k^2 (n_1^2 - n_2^2) = c^2 v^2 \tag{7.239}$$

and hence

$$u\Delta u + t''\Delta t'' = 0. \tag{7.240}$$

With the aid of (7.240), and the usual Bessel function relations (Abramowitz and Stegun, 1964) the l.h.s. of equation (7.238) becomes:

$$\Delta t'' \frac{\partial X_1}{\partial t''} + \Delta u \frac{\partial X_1}{\partial u} = \frac{c^2 v^2 \Delta u}{u t''^2} \left[1 \mp \frac{2l K_l(t'')}{t'' K_{l\pm 1}(t'')} \right]. \tag{7.241}$$

The perturbation $\Delta\beta$ on the propagation constant β is found from the definition (7.210) as $\beta\Delta\beta = -u\Delta u/a^2$, so that (7.241) and (7.238) yield (Kawakami and Nishida, 1975):

$$\beta\Delta\beta = -\frac{\dfrac{I_{l\pm 1}(t'') K_{l\pm 1}(t''d/a)}{I_{l\pm 1}(t''d/a) K_{l\pm 1}(t'')} \dfrac{Y_1(t'', u) Y_2(t'', w)}{X_2(t'', w)}}{\dfrac{c^2 v^2 a^2}{u^2 t''^2} \left[1 \mp \dfrac{2l K_l(t'')}{t'' K_{l\pm 1}(t'')}\right]}. \tag{7.242}$$

For modes above cut-off ($w^2 > 0$) all quantities in (7.242) are real and thus the net effect is a shift of propagation constant β. For modes below cut-off, however, the result is a loss which may be expressed in terms of the power attenuation coefficient α defined, for example, as in (7.186). In order to obtain a relatively simple expression for α we confine attention to the case of relatively well-confined modes with little field intensity at the interface $r = d$. Under these conditions t'' and w are large and the modified Bessel functions appearing in (7.242) may be replaced by their asymptotic forms (Abramowitz and Stegun, 1964):

$$Y_1(t'', u) \sim \frac{2}{t''} \tag{7.243a}$$

$$Y_2(t'', w) \sim \frac{1}{t''} - \frac{1}{w} \tag{7.243b}$$

$$X_2(t'', w) \sim \frac{1}{t''} + \frac{1}{w}. \tag{7.243c}$$

Hence the attenuation coefficient α becomes

$$\alpha \simeq \frac{4u^2 t''}{\beta c^2 v^2 a^2} \operatorname{Im}\left(\frac{w - t''}{w + t''}\right) \frac{\exp[-2t''(d/a - 1)]}{[1 \mp 2l/t'']}. \tag{7.244}$$

For a mode below cut-off, w^2 is negative and hence for large t'' equation (7.244) simplifies to (Kawakami and Nishida, 1975; Arnold, 1977a):

$$\alpha \simeq \frac{8(u^2-v^2)^{1/2}u^2 t''^2}{kn_2 c^2 v^4 a^2 (c^2-1)} \exp[-2t''(d/a-1)]. \qquad (7.245)$$

Equation (7.245) gives an approximate expression for the power attenuation of a leaky mode where u and t'' are found from the solution of the reference fibre eigenvalue equation $X_1(t'', u) = 0$.

Figure 7.27 gives plots of the normalized attenuation versus $ak(n_1^2 - n_2^2)^{1/2}$ for the special case $n_3 = n_1$, as calculated from equation (7.245). For each value of the cladding diameter ratio d/a at the LP_{11} mode loss (solid line) is larger than the corresponding HE_{11} mode loss (broken line). The magnitude of this loss difference increases with increasing values of d/a and can be as large as several orders of magnitude. Hence we conclude that the LP_{11} mode can be made sufficiently lossy for its influence to be negligible whilst the HE_{11} mode propagates with a relatively low loss. The advantage of the W-fibre for single-mode operation lies therefore in the existence of an enhanced single-mode range of operation as compared with the reference SC_1 fibre. This means that the W-fibre core may be made larger than that of SC_1 and hence the tolerance is somewhat relaxed for splicing of these fibres. A further advantage claimed for the W-fibre is that the modal power is relatively more tightly confined to the core than for the reference SC_2 fibre. Once again the situation is similar to that already discussed for the corresponding two-dimensional waveguides in Section 2.6 (see Figure 2.34). The W-fibre also has different dispersion characteristics from the singly-clad fibre and may in fact be operated in a region where the waveguide dispersion (e.g. equation (7.131)) is negative. To see this, note that the shift of LP_{11} mode cut-off is maximized for $c^2 \to \infty$, i.e. $n_1 = n_3$; it follows that for this case $v_0 = u = 0$ and (7.228) reduces to $J_1(v_0) = 0$ for the LP_{11} mode. This means that the cut-off v can be made as high as 3.832 in comparison with 2.405 for the singly-clad

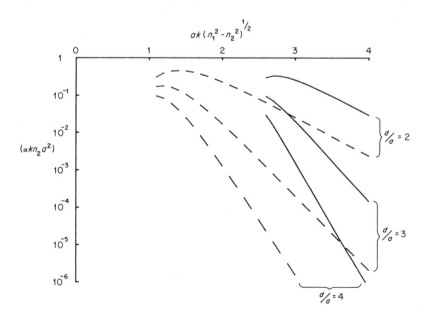

Figure 7.27 Power attenuation coefficient (normalized) versus $ak(n_1^2 - n_2^2)^{1/2}$ for the W-fibre with $n_1 = n_3$, computed from equation (7.245). Broken lines, HE_{11} mode; solid lines, LP_{11} mode

fibre SC_2. The dispersion calculations of Section 7.2 showed that for $v > 3$ the waveguide dispersion term $v\mathrm{d}^2(bv)/\mathrm{d}v^2$ becomes negative. Kawakami and Nishida (1974a, b) have calculated the magnitude of this term for the W-fibre and found that it is sufficiently large to be used to compensate material dispersion under conditions of single-mode operation where the material dispersion remains positive, i.e. for $\lambda \lesssim 1.3\,\mu m$. This represents an alternative possibility for dispersion compensation to that discussed in Section 7.3.4 where negative material dispersion was produced by operation at longer wavelengths.

In Section 7.3.4 we mentioned the necessity for deposited cladding layers in single-mode fibres made by CVD, in order to reduce attenuation due to impurity absorption in the cladding. These layers are usually doped with B_2O_3 in order to depress the index sufficiently below that of the core. Since the substrate tube is composed of SiO_2, the resulting structure is usually of the form shown in Figure 7.23, and there is therefore the possibility of additional attenuation of the HE_{11} mode since it may be a leaky mode. Equation (7.245) may then be used to determine the minimum cladding thickness $(d - a)$ needed to keep the fundamental mode loss to a tolerable level. Using a similar expression due to Gloge (1975), Tasker et al. (1978) have deduced that for the loss to be below 1 dB/km, with $a = 5\,\mu m$, $\lambda = 1.1\,\mu m$, and $vc = 1.8$, the ratio d/a must be at least 9:1. It is worth recalling that equation (7.245) was derived under conditions of weak perturbation by the outer cladding medium. Arnold (1977a) has made a detailed analysis of the attenuation coefficient even in regions where the conventional perturbation analysis fails. The effects of lossy cladding and outer cladding materials have been included in numerical analyses by Roberts (1970, 1975), Clarricoats and Chan (1973), and Kuhn (1975). Marcuse (1971) evaluated the loss by computing the power flow in the transverse direction across the boundary between inner and outer claddings. In all cases the characteristics exponential dependence of attenuation on $(d - a)/a$ exhibited by equation (7.245) is approximately obeyed.

Turning to multimode operation of three-layer circular fibres, we are now concerned with the excess loss introduced by the layer structure as experienced by the total intensity distribution in the fibre, i.e. an average attenuation over all the propagating modes. Whilst equation (7.245) is a valid estimate for those modes whose fields are not too large at $r = d$, it breaks down for modes close to cut-off. Thus a comparison of the results of (7.245) with the measured wavelength-dependence of loss due to optical tunnelling through thinly-clad fibres yields only a general level of agreement with numerous detailed discrepancies (Reeve and Midwinter, 1975). However, the experimental technique (involving an integrating sphere to detect the radiated power) is useful in determining mode cut-offs and permitting an identification of the individual modes of a multimode fibre (Midwinter and Reeve, 1974). For modes which are close to cut-off, including the case of lossy claddings, it may be necessary to solve the eigenvalue equations numerically for all modes of interest (Kuhn, 1975a) in order to estimate the minimum ratio of d/a needed to maintain overall attenuation at a tolerable level. An alternative method, using a ray-tracing technique with reflectivities for multilayer dielectric media, has been given by Cherin and Murphy (1975). This study was specifically concerned with the effect of lossy coatings in decreasing crosstalk between neighbouring fibres and gave conservative loss estimates as a result of only including meridional rays. More recently, an electromagnetic theory approach has been used (Kashima et al., 1977; Kashima and Uchida, 1977) by integrating the Poynting vector at the boundary $r = d$ between the cladding and the outer layer. By using liquid to simulate the outer layer, Kashima and Uchida (1977a) have made measurements of excess loss as a function of cladding thickness, wavelength, outer index n_3, and modal power distribution. The results were found to be in good agreement with the excess loss calculated by this method.

For a multimode fibre the refractive indices of core and cladding are conventionally combined in a single parameter, the numerical aperture (NA), defined as the sine of the maximum angle of acceptance for guided power. Figure 7.28 illustrates this definition in simple ray terms by considering the meridional ray which is just accepted for guidance, i.e. which impinges on the core–cladding interface at the critical angle $\sin^{-1}(n_2/n_1)$. Applying Snell's law at the point of contact of this ray on the fibre end-face leads to the simple relation:

$$\begin{aligned} \text{NA} &= n_1 \cos\left[\sin^{-1}\left(\frac{n_2}{n_1}\right)\right] \\ &= (n_1^2 - n_2^2)^{1/2}. \end{aligned} \qquad (7.246)$$

For a multimode W-fibre this definition of numerical aperture needs to be reconsidered in view of the existence of leaky modes. The definition given in (7.246) is appropriate for the reference fibre SC_1 of Figure 7.24(a) and clearly holds for only a short length of the W-fibre immediately following the input end-face before the leaky modes are significantly attenuated. After an infinite length of fibre the numerical aperture will have reduced to $(n_1^2 - n_3^2)^{1/2}$ which is the appropriate NA for fibre SC_2 of Figure 7.24(b). At lengths between these two limits the numerical aperture will vary according to the overall attenuation of the leaky modes. Measurements of far-field as a function of W-fibre length (Maeda and Yamada, 1977) and of bandwidth and transmission loss under different excitation conditions (Mikoshiba and Kajioka, 1978) all reveal this effect and its dependence on the fibre parameters.

The power fraction confined to the core of a multimode W-fibre has been calculated (Onoda et al., 1976; Kawakami et al., 1976) and shown to be better than that for the reference fibre SC_2. Similarly the bandwidth has been calculated and compared with experimental measurements (Tanaka et al., 1976). The bandwidths are found to be below 300 MHz even for $(n_1 - n_3)/n_1 \simeq 0.03$ per cent, with experimental values lying somewhat below those predicted from theory.

Multimode W-fibres with graded-index cores have also been considered in the literature. The leaky mode attenuation and baseband response have been calculated by the WKB method (Petermann, 1975; Kashima and Uchida, 1978) and confirmed by experimental measurements. In general a thin cladding and large index difference $(n_3 - n_2)$ lead to high bandwidths, but only with a penalty of large excess loss (Kashima et al., 1978). A related topic is the use of mode filters to remove high-order modes and thus improve bandwidths in multimode graded-index fibres (Furuya et al., 1975).

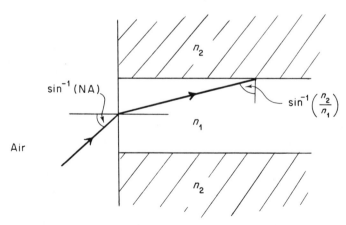

Figure 7.28 Ray diagram illustrating the definition of numerical aperture

A different form of three-layer waveguide may be obtained by again considering the structure of Figure 7.23, but with $n_2 > n_1$ and $n_2 > n_3$. Exact results for the lowest-order modes on such a dielectric-tube waveguide have been reported (Kharadly and Lewis, 1969) and simplified results for the weakly-guiding situation are also available (Marcuse and Mammel, 1973; Miyagi and Nishida, 1977). The theory of dielectric-tube waveguides of elliptical cross-section has been developed by Lewis and Deshpande (1979) and Rengarajan and Lewis (1980). Stolen (1975) has made a detailed study of the solutions for fibres with ring index profiles, with particular reference to the possibility of group velocity matching of modes for nonlinear optics applications. A WKB calculation of dispersive properties for fibres with ring-shaped parabolic index distributions has also been published (Gloge and Marcatili, 1973a).

A multilayer fibre operating on the principle of Bragg reflections has been proposed and analysed by Yeh et al. (1978a). This structure offers the possibility of lossless propagation in a core with a *lower* refractive index than that of the cladding medium. Alternatively, it is possible to envisage a single-mode Bragg fibre whose core diameter is much larger than a wavelength.

Chapter 8
Graded-index Optical Fibres

In this chapter we investigate the properties of circular cylindrical dielectric waveguides with varying refractive index distributions in the core and a uniform cladding assumed to extend to infinity. The general form of the dielectric function to be considered is thus given by

$$n^2(r) = \begin{cases} n_1^2[1 - 2\Delta f(r)], & r < a \quad (8.1a) \\ n_1^2[1 - 2\Delta] \equiv n_2^2, & r \geqslant a \quad (8.1b) \end{cases}$$

where n_1 is the nominal index on axis ($r = a$), and n_2 is the cladding index. The profile grading function $f(r)$ may in general take any form, although power-law or so-called 'alpha-profiles' have been of particular importance for the development of high bandwidth multimode fibres. The alpha-profile grading function is given by (Streifer and Kurtz, 1967):

$$f(r) = \left(\frac{r}{a}\right)^\alpha \quad (8.2)$$

where a is the core radius, and the exponent α may be optimized to minimize intermodal delay differences in a multimode fibre. Plots of $n^2(r)$ versus r for some alpha-profiles according to equations (8.1) and (8.2) are given in Figure 8.1. Note that the special cases $\alpha = \infty$ and $\alpha = 2$ correspond, respectively, to the step-index (uniform core) and parabolic index profiles.

In general, analytic solutions are not available for the fields and propagation constants corresponding to the profiles of equation 8.1 and it is necessary to use approximate methods of solution. However, for the case of the unclad parabolic distribution where we take $\alpha = 2$ in (8.2), remove the condition (8.1b), and permit the profile of (8.1a) to apply for all values of r, then exact solutions are available. These solutions, utilizing either Hermite–Gauss or Laguerre–Gauss functions for the modal fields, are discussed in Section 8.1. Similarly, analytic results are possible for the ray paths in fibres with this profile. These results are also given in Section 8.1 after a brief derivation of the general ray equations of motion in circular graded-index waveguides.

For the analysis of general graded-index multimode fibres the most useful tool has been the WKB approximation and this forms the subject of Section 8.2. Of especial importance here is the derivation of the optimum α-value for the profile of equation (8.2) and a discussion of the effects of material dispersion. The conventional assumption ('linear profile dispersion') is that in equation (8.1a) only n_1 and Δ are functions of wavelength, i.e. that only the 'height' of the index profile changes with wavelength and the 'shape' (i.e. the grading function $f(r)$) does not. It follows that a profile which is optimized for high bandwidth at a specific wavelength will not in general be optimal at any other wavelength, and this effect is also discussed in Section 8.2. A further topic of interest also covered is that of leaky modes; the WKB method provides a useful framework in which to analyse these modes, estimate their attenuation coefficients, and evaluate their effect on the apparent numerical aperture of a graded-index fibre.

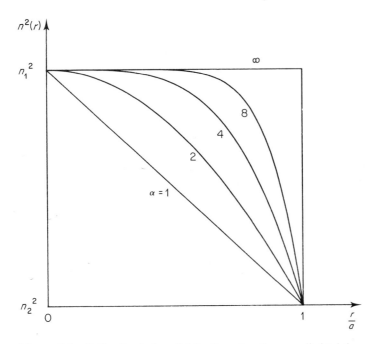

Figure 8.1 Refractive index distributions for the so-called 'alpha-profiles'

Sections 8.3 and 8.4 deal, respectively, with variational and perturbation techniques. Once again the interest centres on obtaining useful approximations for general graded profiles. In addition perturbation theory may also be used to evaluate the effect on fibre bandwidth of small departures from the optimum profile. Section 8.5 deals with other numerical techniques which may be used for arbitrary profiles, in particular those measured on real fibres.

8.1 Parabolic-index Media in Three Dimensions

The parabolic index-profile is worth considering in the context of graded-index fibres since it yields analytic results which facilitate the understanding of more general forms of profile grading. This section therefore presents a discussion of three-dimensional parabolic-index media in analogy with the two-dimensional case considered in Chapter 3. However, before specializing on this profile we give a brief derivation of the ray equations-of-motion for general graded-index media and show how the constants of motion may be related to the familiar modal invariants. For the parabolic profile the ray equations are then solved and ray paths, eigenvalues, and transit times determined. From the modal point of view, the field distributions utilize Hermite–Gauss or Laguerre–Gauss functions. The modal results are derived and compared with the results of geometric optics. A general discussion is given of the relation between vector modes of a weakly-guiding fibre and the solutions of the scalar wave equation. Eigenvalue equations and transit times derived by ray optics are the same as those from solution of the scalar wave equation.

8.1.1 Ray equations for general graded-index media

Consider first the scalar wave equation in the form:

$$\nabla^2 \psi + k^2 n^2(r) \psi = 0 \tag{8.3}$$

where the symbols have their usual meanings. For the ray approximation we are naturally concerned with the case of very small wavelength, i.e. large k, and we seek solutions of the form (in cylindrical polar coordinates):

$$\psi(r, \theta, z) = \psi_0(r, \theta, z)e^{-ikS(r, \theta, z)} \tag{8.4}$$

where ψ_0 and S are assumed slowly-varying over a wavelength of the radiation. Substituting (8.4) into (8.3) yields the equation:

$$\nabla^2 \psi_0 - ik(2\nabla S \cdot \nabla \psi_0 - \psi_0 \nabla^2 S) + k^2 \psi_0 (n^2(r) - \nabla S \cdot \nabla S) = 0. \tag{8.5}$$

For short wavelengths the dominant term in this expression is that with a coefficient of k^2. Hence in the limit of vanishing wavelength we have the eikonal equation (Born and Wolf, 1970; Marcuse, 1972):

$$(\nabla S)^2 = n^2(r). \tag{8.6}$$

This equation determines the propagation in the geometrical optics approximation, since the surfaces $S(r, \theta, z) = $ constant are the phase fronts.

Following Marcuse (1972) we define s as the distance along the ray path to the point r; then the unit vector

$$\mathbf{u} = \frac{d\mathbf{r}}{ds} \tag{8.7}$$

is tangential to the ray and hence normal to the phase fronts. A vector perpendicular to the phase fronts is also found by taking the gradient of S, and from (8.6) the vector $\mathbf{v} \equiv \nabla S$ is of length n. Hence

$$\mathbf{u} = \frac{\mathbf{v}}{n}$$

or

$$n\frac{d\mathbf{r}}{ds} = \nabla S. \tag{8.8}$$

Differentiating both sides of this equation with respect to s yields:

$$\frac{d}{ds}\left(n\frac{d\mathbf{r}}{ds}\right) = \frac{d}{ds}(\nabla S)$$

$$= \frac{d\mathbf{r}}{ds} \cdot \nabla(\nabla S)$$

$$= \frac{\nabla S}{n} \cdot \nabla(\nabla S)$$

$$= \frac{1}{2n}\nabla[(\nabla S)^2]$$

$$= \frac{1}{2n}\nabla n^2.$$

Hence we have finally the ray equation:

$$\frac{d}{ds}\left(n\frac{d\mathbf{r}}{ds}\right) = \nabla n. \tag{8.9}$$

In cylindrical polar coordinates, noting that n is assumed to be a function only of r in accordance with (8.1), the ray equation becomes:

$$\frac{d}{ds}\left[n\left(\frac{dr}{ds}\hat{\mathbf{r}} + r\frac{d\theta}{ds}\hat{\boldsymbol{\theta}} + \frac{dz}{ds}\hat{\mathbf{z}}\right)\right] = \frac{dn}{dr}\hat{\mathbf{r}} \tag{8.10}$$

where $\hat{\mathbf{r}}, \hat{\boldsymbol{\theta}}, \hat{\mathbf{z}}$, are unit vectors in the directions r, θ, z, respectively.

For the z-component of (8.10) we have:

$$\frac{d}{ds}\left(n(r)\frac{dz}{ds}\right) = 0 \tag{8.11}$$

so that a constant of the motion is given by

$$n(r)\frac{dz}{ds} = \text{constant} \equiv \bar{\beta} \text{ (say).} \tag{8.12}$$

The notation $\bar{\beta}$ for the constant (Ankiewicz and Pask, 1977) is chosen to reveal the analogy with the modal propagation constant β, as will become clear later. In terms of the ray path and its direction at point (r, θ, z), Figure 8.2 enables us to evaluate the constant $\bar{\beta}$. From the

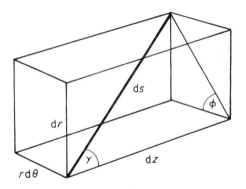

Figure 8.2 Ray components at point (r, θ, z) in a graded-index fibre

figure, the angle between the ray direction and the z-axis is γ, and hence from (8.12) we have (Rawson et al., 1970):

$$\bar{\beta} = n(r) \cos \gamma. \tag{8.13}$$

From (8.12) we also have that

$$\frac{d}{ds} = \frac{\bar{\beta}}{n}\frac{d}{dz}$$

so that equation (8.10) may be re-written as

$$\frac{d}{dz}\left(\frac{dr}{dz}\hat{\mathbf{r}} + r\frac{d\theta}{dz}\hat{\boldsymbol{\theta}}\right) = \frac{n}{\bar{\beta}^2}\frac{dn}{dr}\hat{\mathbf{r}}. \tag{8.14}$$

In order to complete the analysis of equation (8.10) we must evaluate the derivatives with respect to z of the unit vectors $\hat{\mathbf{r}}, \hat{\boldsymbol{\theta}}, \hat{\mathbf{z}}$. In terms of Cartesian unit vectors $\mathbf{i}, \mathbf{j}$, we have:

$$\begin{aligned}\hat{\mathbf{r}} &= \cos\theta\,\mathbf{i} + \sin\theta\,\mathbf{j} \\ \hat{\boldsymbol{\theta}} &= -\sin\theta\,\mathbf{i} + \cos\theta\,\mathbf{j}.\end{aligned} \tag{8.15}$$

Differentiating with respect to z yields the results:

$$\frac{d\hat{r}}{dz} = \frac{d\theta}{dz}(-\sin\theta\mathbf{i} + \cos\theta\mathbf{j}) = \frac{d\theta}{dz}\hat{\theta}$$

$$\frac{d\hat{\theta}}{dz} = \frac{d\theta}{dz}(-\cos\theta\mathbf{i} - \sin\theta\mathbf{j}) = -\frac{d\theta}{dz}\hat{r}. \qquad (8.16)$$

Using these results in (8.14) and denoting differentiation with respect to z by a dot, we find:

$$\ddot{r}\hat{r} + 2\dot{r}\dot{\theta}\hat{\theta} + r\ddot{\theta}\hat{\theta} - r\dot{\theta}^2\hat{r} = \frac{n}{\bar{\beta}}\frac{dn}{dr}\hat{r}. \qquad (8.17)$$

Separating the r- and θ-components of (8.17) gives (Rawson et al., 1970):

$$\ddot{r} - r\dot{\theta}^2 = \frac{n}{\bar{\beta}^2}\frac{dn}{dr} \qquad (8.18)$$

$$2\dot{r}\dot{\theta} + r\ddot{\theta} = 0. \qquad (8.19)$$

Integrating equation (8.19) yields the second ray invariant:

$$r^2\dot{\theta} = \text{constant} \equiv \frac{a\bar{l}}{\bar{\beta}}\,(\text{say}). \qquad (8.20)$$

Once again, using Figure 8.2, we may identify the constant $\bar{l}$. From the figure,

$$r\,d\theta = ds\sin\gamma\cos\phi$$
$$= dz\tan\gamma\cos\phi. \qquad (8.21)$$

Using (8.20), (8.21), and (8.13) yields (Ankiewicz and Pask, 1977):

$$\bar{l} = \frac{r}{a}n(r)\cos\phi\sin\gamma. \qquad (8.22)$$

Having identified the two constants of the motion $\bar{\beta}$, $\bar{l}$ in terms of ray angles on Figure 8.2, it remains to use these in (8.18) to yield a differential equation for the ray trajectory (Jacomme, 1975):

$$\ddot{r} - \left(\frac{\bar{l}a}{\bar{\beta}}\right)^2\frac{1}{r^3} = \frac{1}{2\bar{\beta}^2}\frac{dn^2}{dr}. \qquad (8.23)$$

Integrating with respect to r gives

$$\dot{r}^2 + \left(\frac{\bar{l}a}{\bar{\beta}}\right)^2\frac{1}{r^2} = \frac{n^2}{\bar{\beta}^2} + C \qquad (8.24)$$

where C is a constant of integration. From Figure 8.2 we may evaluate C by noting that

$$dr = ds\sin\gamma\sin\phi$$
$$= dz\tan\gamma\sin\phi. \qquad (8.25)$$

Using the result together with (8.22) and (8.13) in (8.24) yields the value $C = -1$. Hence equation (8.24) may be written in its final form (Jacomme, 1975; Ankiewicz and Pask, 1977; Eve, 1976):

$$\frac{dr}{dz} = \frac{1}{\bar{\beta}}\left[n^2(r) - \bar{\beta}^2 - \left(\frac{a\bar{l}}{r}\right)^2\right]^{1/2}. \qquad (8.26)$$

Combining (8.26) with (8.20) yields:

$$\frac{d\theta}{dr} = \frac{\overline{al}}{r^2}\left[n^2(r) - \overline{\beta}^2 - \left(\frac{\overline{al}}{r}\right)^2\right]^{1/2}. \tag{8.27}$$

Equations (8.26) and (8.27) are the most useful forms of the ray equations for practical calculations of ray paths in graded-index fibres.

Ray equations of the form (8.12), (8.26), and (8.27) may also be derived by the local plane wave analysis which was used for the two-dimensional case in Chapter 3. To follow this approach, we return to the wave equation (8.3) and seek solutions in the form of local plane waves:

$$\psi(r, \theta, z) = \frac{1}{r^{1/2}} \exp(i\beta z \pm il\theta + iqr) \tag{8.28}$$

where β, l are the usual modal invariants in the longitudinal and azimuthal directions, respectively. Substituting (8.28) in (8.3) gives

$$q^2 = k^2 n^2(r) - \beta^2 - \frac{(l^2 - \tfrac{1}{4})}{r^2}$$

$$\simeq k^2 n^2(r) - \beta^2 - \frac{l^2}{r^2} \tag{8.29}$$

for $l \gg \tfrac{1}{2}$, as will be the case for many modes in a multimode fibre. Figure 8.3 illustrates a decomposition of the local plane wave vector based on this analysis (Gloge and Marcatili, 1973) with components q, l/r, β in the directions of $\hat{r}$, $\hat{\theta}$, z, respectively, according to equation

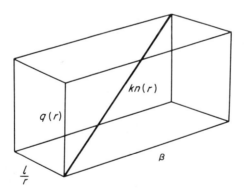

Figure 8.3 Local plane wave decomposition at point (r, θ, z) in a graded-index fibre

(8.29). By comparing Figures 8.3 and 8.2 and recalling that the local plane wave vector is directed tangentially to the ray path, we deduce the following results:

$$\frac{dr}{dz} = \frac{q}{\beta} \tag{8.30}$$

$$\frac{d\theta}{dr} = \frac{l}{r^2 q} \tag{8.31}$$

$$\frac{dz}{ds} = \frac{\beta}{kn(r)}. \tag{8.32}$$

Now equations (8.30), (8.31), and (8.32) are precisely the same as equations (8.26), (8.27), and (8.12), respectively, and by comparing each pair of relations we may identify the ray invariants $\bar{\beta}, \bar{l}$ in terms of the modal constants β, l:

$$\bar{\beta} = \frac{\beta}{k} \tag{8.33}$$

$$\bar{l} = \frac{l}{ak}. \tag{8.34}$$

Hence the reason for the notation used for ray invariants is now clear.

The function q, as given in (8.29), has zeros at r_1 and r_2 as shown in Figure 8.4. These radii define the inner and outer caustics, and hence the projection of the ray path on the (r, θ) plane always remains between r_1 and r_2 for a bound ray, as shown in Figure 8.4. It may also be shown that the function $r(z)$ is periodic in z with period $2\int_{r_1}^{r_2} \beta/q \, dr$, and the function $\theta(z)$ is

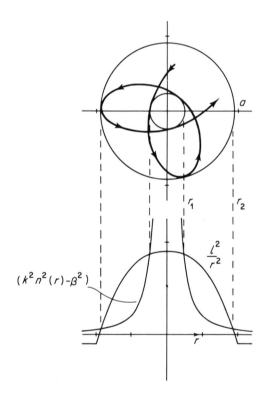

Figure 8.4 Ray path, caustics, and local plane wave components for a graded-index fibre

pseudoperiodic in z (Eve, 1976). To determine the explicit details of these functions it is usually necessary to solve the ray equations numerically. An exception occurs for the parabolic index-profile for which the equations are exactly soluble as will be demonstrated in the next subsection. Two specific ray types are worth mentioning at this point: the meridional rays with $l = 0$ which always pass through the axis of the fibre, and the helical rays with $q = 0$

which always move at a fixed radius from the fibre axis. For the latter ray class, equations (8.30) and (8.31) may be combined to yield the more relevant form:

$$\frac{d\theta}{dz} = \frac{l}{r^2\beta} \tag{8.35}$$

and equation (8.32) still holds. Thus the helical ray path in the θ–z plane may be determined by numerical integration.

Since we are interested in pulse-spreading in graded-index fibres we must derive an expression for the ray transit time τ in a length L of fibre. Recalling that the speed of a ray is given by $c/m(r)$, where $m(r)$ is the group index at radius r, it follows that:

$$\tau = \frac{L}{c} \frac{\oint m(r)\,ds}{\oint dz} \tag{8.36}$$

where

$$m(r) = n(r) + k\frac{dn(r)}{dk} = n(r) - \lambda\frac{dn(r)}{d\lambda} \tag{8.37}$$

and the symbol $\oint$ indicates that the integration is taken over a complete ray period, i.e. an axial distance corresponding to the ray projection leaving the caustic at r_1 and travelling via r_2 back to r_1 (see Figure 8.4). With the aid of equations (8.30) and (8.32), (8.36) becomes:

$$\tau = \frac{L}{c} \frac{\int_{r_1}^{r_2} kn(r)m(r)\,dr/q}{\int_{r_1}^{r_2} \beta\,dr/q}. \tag{8.38}$$

We might proceed direct from this expression to a discussion of intermodal delay time differences and their minimization by suitable choice of the index profile $n(r)$. However, this will be postponed until Section 8.2 where the relation (8.38) derived above will be shown to be the same as that resulting from the WKB approximation.

It remains to give a geometric optics derivation of the eigenvalue equation. Following our earlier practice from previous chapters, this is most easily achieved by calculating the total transverse phase change in one ray period and demanding that this be an integral multiple of 2π:

$$2\int_{r_1}^{r_2} q\,dr - \delta_1 - \delta_2 = 2N\pi \quad (N = 0, 1, 2, \ldots) \tag{8.39}$$

where δ_1, δ_2 are the phase shifts suffered at the caustics r_1, r_2. From Section 3.1.3, we have that $\delta_1 = \delta_2 = \pi/2$, and hence the eigenvalue equation (8.40) becomes:

$$\int_{r_1}^{r_2} \left(k^2 n^2(r) - \beta^2 - \frac{l^2}{r^2}\right)^{1/2} dr = (N + \tfrac{1}{2})\pi \quad (N = 0, 1, 2, \ldots). \tag{8.40}$$

This result will be evaluated for the parabolic index profile in the next subsection.

8.1.2 Geometric optics results for parabolic index-profiles

As a first application of the general results derived in the previous subsection, consider the special case of the parabolic profile. Substituting equations (8.1) and (8.2) with $\alpha = 2$ into (8.30) yields:

$$z - z_0 = \int_{r_0}^{r} \frac{a\beta\,dr}{(u^2 - v^2(r/a)^2 - l^2(a/r)^2)^{1/2}} \tag{8.41}$$

where the initial condition is assumed to be $z = z_0$ at $r = r_0$, and the usual definitions $u^2 = a^2(k^2 n_1^2 - \beta^2)$, $v^2 = a^2 k^2 (n_1^2 - n_2^2)$ apply. For this profile the radii of the caustics are given by

$$\left(\frac{r_{\frac{1}{2}}}{a}\right)^2 = \frac{u^2 \pm (u^4 - 4v^2 l^2)^{1/2}}{2v^2} \tag{8.42}$$

and the condition for bound rays is clearly $u^4 \geqslant 4v^2 l^2$. The integration of (8.41) is straightforward and yields (Jacomme, 1975):

$$\left(\frac{r}{a}\right)^2 = \frac{1}{2v^2}\left[u^2 + (u^4 - 4v^2 l^2)^{1/2} \sin\left(\frac{2v}{a^2\beta}(z - z_0 - z')\right)\right] \tag{8.43}$$

where the length z' is given by $r = r_0$, $z = z_0$:

$$z' = \frac{a^2\beta}{2v} \sin^{-1}\left[\frac{u^2 - 2v^2(r_0/a)^2}{(u^4 - 4v^2 l^2)^{1/2}}\right]. \tag{8.44}$$

Hence the function $r^2(z)$ is periodic with period $(a^2\beta\pi/v)$.

Similarly, equation (8.31) yields for the (r, θ) trajectory:

$$\theta - \theta_0 = \int_{r_0}^{r} \frac{al\, dr}{r^2(u^2 - v^2(r/a)^2 - l^2(a/r)^2)^{1/2}} \tag{8.45}$$

where $\theta = \theta_0$ at $r = r_0$. It follows that (Jacomme, 1975; Matsumura, 1975):

$$\left(\frac{a}{r}\right)^2 = \frac{1}{2l^2}\left[u^2 - (u^4 - 4v^2 l^2)^{1/2} \sin[2(\theta - \theta_0 + \theta')]\right] \tag{8.46}$$

where

$$\theta' = \frac{1}{2}\sin^{-1}\left[\frac{u^2 - 2l^2(a/r_0)^2}{(u^4 - 4v^2 l^2)^{1/2}}\right]. \tag{8.47}$$

Equation (8.46) determines the ray path in the fibre cross-section; to make the path more easily recognizable we change to Cartesian coordinates defined as:

$$X = \frac{r}{a\sqrt{2}}\left[\sin(\theta - \theta_0 + \theta') + \cos(\theta - \theta_0 + \theta')\right]$$

$$Y = \frac{r}{a\sqrt{2}}\left[\sin(\theta - \theta_0 + \theta') - \cos(\theta - \theta_0 + \theta')\right]. \tag{8.48}$$

With this change of variables, equation (8.46) becomes:

$$\left(\frac{X}{X_0}\right)^2 + \left(\frac{Y}{Y_0}\right)^2 = 1 \tag{8.49}$$

where

$$\left.\begin{array}{c}X_0^2 \\ Y_0^2\end{array}\right\} = \frac{2l^2}{u^2 \mp (u^4 - 4v^2 l^2)^{1/2}}. \tag{8.50}$$

Thus it follows that the projection of the ray path in the fibre cross-section is an ellipse whose major axis makes an angle $(\theta_0 - \theta' + \pi/4)$ with the x-axis. The projection is illustrated in Figure 8.5; it should be emphasized that the angle of the ellipse is a function of the input conditions (r_0, θ_0) of the ray considered.

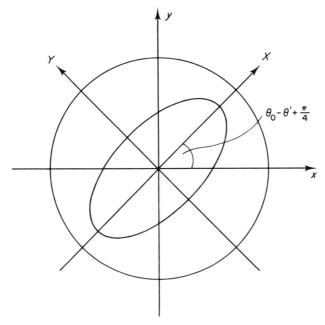

Figure 8.5 Projection of ray path in fibre cross-section for a skew ray in a parabolic index fibre

If the ray path is to remain inside the fibre core then it is necessary that the major axis of the ellipse shown in Figure 8.5 be less than the core radius. Imposing this condition yields, with the aid of (8.50):

$$u^2 - v^2 \leq l^2. \tag{8.51}$$

This relation may be re-cast into ray angles by means of equations (8.13), (8.22), (8.33), and (8.34). The result then becomes:

$$\sin^2\gamma \leq \frac{2\Delta[1 - (r/a)^2]}{[1 - (r/a)^2 \cos^2\phi][1 - 2\Delta(r/a)^2]}. \tag{8.52}$$

Consider now the input face of the fibre with the ray impinging at the point (r_0, θ_0) with $\gamma = \gamma_0$ and $\phi = \phi_0$. If I is the angle of incidence from the outer medium (assumed to be air) then Snell's law ($\sin I = n(r) \sin \gamma_0$) may be combined with (8.52) to yield (Matsumura, 1975):

$$\sin^2 I \leq \frac{2\Delta n_1^2 [1 - (r_0/a)^2]}{[1 - (r_0/a)^2 \cos^2 \phi_0]}. \tag{8.53}$$

That is, the angle of acceptance, or local numerical aperture, on the fibre end-face varies as a function of the launch radius r_0 and the projected angle ϕ_0. The acceptance angle is always less than or equal to that experienced at the core centre ($r_0 = 0$). For $\phi_0 = 0$ or π the angle is always the same as that at $r_0 = 0$, whilst for $\phi_0 = \pi/2$ or $3\pi/2$ the angle assumes its minimum value for each radius r_0. The functional form taken by the local numerical aperture as ϕ_0 varies can be easily visualized by a change to Cartesian coordinates ξ_0, η_0 defined as:

$$\left. \begin{array}{l} \xi_0^2 + \eta_0^2 = \dfrac{\sin^2 I}{2\Delta n_1^2} \\ \xi_0 = \eta_0 \tan \phi_0 \end{array} \right\}. \tag{8.54}$$

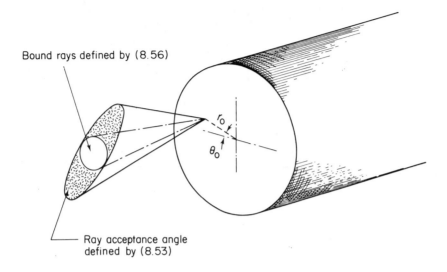
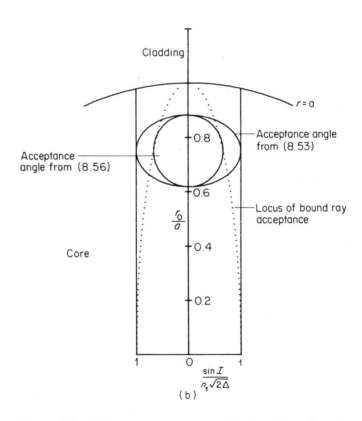

Figure 8.6 (a) Acceptance cones at position (r_0, θ_0) on the end-face of a fibre with parabolic index-profile. (b) Cross-section through acceptance cones on the fibre end face of (a), showing locus of major axis for elliptical cone and that of diameter for circular cone. (From Adams *et al.* (1976). Reproduced by permission of the American Institute of Physics)

With the aid of this change of variables, the relation (8.53) becomes:

$$\xi_0^2 + \frac{\eta_0^2}{[1-(r_0/a)^2]} \leq 1. \tag{8.55}$$

This result shows that the local acceptance cone on the fibre end-face has an elliptical cross section with ratio of minor to major axes given by $[1-(r_0/a)^2]^{1/2}$.

It might at first sight appear that the condition (8.53), or equivalently (8.55), defines the local acceptance angle for bound rays. In fact this is not the case, since inspection of the modal form (8.51) reveals that modes below cut-off are included within the allowed limit. In fact the well-established cut-off condition $u = v$ yields in ray terms for the parabolic profile the result:

$$\sin^2 I \leq 2\Delta n_1^2 [1-(r_0/a)^2]. \tag{8.56}$$

Clearly this defines a cone of circular cross-section which has its radius of the same length as the semi-minor axis of the elliptical cross-section referred to above. The situation is illustrated schematically in Figure 8.6(a) whilst Figure 8.6(b) shows the locus of the major axis of the elliptical cone cross-section and that of the diameter for the circular cone as the launch radius r_0 is varied. In fact the region between the elliptical and circular acceptance cones is occupied by leaky rays (Adams et al., 1976); a formal proof of this will be given in Section 8.2 where a more general discussion of leaky rays/modes will also be found.

The ray transit time τ in a length L of parabolic-profile fibre is given by equation (8.38) with $n(r)$ defined as in (8.1) and (8.2) for $\alpha = 2$. If we ignore material dispersion for the present and concentrate on waveguide effects $(m(r) = n(r))$, then the result is:

$$\begin{aligned}\tau &= \frac{Ln_1}{c}\left(\frac{kn_1}{\beta}\right)\left[1 - \Delta\frac{u^2}{v^2}\right] \\ &= \frac{Ln_1}{c}\left(\frac{kn_1}{2\beta}\right)\left[1 + \left(\frac{\beta}{kn_1}\right)^2\right].\end{aligned} \tag{8.57}$$

Equation (8.57) is the same as (3.16) and (3.25) for parabolic profiles in two-dimensions. It is

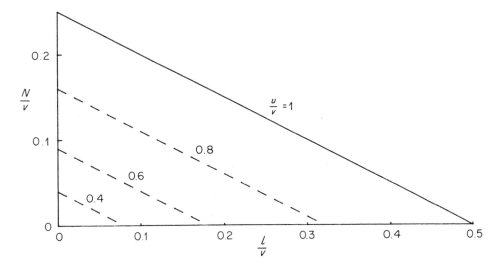

Figure 8.7 Radial mode number N versus azimuthal mode number l for different values of u/v in the parabolic profile fibre. The plot was calculated from equation (8.58) assuming $v \gg 1$

noteworthy that this relation for τ depends only on β and not explicitly on the azimuthal invariant l; we shall see later that this property is characteristic of all α-profiles defined as in equation (8.2). Equation (8.57) also shows that within the paraxial approximation ($\beta \simeq kn_1$) all the ray transit times are the same for the parabolic profile. This implies that the optimum profile for equalizing ray transit times in multimode fibres is close to parabolic, and this will also be confirmed in Section 8.2.

The eigenvalue equation is given by equation (8.40) with the integration limits r_1, r_2 given by the caustics as in (8.42) for the parabolic index-profile. Performing the integration (Gradshteyn and Ryzhik, 1965), we find:

$$u^2 = 2v(2N + l + 1) \tag{8.58}$$

where $N = 0, 1, 2, \ldots$ and $l = 0, 1, 2, \ldots$. Following the procedure used to estimate total mode numbers for step-index fibres in Chapter 7, we may represent the results of equation (8.58) as a plot of N/v versus l/v as in Figure 8.7. To find the total number of guided modes the area of the figure enclosed by the cut-off line ($u/v = 1$) must be determined. Allowing for the four-fold degeneracy of each mode ($l \neq 0$) the total mode number is thus given approximately by $4 \times v^2/16$, i.e. $v^2/4$. Thus the number of guided modes on the parabolic-profile fibre is one half that for the step-index fibre as given in equation (7.136).

8.1.3 Vector field theory for general graded-index media

Rather than proceeding directly to solutions of the scalar wave equation for parabolic-index media, we give first the vector field theory for general graded index media under the conditions of weak guidance. The results will be useful not only in the context of parabolic index-profiles but also later in the chapter when we consider approximate methods of solution for arbitrary index-profiles. We commence therefore with Maxwell's equations in cylindrical coordinates (cf. equations (7.10)–(7.15) for uniform-core waveguides):

$$\frac{v}{r} H_z - \beta H_\theta = -n^2(r)\varepsilon_0 \omega E_r \tag{8.59}$$

$$i\beta H_r - \frac{\partial H_z}{\partial r} = -in^2(r)\varepsilon_0 \omega E_\theta \tag{8.60}$$

$$\frac{1}{r}\frac{\partial}{\partial r}(rH_\theta) - \frac{iv}{r}H_r = -in^2(r)\varepsilon_0 \omega E_z \tag{8.61}$$

$$\frac{v}{r} E_z - \beta E_\theta = \omega\mu_0 H_r \tag{8.62}$$

$$i\beta E_r - \frac{\partial E_z}{\partial r} = i\omega\mu_0 H_\theta \tag{8.63}$$

$$\frac{1}{r}\frac{\partial}{\partial r}(rE_\theta) - \frac{iv}{r}E_r = i\omega\mu_0 H_z. \tag{8.64}$$

In equations (8.59)–(8.64) we have assumed the usual time-, z-, and θ-dependence in the form $\mathbf{E} = (E_r, E_\theta, E_z)e^{i\beta z - i\omega t + iv\theta}$, $\mathbf{H} = (H_r, H_\theta, H_z)e^{i\beta z - i\omega t + iv\theta}$.

In the usual way, equations (8.59)–(8.64) may be rearranged in order to express the axial field components in terms of the longitudinal components. Thus, using the profile of (8.1), we find for $r \leqslant a$:

$$E_r = \frac{a^2}{[u^2 - v^2 f(r)]} \left[i\beta \frac{\partial E_z}{\partial r} - \frac{v\omega\mu_0}{r} H_z \right] \tag{8.65}$$

$$H_r = \frac{a^2}{[u^2 - v^2 f(r)]} \left[i\beta \frac{\partial H_z}{\partial r} + \frac{n^2(r)v\omega\varepsilon_0}{r} E_z \right] \tag{8.66}$$

$$E_\theta = \frac{-a^2}{[u^2 - v^2 f(r)]} \left[i\omega\mu_0 \frac{\partial H_z}{\partial r} + \frac{v\beta}{r} E_z \right] \tag{8.67}$$

$$H_\theta = \frac{a^2}{[u^2 - v^2 f(r)]} \left[i n^2(r)\omega\varepsilon_0 \frac{\partial E_z}{\partial r} - \frac{v\beta}{r} H_z \right]. \tag{8.68}$$

Further manipulation yields two coupled second-order equations for the longitudinal fields ($r \leqslant a$):

$$\frac{\partial^2 H_z}{\partial r^2} + \left[\frac{1}{r} + \frac{v^2 f'}{(u^2 - v^2 f)} \right] \frac{\partial H_z}{\partial r} + \left[\frac{(u^2 - v^2 f)}{a^2} - \frac{v^2}{r^2} \right] H_z = \frac{i v \beta v^2 f'}{r\omega\mu_0 (u^2 - v^2 f)} E_z \tag{8.69}$$

$$\frac{\partial^2 E_z}{\partial r^2} + \left[\frac{1}{r} + \frac{v^2 f'}{(u^2 - v^2 f)} \left(\frac{\beta}{kn(r)} \right)^2 \right] \frac{\partial E_z}{\partial r} + \left[\frac{(u^2 - v^2 f)}{a^2} - \frac{v^2}{r^2} \right] E_z = \frac{-i v \beta v^2 f'}{r n^2(r) \omega\varepsilon_0 (u^2 - v^2 f)} H_z \tag{8.70}$$

where $f' = df/dr$.

Next we define new functions $\phi(r)$ and $\psi(r)$ proportional to the fields E_z, H_z (Kurtz and Streifer, 1969):

$$\phi = n_1^{1/2} E_z \tag{8.71}$$

$$\psi = -i \left(\frac{\mu_0}{\varepsilon_0 n_1} \right)^{1/2} H_z. \tag{8.72}$$

Using these definitions together with the usual notation $b = 1 - u^2/v^2$, equations (8.69) and (8.70) reduce to:

$$\frac{\partial^2 \psi}{\partial r^2} + \left[\frac{1}{r} + \frac{f'}{(1-b-f)} \right] \frac{\partial \psi}{\partial r} + \left[\frac{v^2}{a^2}(1-b-f) - \frac{v^2}{r^2} \right] \psi = \frac{vf'[1 - 2\Delta(1-b)]^{1/2}}{r(1-b-f)} \phi \tag{8.73}$$

$$\frac{\partial^2 \phi}{\partial r^2} + \left[\frac{1}{r} + \frac{f'[1 - 2\Delta(1-b)]}{(1-b-f)(1-2\Delta f)} \right] \frac{\partial \phi}{\partial r} + \left[\frac{v^2}{a^2}(1-b-f) - \frac{v^2}{r^2} \right] \phi = \frac{vf'[1 - 2\Delta(1-b)]^{1/2}}{r(1-b-f)(1-2\Delta f)} \psi. \tag{8.74}$$

Note that the coupling terms on the right of (8.73) and (8.74) vanish for the cases $v = 0$ and $f' = 0$. For the latter case, the core index is constant and we return to the uniform-core step-index fibres discussed in Chapter 7; the equations (8.73) and (8.74) then reduce to the wave equation (7.45) for $f = 0$ and (7.46) for $f = 1$. For the case $v = 0$, the equations are again decoupled and we may determine the TE and TM modes by solving the resulting wave equations.

To discuss the general vector theory further we limit attention to the weakly-guiding case $\Delta \ll 1$. Applying this condition consistently to equations (8.73) and (8.74) by ignoring terms in Δ compared with those of order 1, we find (Kurtz and Streifer, 1969):

$$M\psi = \phi \tag{8.75a}$$
$$M\phi = \psi \tag{8.75b}$$

where

$$M = \frac{r(1-b-f)}{vf'} \left\{ \frac{d^2}{dr^2} + \left[\frac{1}{r} + \frac{f'}{(1-b-f)} \right] \frac{d}{dr} + \left[\frac{v^2}{a^2}(1-b-f) - \frac{v^2}{r^2} \right] \right\}. \quad (8.76)$$

From equations (8.75a) and (8.75b) it is clear that ϕ, ψ satisfy the fourth-order equations:

$$MM\psi = \psi \quad (8.77a)$$
$$MM\phi = \phi. \quad (8.77b)$$

The functions G_j ($j = 1-4$) which satisfy the corresponding second-order equations:

$$MG_j = +G_j, \quad j = 1, 3 \quad (8.78a)$$
$$MG_j = -G_j, \quad j = 2, 4 \quad (8.78b)$$

are therefore also solutions of (8.77a) and (8.77b). Since these solutions G_j are linearly independent, ϕ and ψ are linear combinations of the set $\{G_j\}$. In particular, if

$$\psi = A_1 G_1 + A_2 G_2 + A_3 G_3 + A_4 G_4 \quad (8.79)$$

where A_j ($j = 1-4$) are arbitrary constants, then it follows that

$$\phi = A_1 G_1 - A_2 G_2 + A_3 G_3 - A_4 G_4. \quad (8.80)$$

Hence equations (8.79) and (8.80) may be used to find the functions ψ, ϕ provided that the solutions G_j are known from the wave equation

$$G_j'' + \left[\frac{1}{r} + \frac{f'}{(1-b-f)} \right] G_j' + \left[\frac{v^2}{a^2}(1-b-f) - \frac{v^2}{r^2} \mp \frac{vf'}{r(1-b-f)} \right] G_j = 0 \quad (8.81)$$

where the primes denote differentiation with respect to r, the upper sign corresponds to $j = 1$, 3, and the lower sign to $j = 2, 4$. It should be noted that the coupling of the longitudinal fields, included in the term $\mp vf'G_j/r(1-b-f)$ has been retained in this reduction to a second-order differential equation. Note also in general that one solution of each of equations (8.78a) and (8.78b) will have a singularity at the origin and may therefore be discarded as unphysical for the representation of waveguide modes. If these singular solutions are G_3 and G_4 (say), then the longitudinal fields are, from (8.71), (8.72), (8.79), and (8.80):

$$E_z = n_1^{-1/2}(A_1 G_1 - A_2 G_2) \quad (8.82)$$

$$H_z = i\left(\frac{\varepsilon_0 n_1}{\mu_0}\right)^{1/2} (A_1 G_1 + A_2 G_2). \quad (8.83)$$

For $v = 0$ the TE and TM modes are found by noting that $G_1 = G_2$ (from (8.81)) and setting $A_1 = A_2$ or $A_1 = -A_2$, respectively.

The transverse field components are most conveniently dealt with by setting either $A_1 = 0$ and $A_2 = 1$, or $A_1 = 1$, $A_2 = 0$ in order to obtain two independent sets of solutions. Then, using equations (8.82), (8.83) in the expressions for transverse fields (8.65)–(8.68), and consistently ignoring terms of order Δ compared with those of order unity, we find:

$$E_r = \pm \frac{i}{2\Delta k n_1^{3/2}(1-b-f)} \left[\frac{\partial G_j}{\partial r} \mp \frac{vG_j}{r} \right] \quad (8.84)$$

$$E_\theta = \frac{1}{2\Delta k n_1^{3/2}(1-b-f)} \left[\frac{\partial G_j}{\partial r} \mp \frac{vG_j}{r} \right] \quad (8.85)$$

$$H_r = -n_1 \left(\frac{\varepsilon_0}{\mu_0}\right)^{1/2} \frac{1}{2\Delta k n_1^{3/2}(1-b-f)} \left[\frac{\partial G_j}{\partial r} \mp \frac{vG_j}{r}\right] \tag{8.86}$$

$$H_\theta = \pm n_1 \left(\frac{\varepsilon_0}{\mu_0}\right)^{1/2} \frac{i}{2\Delta k n_1^{3/2}(1-b-f)} \left[\frac{\partial G_j}{\partial r} \mp \frac{vG_j}{r}\right] \tag{8.87}$$

where the upper signs refer to $j = 1$ and the lower signs refer to $j = 2$. It follows that the transverse fields $\mathbf{E}_t^{(j)}$, $\mathbf{H}_t^{(j)}$ may be expressed as (Kurtz and Streifer, 1969):

$$\mathbf{E}_t^{(j)} = (\pm i\hat{\mathbf{r}} + \hat{\boldsymbol{\theta}}) \frac{\Phi_j(r)}{2\Delta k n_1^{3/2}} \tag{8.88}$$

$$\mathbf{H}_t^{(j)} = n_1 \left(\frac{\varepsilon_0}{\mu_0}\right)^{1/2} \hat{\mathbf{z}} \times \mathbf{E}_t^{(j)} \tag{8.89}$$

where

$$\Phi_j(r) = \frac{1}{(1-b-f)} \left[\frac{\partial G_j}{\partial r} \mp \frac{vG_j}{r}\right]. \tag{8.90}$$

The 'transverse field function' Φ_j is related to the 'longitudinal field function' G_j also by the equation:

$$G_j = \mp \frac{a^2}{v^2} \left[\frac{\partial \Phi_j}{\partial r} \pm \frac{(v \pm 1)}{r} \Phi_j\right]. \tag{8.91}$$

The wave equation satisfied by Φ_j is:

$$\frac{d^2\Phi_j}{dr^2} + \frac{1}{r}\frac{d\Phi_j}{dr} + \left[\frac{v^2}{a^2}(1-b-f) - \frac{(v \pm 1)^2}{r^2}\right]\Phi_j = 0. \tag{8.92}$$

In all these equations the upper (lower) signs refer to $j = 1$ (2).

8.1.4 Modal results for parabolic index-profiles

Consider now the application of the general theory from the previous section to the specific profile of interest here, i.e. $f(r) = (r/a)^2$, for all r. Equation (8.92) for the transverse field function then becomes:

$$\frac{d^2\Phi_j}{dr^2} + \frac{1}{r}\frac{d\Phi_j}{dr} + \left[\frac{v^2}{a^2}(1-b) - \frac{v^2 r^2}{a^4} - \frac{(v \pm 1)^2}{r^2}\right]\Phi_j = 0. \tag{8.93}$$

Making the change of variables $x = v(r/a)^2$, this equation may be re-written as:

$$x \frac{d^2\Phi_j}{dx^2} + \frac{d\Phi_j}{dx} + \left[v(1-b) - x - \frac{(v \pm 1)^2}{x}\right]\frac{\Phi_j}{4} = 0. \tag{8.94}$$

Solutions of (8.94) are sought in the form:

$$\Phi_j = x^{(v \pm 1)/2} e^{-x/2} X(x). \tag{8.95}$$

Substituting (8.95) into (8.94) yields:

$$x \frac{d^2 X}{dx^2} + (v \pm 1 - x + 1)\frac{dX}{dx} + [v(1-b) - 2 - 2(v \pm 1)]\frac{X}{4} = 0 \tag{8.96}$$

and the solutions of this equation are the generalized Laguerre polynomials (Abramowitz

and Stegun, 1964):
$$X = L_{M-\frac{1}{2}}^{\nu \pm 1}{}_{\frac{1}{2}}(x) \tag{8.97}$$
where
$$v(1-b) = 2(2M+\nu). \tag{8.98}$$

Equation (8.98) is the eigenvalue equation for vector solutions of the parabolic-index medium with $\Delta \ll 1$, and is to be interpreted with $M = 1, 2, 3, \ldots$ for $j = 1$ (upper signs in all equations) and $M = 0, 1, 2, \ldots$ for $j = 2$ (lower signs) (Kurtz and Streifer, 1969).

It is of interest to compare these results with those for the familiar scalar wave equation in the parabolic-index medium. Written in cylindrical polar coordinates, and assuming a field variation as $e^{il\theta + i\beta z - i\omega t}$, the scalar wave equation for any Cartesian field component ψ reads:

$$\frac{d^2\psi}{dr^2} + \frac{1}{r}\frac{d\psi}{dr} + \left[\frac{v^2}{a^2}(1-b) - \frac{v^2 r^2}{a^4} - \frac{l^2}{r^2}\right]\psi = 0. \tag{8.99}$$

This equation is of identical form to (8.93) with $(\nu \pm 1)$ replaced by l. Hence the solutions are of the form
$$\psi = x^{l/2} e^{-x/2} L_N^l(x) \tag{8.100}$$
where $x = v(r/a)^2$, and
$$v(1-b) = 2(2N+l+1). \tag{8.101}$$

Equation (8.101) is recognized as being identical to the eigenvalue equation (8.58) derived by ray-optics methods. Comparing the eigenvalue equations (8.98) and (8.101) we see that if the scalar and vector fields are to have the same phase and group velocity, then we must demand that (Kurtz, 1975):
$$2M + \nu = 2N + l + 1. \tag{8.102}$$

Further comparison of the scalar and vector solutions requires that they have the same variation in the r-direction, i.e. that the subscripts and superscripts of the Laguerre functions in (8.97) and (8.100) be the same. For $j = 1$ this implies that $\psi_{l,N}$ and $\Phi_{1\nu,M}$ match, with the result (Kurtz, 1975; Yip and Nemoto, 1975):

$$j = 1: \quad l = \nu + 1, \quad N = M - 1. \tag{8.103a}$$

Similarly, for $j = 2$, matching $\psi_{l,N}$ and $\Phi_{2\nu,M}$ yields

$$j = 2: \quad l = \nu - 1, \quad N = M. \tag{8.103b}$$

We thus have the explicit relations between the cylindrical vector solutions and the Cartesian scalar functions:
$$\psi_{l,N} = \Phi_{1l-1,N+1} = \Phi_{2l+1,N}. \tag{8.104}$$

Kurtz (1975) has shown that the results (8.102)–(8.104) proved here for parabolic profiles remain true for arbitrary index distributions, subject only to the weak guidance condition $\Delta \ll 1$.

The relations between the Cartesian field components from the scalar solutions and the cylindrical components from the vector solutions may now be demonstrated. Using (8.88) we form the transverse field combination:

$$\mathbf{E}_{tl-1,N+1}^{(1)} - \mathbf{E}_{tl+1,N}^{(2)} \propto (i\hat{\mathbf{r}} + \hat{\boldsymbol{\theta}})e^{i(l-1)\theta}\Phi_{1l-1,N+1} - (-i\hat{\mathbf{r}} + \hat{\boldsymbol{\theta}})e^{i(l+1)\theta}\Phi_{2l+1,N}$$

$$\propto (\hat{\mathbf{r}}\cos\theta - \hat{\boldsymbol{\theta}}\sin\theta)e^{il\theta}\psi_{l,N}$$

$$= i e^{il\theta}\psi_{l,N} \tag{8.105}$$

which represents a wave linearly polarized in the x-direction. Similarly, using (8.88) and (8.104) again yields a wave linearly polarized in the y-direction:

$$\mathbf{E}^{(1)}_{tl-1,N+1} + \mathbf{E}^{(2)}_{tl+1,N} \propto \mathbf{j} e^{il\theta} \psi_{l,N}. \tag{8.106}$$

Thus we have demonstrated that we can construct linearly-polarized modes for the weakly-guiding parabolic index-profile in an analogous fashion to the LP modes of the step-index fibre discussed in Chapter 7. Similarly it is possible to construct the cylindrical solutions from combinations of the linearly-polarized scalar wave solutions. The results, which are easily verified, are (Kurtz, 1975):

$$\mathbf{E}^{(1)}_{tl-1,N+1} \propto (i\hat{\mathbf{r}} + \hat{\boldsymbol{\theta}})\Phi_{1l-1,N+1} e^{i(l-1)\theta} = (i\mathbf{i} + \mathbf{j})\psi_{l,N} e^{il\theta} \tag{8.107}$$

$$\mathbf{E}^{(2)}_{tl+1,N} \propto (-i\hat{\mathbf{r}} + \hat{\boldsymbol{\theta}})\Phi_{2l+1,N} e^{i(l+1)\theta} = (-i\mathbf{i} + \mathbf{j})\psi_{l,N} e^{il\theta} \tag{8.108}$$

where $\mathbf{i}$, $\mathbf{j}$ are unit vectors in the x, y directions and $i = (-1)^{1/2}$. Reference to equations (8.82) and (8.83) shows that the cases $A_2 = 0$, $A_1 = 0$ correspond, respectively, to $E_z/H_z = -i(\mu_0/\varepsilon_0)^{1/2}/n_1$ and $E_z/H_z = +i(\mu_0/\varepsilon_0)^{1/2}/n_1$. Hence from the definition of mode nomenclature given in Section 7.2 (Snitzer, 1961), it follows that the former case corresponds to EH and the latter to HE modes. In addition we know from the comments after (8.83) that TE and TM modes occur for $v = 0$ with $A_1 = A_2$ or $A_1 = -A_2$, respectively. Thus equations (8.107), (8.108) enable us to construct the EH, HE, TE, and TM modes from the LP modes for parabolic index profiles in a similar fashion to that discussed for step-index guides in Section 7.2. Snyder and Young (1978) have shown that these results are merely special cases of more general rules for construction of vector modal fields for the weakly-guiding case ($\Delta \ll 1$) of any optical waveguide. Morishita et al. (1980) have examined numerically the accuracy of the scalar approximation for a range of fibre profiles.

For fibres with a parabolic-index core and a homogeneous cladding, the Kurtz-Streifer approach discussed above has been extended to include the effects of the cladding (Kirchhoff, 1970; Yamada and Inabe, 1974). For this case the wave equation in the core may be transformed to Whittaker's equation whose solution may be expressed as a Kummer function (Abramowitz and Stegun, 1964). In the cladding the fields are described by modified Bessel functions $K_l(wr/a)$ as in Chapter 7, and the eigenvalue equation is derived in the usual way by matching the tangential field components at the core–cladding interface. Cut-off conditions for the modes may then be derived and solved numerically (Lukowski and Kapron, 1977). Similarly the mode equations far from cut-off may be obtained and used to establish a consistent mode designation scheme (Lim et al., 1979) in a similar manner to that for step-index fibres as discussed in Chapter 7. For weakly-guiding fibres the degeneracy of the $HE_{v+1,N}$ and $EH_{v-1,N}$ modes may be explicitly demonstrated and the LP modes derived (Garside et al., 1980). Analytic approximations for the eigenvalues may be found from a field decomposition technique (Hashimoto et al., 1977) where the transverse field function is expressed in terms of the Laguerre function $L^l_{M-1/2\mp 1/2}$ where M is no longer an integer as a consequence of the effects of the cladding. The approach is similar to that discussed in Section 5.4.4 for the two-dimensional waveguide with cladded-parabolic profile. Whilst the original field decomposition technique applies only to modes far from cut-off, the idea of allowing all cladding effects to be contained in a non-integer M has also been used in a treatment of mode cut-offs (Arnold, 1977b). Another extension of the method is its application in the case of lossy claddings (Masaki et al., 1978) where it was found that the intermodal dispersion may be somewhat improved without significant increase of the total transmission loss.

For the sake of completeness we conclude this section with a discussion of solutions of the scalar wave equation for parabolic-index media when Cartesian, rather than cylindrical polar, coordinates are used. The scalar wave equation satisfied by any Cartesian field component $\psi(x, y)$ may be written in Cartesian coordinates as:

$$\frac{\partial^2 \psi}{\partial x^2} + \frac{\partial^2 \psi}{\partial y^2} + \frac{v^2}{a^2}\left(1 - b - \frac{(x^2 + y^2)}{a^2}\right)\psi = 0. \tag{8.109}$$

In the usual way we seek separable solutions of the form:

$$\psi(x, y) = X(x) Y(y). \tag{8.110}$$

Insertion of (8.110) in (8.109) yields:

$$\left[\frac{1}{X}\frac{d^2 X}{dx^2} + \frac{v^2}{a^2}\left(1 - b - \frac{x^2}{a^2}\right)\right] + \left[\frac{1}{Y}\frac{d^2 Y}{dy^2} - \frac{v^2 y^2}{a^4}\right] = 0. \tag{8.111}$$

Separating the x-and y-dependent parts of this equation with the aid of a separation constant K^2 yields:

$$\frac{d^2 X}{dx^2} + \left[\frac{v^2}{a^2}\left(1 - b - \frac{x^2}{a^2}\right) - K^2\right]X = 0 \tag{8.112}$$

$$\frac{d^2 Y}{dy^2} + \left(K^2 - \frac{v^2 y^2}{a^4}\right)Y = 0. \tag{8.113}$$

Equations (8.112) and (8.113) are familiar since they are the relations governing TE mode propagation in two-dimensional parabolic-index media. Their solutions are the familiar Hermite–Gauss functions, so that the total field variation is of the form (as yet not normalized):

$$\psi(x, y) = H_P\left(\frac{x\sqrt{2}}{w_0}\right) H_Q\left(\frac{y\sqrt{2}}{w_0}\right) \exp\left[-\frac{(x^2 + y^2)}{w_0^2}\right] \quad (P, Q = 0, 1, 2 \ldots) \tag{8.114}$$

where

$$w_0^2 = \frac{2a^2}{v}. \tag{8.115}$$

Substituting the x- and y-dependent parts of (8.114) into (8.112) and (8.113), respectively, yields:

$$2P = v(1 - b) - \frac{K^2 a^2}{v} - 1 \tag{8.116}$$

$$2Q = \frac{K^2 a^2}{v} - 1. \tag{8.117}$$

Combining these equations so as to eliminate the constant K^2 gives (Marcuse, 1973):

$$v(1 - b) = 2(P + Q + 1) \quad (P, Q = 0, 1, 2, \ldots). \tag{8.118}$$

Equation (8.118) is the eigenvalue for parabolic-index media when the profile is expressed in Cartesian coordinates and the mode integers P, Q define the field variations in the x, y directions, respectively. Note that equation (8.118) may be viewed as the sum of two 2-dimensional eigenvalue equations each of the form of (3.51) or (3.52).

One might ask at this point why there are two sets of solutions to the scalar wave equation in parabolic-index media, viz. the Laguerre–Gaussian modes with azimuthal and radial periodicity, and the Hermite–Gaussian modes with periodicity in two Cartesian coordinates.

The solution to this apparent paradox lies in the fact that we are considering unbounded parabolic-index media, i.e. the refractive index goes on decreasing as r^2 for all r. Hence the only boundary condition which must be satisfied by the fields of guided modes is that of vanishing as $r \to \infty$. This condition is not sufficiently strict to make the choice of mode family unique. However, as soon as a boundary condition of a more demanding type is imposed, e.g. a cladding in a real fibre, then the choice of the appropriate set of functions to describe the modal fields becomes clearer. Further discussion of the Laguerre–Gauss and Hermite–Gauss modes, including the cases of complex permittivity and off-axis beams, has been given by Casperson (1976). Petermann (1977) has used an expansion of the Hermite–Gaussian modes in terms of the Laguerre–Gaussian modes in order to estimate leaky-mode attenuation in parabolic-profile fibres.

8.2 The WKB Approximation in Graded-index Fibre Analysis

In the analysis and optimization of multimode fibres, the WKB approximation has played an especially important role. Although the results, as stressed earlier, are identical with those from geometric optics, the attraction of WKB theory lies in its ability also to provide reasonably accurate field distributions without the necessity for lengthy numerical computations. We shall present here a straightforward extension of the WKB principles covered in Chapter 5 to the three-dimensional case of graded-index fibres. This approach leads to optimum index-profiles of the α-type already mentioned, and to an understanding of the behaviour of multimode fibre bandwidth with variation of operating wavelength. The approach is then extended to permit a detailed discussion of the general class of leaky modes on graded-index fibres. The attenuation coefficients of the leaky modes may easily be evaluated using WKB theory, and thus the effect of these modes on the fibre near-field distribution and effective numerical aperture may be calculated.

8.2.1 Formal theory

We seek approximate solutions to the scalar wave equation in cylindrical polar coordinates (neglecting the usual axial-, azimuthal-, and time-dependences):

$$\frac{d^2\psi}{dr^2} + \frac{1}{r}\frac{d\psi}{dr} + \left(k^2 n^2(r) - \beta^2 - \frac{l^2}{r^2}\right)\psi = 0. \tag{8.119}$$

Following the same approach as in Section 5.2, a trial solution of the form $\psi = \psi_0 \exp[ikS(r)]$ is assumed, and S is expanded in powers of $1/k$:

$$S(r) = S_0(r) + \frac{1}{k}S_1(r) + \ldots \tag{8.120}$$

so that

$$\psi = \psi_0 \exp[ikS_0(r) + iS_1(r) + \ldots]. \tag{8.121}$$

Inserting (8.121) in (8.119) and equating terms of similar order in k, the zero and first-order WKB approximations are obtained:

$$S_0(r) = \frac{1}{k}\int \left(k^2 n^2(r) - \beta^2 - \frac{l^2}{r^2}\right)^{1/2} dr \tag{8.122}$$

$$S_1(r) = \frac{i}{4}\ln\left(r^2 n^2(r) - \frac{\beta^2 r^2}{k^2} - \frac{l^2}{k^2}\right) \tag{8.123}$$

where the integral limits in (8.122) will be discussed later. It follows that we have the usual oscillatory or evanescent field solutions, depending on the sign of the radical in (8.122):

$$\psi(r) = \frac{\psi_0}{(rq)^{1/2}} \exp\left(\pm i \int q \, dr\right), \quad q^2 > 0 \tag{8.124}$$

$$\psi(r) = \frac{\psi_0}{(rp)^{1/2}} \exp\left(\pm \int p \, dr\right), \quad p^2 > 0 \tag{8.125}$$

where

$$q^2 = -p^2 = k^2 n^2(r) - \beta^2 - \frac{l^2}{r^2}. \tag{8.126}$$

For a guided mode in a general graded-index fibre, q^2 will be positive between the caustics r_1, r_2 where $q^2 = 0$; for $r < r_1$ and $r > r_2$ then p^2 is positive. Hence the field distribution is as schematically indicated in Figure 8.8, with oscillatory variation for $r_1 < r < r_2$ and evanescent behaviour outside this range. In fact, for radii close to the caustics r_1, r_2 the simple WKB approximation fails (cf. Section 5.2) and it becomes necessary to find correction formulae for the field functions in the regions of r_1 and r_2 (Gordon, 1966).

For values of r close to the caustics a change of variables $r = e^x$ (Timmermann, 1975) transforms the scalar wave equation (8.119) to the form:

$$\frac{d^2\psi}{dx^2} + Q^2(x)\psi = 0 \tag{8.127}$$

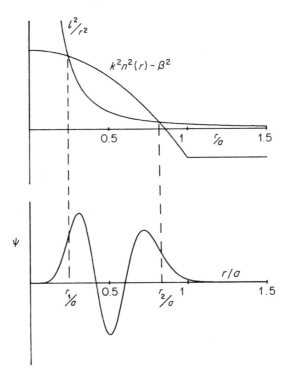

Figure 8.8 Schematic field distribution $\psi(r)$, showing the caustics at radii r_1, r_2 as given by the zeros of q^2 in (8.126) for the WKB approximation

where

$$Q^2(x) = r^2 q^2 = r^2 \left(k^2 n^2(r) - \beta^2 - \frac{l^2}{r^2} \right). \tag{8.128}$$

Equation (8.127) is now in the same form as the two-dimensional wave equation considered in Chapter 5. Hence we may use the results of the WKB analysis given in Section 5.2 directly here with the appropriate definition of Q as given in (8.128). In particular, for the approximate solution near a caustic x_c we may use the results (5.72) and (5.73) for a wave which decays as $x \to \infty$, and (5.75), (5.76) for an exponentially growing wave. From the discussion of the caustics given above, it is clear that the former case should apply at both the caustics r_1, r_2. Noting also, from the definitions of $Q(x)$ and $q(r)$, that:

$$Q(x)\,dx = q(r)\,dr \tag{8.129}$$

then it follows that the approximate field distributions which satisfy the appropriate connection formulae are:

$$\psi(r) \propto \begin{cases} \dfrac{1}{2(rp)^{1/2}} \exp\left[-\int_r^{r_1} p\,dr \right], & 0 \leqslant r < r_1 & (8.130a) \\[4pt] \dfrac{1}{(rq)^{1/2}} \cos\left[\int_{r_1}^r q\,dr - \dfrac{\pi}{4} \right], & r_1 < r < r_2 & (8.130b) \\[4pt] \dfrac{1}{2(rp)^{1/2}} \exp\left[-\int_{r_2}^r p\,dr \right], & r_2 < r. & (8.130c) \end{cases}$$

Once again it follows from the arguments of Section 5.2 that the presence of the caustics at r_1 and r_2 implies a phase matching condition on the oscillatory solution. In analogy with equation (5.79), the appropriate condition is given by:

$$\int_{r_1}^{r_2} q\,dr = (N + \tfrac{1}{2})\pi \quad (N = 0, 1, 2, \ldots). \tag{8.131}$$

Inspection of equation (8.131) confirms that it is identical to the result (8.40) obtained from the geometric optics approach.

As an example of the use of equation (8.130) to provide approximate field distributions, consider again the parabolic index-profile given by equations (8.1) and (8.2) with $\alpha = 2$. For this profile the integrals in equation (8.130) may be evaluated analytically and the resultant oscillatory field distribution is (Hartog and Adams, 1977):

$$\psi(r) = \frac{1}{Q'^{1/2}} \cos\left\{ \frac{Q'}{2} + \frac{u^2}{4v} \cos^{-1}\left[\frac{u^2 - 2v^2(r/a)^2}{(u^4 - 4v^2 l^2)^{1/2}} \right] - \frac{l}{2}\cos^{-1}\left[\frac{2l^2(a/r)^2 - u^2}{(u^4 - 4v^2 l^2)^{1/2}} \right] - \frac{\pi}{4} \right\} \tag{8.132}$$

for $r_1 < r < r_2$, where

$$Q' = \left(u^2 \left(\frac{r}{a}\right)^2 - l^2 - v^2 \left(\frac{r}{a}\right)^4 \right)^{1/2}. \tag{8.133}$$

The evanescent field distribution for $r > r_2$ is given by (Hartog and Adams, 1977):

$$\psi(r) = \frac{1}{2P'^{1/2}} \left| \frac{2P' + 2l^2 - u^2(r/a)^2}{(r/a)^2(u^4 - 4v^2 l^2)^{1/2}} \right|^{u^2/4v} \left| \frac{2vP' + 2v^2(r/a)^2 - u^2}{(u^4 - 4v^2 l^2)^{1/2}} \right|^{1/2} \exp\left(-\frac{P'}{2} \right) \tag{8.134}$$

where

$$P'^2 = -Q'^2. \tag{8.135}$$

A similar expression for $\psi(r)$ in the region $0 \leqslant r < r_1$ is easily achieved by changing the signs of all the major exponents, viz. $u^2/4v$, $1/2$, $-P'/2$, in equation (8.134).

Comparisons of the fields given by equations (8.132)–(8.135) with those from the solutions of the scalar wave equation, i.e. the Laguerre–Gauss functions as in (8.100), are shown in Figures 8.9 and 8.10. The zero-order WKB approximation, obtained by omitting the denominators $Q'^{1/2}$ and $2P'^{1/2}$ in (8.132) and (8.134), respectively, is also shown in the figures. The parameters used were $v = 20$, $N = 3$, $l = 0$ for Figure 8.9 and $l = 1$ for Figure 8.10; the fields are normalized to carry the same power in each case. It is seen that the zeros of the WKB eigenfunctions are in good agreement with those of the scalar mode fields and that, sufficiently far from the caustics, the first-order WKB method gives an excellent approximation to the Laguerre–Gaussian modes. The slight error observed at the peaks is believed to be due to the problem of normalization of the WKB fields, rather than to an inaccuracy of the method. The first-order WKB approximation cannot be very accurate in the vicinity of the turning points, particularly between the outer caustic and the cladding; in this region other asymptotic expressions based on the use of Airy functions may be used, as outlined in Chapter 5 (Timmermann, 1975). The field comparisons given in Figures 8.9 and 8.10 may be viewed as an extension of the analogous two-dimensional comparison carried out by Arnaud (1973).

Returning now to the general WKB eigenvalue equation (8.131) we may use this (a) to obtain an expression for the field transit time in a fibre of length L, and (b) to calculate the number of guided modes in a fibre with a given index profile. To find the transit time

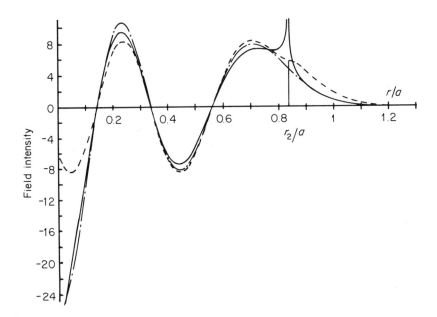

Figure 8.9 Field distributions in the infinite parabolic-index medium calculated by (a) zero-order WKB approximation (broken line), (b) first-order WKB approximation (solid line), and (c) scalar mode theory (dot–dash line). The parameters used were $v = 20$, $l = 0$, $N = 3$. (From Hartog and Adams (1977). Reproduced by permission of Chapman and Hall)

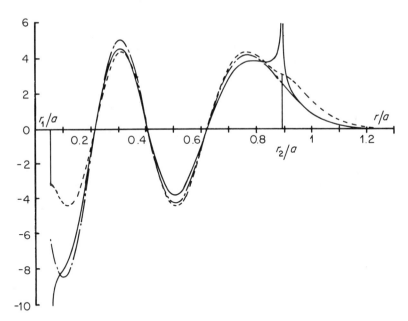

Figure 8.10 As Figure 8.9, except that here $l = 1$. (From Hartog and Adams (1977). Reproduced by permission of Chapman and Hall)

$\tau = (L/c) \, d\beta/dk$, it is a straightforward matter to differentiate (8.131) with respect to k whilst keeping the mode numbers l, N fixed. The result is:

$$\tau = \frac{L}{c} \frac{\int_{r_1}^{r_2} kn(r)m(r)dr/q}{\int_{r_1}^{r_2} \beta \, dr/q} \tag{8.136}$$

which is exactly the same as equation (8.38) which was derived by the ray technique. Error estimates for the values of τ given by (8.136) have been made by Jacobsen and Ramskov Hansen (1980) using an improved evanescent wave theory to provide very accurate upper and lower limits for the exact results. For near-parabolic profiles they found errors between 3 and 10 ps/km, whilst for a profile far from parabolic the group delay error is no greater than 30 ps/km, the value found for the fundamental mode.

To evaluate the total number M of guided modes we follow the process used in Chapter 7 for step-index fibres and in Section 8.1 for parabolic-index fibres. Using the cut-off condition $\beta = kn_2$, we sum the radial mode number N given by (8.131) over all permitted values of the azimuthal mode number l from 0 to l_{max}. Since l_{max} is usually a large number for multimode fibres, the summation is replaced by an integral. Ignoring the term $\frac{1}{2}$ in (8.131), and allowing for the four-fold degeneracy of each mode group designated by N, l, we obtain (Gloge and Marcatili, 1973):

$$M = \frac{4}{\pi} \int_0^{l_{max}} \int_{r_1}^{r_2} \left(k^2(n^2(r) - n_2^2) - \frac{l^2}{r^2} \right)^{1/2} dr \, dl. \tag{8.137}$$

Changing the order of integration gives:

$$M = \frac{4}{\pi} \int_0^a \int_0^{rk(n^2(r) - n_2^2)^{1/2}} \left(k^2(n^2(r) - n_2^2) - \frac{l^2}{r^2} \right)^{1/2} dl \, dr$$

$$= \int_0^a k^2(n^2(r) - n_2^2) r \, dr. \tag{8.138}$$

Note that for small index differences the integral in equation (8.138) represents the volume of rotation defined by the refractive index profile. The procedure leading to the result (8.138) may be repeated for any arbitrary value of propagation constant β, not just kn_2 as above, and the result will be the number of guided modes, $\mu(\beta)$ say, having a propagation constant larger than β. The result is:

$$\mu(\beta) = \int_0^{r'} (k^2 n^2(r) - \beta^2) r \, dr \tag{8.139}$$

where r' is the radius at which $\beta = kn(r')$.

Thus far the results for eigenvalues, mode numbers, and transit times have applied to arbitrary (circularly symmetric) refractive index distributions. In the next subsection we limit the discussion to α-profiles of the type described by equations (8.1) and (8.2).

8.2.2 Alpha-profiles and their optimization

Inserting the profile of (8.1) and (8.2) in equation (8.139) yields:

$$\mu(\beta) = \left(\frac{\alpha}{\alpha+2}\right) \frac{v^2}{2} \left(\frac{u^2}{v^2}\right)^{2/\alpha + 1}. \tag{8.140}$$

Similarly, (8.138) gives for the total number of guided modes:

$$M = \left(\frac{\alpha}{\alpha+2}\right) \frac{v^2}{2}. \tag{8.141}$$

Note that for the step-index case ($\alpha = \infty$) then (8.141) gives $M = v^2/2$ in agreement with equation (7.136), whilst for parabolic-index fibres ($\alpha = 2$) (8.141) gives $M = v^2/4$ in agreement with the result derived in Section 8.1.2.

For future use we may combine equations (8.140) and (8.141) to yield u/v in terms of μ/M:

$$\left(\frac{u}{v}\right)^2 = \left(\frac{\mu}{M}\right)^{\alpha/(\alpha+2)}. \tag{8.142}$$

The mode transit time τ in length L of fibre may now be calculated either by differentiating (8.142) or by using the integral form (8.136); we shall follow here the latter approach. For the α-profile, assuming n_1 and Δ are functions of λ but α is not, the phase/group index product required takes the form:

$$n(r) m(r) = n_1 m_1 - \left(2\Delta n_1 m_1 + k n_1^2 \frac{d\Delta}{dk}\right) \left(\frac{r}{a}\right)^\alpha. \tag{8.143}$$

Putting this expression in (8.136) and substituting $y = (r/a)^2$ gives:

$$\tau = \frac{Lk}{c\beta I} \int_{y_1}^{y_2} \frac{[n_1 m_1 - (2\Delta n_1 m_1 + k n_1^2 d\Delta/dk) y^{\alpha/2}]}{Q} dy \tag{8.144}$$

where

$$I = \int_{y_1}^{y_2} \frac{dy}{Q} \tag{8.145}$$

and

$$Q^2 = u^2 y - v^2 y^{(\alpha+2)/2} - l^2. \tag{8.146}$$

Using the fact that

$$2Q\frac{dQ}{dy} = u^2 - \left(\frac{\alpha+2}{2}\right)v^2 y^{\alpha/2} \tag{8.147}$$

we then find (Sammut, 1977):

$$\int_{y_1}^{y_2} \frac{y^{\alpha/2} dy}{Q} = \left(\frac{2}{\alpha+2}\right)\frac{u^2}{v^2} I \tag{8.148}$$

since Q vanishes at each limit $y = y_1$ and $y = y_2$. Finally, using (8.148) in (8.144) gives:

$$\tau = \left(\frac{Lm_1}{c}\right)\left(\frac{kn_1}{\beta}\right)\left[1 - 2\Delta\left(\frac{u}{v}\right)^2\left(1 + \frac{n_1 k}{2m_1 \Delta}\frac{d\Delta}{dk}\right)\left(\frac{2}{\alpha+2}\right)\right]. \tag{8.149}$$

Equation (8.149) has also been derived by Pask (1979), without the aid of WKB theory, by the use of a scaling argument on the scalar wave equation. Note that if we neglect material dispersion effects ($m_1 = n_1$, $d\Delta/dk = 0$) then for $\alpha = 2$ equation (8.149) reduces to (8.57) which is the transit time for parabolic-index media derived earlier.

We now define a profile dispersion parameter ε (Olshansky and Keck, 1976):

$$\varepsilon = \frac{2n_1 k}{m_1 \Delta}\frac{d\Delta}{dk}. \tag{8.150}$$

In terms of ε, equation (8.149) may be re-written as:

$$\tau = \left(\frac{Lm_1}{c}\right)\left[1 - 2\Delta\left(\frac{u}{v}\right)^2\left(\frac{2}{\alpha+2}\right)\left(1 + \frac{\varepsilon}{4}\right)\right]\left[1 - 2\Delta\left(\frac{u}{v}\right)^2\right]^{-1/2}. \tag{8.151}$$

Recalling that for weakly-guiding fibres $\Delta \ll 1$, it is clear that the terms in $2\Delta(u/v)^2$ in (8.151) are very small by comparison with terms of order unity. It is therefore appropriate to expand this expression for τ in ascending powers of $2\Delta(u/v)^2$. Using the relation (8.142) to eliminate (u/v), this expansion yields:

$$\tau = \frac{Lm_1}{c}\left[1 + \Delta\left(\frac{\alpha-2-\varepsilon}{\alpha+2}\right)\left(\frac{\mu}{M}\right)^{\alpha/(\alpha+2)} + \frac{\Delta^2}{2}\left(\frac{3\alpha-2-2\varepsilon}{\alpha+2}\right)\left(\frac{\mu}{M}\right)^{2\alpha/(\alpha+2)}\right.$$
$$\left. + \frac{\Delta^3}{2}\left(\frac{5\alpha-2-3\varepsilon}{\alpha+2}\right)\left(\frac{\mu}{M}\right)^{3\alpha/(\alpha+2)}\right] + O(\Delta^4). \tag{8.152}$$

Now for a high bandwidth fibre we require that intermodal transit time differences shall be minimized. Using (8.152) it is clear that values of τ for all modes will be the same to order Δ provided we choose (Olshansky and Keck, 1976):

$$\alpha = 2 + \varepsilon. \tag{8.153}$$

Since $0 \leq \mu \leq M$, for this choice of α it follows that the total spread of transit times will be $Lm_1\Delta^2/(2c)$, ignoring higher order terms. However, it is possible to find a true minimum in the spread of transit times by the choice

$$\alpha = 2 + \varepsilon - 2\Delta \tag{8.154}$$

in which case the modes with $\mu = 0$ arrive at the same time as those with $\mu = M$, and the overall spread in times is reduced to $Lm_1\Delta^2/(8c)$. These results are illustrated in Figure 8.11 where, for clarity, the case $\varepsilon = 0$ is taken. In Figure 8.11(a) the relative mode transit time is plotted versus $(\mu/M)^{\alpha/(\alpha+2)}$ for some α-profiles which are non-optimal. In general the spread of delays is of order $\Delta Lm_1/c$; in fact for the step-index fibre ($\alpha = \infty$) the spread takes exactly

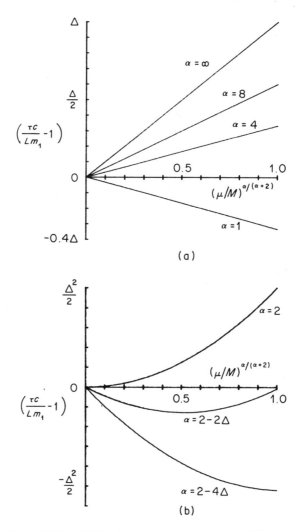

Figure 8.11 (a) Relative transit times (to order Δ) for non-optimal α-profiles, from equation (8.152). (b) Relative transit times (to order Δ^2) for near-optimal α-profiles, from equation (8.152)

this value, which is approximately the same as that deduced in Section 7.2. In Figure 8.11(b) the relative mode transit times are plotted versus $(\mu/M)^{\alpha/(\alpha+2)}$ again using equation (8.152), but this time for profiles close to optimal. In this region the variation of $(\tau c/Lm_1 - 1)$ is no longer monotonic with $(\mu/M)^{\alpha/(\alpha+2)}$ and thus there exists the optimal choice (8.154) where the mode at cut-off ($\mu = M$) has the same group delay as the mode furthest from cut-off ($\mu = 0$).

How accurately must an α-profile be achieved in order to ensure optimal bandwidth? Figure 8.12 plots overall intermodal delay spread $\Delta\tau$ in ns/km versus the α-value for the case $\Delta = 0.01$ and $\varepsilon = 0$. At the optimum α (1.98) the value of $\Delta\tau$ is about 0.06 ns/km. However, for $\alpha = 1.90$ and 2.05, i.e. departures of only about 3–4 per cent from optimum, this value becomes approximately 1 ns/km. Clearly there are rather stringent requirements in achieving the optimum α-value if high bandwidths are to be guaranteed. In addition the theory assumes profiles that are of the form given by equations (8.1) and (8.2); in practice most real fibre index

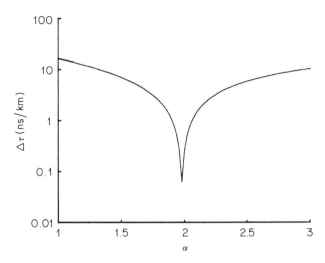

Figure 8.12 Intermodal delay spread $\Delta\tau$ versus α for graded-index fibres with $\Delta = 0.01$ (numerical aperture $\simeq 0.21$) and $\varepsilon = 0$

profiles show marked departures from this ideal. Profiles of fibres fabricated by the CVD process frequently show an axial depression in the index profile (Presby, 1977) which occurs as a result of dopant evaporation during the high-temperature collapse stage of the process. The effects of this index dip on modal transit times has been calculated by the WKB method by Behm (1977) and Ramskov Hansen (1978); we shall return to this topic in a later section. WKB results for index profiles of the form

$$n^2(r) = n_1^2 \left[1 - 2\Delta \left(\frac{r}{a}\right)^2 + 2K\Delta \left(\frac{r}{a}\right)^4 \right]$$

have been obtained by Timmerman (1974a, 1974b) with the object of optimizing fibre bandwidths by suitable choice of the coefficient K. Fourth-order profiles of this type are, however, somewhat easier to treat using perturbation theory, and we therefore postpone their discussion until Section 8.4.

8.2.3 Wavelength-dependence of multimode fibre bandwidths

The optimum α-value as given by (8.153) or (8.154) is clearly a material property which is a function of operating wavelength. The variation of the profile dispersion parameter ε as a function of wavelength may be found either from determination of the refractive index of bulk samples (see, for example, the results in Table 7.3) or by experiments designed to measure the relative index difference Δ in the fibre. A review of available data for ε in germania-doped fibres has been given by Adams et al. (1978a). A measurement of ε made directly on the fibre has been reported by Sladen et al. (1977, 1978, 1979). In this case the variation of numerical aperture squared, $(NA)^2$, with wavelength is determined by measuring the total power transmitted through a short length of fibre excited by a small spot of focused light on the end-face; a reference signal is obtained by the use of an angular aperture at the launching end of the fibre. From (7.246) the definition of NA for a graded-index fibre is assumed to be

$$(NA)^2 = n_1^2 - n_2^2 \tag{8.155}$$

and the power transmitted for a source which excites all modes equally is directly proportional to NA^2. The refractive indices n_1, n_2 may each be expressed in terms of a Sellmeier equation of the form (7.126). Hence the variation of $(NA)^2$ with wavelength is given by the difference between two three-term Sellmeier equations. This somewhat cumbersome expression may be simplified by noting (see, for example, Table 7.3) that two terms have resonance wavelengths in the infrared (typically 8 to 12 μm) and four terms are centred in the ultraviolet (of order 0.1 μm or less). Thus a simple expansion can be used in the form (Sladen et al., 1977):

$$(NA)^2 = 2\Delta n_1^2 = A + B\lambda^2 + \frac{C}{\lambda^2}. \tag{8.156}$$

Equation (8.156) may be used in conjunction with experimental measurements of $(NA)^2$ to find the coefficients A, B, C by a least-squares fitting procedure. Values of these coefficients for some commonly used dopants are given in Table 8.1 (Sladen, 1978); in all cases the other material component in the fibre was silica.

Table 8.1 Coefficients of the equation $(NA)^2 = A + B\lambda^2 + c\lambda^{-2}$ where λ is in nm. (From Sladen (1978). (Reproduced by permission of F.M.E. Sladen)

Dopant material	$A \times 10^{-2}$	$B \times 10^{-10}$	$C \times 10^2$	NA at 550 nm
Germania 8.1 m/o	3.90341	8.18350	8.31169	0.205
Germania 11.0 m/o	5.20366	11.2403	11.0722	0.236
Germania 13.1 m/o	5.63887	12.0034	11.7935	0.247
Boric oxide	1.74566	8.65049	−2.47673	0.130
Fluorine	1.76075	4.90045	1.25199	0.135
Phosphorus pentoxide	1.35802	0.735761	−0.367621	0.116

The profile dispersion parameter ε may be found by differentiation of equation (8.156), using the fact that:

$$\varepsilon = -\frac{2n_1 \lambda}{m_1 \Delta} \frac{d\Delta}{d\lambda} = \frac{2n_1}{m_1} \left[\frac{2\lambda}{n_1} \frac{dn_1}{d\lambda} - \frac{\lambda}{(NA)^2} \frac{d(NA)^2}{d\lambda} \right] \tag{8.157a}$$

$$\simeq -\frac{2\lambda}{(NA)^2} \frac{d(NA)^2}{d\lambda}. \tag{8.157b}$$

In practice the approximate form (8.157b) yields quite accurate results, but if the index of the core (or of the cladding) is known then the full equation (8.157a) may be used to obtain improved values for ε. Figure 8.13 shows the results for the phosphorus-doped fibre and the three germania-doped fibres of Table 8.1; the plot shows optimum α as given by (8.153) versus wavelength λ (Sladen et al., 1978, 1979). The results for the three GeO_2 samples are very nearly identical, being within the thickness of the line. The curves clearly indicate that for the range of compositions tested, the optimum α-value is independent of germania concentration. This result confirms the validity of ignoring the wavelength dependence of α in (8.143). The GeO_2 curves also show considerable variation with wavelength, giving optimum α's of about 2.0 at the GaAs emission wavelength and 1.87 at $\lambda = 1.3$ μm. This emphasizes the fact that a profile designed to be optimal at a specific operating wavelength may well be far from optimal at another operating wavelength. It follows that multimode fibres presently operated at 0.85 μm will not necessarily give higher bandwidth when operated at 1.3 μm as considerations based solely on material dispersion would indicate. By contrast the P_2O_5

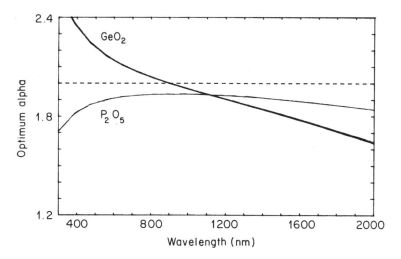

Figure 8.13 Optimal α-value (2+ε) for phosphorus and germania-doped silica fibres as a function of wavelength. (From Sladen et al. (1978). Reproduced by permission of the Istituto Internazionale delle Communicazioni)

curve shows relatively little change with wavelength over the range 800–1500 nm. In practice the complex interaction of intermodal group delay differences and pulse-spreading resulting from material dispersion must be carefully considered in designing fibres for specific operating wavelengths.

In order to investigate multimode pulse dispersion effects further, we use the same expansion of $\tau(\lambda)$ about the mean source wavelength λ_s as was used earlier in Section 7.2, equation (7.118); we repeat the equation here for convenience:

$$\tau(\lambda) = \tau(\lambda_s) + (\lambda - \lambda_s)\tau'(\lambda_s) + \frac{(\lambda - \lambda_s)^2}{2}\tau''(\lambda_s) + \ldots \qquad (8.158)$$

where the primes denote differentiation with respect to λ. Let us assume a source whose spectral distribution is $S(\lambda)$, normalized such that

$$\int_0^\infty S(\lambda)\,d\lambda = 1 \qquad (8.159)$$

and with the mean wavelength λ_s given by:

$$\lambda_s = \int_0^\infty \lambda S(\lambda)\,d\lambda. \qquad (8.160)$$

The root mean square (r.m.s.) spectral width σ_s is defined as:

$$\sigma_s^2 = \int_0^\infty (\lambda - \lambda_s)^2 S(\lambda)\,d\lambda \qquad (8.161)$$

and higher-order moments $\eta_k (k = 3, 4, \ldots)$ are defined, for later use, by:

$$\eta_k = \int_0^\infty (\lambda - \lambda_s)^k S(\lambda)\,d\lambda. \qquad (8.162)$$

Now in order to characterize the dispersion behaviour of the fibre completely we need to calculate the impulse response, i.e. the output after a length L of fibre which has been excited by a delta-function pulse in the time domain. Clearly this will be a function of the relative amounts of power carried by each mode of a multimode fibre, so we need to define the modal power distribution function, p_{Nl}, for the mode denoted by integers N and l. In fact, the impulse response may be adequately described by its moments w_j ($j = 0, 1, 2, \ldots$) which are defined as (Olshansky and Keck, 1976; Adams et al., 1978):

$$w_j = \sum_{N,l} \left[p_{Nl} \int_0^\infty \tau_{Nl}^j(\lambda) S(\lambda) d\lambda \right] \qquad (8.163)$$

where τ has been ascribed subscripts N, l for the mode (N, l). The lowest order moment w_0 gives by definition the total power arriving after length L of fibre. The mean delay τ_0 of the pulse is given by:

$$\tau_0 = \frac{w_1}{w_0} \qquad (8.164)$$

and the r.m.s. pulse width σ is given by:

$$\sigma^2 = \frac{w_2}{w_0} - \left(\frac{w_1}{w_0}\right)^2. \qquad (8.165)$$

For specifying fibre bandwidth in digital systems, Personick (1973) has shown that σ is a necessary pre-requisite. Hence σ, or an associated bandwidth, is conventionally considered as adequate characterization of a fibre, whilst the impulse response, or its Fourier transform, is needed for a full description. The r.m.s. pulse width is separable into intermodal terms resulting from delay differences between modes, and intra-modal terms which include averages of the pulse broadening within each mode as well as the purely material effect:

$$\sigma^2 = \sigma^2_{\text{intermodal}} + \sigma^2_{\text{intramodal}}. \qquad (8.166)$$

Using equations (8.158)–(8.166) and retaining terms up to fourth order yields (Adams et al., 1978):

$$\sigma^2_{\text{intermodal}} = \left[\langle \tau^2 \rangle - \langle \tau \rangle^2 \right] + \sigma_s^2 \left[\langle \tau'' \tau \rangle - \langle \tau \rangle \langle \tau'' \rangle \right]$$
$$+ \frac{\eta_3}{3} \left[\langle \tau''' \tau \rangle - \langle \tau \rangle \langle \tau''' \rangle \right] + \frac{\eta_4}{12} \left[\langle \tau^{iv} \tau \rangle - \langle \tau \rangle \langle \tau^{iv} \rangle \right] \qquad (8.167)$$

$$\sigma^2_{\text{intramodal}} = \sigma_s^2 \langle \tau'^2 \rangle + \eta_3 \langle \tau'' \tau' \rangle + \frac{\eta_4}{3} \langle \tau''' \tau' \rangle + \frac{1}{4} \left[\eta_4 \langle \tau''^2 \rangle - \sigma_s^4 \langle \tau'' \rangle^2 \right] \qquad (8.168)$$

where $\langle \ \rangle$ denotes an average over the power distribution p_{Nl}. The amendments to these expressions which result from considering a Gaussian temporal distribution of the input pulse (rather than a delta function as here) have been calculated by Thyagarajan and Ghatak (1977): in practice, however, the magnitudes of these corrections are sufficiently small to be negligible.

Let us consider first the intermodal contribution to pulse broadening as given by (8.167); the dominant term here is the first one on the r.h.s., and the other terms which depend on the source moments may be viewed as small corrections which are usually negligible. For the α-profiles the expression for τ in (8.152) may be used to evaluate the dominant term in (8.167). If we assume that all modes carry equal power ($p_{Nl} = 1$) and approximate the summation in each $\langle \ \rangle$ by an integral over mode group number μ, then the result to order Δ^2 is given by

(Olshansky and Keck, 1976):

$$\sigma_{\text{intermodal}} = \frac{Lm_1\Delta}{2c}\left(\frac{\alpha}{\alpha+1}\right)\left(\frac{\alpha+2}{3\alpha+2}\right)^{1/2}\left[C_1^2 + \frac{4C_1C_2\Delta(\alpha+1)}{(2\alpha+1)} + \frac{4\Delta^2C_2^2(2\alpha+2)^2}{(5\alpha+2)(3\alpha+2)}\right]^{1/2} \quad (8.169)$$

where

$$C_1 = \frac{\alpha-2-\varepsilon}{\alpha+2} \quad (8.170)$$

and

$$C_2 = \frac{3\alpha-2-2\varepsilon}{2(\alpha+2)}. \quad (8.171)$$

Equation (8.169) has been verified by Feit (1979) on the basis of the Hellman–Feynman theorem and the virial theorem of quantum mechanics, rather than via WKB theory.

The minimum of the intermodal r.m.s. pulse width occurs for

$$\alpha_{\text{opt}} = 2+\varepsilon-\Delta\frac{(4+\varepsilon)(3+\varepsilon)}{(5+2\varepsilon)} \quad (8.172)$$

when the value of pulse width is

$$(\sigma_{\text{intermodal}})_{\text{opt}} = \frac{Lm_1\Delta^2}{20c\sqrt{3}}. \quad (8.173)$$

If we compare this minimum value with the intermodal r.m.s. pulse width for a step-index fibre ($\alpha = \infty$), we find from (8.169):

$$(\sigma_{\text{intermodal}})_{\text{step}} = \frac{Lm_1\Delta}{2c\sqrt{3}} \quad (8.174)$$

so that

$$\frac{(\sigma_{\text{intermodal}})_{\text{opt}}}{(\sigma_{\text{intermodal}})_{\text{step}}} = \frac{\Delta}{10}. \quad (8.175)$$

Thus we see that for a Δ of 0.01 the intermodal r.m.s. pulse spreading for the optimum profile is 1000 times better than that for a step-index fibre; for this value of Δ the pulse spreading given by (8.173) is about 15 ps/km. Note that the results (8.169)–(8.175) all assume equal power in all modes ($p_{Nl} = 1$) as would be the case for excitation by a Lambertian source. WKB results for partially-excited graded-index fibres have been presented by Sammut (1977).

Turning now to the intramodal contribution to pulse spreading as given by (8.168), we note that the dominant term here is normally that due to material dispersion (first term on the r.h.s. of (8.168)) which has already been discussed at some length in Section 7.2. If we include also the first-order waveguide terms in (8.168), then for the α-profile, assuming all modes carry equal power and integrating rather than summing over μ, we find (Olshansky and Keck, 1976):

$$\sigma_{\text{intramodal}}^2 \simeq \sigma_s^2\left[M_1^2 - \frac{2M_1m_1\Delta}{\lambda c}\left(\frac{\alpha-2-\varepsilon}{\alpha+2}\right)\left(\frac{2\alpha}{2\alpha+2}\right)\right.$$
$$\left. + \left(\frac{m_1\Delta}{\lambda c}\right)^2\left(\frac{\alpha-2-\varepsilon}{\alpha+2}\right)^2\left(\frac{2\alpha}{\alpha+2}\right)\left(\frac{2\alpha}{3\alpha+2}\right)\right] \quad (8.176)$$

where we have used the definition, familiar already from (7.125):

$$M_1 = \frac{\lambda}{c}\frac{d^2 n_1}{d\lambda^2}. \tag{8.177}$$

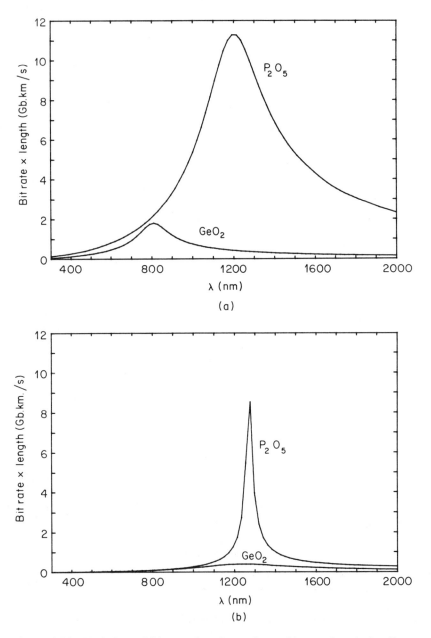

Figure 8.14 Variation of bit rate–length product with wavelength for fibres having a profile designed for operation at 850 nm. The curves are plotted for fibres having P_2O_5 and GeO_2 content at core centre of $12^m/_0$ and $13.1^m/_0$, respectively (see Table 8.1 and Figure 8.13): (a) gives result for a laser of 2 nm spectral spread, and (b) that for an LED of 40 nm. (From Sladen et al. (1978). Reproduced by permission of the Instituto Internazionale delle Communicazioni)

In (8.176) the derivative of ε with respect to wavelength has been ignored; we shall return to this point later. For profiles close to optimal the material dispersion term dominates $\sigma_{\text{intramodal}}$ and it may further be necessary to include higher-order terms in source spectral width. In this case for a Gaussian spectral distribution with $\eta_3 = 0$ and $\eta_4 = 3\sigma_s^4$, we find (Gloge, 1979):

$$\sigma_{\text{intramodal}}^2 \simeq \sigma_s^2 M_1^2 + \frac{\sigma_s^4}{2}\left(\frac{dM_1}{d\lambda}\right)^2. \tag{8.178}$$

We may use equation (8.178) together with (8.169) for $\sigma_{\text{intermodal}}$ in equation (8.166) to determine the variation of σ with wavelength for near-optimal profiles. Some results are shown in Figure 8.14 for GeO_2- and P_2O_5-doped fibres using the data from Table 8.1 and Figure 8.13. The plots are given as (bit rate × length) versus wavelength, using the approximate relation (Personick, 1973): bit rate = $1/4\sigma$. Using the values of optimum α given in Figure 8.13, the curves are plotted for P_2O_5 and GeO_2-doped fibres which have been designed to operate at 850 nm. Figure 8.14(a) gives the curves for a laser of 2 nm spectral spread whilst (b) gives the equivalent for an LED of 40 nm. We see that in the case of the laser operating in conjunction with the GeO_2-doped fibre, the bandwidth actually decreases as we move to longer wavelengths. This is because the departure of the profile from optimum outweighs the advantage of lower material dispersion at these wavelengths. For P_2O_5-doped fibres with their characteristically low wavelength-variation of optimum α, we obtain an improvement of bandwidth at longer wavelengths due to reduced material dispersion. These effects are equally marked for LED operation where for the GeO_2-doped fibre very little improvement in bandwidth is gained by moving the operating wavelength to the 1.3 μm region. In contrast, the P_2O_5-doped fibre exhibits an increase of nearly two orders of magnitude.

Consider now the case of multimode fibres specifically designed for operation at wavelengths close to the zero of material dispersion. In this case we must return to the full expression for intramodal r.m.s. pulsewidth given by equation (8.168). It is possible to eliminate the leading term on the r.h.s. of (8.168) by operating at the wavelength of zero material dispersion only if this wavelength does not vary for the range of compositions used to grade the core refractive index. In fact, however, the zero dispersion wavelength depends on the glass composition, particularly for fibres containing germania (Payne and Hartog, 1977; Adams et al., 1978a); thus total elimination of the first term of (8.168) is not possible, although we will show that minimization is possible. Similar remarks apply to terms 2 and 3 of (8.168), which together with the final term represent residual higher-order chromatic dispersion; moreover, term 2 vanishes for sources symmetric about the mean. The final term is the second-order effect investigated by Kapron (1977). For a Gaussian spectral intensity distribution ($\eta_3 = 0$, $\eta_4 = 3\sigma_s^4$), using the WKB approximation (8.152) for τ, and assuming all modes equally excited ($p_{Nl} = 1$), the averages in equation (8.168) may be computed, retaining terms to order Δ^3. Let us take for example a fibre having a germania concentration of 13.1$^m/_0$ (see Table 8.1) at the core centre and a silica cladding, and assume that at each wavelength the optimal index profile is achieved (see Figure 8.13). The results are given in Figure 8.15 in the form $\sigma_{\text{intramodal}}/\sigma_s$ for values of $2\sigma_s$ from 1 to 100 nm; also shown are measured values of $|M_1|$ for the peak GeO_2 content (Payne and Hartog, 1977) and the calculated values of $|M_2|$ for the silica cladding (see Table 7.3 and Mallitson (1965)). The curves shown are computed for source spectral width $2\sigma_s$ of 1, 20, 40, 60, 80, and 100 nm (Adams et al., 1978). The lowest curve thus indicates the limitation imposed by the first-order term of (8.168) and clearly shows that although this achieves a minimum at a wavelength $\hat{\lambda}_0$ between those of the constituent materials, it cannot be eliminated. The remaining curves are increasingly dominated by the second-order term in source spectral width but continue to

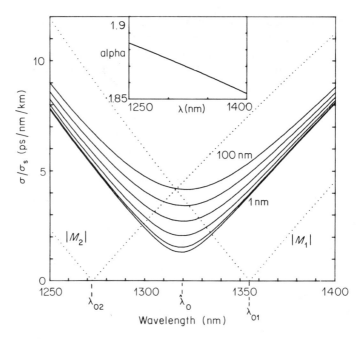

Figure 8.15 Intramodal dispersion versus wavelength for optimal profiles in the region of minimum material dispersion. Solid lines, effective material dispersion parameter calculated for graded-index fibre having $13.1^m/_0$ GeO_2 at core centre and silica cladding; curves shown for source spectral width $2\sigma_s$ of 1, 20, 40, 60, 80, and 100 nm. Dotted lines, material dispersion parameter $|M_1|$ measured for a $13.1^m/_0$ GeO_2/SiO_2 composition and $|M_2|$ calculated for SiO_2. Inset: measured wavelength dependence of optimum $\alpha = 2 + \varepsilon$ for germania-doped fibres. (From Adams et al. (1978). Reproduced by permission of the IEE)

exhibit a minimum close to $\hat{\lambda}_0$. The inset on the figure shows the relevant part of the optimum α versus wavelength plot of Figure 8.13 used for this calculation.

The relative magnitudes of the residual first- and second-order material dispersion effects may be seen more easily in Figure 8.16 where the optimized intermodal r.m.s. pulse width for this fibre is included, according to equations (8.166) and (8.167). The left axis indicates the (data rate × distance) product and the right axis gives the r.m.s. pulse dispersion 2σ. Operation at the optimum wavelength $\hat{\lambda}_0$ produces a bandwidth capability which differs only marginally from that predicted by consideration of the second-order limitation above. However, operation at the wavelengths λ_{01} or λ_{02} corresponding to the zeros of material dispersion of the core centre and the cladding is dominated by first-order material dispersion and results in a curtailment of bandwidth.

In calculating the results of Figures 8.15 and 8.16 it was necessary to use numerical computation owing to the excessive complexity of the terms involved in equations (8.167) and (8.168). Since the material dispersion properties of core and cladding materials are different it is also necessary to include the derivative $d\varepsilon/d\lambda$ as this can no longer be ignored. The exact expression for this derivative is straightforward to evaluate but takes a rather complicated form and a useful approximation which is usually sufficiently accurate is given by:

$$\frac{d\varepsilon}{d\lambda} \simeq \frac{\varepsilon}{\lambda} - \frac{2c(M_1 - M_2)}{\Delta n_1}. \tag{8.179}$$

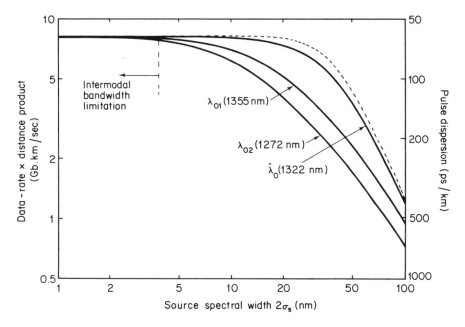

Figure 8.16 Dependence of r.m.s. pulse dispersion 2σ and data-rate × distance product on r.m.s. source spectral width $2\sigma_s$ for fibre of Figure. 8.15. Curves indicate the effect of operating at wavelengths λ_{01}, λ_{02}, and $\hat{\lambda}_0$ (see Figure 8.15); dotted curve shows result for $M_1 = M_2$. *Note*: f.w.h.m. spectral width = $1.18 \times (2\sigma_s)$. (From Adams *et al.* (1978). Reproduced by permission of the IEE)

With the aid of this expression a useful approximation for the intramodal pulse broadening may be obtained (Adams *et al.*, 1978):

$$\left(\frac{\sigma_{\text{intramodal}}}{\sigma_s}\right)^2 \simeq \left(\frac{2M_1 + M_2}{3} + \frac{\varepsilon \Delta n_1}{6c\lambda}\right)^2 + \frac{1}{8}\left(\frac{dM_1}{d\lambda}\right)^2 (2\sigma_s)^2 + O\left(\frac{\Delta^2 M_1}{\lambda_c}\right) \quad (8.180)$$

where M_1 and M_2 are respectively the material dispersion parameters of core centre and cladding. The expression is normally dominated by the first term on the r.h.s.; the zero of this term corresponds approximately to the optimum wavelength $\hat{\lambda}_0$. For the relatively small values of ε usually found, we see that this wavelength is given simply by the zero of the weighted mean $(2M_1 + M_2)$, i.e. roughly $\frac{2}{3}$ of the way between λ_{02} and λ_{01}. Note, however, that this wavelength will depend on the modal power distribution within the fibre. The second term in equation (8.180) is the r.m.s. form of the second-order contribution which dominates in practice at the optimum wavelength $\hat{\lambda}_0$ (Muska *et al.*, 1977) as discussed above. The remaining term gives the order of the residual first-order effect at the optimum wavelength and represents the ultimate bandwidth limitation for multimode fibres optimized for operation with narrow-linewidth sources.

It is possible to determine optimum α-values for multimode fibres by direct measurements of pulse dispersion in the fibres. The measurements may be made either in the time domain where the fibre impulse response may be determined by deconvolution of input and output pulses, or in the frequency domain where the modulation transfer function is determined. These techniques are equivalent in the sense that the transfer function is the Fourier transform of the impulse response (Personick, 1973a), although in practice only the amplitude of the transfer function is usually measured and hence phase information is lost.

Optimal α-values may be estimated by measuring r.m.s. pulse-spreading or bandwidth as a function of α on a number of fibres (Cohen, 1976).

An alternative approach is to find the wavelength at which a given α is optimal by making measurements as a function of wavelength for one specific fibre (Keck and Bouillie, 1978; Eve et al., 1978; Matsumoto and Nakagawa, 1979; Horiguchi et al., 1980). In some cases measurements of both types provide a further check (Cohen et al., 1978; Ohmori et al., 1978; Kitayama et al., 1979), although all these methods suffer from the disadvantage of requiring a very accurate determination of the fibre refractive index profile and hence α. The problem of accurately determining α is extremely difficult and we shall return to this topic in a later section. Most of the measurements referred to above were performed on germania-doped fibres and give reasonable confirmation of the plot of optimum α versus wavelength for GeO_2 given in Figure 8.13. As regards the achievement of high bandwidths, a fibre of NA $\simeq 0.2$ made by the plasma-activated CVD process has been reported by Hazan et al. (1977) to have a f.w.h.m. intermodal pulse-broadening at $\lambda = 905$ nm of about 150 ps/km. This figure is within a factor 5–6 of that predicted as the ultimate achievable limit set by equation (8.173) and assuming a Gaussian frequency response. The highest 6 dB bandwidth reported to date is 6.53 GHz.km at $\lambda = 1.31$ μm for a fibre fabricated by the VAD process (Nakahara et al., 1980). Again assuming a Gaussian response, this value is within a factor of about 2 of the limit set by (8.173) with $\Delta = 0.01$.

The variation of fibre bit rate–length product with wavelength of the forms shown in Figure 8.14 is a cause of some concern for system designers who would possibly wish to operate a high-capacity long-length system over a range of wavelengths. Such considerations arise for instance if one wishes to switch from present sources based on GaAs to future sources in the 1300 nm range without reduction of bandwidth due to profile degradation. Another possibility is the use of wavelength multiplexing (Tomlinson, 1977) to up-rate existing systems. It is therefore highly desirable to be able to design fibres with a small r.m.s. pulse dispersion which shows little change over a broad wavelength range, or alternately which peaks at two or more selected wavelengths. One way in which this might be achieved is the use of multicomponent glass fibres whose component concentration profiles are specified independently. Kaminow and Presby (1977), using their measurements of ε versus λ in several binary silica fibres (Presby and Kaminow, 1976) and assuming a linear relationship between concentration and refractive index, have computed concentration profiles required for a SiO_2–GeO_2–P_2O_5 fibre with minimal intermodal dispersion over a wide range of wavelengths. Experimental confirmation of this effect has been reported by Kaminow et al. (1977). Marcuse and Presby (1979) have made further calculations for phosphorus and germania-doped ternary glass fibres with bandwidths optimized at two specified wavelengths.

The 'concatenation' of several graded-index fibres jointed together to form a link may lead to a form of optical equalization (Eve 1977, 1978). Consider, for example, the case of two fibres whose α-values lie on opposite sides of the optimum α given by (8.172). Reference to Figure 8.11 shows that slow modes in the first may become fast modes in the second and vice versa, and thus provided that the modes propagate undisturbed through the joint the total delays may be equalized. Eve (1978) has made a detailed study of the effects of imperfect joints, mode-mixing, etc. and has demonstrated a procedure for the optimization of the bandwidth of a link in this way. An extension of this idea is to the case where the α-value varies slowly along the length of an individual fibre. Marcuse (1979a) has shown that the optimum r.m.s. impulse width is achievable provided that α deviates on average by equal amounts to either side of its optimum value. Pask (1980) has shown that if $\alpha = \alpha_0 + \delta g(z)$ where δ is small and $g(z)$ ($\ll 1$) is smoothly varying with distance z, then the fibre may be described by an equivalent exponent α_e which depends on ε, $g(z)$, and the fibre length. Another corollary of equalization in a jointed link is that alternately jointing under- and over-compensated fibres can produce a high bandwidth over a range of wavelengths (Eve et al., 1978; Matsumoto and Nakagawa, 1979). This is therefore another possible technique for stabilizing the wavelength-dependence of intermodal pulse dispersion.

8.2.4 More complicated index profiles

For the refractive index measurements summarized in Table 7.3 it is sometimes the case that $n\,dn/d\lambda$ does not vary linearly with n^2 as the dopant concentration varies. The requirement that $ndn/d\lambda$ should vary linearly with n^2 is implicit in the theory of Sections 8.2.2 and 8.2.3, since it was assumed there that only Δ is a function of wavelength for α-profiles and hence the shape of the profile is independent of λ. Since this is not the case, for example, for boron doping (Fleming, 1976; Presby and Kaminow, 1976) and, according to Fleming's bulk measurements, also for germania, it is necessary to develop more general theories for calculating pulse dispersion under these conditions (see, for example, Arnaud 1975; Arnaud and Fleming, 1976; Arnaud, 1976, 1977). We will discuss here an approach due to Marcatili (1977) which is based on the WKB approximation.

Let us commence with the WKB resonance condition (8.131) and use the profile of equation (8.1) so that the quantity $q(r)$ becomes:

$$q = \left(2\Delta k^2 n_1^2(1-b-f) - \frac{l^2}{r^2}\right)^{1/2}. \tag{8.181}$$

Taking the derivative of both sides of equation (8.131) with this definiton of q, whilst keeping the mode numbers N, l fixed, yields:

$$\int_{r_1}^{r_2} \left\{\frac{d}{dk}[\Delta k^2 n_1^2(1-b)] - 2\Delta k n_1 m_1 f\left(1 - \frac{P}{2}\right)\right\} \frac{dr}{q} = 0 \tag{8.182}$$

where

$$P = -\frac{n_1}{m_1} \frac{k}{\Delta f} \frac{\partial(\Delta f)}{\partial k} = \frac{n_1}{m_1} \frac{\lambda}{\Delta f} \frac{\partial(\Delta f)}{\partial \lambda} \tag{8.183}$$

is related to a generalized version of the ε defined for α-profiles in (8.150) and is cast in this form as a generalization of the profile dispersion parameter introduced by Gloge et al. (1975); for α-profiles (8.183) reduces to $P = -\varepsilon/2$. The parameter b in (8.182) is defined in the usual way by the relation:

$$\beta^2 = k^2 n_1^2[1 - 2\Delta(1-b)]. \tag{8.184}$$

Differentiation of this equation with respect to k gives the result for the transit time τ:

$$m_1 \beta \frac{\tau}{T} = k n_1 m_1 - \frac{d}{dk}[\Delta k^2 n_1^2(1-b)] \tag{8.185}$$

where T is the on-axis time of flight Lm_1/c. Substituting (8.185) and (8.184) into (8.182) gives:

$$\int_{r_1}^{r_2} \left\{1 - \frac{\tau}{T}[1 - 2\Delta(1-b)]^{1/2} - 2\Delta f\left(1 - \frac{P}{2}\right)\right\} \frac{dr}{q} = 0. \tag{8.186}$$

In order to proceed further we use the self-evident relation:

$$\int_{r_1}^{r_2} \frac{d}{dr}(qr)\,dr = r_2 q(r_2) - r_1 q(r_1) = 0. \tag{8.187}$$

With the definition of q from (8.181), this yields a more useful form:

$$\int_{r_1}^{r_2} \left(1 - b - f - \frac{r}{2}\frac{\partial f}{\partial r}\right)\frac{dr}{q} = 0. \tag{8.188}$$

Combining (8.188) with (8.186) we arrive at the general expression:

$$\frac{1 - \frac{\tau}{T}[1 - 2\Delta(1-b)]^{1/2}}{2\Delta(1-b)} = \frac{\int_{r_1}^{r_2}\left(1 - \frac{P}{2}\right)\frac{f}{q}dr}{\int_{r_1}^{r_2}\left(1 + \frac{r}{2f}\frac{\partial f}{\partial r}\right)\frac{f}{q}dr}. \tag{8.189}$$

Equation (8.189) relates the group delay τ of a mode characterized by the propagation parameter b and azimuthal mode number l (contained in q) to the profile function f and profile dispersion P. If we consider a class of fibres which satisfy the condition (Marcatili, 1977):

$$\frac{1 + (r/2f)\partial f/\partial r}{1 - P/2} = D(\lambda) \tag{8.190}$$

where D is an arbitrary function of λ, then the r.h.s. of (8.189) reduces to $1/D$ and the group delay is given simply by:

$$\frac{\tau}{T} = \frac{1 - 2\Delta(1-b)/D}{[1 - 2\Delta(1-b)]^{1/2}}. \tag{8.191}$$

Equation (8.191) demonstrates that fibres of the class defined by (8.190) have group delays which are independent of the mode number and functions only of the mode parameter b and the dispersion parameter D. Equation (8.190), on the other hand, is a design formula for fibres which may be solved in several ways, depending on the degree of control which is available for the profile function f and the profile dispersion P. This result has also been derived from Hamilton's equations for ray optics by Geckeler (1978) using the formulation of Arnaud (1975), and from more straightforward geometric optics techniques (Pask, 1978; Jacomme, 1978).

The profiles specified by (8.190) may be optimized for minimum pulse-spreading by consideration of the fastest and slowest group delays as given by (8.191). In fact modes for which $b = 0$ and $b = 1$ are the slowest and have the same delay ($\tau = T$) and modes characterized by $b = 1 - (2 - D)/2\Delta$ are the fastest.

The choice

$$D = 1 + (1 - 2\Delta)^{1/2} \tag{8.192}$$

gives the minimum time interval $\Delta\tau$ between fastest and slowest modes, namely:

$$\frac{\Delta\tau}{T} = \frac{[1 - (1 - 2\Delta)^{1/4}]^2}{1 + (1 - 2\Delta)^{1/2}}. \tag{8.193}$$

Clearly, for weakly-guiding fibres with $\Delta \ll 1$, equation (8.192) reduces to $D \simeq 2 - \Delta$, and (8.193) reduces to $\Delta\tau = \Delta^2 T/8$ which is the same as the minimum pulse-spreading calculated for α-profiles using the optimization condition (8.154). Thus an optimized fibre with arbitrary profile dispersion P has the same minimum impulse-response width as an optimized α-profile fibre with the same Δ having profile dispersion independent of position r in the core. The main results derived so far in this subsection have been extended by Arnaud (1977) to include the case of noncircular fibres. A number of profiles obeying the Marcatili condition (8.190) have been proposed (Marcatili, 1977; Geckeler, 1979; Weierholt, 1979). We shall discuss here one of these which may hold promise for the future of high-bandwidth multimode fibres, namely the multiple-α profile (Olshansky, 1978, 1979).

The Marcatili profile condition (8.190) may be formulated as a partial differential equation (Jacomme, 1978; Olshansky, 1978, 1979):

$$r\frac{\partial(\Delta f)}{\partial r} + D\frac{n_1}{m_1}\lambda\frac{\partial(\Delta f)}{\partial \lambda} - 2(D-1)\Delta f = 0. \tag{8.194}$$

A solution of this equation is the multiple-α profile proposed by Olshansky (1978, 1979):

$$\Delta f = \sum_{j=1}^{M} \Delta_j \left(\frac{r}{a}\right)^{\alpha_j} \tag{8.195}$$

where

$$\alpha_j = 2(D-1) - D\frac{n_1}{m_1}\frac{\lambda}{\Delta_j}\frac{\partial \Delta_j}{\partial \lambda} \quad (j = 1, 2, \ldots, M) \tag{8.196}$$

and

$$\Delta = \sum_{j=1}^{M} \Delta_j. \tag{8.197}$$

These solutions represent a generalization of the single-α profile discussed earlier; they have the advantage of further degrees of freedom in designing the profile so that, for example, high bandwidths may be obtained over a range of wavelengths. The modal delay time given by (8.191) may be expanded in a Taylor series:

$$\frac{\tau}{T} \simeq 1 + \left(\frac{1}{2} - \frac{1}{D}\right)2\Delta(1-b) + \left(\frac{3}{8} - \frac{1}{2D}\right)4\Delta^2(1-b)^2 + O(\Delta^3) \tag{8.198}$$

from which it is clear that the lowest-order term vanishes if $D = 2$. If Δf is approximately parabolic as a function of r then we may use the result (8.172) which implies that the r.m.s intermodal pulse-width is minimized for $D = 2 - 6\Delta/5$. Using this result in (8.196) yields for the optimal profile:

$$\alpha_j \simeq 2 - \frac{12\Delta}{5} - \frac{2n_1}{m_1}\frac{\lambda}{\Delta_j}\frac{\partial \Delta_j}{\partial \lambda} \quad (j = 1, 2, \ldots, M). \tag{8.199}$$

This result represents a generalization of equation (8.172) for the single-α profile, and clearly reduces to (8.172) for the case $M = 1$.

Olshansky (1979) has applied this approach to the case of a double-α profile and has shown how to choose the α's and the core dopants to impose extra conditions on the dispersion behaviour of the fibre. The double-α profile is given by:

$$n^2(r) = n_1^2\left[1 - 2\Delta_1\left(\frac{r}{a}\right)^{\alpha_1} - 2\Delta_2\left(\frac{r}{a}\right)^{\alpha_2}\right] \tag{8.200}$$

where

$$\Delta = \Delta_1 + \Delta_2. \tag{8.201}$$

Let us consider the specific case of a GeO_2–B_2O_3–SiO_2 fibre and assume that at $r = 0$ the

GeO$_2$ concentration is $C_G(0)$ and there is no B$_2$O$_3$, whilst at $r = a$ the B$_2$O$_3$ concentration is $C_B(a)$ and there is no GeO$_2$. The two new degrees of freedom which are introduced by the use of the double-α profile with two dopants may be used to impose conditions such as:

$$\left.\frac{d\alpha_j}{d\lambda}\right|_{\lambda=\lambda_0} = 0 \qquad (j = 1, 2) \qquad (8.202)$$

or

$$\alpha_j(\lambda_1) = \alpha_j(\lambda_2) \qquad (j = 1, 2). \qquad (8.203)$$

The application of condition (8.202) would result in pulse dispersion which is independent of wavelength in the vicinity of λ_0, whilst condition (8.203) would mean that the dispersion is minimized at two wavelengths λ_1 and λ_2.

To illustrate the approach to designing a double-α, GeO$_2$–B$_2$O$_3$–SiO$_2$ fibre we will consider here the case where condition (8.203) is required to be obeyed. Let us assume that the square of refractive index is proportional to dopant concentration; this is almost certainly a good approximation for GeO$_2$–SiO$_2$ compositions but may not be so good for B$_2$O$_3$–SiO$_2$ glasses at high dopant concentrations. Nevertheless the assumption is a reasonable first approximation and small deviations from linearity can be treated as perturbations of the linear solutions discussed here. We define Δ_1 and Δ_2 in terms of parameters X_G and X_B denoting the relative amounts of GeO$_2$ and B$_2$O$_3$, respectively:

$$\left.\begin{array}{l}2n_1^2\Delta_1 = \delta_G(1-X_G)+\delta_B X_B \\ 2n_1^2\Delta_2 = \delta_G X_G + \delta_B(1-X_B)\end{array}\right\} \qquad (8.204)$$

where

$$\left.\begin{array}{l}\delta_G = n_1^2 - n_s^2 \\ \delta_B = n_s^2 - n_2^2\end{array}\right\} \qquad (8.205)$$

and n_s is the refractive index of silica; as usual n_1 is the index on-axis and n_2 that at $r = a$. From equations (8.200) and (8.204) the dopant concentration profiles for GeO$_2$ and B$_2$O$_3$ are, respectively:

$$\left.\begin{array}{l}C_G(r) = C_G(0)\left[1 - (1-X_G)\left(\dfrac{r}{a}\right)^{\alpha_1} - X_G\left(\dfrac{r}{a}\right)^{\alpha_2}\right] \\ C_B(r) = C_B(a)\left[X_B\left(\dfrac{r}{a}\right)^{\alpha_1} + (1-X_B)\left(\dfrac{r}{a}\right)^{\alpha_2}\right]\end{array}\right\} \qquad (8.206)$$

whence it is clear that X_G is proportional to the amount of GeO$_2$ following profile α_2, and X_B is proportional to the amount of B$_2$O$_3$ following profile α_1. The condition (8.203) that each α shall be the same at λ_1 and λ_2 can be evaluated using the expression for optimum α's from (8.199) with $M = 2$, and the result may be expressed as:

$$\left.\begin{array}{l}(1-X_G)^2 A_{GG} + 2(1-X_G) X_B A_{GB} + X_B^2 A_{BB} = 0 \\ (1-X_B)^2 A_{BB} + 2(1-X_B) X_G A_{GB} + X_G^2 A_{GG} = 0\end{array}\right\} \qquad (8.207)$$

where the coefficients are given approximately by:

$$\left.\begin{array}{l}A_{GG} = \delta'_G \overline{\delta}_G - \overline{\delta'}_G \delta_G \\ A_{BB} = \delta'_B \overline{\delta}_B - \overline{\delta'}_B \delta_B \\ 2A_{GB} = \delta'_G \overline{\delta}_B - \overline{\delta'}_G \delta_B + \overline{\delta}_G \delta'_B - \delta_G \overline{\delta'}_B\end{array}\right\} \qquad (8.208)$$

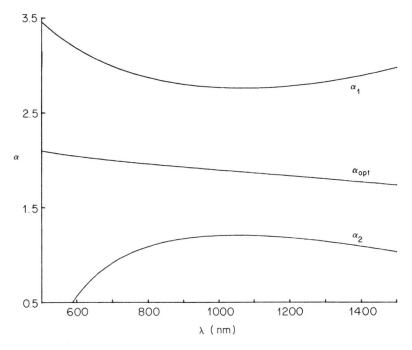

Figure 8.17 Optimum α's versus wavelength for GeO_2–B_2O_3–SiO_2 fibres with 8.1^{m}/$_0$ GeO_2 on axis (data from Table 8.1). For a single-α profile, α_{opt} is the value given by equation (8.172). For a double-α profile, α_1 and α_2 are the values given by (8.199) and chosen such that α_j (850 nm) = α_j (1300 nm), for $j = 1, 2$

where the primes denote the operation $\lambda \, (d/d\lambda)$. The quantities denoted by a bar are evaluated at λ_2 and the other quantities are evaluated at λ_1. In calculating the coefficients in (8.208) terms involving the derivative of n_1 have been ignored by comparison with those involving the derivatives of Δ_1, Δ_2, and m_1 has been replaced by n_1 throughout; this approximation is of the same type as that involved in (8.157b). The solutions of equations (8.207) are:

$$X_G = \frac{(A_{GB} + B + A_{BB})}{2B}$$
$$X_B = \frac{(A_{GB} + B + A_{GG})}{2B}$$
(8.209)

where

$$B = (A_{GB}^2 - A_{BB} A_{GG})^{1/2}.$$
(8.210)

These results, together with (8.206), enable us to compute dopant concentration profiles which should produce optimized pulse dispersion at wavelengths λ_1, λ_2.

As a specific example of the use of this theory, let us use the data of Table 8.1 for the variation of δ_B and δ_G with wavelength. Using the data for 8.1^{m}/$_0$ germania on-axis, and setting $\lambda_1 = 850$ nm, $\lambda_2 = 1300$ nm, the solutions of equation (8.209) are $X_G = -1.68$, $X_B = -4.60$. The corresponding α-values given by equation (8.199) are $\alpha_1 = 2.83$ and $\alpha_2 = 1.14$. The variation of α_1, α_2 with wavelength is plotted in Figure 8.17 where it is clear that we have achieved the condition (8.203) that α_j(850 nm) = α_j(1300 nm) ($j = 1, 2$). Also plotted on the figure is the variation of a single optimum α for the conventional α-profile optimized

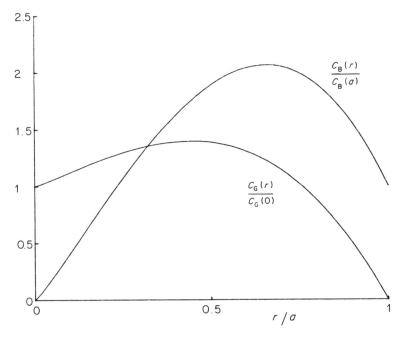

Figure 8.18 Dopant profiles for germania and boric oxide as given by equation (8.206) for the optimized double-α profile of Figure 8.17

according to (8.172) and using the same data for GeO_2–B_2O_3–SiO_2 from Table 8.1. The dopant profiles for the optimized double-α profile as given by equation (8.206) are plotted in Figure 8.18. It is clear that these profiles are vastly different from those required for a single α-profile. The results given in Figures 8.17 and 8.18 provide design data for fibres optimized to have the same intermodal pulse-spreading at 850 nm and 1300 nm. Using bulk index data from Table 7.3, rather than fibre data from Table 8.1 as here, Olshansky (1979) has given similar results for GeO_2–SiO_2–B_2O_3 double-fibres optimized according to either (8.202) or (8.203).

8.2.5 Leaky modes on fibres of arbitrary index profile

Leaky rays or modes correspond to discrete solutions of the eigenvalue equation for the waveguide which are below cut-off, i.e. $\beta < kn_2$. In general the propagation constant β for a leaky mode is complex and hence the mode suffers attenuation which may range from zero for the mode at cut-off to infinity for the most leaky mode. In ray terms we may think of a skew ray which satisfies the usual condition for a trapped ray in terms of the critical angle of incidence, and yet as a consequence of its skewness may correspond to the below cut-off criterion $\beta < kn_2$. In wave terms the cladding supports evanescent fields which may apparently be forced to travel faster than the velocity of light in the medium as they try to keep pace with the oscillatory fields and simultaneously rotate about the fibre axis. This apparent contradiction leads to the mode radiating from a caustic in the cladding. The phenomenon is similar to tunnelling in quantum mechanics in that an evanescent field occurs between the guided power in the core and the radiated power in the cladding beyond the outer caustic. A review of the literature for leaky rays and modes on circular step-index fibres has been given by Snyder (1974). Here, however, we wish to develop an approach based on the

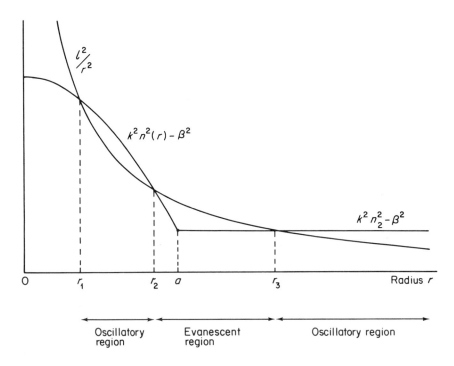

Figure 8.19 Squared magnitude of local plane wave components for a mode below cut-off. (From Adams et al. (1975). Reproduced by permission of the IEE)

WKB method which is appropriate for fibres of arbitrary index profile (Adams et al., 1975).

From our earlier considerations of the local plane-wave decomposition we know that the radical component of the wave vector at radius r is given by q as defined in equation (8.126). In Figure 8.19 the squared magnitudes of the various components of q are shown as functions of radius r; the figure is drawn for a fibre core of arbitrary index profile and radius a, surrounded by a uniform cladding of index n_2. Figure 8.19 differs from the corresponding plots of Figures 8.4 and 8.8 in that the mode here is below cut-off, i.e. $\beta < kn_2$. It can be seen that this mode has:

(a) a region of radial periodicity within the core ($r_1 \leq r \leq r_2$), representing bound power,
(b) a region of evanescent field extending into the cladding ($r_2 \leq r \leq r_3$), and
(c) a further region having an oscillatory field solution within the cladding ($r \geq r_3$), representing radiated power.

The mode may therefore be identified as a leaky tunnelling mode. The limiting values for the propagation constant between which these modes may exist are given from Figure 8.19 by:

$$k^2 n_2^2 - \frac{l^2}{a^2} \leq \beta^2 < k^2 n_2^2. \tag{8.211}$$

From the relations (8.13), (8.22), (8.33), and (8.34) between modal parameters β, l and ray angles γ, ϕ, applied at radius r_0, we find the relations:

$$\left. \begin{array}{l} \beta = kn(r_0)\cos\gamma \\ l = rkn(r_0)\cos\phi \sin\gamma \end{array} \right\}. \tag{8.212}$$

These results are confirmed by the local plane wave decomposition diagram which is repeated

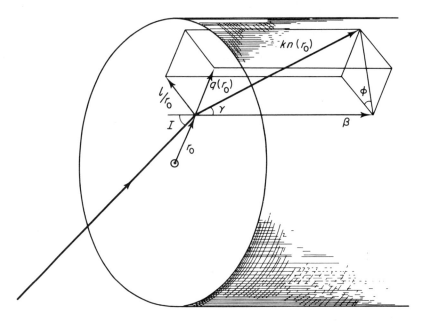

Figure 8.20 Local wave-vector diagram for ray incident on fibre face at radius r_0 and with angle I. (From Adams et al. (1975). Reproduced by permission of the IEE)

for convenience in Figure 8.20. The figure is drawn for incidence at angle I from an external medium on the end-face of the fibre at radius r_0. If the external medium is air then Snell's law yields:

$$\sin I = n(r_0) \sin \gamma. \tag{8.213}$$

Inserting equations (8.212) and (8.213) into (8.211) gives the condition on acceptance angle I for leaky rays (Adams et al., 1975):

$$\frac{n^2(r_0) - n_2^2}{1 - (r_0/a)^2 \cos^2 \phi} \geq \sin^2 I > n^2(r_0) - n_2^2. \tag{8.214}$$

Equation (8.214) defines the angular region in which leaky rays are found at a given radius on the endface of a fibre with an arbitrary circularly-symmetric index profile. The corresponding region for guided rays is given by substituting (8.213) and (8.212) into the condition $\beta \geq kn_2$, namely:

$$\sin^2 I \leq n^2(r_0) - n_2^2. \tag{8.215}$$

The r.h.s. of equation (8.215) is the square of the local numerical aperture (Gloge and Marcatili, 1973) for a graded-index fibre, defined in analogy with the conventional step-index numerical aperture as given by equation (7.246). Thus the l.h.s. of (8.214) defines a local acceptance angle that includes both bound and leaky rays. As expected, no leaky-ray region exists for $\phi = \pi/2$, as this defines a meridional ray.

Equation (8.214) forms a generalization of results which were derived earlier by the methods of geometric optics. For example, leaky ray conditions for step-index fibres (Snyder et al., 1974) and fourth-order profiles (Ikeda, 1974) are available and confirm the validity of (8.214). For the parabolic index profile the l.h.s. of (8.214) reproduces equation (8.53) with $\phi_0 = \phi$, and hence the arguments presented here confirm the remarks concerning leaky rays on parabolic profiles in Section 8.1.

Consider now a multimode fibre core cross-section which is illuminated by an incoherent source (exciting all modes uniformly) so that the power incident per unit solid angle at any point in the cross-section is constant. To calculate the power accepted by the fibre at radius r_0 we must sum over the solid angle available for guided or leaky modes. Allowing for the cosine distribution of power from such a Lambertian source (Carpenter and Pask, 1977; Pask, 1978a) it follows that the intensity distribution $P(r_0)$ is given by:

$$\frac{P(r_0)}{P(0)} = \frac{\int_0^{\pi/2} d\phi \int_0^{I_{max}(r_0)} \sin I \cos I \, dI}{\int_0^{\pi/2} d\phi \int_0^{I_{max}(0)} \sin I \cos I \, dI}$$

$$= \frac{\int_0^{\pi/2} \sin^2 (I_{max}(r_0)) \, d\phi}{\int_0^{\pi/2} \sin^2 (I_{max}(0)) \, d\phi}. \tag{8.216}$$

In writing (8.216) in this form we are ignoring attenuation of any modes; in general this will only be true for guided modes or for leaky modes at the input face of the fibre. For bound modes the value of I_{max} is given by (8.215) and we find the result (Gloge and Marcatili, 1973):

$$\frac{P(r_0)}{P(0)} = \frac{n^2(r_0) - n_2^2}{n_1^2 - n_2^2}. \tag{8.217}$$

For leaky plus guided modes the appropriate expression for I_{max} is given by the l.h.s. of (8.214). Using this relation in (8.216) yields (Adams et al., 1975):

$$\frac{P(r_0)}{P(0)} = \frac{n^2(r_0) - n_2^2}{n_1^2 - n_2^2} \frac{1}{[1 - (r_0/a)^2]^{1/2}}. \tag{8.218}$$

However, equation (8.218) may only be generally valid up to some value of r_0 ($\leq a$), since a condition which we have not imposed is the requirement that the ray acceptance angle I does not exceed $\pi/2$:

$$\sin^2 I \leq 1$$

or, in terms of the profile $n(r_0)$ and the projected angle ϕ:

$$\left(\frac{r_0}{a}\right)^2 \cos^2 \phi \leq 1 - n^2(r_0) + n_2^2. \tag{8.219}$$

The condition (8.219) may be built-in to the integration over ϕ in (8.216) in order to correct (8.218) for this feature at values of r_0/a close to unity. In practice it is usually only significant for profiles which are close to step-index. For the step-index case, use of (8.219) results in the replacement of (8.218) by:

$$\frac{P(r_0)}{P(0)} = \begin{cases} [1 - (r_0/a)^2]^{-1/2}, & \text{for } (r_0/a)^2 < 1 - NA^2 \tag{8.220a} \\ [1 - (r_0/a)^2]^{-1/2} \left\{1 - \frac{2}{\pi}\tan^{-1}\left[\left(\frac{NA^2 - 1 + (r_0/a)^2}{(1 - NA^2)(1 - (r_0/a)^2)}\right)^{1/2}\right]\right\} \\ + \frac{2}{\pi NA^2} \cos^{-1}\left[\frac{a}{r_0}(1 - NA^2)^{1/2}\right], & \text{for } \left(\frac{r_0}{a}\right)^2 \geq 1 - NA^2. \tag{8.220b} \end{cases}$$

In cases where truncation of the angle I occurs as a result of the use of a lens, this may be included in (8.220) merely by replacing NA by the ratio of NA to the numerical aperture of the lens. Plots of the intensity distribution for several α-profiles (see equations (8.1) and (8.2) for definition) are given in Figure 8.21, as calculated from equations (8.218) and (8.220). Note that if only bound rays were included in the calculation, then equation (8.217) would apply and the

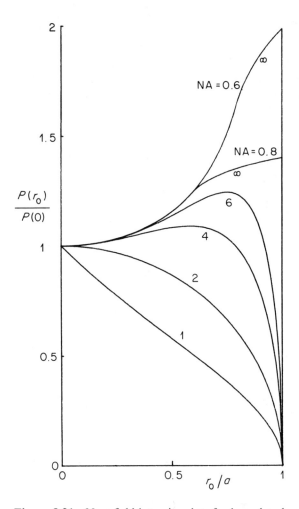

Figure 8.21 Near-field intensity plots for bound and all leaky rays, from equations (8.218) and (8.220). Parameter gives the value of exponent α for α-profiles. (From Adams et al. (1975). Reproduced by permission of the IEE)

plots of Figure 8.1 for α-profiles would be appropriate. The inclusion of the additional acceptance angle defined by the leaky-ray region thus causes a large deviation from the near-field intensity distribution predicted for guided modes.

In order to calculate the number of leaky modes for a given profile, consider again the local plane wave component plot given in Figure 8.19. It is clear that the allowed values of azimuthal mode number l are bounded by the values l_{min}, l_{max}, where l_{min} is given by (8.211) which may be re-written as:

$$l_{min}^2 = u^2 - v^2. \tag{8.221}$$

The value of l_{max} is given by the condition that the caustics r_1, r_2 shown in Figure 8.19 should be the same, i.e. the azimuthal wave vector component $(l/r)^2$ shall intersect the curve of $(k^2 n^2(r) - \beta^2)$ at only one point. For an α-profile this condition can be expressed by demanding that at the point of intersection the slope of the two curves is the same:

$$u^2 - v^2 \left(\frac{r}{a}\right)^\alpha - \left(\frac{l_{\max} a}{r}\right)^2 = 0 \tag{8.222}$$

and

$$v^2 \left(\frac{r}{a}\right)^{\alpha-1} \alpha = 2 l_{\max}^2 \left(\frac{a}{r}\right)^3. \tag{8.223}$$

Eliminating (r/a) from these equations yields (Stewart, 1975):

$$l_{\max} = \left(\frac{2}{\alpha+2}\right)^{1/\alpha} \left(\frac{\alpha}{\alpha+2}\right)^{1/2} u \left(\frac{u}{v}\right)^{2/\alpha}. \tag{8.224}$$

Leaky modes for a given β have $l_{\min} \leqslant l \leqslant l_{\max}$ where $l_{\min}$, $l_{\max}$ are given by (8.221) and (8.224) for α-profiles. If, on the other hand, we consider guided modes with a given β, similar consideration of Figure 8.8 shows that in this case $0 \leqslant l \leqslant l_{\max}$ with $l_{\max}$ from (8.224). Thus the number $\mu_1(\beta)$ of leaky modes for a given β is given by a double integral of the form (8.137) but with the lower limit on the l-integration replaced by $l_{\min}$. For α-profiles we have therefore:

$$\mu_1(\beta) = \frac{4}{\pi} \int_{l_{\min}}^{l_{\max}} \int_{r_1}^{r_2} \left[u^2 - v^2 \left(\frac{r}{a}\right)^\alpha - \left(\frac{la}{r}\right)^2\right]^{1/2} \frac{dr}{a} dl. \tag{8.225}$$

Unfortunately, there is no analytic solution for the r-integral in (8.225) for arbitrary α, and to proceed further we need to either specify an α-value for which the integral may be performed or use some approximation. Gloge (1975) has suggested the useful approximation:

$$\frac{4}{\pi} \int_{r_1}^{r_2} \left[u^2 - v^2 \left(\frac{r}{a}\right)^\alpha - \left(\frac{la}{r}\right)^2\right]^{1/2} \frac{dr}{a} \simeq \left(\frac{2+\alpha}{2}\right)^{2/\alpha} (l_{\max} - l). \tag{8.226}$$

This approximation is exact for $\alpha = 2$ and very good provided α is not too large; the maximum error occurs for $\alpha \to \infty$. Note that this approximation, when introduced into the expression (8.137) for number of guided modes, yields the result (8.140) which was derived for α-profiles by a different method. Using (8.226) in (8.225) gives the result:

$$\mu_1(\beta) \simeq \frac{1}{2} \left(\frac{2+\alpha}{2}\right)^{2/\alpha} \left[\left(\frac{2}{\alpha+2}\right)^{1/\alpha} \left(\frac{\alpha}{\alpha+2}\right)^{1/2} u \left(\frac{u}{v}\right)^{2/\alpha} - (u^2 - v^2)^{1/2}\right]^2. \tag{8.227}$$

Note that μ_1 varies from the value M given by (8.141) at $u = v$ to 0 at the u corresponding to the condition $l_{\max} = l_{\min}$. That is just at cut-off all the guided modes become leaky, and as u increases further the number of leaky modes for given u decreases to 0 at the upper limit u_m given by $l_{\max} = l_{\min}$. This latter condition may be solved with the aid of (8.221) and (8.224) to yield (Adams et al., 1975a):

$$u_m = \left(\frac{\alpha+2}{2}\right)^{1/2} v. \tag{8.228}$$

For example for a parabolic profile ($\alpha = 2$), μ_1 varies from $v^2/4$ at $u = v$ to 0 at $u_m = v\sqrt{2}$.

We have stressed above the fact that $\mu_1(\beta)$ only counts the number of modes which are leaky at a given β. To find the total number of modes (guided and leaky) with propagation constants greater than a specified β we must add the number $\mu_2(\beta)$ of all modes which propagate with higher β-values. This may be calculated by a similar double integral to (8.225) but imposing the condition that $l = l_{\min}$ in the integrand, so that the l-integral is replaced by an integration over u. Thus, for α-profiles with the aid of the approximation (8.226), we find:

$$\mu_2(\beta) \simeq v^2 \left(\frac{\alpha+2}{2}\right)^{2/\alpha} \int_1^{u/v} \left[\left(\frac{\alpha}{\alpha+2}\right)^{1/2} \left(\frac{2}{\alpha+2}\right)^{1/\alpha} \frac{(u/v)^{2/\alpha+2}}{[(u/v)^2-1]^{1/2}} - \left(\frac{u}{v}\right)\right] d\left(\frac{u}{v}\right). \tag{8.229}$$

For $\alpha = 2$ the integral in (8.229) is considerably simplified and we find the result:

$$\mu_2(\beta) \simeq v^2 \left\{\frac{1}{3}\left[\left(\frac{u}{v}\right)^2-1\right]^{3/2} - \left[\left(\frac{u}{v}\right)^2-1\right] + \left[\left(\frac{u}{v}\right)^2-1\right]^{1/2}\right\}. \tag{8.230}$$

Combining this result with the $\alpha = 2$ case for $\mu_1(\beta)$ from (8.227) yields for the total number of guided and leaky modes with propagation constants greater than β:

$$\mu_1(\beta) + \mu_2(\beta) \simeq v^2 \left\{\frac{1}{4}\left(\frac{u}{v}\right)^4 - \frac{2}{3}\left[\left(\frac{u}{v}\right)^2-1\right]^{3/2}\right\}. \tag{8.231}$$

Thus the number of modes given by this expression varies from $v^2/4$ at $u = v$ to $v^2/3$ at $u_m = v\sqrt{2}$. Hence the total number of guided and leaky modes on a parabolic-profile fibre is $v^2/3$, whilst the total number of guided modes is $v^2/4$; i.e. the number of leaky modes is $v^2/12$ (Adams et al., 1975a; Stewart, 1975a; Arnaud, 1975). Figure 8.22 shows a plot of $\mu_1(\beta) + \mu_2(\beta)$ versus $(u/v)^2$ from equation (8.231), together with the corresponding plot for guided modes $(u < v)$ from (8.140) with $\alpha = 2$. The total number of modes below cut-off includes tunnelling

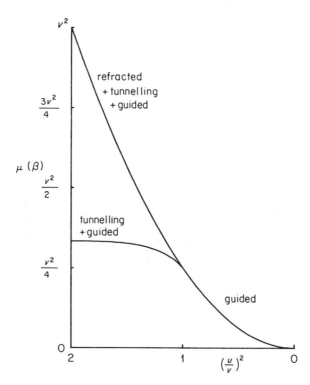

Figure 8.22 Number of modes $\mu(\beta)$ having propagation constant greater than β as specified by u/v in abscissa. (From Adams et al. (1975a). Reproduced by permission of the IEE)

leaky modes and refracted (radiation) modes and is given by the analytic continuation of (8.140) for $u > v$; this curve is also included in Figure 8.22.

For step-index fibres the above argument must be modified slightly in order to obtain an expression for the number of leaky modes. The upper limit for azimuthal mode number $l_{\max}$ is given from (8.224) as u for $\alpha \to \infty$. For this case both integrals in equation (8.225) may be performed exactly, with the result:

$$\mu_1(\beta) = \frac{4}{\pi} \int_{(u^2-v^2)^{1/2}}^{u} \int_{la/u}^{a} \left[u^2 - \left(\frac{la}{r}\right)^2\right]^{1/2} \frac{dr}{a} dl$$

$$= \frac{4}{\pi} \int_{(u^2-v^2)^{1/2}}^{u} \left[(u^2-l^2)^{1/2} - l \cos^{-1}\left(\frac{l}{u}\right)\right] dl$$

$$= \frac{v^2}{\pi} \left[\left(\frac{3u^2}{v^2} - 2\right) \sin^{-1}\left(\frac{v}{u}\right) - 3\left(\frac{u^2}{v^2} - 1\right)^{1/2}\right]. \qquad (8.232)$$

Similarly, for the step-index profile the integrals involved in $\mu_2(\beta)$ are straightforward to evaluate, with the result:

$$\mu_2(\beta) = \frac{2v^2}{\pi} \left[\left(\frac{u^2}{v^2} - 1\right)^{1/2} + \frac{\pi}{2} - \frac{u^2}{v^2} \sin^{-1}\left(\frac{v}{u}\right)\right]. \qquad (8.233)$$

Combining these two results we obtain finally:

$$\mu_1(\beta) + \mu_2(\beta) = \frac{v^2}{\pi} \left[\pi + \left(\frac{u^2}{v^2} - 2\right) \sin^{-1}\left(\frac{v}{u}\right) - \left(\frac{u^2}{v^2} - 1\right)^{1/2}\right]. \qquad (8.234)$$

The number of modes given by this expression for step-index fibres is $v^2/2$ at $u = v$, corresponding to the total number of guided modes. For this profile leaky modes occur for all β-values down to $\beta = 0$, so that the total number of leaky modes is given by (Pask et al., 1975):

$$\mu_1(0) + \mu_2(0) - \frac{v^2}{2} = \frac{v^2}{\pi} \left[\frac{\pi}{2} + \left(\frac{1}{2\Delta} - 2\right) \sin^{-1}(\sqrt{2\Delta}) - \left(\frac{1}{2\Delta} - 1\right)^{1/2}\right]. \qquad (8.235)$$

For most practical cases of interest, the weakly-guiding approximation $\Delta \ll 1$ means that the number of leaky modes on a step-index fibre is given approximately by $v^2/2$ from (8.235), i.e. the same as the number of bound modes.

The results derived above for total numbers of leaky modes may also be verified in terms of total powers directly from the near-field intensity distributions calculated earlier. For α-profiles, the results (8.217) and (8.218) may be integrated over the core cross-section to give the ratio of powers in the guided and leaky modes. Denoting guided power by S_B and leaky power by S_L, we find:

$$\frac{S_L + S_B}{S_B} = \frac{\int_0^1 \frac{(1-x^\alpha)}{(1-x^2)^{1/2}} x\, dx}{\int_0^1 \frac{x\, dx}{(1-x^2)^{1/2}}}. \qquad (8.236)$$

With the aid of tables of integrals (Gradshteyn and Ryzhik, 1965) and relationships between gamma functions (Abramowitz and Stegun, 1964), the integrals in (8.236) may be performed, with the result (Ankiewicz and Pask, 1977):

$$\frac{S_L}{S_B} = \frac{\alpha+2}{\alpha} \left\{\frac{\alpha+4}{\alpha+2} - \sqrt{\pi} \frac{\Gamma[(\alpha+2)/2]}{\Gamma[(\alpha+3)/2]}\right\} \qquad (8.237)$$

where Γ is the gamma function. S_L/S_B increases monotonically from 0.288 at $\alpha = 1$, to $\frac{1}{3}$ at $\alpha = 2$, to 0.370 at $\alpha = 3$. This result confirms that found above for ratio of mode numbers in the case $\alpha = 2$; the ratios of powers and mode numbers are the same since a Lambertian source was assumed and such a source launches equal power into all modes (Pask, 1978). For step-index fibres a careful integration of (8.220) over the fibre core confirms the form of the result (8.235). The effects of non-uniform sources have been investigated by Ankiewicz and Pask (1977, 1978).

8.2.6 Impulse response

For equal power in all modes the impulse response is given by the mode density per unit time interval. Within the limits of the present treatment the impulse response is therefore given by $(T/M)\,d\mu/dt$, where μ, M are mode counts as defined earlier and T is the on-axis transit time Lm_1/c. For α-profiles, the guided-mode results for μ and M are given in equations (8.140) and (8.141), respectively, whilst the connection between time t and mode parameter (u/v) is given by (8.151). With the aid of these results we find for guided modes (Gloge and Marcatili, 1973):

$$\frac{T}{M}\frac{d\mu}{dt} = \frac{T}{M}\frac{\partial \mu}{\partial (u/v)}\frac{\partial (u/v)}{\partial t} = \frac{\left(\frac{2+\alpha}{\alpha}\right)\left(\frac{u}{v}\right)^{4/\alpha}\left[1-2\Delta\left(\frac{u}{v}\right)^2\right]^{3/2}}{\Delta\left|1-\left(\frac{4+\varepsilon}{\alpha+2}\right)\left[1-\Delta\left(\frac{u}{v}\right)^2\right]\right|} \quad (8.238a)$$

$$\simeq \begin{cases} \left(\frac{2+\alpha}{\alpha}\right)\left|\frac{\alpha+2}{\alpha-2-\varepsilon\Delta}\right|\frac{1}{T^{|2/\alpha+1|}}\left|\frac{t}{T}-1\right|^{2/\alpha}, & \text{except for } \alpha \simeq 2+\varepsilon \quad (8.238b) \\ \left(\frac{4+\varepsilon}{2+\varepsilon}\right)\frac{1}{\Delta^2}\left|\frac{2}{\Delta^2}\left(\frac{t}{T}-1\right)\right|^{-\varepsilon/(4+2\varepsilon)}, & \text{for } \alpha = 2+\varepsilon. \quad (8.238c) \end{cases}$$

Note that for $\varepsilon = 0$, equation (8.238c) gives a square impulse response of height $2/\Delta^2$.

For leaky and guided modes the impulse response may be calculated for α-profiles by a similar procedure to that outlined above. In this case the appropriate result for μ is given by combining equation (8.227) for μ_1 and equation (8.229) for μ_2, so that we find:

$$\frac{T}{M}\frac{\partial \mu}{\partial t} = \frac{T}{M}\left[\frac{\partial \mu_1}{\partial (u/v)} + \frac{\partial \mu_2}{\partial (u/v)}\right]\frac{\partial (u/v)}{\partial t}$$

$$= \frac{F(u/v)\left[1-2\Delta\left(\frac{u}{v}\right)^2\right]^{3/2}}{\Delta\left|1-\left(\frac{4+\varepsilon}{\alpha+2}\right)\left[1-\Delta\left(\frac{u}{v}\right)^2\right]\right|\left\{2-\sqrt{\pi}\,\frac{\Gamma[(\alpha+2)/2]}{\Gamma[(\alpha+3)/2]}\right\}} \quad (8.239)$$

where

$$F(u/v) = \begin{cases} \left(\frac{u}{v}\right)^{4/\alpha}, & \text{for } u \leqslant v \quad (8.240a) \\ \left[\left(\frac{u}{v}\right)^{4/\alpha} - \left(\frac{\alpha+2}{2}\right)^{1/\alpha}\left(\frac{\alpha+2}{\alpha}\right)^{1/2}\left(\frac{u}{v}\right)^{2/\alpha}\left(1-\frac{v^2}{u^2}\right)^{1/2}\right], & \text{for } v \leqslant u \leqslant u_m \quad (8.240b) \end{cases}$$

where u_m is given by equation (8.228). The value of M used for normalization here is that appropriate to leaky and guided modes and has been obtained from the result (8.237). Note that equations (8.239) and (8.240a) together give the same result for guided modes as (8.238a) except for the different normalization factor M. As u/v varies from 0 to 1 for guided modes, the value of normalized time $(t/T-1)$ varies from 0 to

$$\Delta\left(\frac{\alpha-2-\varepsilon}{\alpha+2}\right)+\frac{\Delta^2}{2}\left(\frac{3\alpha-2-2\varepsilon}{\alpha+2}\right)+O(\Delta^3)$$

and we have already seen in Section 8.2.2 that this leads to an optimum given by equation (8.154) and a maximum pulse width of $T\Delta^2/8$. If we include all the leaky and guided modes, then u/v varies to its maximum value of $((\alpha+2)/2)^{1/2}$, and the value of $(t/T-1)$ becomes

$$\Delta\left(\frac{\alpha-2-\varepsilon}{2}\right)+\frac{\Delta^2}{8}(\alpha+2)(3\alpha-2-2\varepsilon)+O(\Delta^3)$$

which leads to a new optimum profile exponent of $\alpha=2+\varepsilon-4\Delta$ and a minimum pulse width of $T\Delta^2/2$ (Adams et al., 1975a), a result which was also found for helical rays only by Geckeler (1975). However, this treatment assumes all the leaky modes propagate unattenuated, which will not be the case in practice. Thus a true optimum value of α probably lies somewhere between the value given here and that of (8.154): the difference is rather small in any case for $\Delta \ll 1$.

For a given α-profile impulse responses for guided and unattenuated leaky modes may be calculated from (8.239) and (8.240), together with the time relationship (8.151). For the case of zero profile dispersion ($\varepsilon=0$) and near-parabolic profiles, these results may be simplified by the assumption:

$$\alpha = 2 - \kappa\Delta \tag{8.241}$$

where κ is a variable parameter and, as usual, $\Delta \ll 1$. The resultant impulse response is given by:

$$\frac{T}{M}\frac{d\mu}{dt} = \begin{cases} \dfrac{6(u/v)^2}{\Delta^2|4(u/v)^2-\kappa|}, & \text{for } u \leqslant v \tag{8.242a} \\[2ex] \dfrac{6[(u/v)^2-2(u^2/v^2-1)^{1/2}]}{\Delta^2|4(u/v)^2-\kappa|}, & \text{for } v \leqslant u \leqslant v\sqrt{2} \tag{8.242b} \end{cases}$$

where time t is given by:

$$\left(\frac{t}{T}-1\right)=\frac{\Delta^2}{4}\left(\frac{u}{v}\right)^2\left[2\left(\frac{u}{v}\right)^2-\kappa\right]. \tag{8.243}$$

Results are shown in Figure 8.23 for κ of 0, 2, and 4 (Adams et al., 1975a) corresponding, respectively, to the perfect parabolic profile, the optimized distribution for guided modes, and the optimized distribution for guided-plus-tunnelling modes. It is seen that, for $\kappa = 0$ and 2, leaky modes have the effect of adding a tail at the end of the pulse, since their transit times are all greater than that of the slowest bound mode. For the perfect parabolic profile, the r.m.s. width of the response is increased from $T\Delta^2/(4\sqrt{3}) = 0.144\Delta^2 T$ to $T\Delta^2\sqrt{(2963/33)}/35 = 0.271\Delta^2 T$ when all the leaky modes are present. For the new optimum condition $\kappa = 4$ the effect of leaky modes is to increase the amplitude of the response, rather than the overall width, since this choice of profile has the effect of adjusting the transit times of the leaky modes to lie within the normal range of those of the guided modes. Note that it is this effect which gives the unambiguous pulse dispersion shown by the broken line in Figure 8.23. The complete overlap of the transit times of both leaky and guided modes is only possible in fibres with close-to-parabolic index profiles.

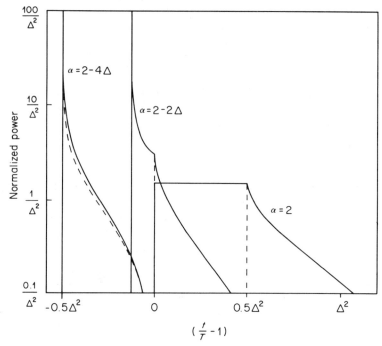

Figure 8.23 Impulse response for index profiles described by $\alpha = 2, 2 - 2\Delta$, and $2 - 4\Delta$. The solid lines show the pulse dispersion with both guided and leaky modes present (assumed unattenuated) and the broken lines show the effect of bound modes only. (From Adams *et al.* (1975a). Reproduced by permission of the IEE)

It should be emphasized that the results derived above ignore the attenuation of leaky modes, so that they represent a worse-case estimate. Impulse response calculations for near-parabolic profiles including leaky-mode attenuation have been presented by Ankiewicz and Pask (1978); attenuation reduces the pulse tails present in Figure 8.23 by varying amounts depending on the length of fibre, core radius, numerical aperture, and v-value. For a fibre with $v = 32.4$, $a = 25\ \mu\text{m}$, $\Delta = 0.01$, considering only leaky modes with loss below 1 dB/km, Gloge *et al.* (1979) have shown that the r.m.s. width of the impulse response is $0.15\Delta^2 T$ (see also the figures given in previous paragraph for r.m.s. widths). The effect of cladding loss on the impulse response has been investigated by Barrell and Pask (1978) who found that most of the leaky mode power is lost through radiation before cladding absorption becomes significant. These and other authors (Barrell and Pask, 1979; Jacomme, 1977; Jacomme and Rossier, 1977) have also calculated impulse responses for parabolic-index fibres excited by parallel beams rather than by Lambertian sources as treated here. Corresponding results for impulse responses of step-index fibres are also available in the literature (Albertin *et al.*, 1974; Ankiewicz and Pask, 1978; Barrell and Pask, 1978).

8.2.7 Leaky mode attenuation

Attenuation coefficients for leaky modes on graded-index fibres may be calculated from the WKB expressions for scalar field distributions. Following the theoretical formalism developed in Section 8.2.1 and using the appropriate connection formulae from Chapter 5, the forms of the leaky-mode fields in the core are given by (Petermann, 1975a; Olshansky,

1976a):

$$\psi(r) = \begin{cases} \dfrac{1}{2(rp_1)^{1/2}} \exp\left[-\int_r^{r_1} p_1 \, dr\right], & 0 \leqslant r < r_1 \quad (8.244\text{a}) \\[1em] \dfrac{1}{(rq_1)^{1/2}} \cos\left[\int_{r_1}^r q_1 \, dr - \dfrac{\pi}{4}\right], & r_1 < r < r_2 \quad (8.244\text{b}) \\[1em] \dfrac{1}{2(rp_1)^{1/2}} \left\{\sin \phi_1 \exp\left[-\int_{r_2}^r p_1 \, dr\right] + 2\cos \phi_1 \exp\left[\int_{r_2}^r p_1 \, dr\right]\right\}, & \\ & r_2 < r \leqslant a \quad (8.244\text{c}) \end{cases}$$

and in the cladding region by:

$$\psi(r) = \begin{cases} \dfrac{1}{2(rp_2)^{1/2}} \left\{\sin \phi_1 \exp(-\phi_2) \exp\left[\int_r^{r_3} p_2 \, dr\right] \\ + 2\cos \phi_1 \exp(\phi_2) \exp\left[-\int_r^{r_3} p_2 \, dr\right]\right\}, & a \leqslant r < r_3 \quad (8.244\text{d}) \\[1em] \dfrac{1}{2(rq_2)^{1/2}} \left\{4\cos \phi_1 \exp(\phi_2) \cos\left[\int_{r_3}^r q_2 \, dr - \dfrac{\pi}{4}\right] \\ - \sin \phi_1 \exp(-\phi_2) \sin\left[\int_{r_3}^r q_2 \, dr - \dfrac{\pi}{4}\right]\right\}, & r_3 < r \quad (8.244\text{e}) \end{cases}$$

where

$$\left. \begin{array}{l} q_1^2 = -p_1^2 = k^2 n^2(r) - \beta^2 - \dfrac{l^2}{r^2} \\[0.5em] q_2^2 = -p_2^2 = k^2 n_2^2 - \beta^2 - \dfrac{l^2}{r^2} \end{array} \right\} \quad (8.245)$$

$$\phi_1 = \int_{r_1}^{r_2} q_1 \, dr \quad (8.246)$$

$$\phi_2 = \int_{r_2}^{a} p_1 \, dr + \int_{a}^{r_3} p_2 \, dr. \quad (8.247)$$

The caustics r_1, r_2, r_3 are defined by the zeros of q_1 and q_2 as shown in Figure 8.19; at each of these caustics the appropriate connection formula is obeyed by the fields of (8.244). In addition the fields and their derivatives are continuous at the core–cladding boundary $r = a$.

For the field distributions of (8.244) to fully represent a leaky mode, the solution (8.244e) must represent an outgoing wave. This requirement leads to the eigenvalue equation (Olshansky, 1976a):

$$\cot \phi_1 \exp(2\phi_2) = i/4. \quad (8.248)$$

Using the fact that the real part of β is of order $n_2 k$, or about 10^7 m^{-1}, it is safe to assume $\beta_r \gg \beta_i$ for the leaky modes of interest, where $\beta = \beta_r + i\beta_i$. Expanding the l.h.s. of equation (8.248) in a Taylor series about β_r, we find from the lowest-order term the result:

$$\cot \phi_1 = 0$$

or

$$\phi_1 = (N + \tfrac{1}{2})\pi \quad (N = 0, 1, 2, \ldots) \quad (8.249)$$

which from (8.246) is seen to be an identical result to that for guided mode eigenvalues as in (8.131). The next term in the Taylor series gives the result for β_i, or equivalently, attenuation coefficient per unit length α_L, as (Petermann, 1975a; Olshansky, 1976a):

$$\alpha_L = -\frac{1}{2}\exp(-2\phi_2)\left[\frac{d\phi_1}{d\beta}\bigg|_{\beta=\beta_r}\right]^{-1}. \tag{8.250}$$

With the aid of the definition (8.246), this result may also be re-written in the form (Adams et al., 1976a; Snyder and Love, 1976):

$$\alpha_L = \frac{\exp(-2\phi_2)}{Z_p} \tag{8.251}$$

where

$$Z_p = 2\int_{r_1}^{r_2} \frac{\beta}{q_1}\,dr. \tag{8.252}$$

In this form, the expression for attenuation coefficient α_L may be interpreted as the ratio of a transmission coefficient $\exp(-2\phi_2)$ to a characteristic axial length Z_p. The transmission coefficient gives the probability of a photon at the central caustic r_2 emerging at the outer caustic r_3 by a process analogous to quantum-mechanical tunnelling (Stewart, 1975b). The axial length Z_p is recognized from equation (8.26) and the ensuing discussion as the distance along the fibre axis between successive inner caustics of the ray path. The result (8.251) may also be derived directly from ray optics with the inclusion of the concept of a tunnelling coefficient (Adams et al., 1976a, b; Snyder and Love, 1976; Love and Winkler, 1977). The tunnelling coefficient concept may also be extended to include the effects of absorbing claddings (Snyder and Love, 1976a). The tunnelling property of leaky modes has been used by Stewart (1975c) and others (Zemon and Fellows, 1976; Szczepanek and Berthold, 1978; Ankiewicz et al., 1979) to launch individual modes by a prism-coupling technique. Observation of the intensity distribution of leaky modes launched in this way provides useful confirmation of the WKB theory presented here.

For α-profiles the value of Z_p may be easily evaluated with the aid of the approximation (8.226); the result is (Adams et al., 1976b):

$$Z_p \simeq \frac{a^2\beta\pi}{2u}\left(\frac{2+\alpha}{2}\right)^{1/\alpha}\left(\frac{\alpha+2}{\alpha}\right)^{1/2}\left(\frac{u}{v}\right)^{2/\alpha}. \tag{8.253}$$

For parabolic profiles this reduces to the exact result $Z_p = a^2\beta\pi/v$ (Snyder and Love, 1976; Olshansky, 1976a; Love and Winkler, 1977), whilst it becomes less accurate as α becomes large. The evaluation of ϕ_2 as defined in (8.247) is in general not possible analytically and recourse must be made to numerical integration. For modes whose central caustic r_2 is close to the interface $r = a$, it may sometimes be a valid approximation to neglect the first term on the l.h.s. of (8.247) by comparison with the second term; the result is (Adams et al., 1976b):

$$\phi_2 \simeq -y - \frac{l}{2}\ln\left[\frac{u^2-v^2}{(l+y)^2}\right] \tag{8.254}$$

where $y^2 = l^2 + v^2 - u^2$. Another approximation for ϕ_2 in the case of high-loss leaky-modes has been suggested by Snyder and Love (1976).

For the parabolic index-profile the expression (8.247) for ϕ_2 may be evaluated exactly and gives (Adams et al., 1976a; Olshansky, 1976a; Love and Winkler, 1977):

$$\phi_2 = -\frac{y}{2} - \frac{l}{2}\ln\left[\frac{(u^2 - v^2)|2ly + 2l^2 - u^2|}{(l+y)^2}\right] - \frac{u^2}{4v}\ln|2vy + 2v^2 - u^2|$$
$$+ \frac{1}{4}\left(\frac{u^2}{2v} + l\right)\ln(u^4 - 4v^2l^2). \tag{8.255}$$

For the parabolic index-profile the real part β_r of the propagation constant may be evaluated from (8.249) (or, equivalently, (8.131) or (8.40)); the result, $u^2 = 2v(2N + l + 1)$, has already been given in equation (8.58). With the aid of this relation it is a simple matter to evaluate the attenuation coefficient α_L using the expression for Z_p given earlier with ϕ_2 from (8.255). Numerical results are shown in Figure 8.24 for a parabolic-profile fibre with $v = 50$, core radius 30 μm, and $\Delta = 0.01$. The loss is plotted versus radial mode number N, and lines have been drawn through points with the same value of principal mode number $M = 2N + l + 1$; for clarity only results corresponding to even values of M have been included on the figure. The functional behaviour of the attenuation coefficient may easily be understood in terms of the local plane wave components of Figure 8.19. Fixed values of M imply constant β_r, and hence as N increases the azimuthal mode number l decreases; from Figure 8.19 this means that the evanescent field region becomes narrower, the wave has less distance to tunnel, and hence the loss increases sharply.

For step-index fibres we have already noted that the approximation (8.253) for Z_p will not

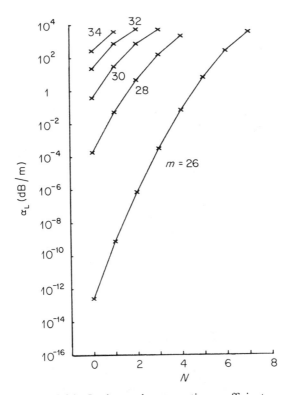

Figure 8.24 Leaky mode attenuation coefficient α_L for parabolic index-profile fibre with $v = 50$, core radius $a = 30$ μm, and $\Delta = 0.01$. Labelling parameter gives the value of principal mode number $m = 2N + l + 1$ where N = radial mode number and l = azimuthal mode number

be very accurate. Instead we may evaluate Z_p either directly from (8.252) with $r_1 = al/u$ and $r_2 = a$, or by geometrical considerations as in Section 7.1. The result obtained from either method is:

$$Z_p(\text{step}) = \frac{2a^2\beta(u^2 - l^2)^{1/2}}{u^2}. \tag{8.256}$$

A further amendment is necessary to the above theory when applied to calculation of leaky mode loss in step-index fibres. This is because of the occurrence of an index discontinuity at $r = a$, the core–cladding interface, which forms the central caustic r_2 for leaky modes on step-index fibres (Love and Winkler, 1978). The necessary correction is to multiply the WKB tunnelling coefficient $\exp(-2\phi_2)$ by a factor which is equal to the generalized Fresnel transmission coefficient at the interface (Love and Winkler, 1977); an analogous result for WKB tunnelling theory at index discontinuities in slab waveguides was derived in Section 5.2. The appropriate Fresnel transmission coefficient is given in equations (2.13) or (2.21) (which are the same for $n_1 \simeq n_2$) and may be expressed in the usual normalized variables via the established ray-mode notational connections. For this case ϕ_2 is given by (8.254), and using Z_p from (8.256) the step-index leaky-mode attenuation coefficient is given by:

$$\alpha_L(\text{step}) = \frac{2y}{a^2\beta}\left(\frac{u}{v}\right)^2 \exp\left[-2y - 2l\cosh^{-1}\left(\frac{l}{(u^2-v^2)^{1/2}}\right)\right]. \tag{8.257}$$

This result agrees with the asymptotic form of the attenuation coefficient found from the solution of the weakly-guiding eigenvalue equation for leaky modes (Snyder and Mitchell, 1974).

Once the attenuation coefficient α_L is known for a given mode denoted by (u, l), then the total power $S_L(z)$ in the leaky modes after a distance z may be calculated by summing the source distribution, appropriately weighted by the attenuation factor $\exp(-\alpha_L z)$, over all the leaky modes. For large values of v the summation may be replaced by a double integral:

$$S_L(z) = \sum_{N,l} J(u,l)e^{-\alpha_L z} \int_v^{u_m} du \int_{l_{min}}^{l_{max}} J(u,l)e^{-\alpha_L z} dl \tag{8.258}$$

where $J(u, l)$ is the source intensity distribution. Restricting attention to α-profiles, u_m is given by (8.228), l_{min} is given by (8.221), and l_{max} is given by (8.224). For $z = 0$ all leaky modes are present unattenuated in (8.258) and for a Lambertian source the integrals may be converted from functions of u, l to ray parameters γ, ϕ, r as in the argument leading to (8.236) and (8.237). For $z \to \infty$ all the leaky modes are strongly attenuated and $S_L(z) \to 0$ in (8.258). Clearly the least leaky modes will contribute most to the integrals in the general case, and for these modes we may expand the exponential factor as $e^{-\alpha_L z} \simeq 1 - \alpha_L z$. This observation leads naturally to the suggestion (Adams et al., 1976a) that the integrals in (8.258) be truncated at values corresponding to $\alpha_L z = 1$. That is, modes for which $\alpha_L z < 1$ might be assumed to propagate unattenuated whilst those with $\alpha_L z > 1$ are totally attenuated (Pask, 1978b). Using the approximations (8.253) for Z_p and (8.254) for ϕ_2, it follows from (8.251) that the condition $\alpha_L z = 1$ may be written in the form:

$$\frac{2vz}{a^2\beta\pi}\left(\frac{2}{2+\alpha}\right)^{1/\alpha}\left(\frac{\alpha}{\alpha+2}\right)^{1/2}\left(\frac{u}{v}\right)^{(\alpha-2)/\alpha} = \exp\left[2y + 2l\cosh^{-1}\left(\frac{l}{(u^2-v^2)^{1/2}}\right)\right]. \tag{8.259}$$

We note that the l.h.s. of this equation contains the dependence on length z and profile exponent α, together with a weak dependence on waveguide parameters for α close to 2, whilst the r.h.s. contains a rather strong dependence on waveguide parameters. Hence it seems reasonable, for profiles not too far from parabolic, to define a normalization parameter $D(\alpha)$ as follows (Adams et al., 1976b):

$$D(\alpha) = \frac{1}{v}\left\{\ln\left(\frac{(2\Delta)^{1/2}}{\pi}\right) + \ln\left[2\left(\frac{2}{2+\alpha}\right)^{1/\alpha}\left(\frac{\alpha}{\alpha+2}\right)^{1/2}\right] + \ln\left(\frac{z}{a}\right)\right\}. \qquad (8.260)$$

Hence for $\alpha \simeq 2$ and $\beta \simeq kn_1$, equation (8.259) may be re-written as

$$\frac{D(\alpha)}{2} \simeq \left[\left(\frac{l}{v}\right)^2 + b\right]^{1/2} + \left(\frac{l}{v}\right)\cosh^{-1}\left(\frac{(l/v)}{(-b)^{1/2}}\right). \qquad (8.261)$$

For each value of $D(\alpha)$ equation (8.261) defines a line in the $(l/v)^2$, b plane separating leaky modes which suffer little attenuation from those which are heavily attenuated; numerical results are given in Figure 8.25. The figure is drawn for a parabolic-index fibre and shows the domains occupied by guided and leaky modes in the $(l/v)^2$, b plane. The upper limit for each b is given by the line $l = l_{max}$ where l_{max} for $\alpha = 2$ from (8.224) is given by $v(1-b)/2$. For guided modes the lower limit for each b is $l = 0$, whilst that for leaky modes is $l = l_{min} = v(-b)^{1/2}$. The line for $D = 0.1$ is calculated from (8.261) and has the property that modes lying to the left of this line are strongly attenuated, whilst those to the right of the line suffer only very low losses. Lower values of D shift the demarcation line to the left, this giving more low-loss leaky modes; higher values of D rotate the line further to the right thus giving fewer low-loss leaky modes. Note that the lines of constant D are independent of the profile exponent α; as α varies

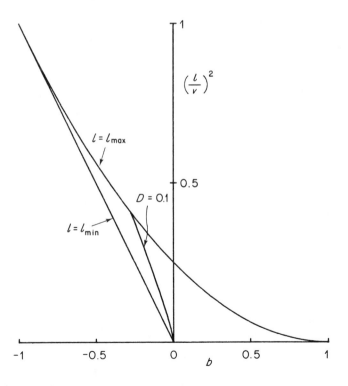

Figure 8.25 Tunnelling and guided mode domains for parabolic-index fibres. The line $D = 0.1$ separates low-loss from high-loss leaky modes, according to equation (8.261)

the upper limit $l = l_{max}$ varies according to (8.224). For $\alpha = 2$ the normalization parameter given by (8.260) becomes

$$D(2) = \frac{1}{v}\ln\left(\frac{(2\Delta)^{1/2}z}{\pi a}\right). \tag{8.262}$$

Similar considerations based on the step-index attenuation coefficient (8.257) lead to a normalization parameter $D(\text{step})$ given by:

$$D(\text{step}) = \frac{1}{v}\ln\left(\frac{(2\Delta)^{1/2}2z}{a}\right). \tag{8.263}$$

The normalization parameters $D(2)$ and $D(\text{step})$ given in (8.262) and (8.263) were first discovered by Love and Pask (1976). Pask (1978b) has also derived normalization parameters $D(\alpha)$ of a slightly different form from those in equation (8.260); the difference stems from a slightly different approximation for the analytic form of periodic distance Z_p.

Some idea of the overall power attenuation for leaky modes may be obtained from Figure 8.26 which plots $S_L(z)/S_L(0)$ versus length z for a parabolic index fibre with $v = 50$, core radius $a = 30\,\mu\text{m}$, and $\Delta = 0.01$. The solid line was calculated using the modal sum in equation (8.258) and assuming a Lambertian source (all modes excited equally). It is thought that the ripples in the line are associated with the relatively small number of leaky modes included in the sum. Attenuation coefficients were calculated from (8.251) using ϕ_2 from (8.255). It is seen that the value of $S_L(z)/S_L(0)$ falls from 0.31 at $z = 1\,\text{m}$ to 0.16 at $z = 1\,\text{km}$. Expressed as percentages of the total guided mode power S_G the leaky mode contributions for this fibre

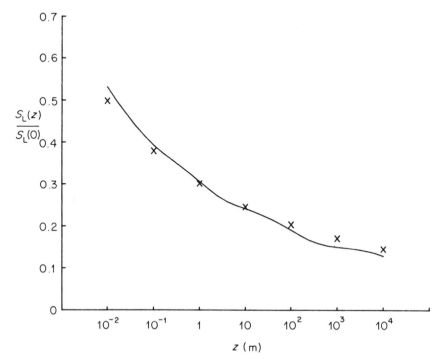

Figure 8.26 Ratio of leaky mode power $S_L(z)$ after length z to the initial power $S_L(0)$ for a parabolic index fibre with $v = 50$, $a = 30\,\mu\text{m}$, and $\Delta = 0.01$. The solid line is calculated from the modal sum in equation (8.258), whilst the crosses give the result of using the truncation procedure discussed in the text

would be 10 per cent after 1 m and 5 per cent after 1 km (as compared with an initial launched value of 33.3 per cent). If an attenuation measurement were carried out on a 1 km length of such a fibre using a 1 m reference length, the leaky modes would account for an excess loss of about 0.2 dB/km. This is the excess loss over that which would be measured if only guided modes were excited. The crosses on Figure 8.26 are the result of using the normalized parameter $D(2)$ from (8.262) and truncating the integrals in (8.258) so as to include only values of u, l lying in the low-loss domain defined by (8.261). The results are in good agreement with those from the modal sum using the full expression for attenuation coefficient. Some of the discrepancy must be attributed to the difference between the use of the modal sum and the double integral in (8.258). Universal power attenuation curves for parabolic-profile and step index fibres have been given by Love and Pask (1976), using the normalized parameters of (8.262) and (8.263).

For fibres with random diameter perturbations whose longitudinal spatial frequencies are below the range required for mode coupling, an excess loss may be produced as a consequence of guided modes being converted to leaky modes. Using the WKB approach presented above for the calculation of α_L, Olshansky and Nolan (1976, 1977) have calculated the magnitude of this excess loss for both parabolic and step-index fibres. The total losses are found to be essentially the same for these profiles and to be approximately independent of wavelength, although strongly dependent on the r.m.s. diameter variation assumed.

In the same way that the overall power attenuation varies with fibre length, there is an associated change of the fibre near-field intensity distribution $P(r)$. This may be calculated in the same way as that for unattenuated rays as in (8.216), except that now we must include the attenuation factor $\exp(-\alpha_L z)$. The measurement of near-field intensity distribution as a function of radius has been used as a method of determining the refractive-index profile of a fibre (Sladen et al., 1976). If no leaky modes were present, then the near-field distribution would be directly proportional to the index profile as in (8.217). On the other hand, if all leaky modes propagated unattenuated a correction of the form $[1-(r/a)^2]^{-1/2}$ would have to be applied at radius r, according to equation (8.218). Since, in reality, the attenuation of the leaky modes renders the near-field length-dependent then a correction factor $C(r,z)$ is needed after length z:

$$\frac{n^2(r)-n_2^2}{n^2(0)-n_2^2} = \frac{P(r)}{P(0)}\frac{1}{C(r,z)}. \tag{8.264}$$

For a Lambertian source the correction factor may be calculated as (Sladen et al., 1976):

$$C(r,z) = \frac{4}{\pi(n^2(r)-n_2^2)} \int_0^{\pi/2} d\phi \int_0^{\pi/2} \sin I \cos I \exp(-\alpha_L z) dI \tag{8.265}$$

where α_L is given by (8.250) and is a function of ϕ, I, r (related to modal parameters u, l via equations (8.212) and (8.213)). In evaluating the correction factors $C(r,z)$, Adams et al. (1976a, b) have defined a new normalization parameter X by retaining only the third term of equation (8.260) for the general normalization:

$$X = \frac{1}{v}\ln\left(\frac{z}{a}\right) \tag{8.266}$$

which is generally valid for fibres with $z/a > 10^3$.

The numerical evaluation of $C(r,z)$ for a wide range of index-profiles has shown that the correction factors are very similar up to a normalized radius ~ 0.8, after which a divergence of ± 6 per cent may occur. Figures 8.27 and 8.28 show computed correction factors for $\alpha = 2$ and 3, respectively, for X-values from 0.05 to 0.5. In practice the curves for $\alpha = 3$ are a good

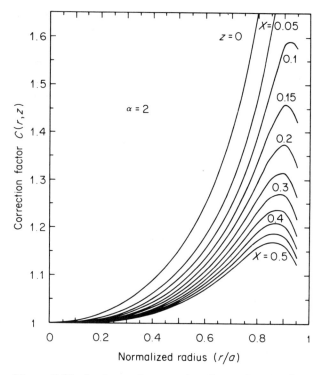

Figure 8.27 Leaky mode correction factor $C(r, z)$ calculated from equation (8.265) for parabolic index profiles using the normalization parameter X defined in (8.266). The line for $z = 0$ is given from (8.218) as $[1 - (r/a)^2]^{-1/2}$. (From Adams et al. (1976a). Reproduced by permission of the IEE)

median value for the range of profiles normally encountered. An error of 8 per cent at a normalized radius of 0.9 would result if these curves were used to correct an $\alpha = 1.5$ profile for a typical fibre having $X = 0.5$. This represents a deviation of about 1 per cent of the index difference at the core centre, an insignificant error. Numerical values for $C(r,z)$ corresponding to Figures 8.27 and 8.28 are given in Tables 8.2 and 8.3. The correction factors are required not only for the near-field scanning method but also for its inverse, i.e. a method for profile determination where a small spot of light is traversed across the end-face and the total transmitted power measured as a function of position (Arnaud and Derosier, 1976; Sumner, 1977).

As first pointed out by Petermann (1977, 1977a), a small ellipticity of the core may cause the leaky modes to carry a smaller amount of power in an optical fibre with a near-parabolic index profile than in the corresponding circular fibre. Ankiewicz (1979, 1979a) has shown that this is because the caustics of the leaky modes in a general class of near-parabolic index profiles are confocal ellipses. The focal points approach infinity when the profile approaches a parabolic shape and as leaky modes can no longer be contained in the core they become refracting modes. For the perfectly parabolic profile in an elliptical fibre, all rays which are not bound will eventually become refracting (Barrell and Pask, 1979a) and this fact enables the ray attenuation and hence the modified correction factors to be calculated by geometric optics (Barrell and Pask, 1980). For this profile the correction factors are very sensitive to fibre noncircularity, although this sensitivity decreases rapidly as the power-law exponent deviates from two. For a generalized class of noncircular graded-index profiles the leaky-

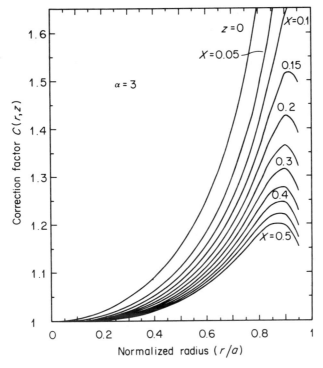

Figure 8.28 As in Figure 8.27 but for α-profiles with α = 3. (From Adams *et al.* (1976a). Reproduced by permission of the IEE)

mode attenuation coefficient may be derived either from ray optics or the WKB method (Ramskov Hansen *et al.*, 1980). Correction factors calculated for these profiles (Ramskov Hansen *et al.*, 1980a) have an azimuthal dependence which is in qualitative agreement with near-field intensity distributions measured along major and minor axes of some fibres.

8.3 Variational Methods

We are concerned here with extending the variational methods already discussed in Section 5.3, to deal with the analysis of graded-index optical fibres. First we discuss the fairly conventional approach to calculating eigenvalues and modal dispersion by this method and then the newer method of Okoshi and Okamoto (1974) will be described. The latter is of especial interest since for α-profiles it permits a number of useful analytic approximations to be derived (Okamoto and Okoshi, 1976; Okamoto *et al.*, 1979; Krumbholz *et al.*, 1980); these results differ from those of the WKB theory in that cladding effects are included. In addition the method leads to computer-aided synthesis of optimal profiles for multimode fibres (Okamoto and Okoshi, 1977) and to a finite element analysis which is applicable to both multimode and single-mode fibres (Okamoto and Okoshi, 1978; Okamoto, 1979).

8.3.1 The conventional approach

The general form of the variational principle for eigenvalues has been given in equation (5.110) for two-dimensional graded-index waveguides, and extended to the case of three-dimensional guides in Cartesian coordinates in equation (6.139). In terms of an arbitrary

Table 8.2 Leaky-mode correction factors for fibres with a parabolic index profile

X: r/a	0.05	0.10	0.15	0.20	0.25	0.30	0.35	0.40	0.45	0.50	Z = 0
0.00	1.00000	1.00000	1.00000	1.00000	1.00000	1.00000	1.00000	1.00000	1.00000	1.00000	1.00000
0.05	1.00028	1.00012	1.00006	1.00003	1.00002	1.00002	1.00002	1.00002	1.00002	1.00002	1.00125
0.10	1.00001	1.00123	1.00082	1.00057	1.00041	1.00030	1.00022	1.00016	1.00012	1.00010	1.00504
0.15	1.00571	1.00397	1.00294	1.00225	1.00177	1.00141	1.00114	1.00094	1.00078	1.00065	1.01144
0.20	1.01164	1.00865	1.00677	1.00546	1.00449	1.00375	1.00316	1.00270	1.00231	1.00199	1.02062
0.25	1.02001	1.01550	1.01258	1.01047	1.00887	1.00760	1.00658	1.00574	1.00504	1.00445	1.03280
0.30	1.03109	1.02479	1.02061	1.01752	1.01512	1.01320	1.01162	1.01031	1.00919	1.00824	1.04829
0.35	1.04521	1.03681	1.03112	1.02687	1.02352	1.02081	1.01854	1.01663	1.01500	1.01359	1.06752
0.40	1.06283	1.05194	1.04448	1.03883	1.03434	1.03066	1.02758	1.02494	1.02268	1.02071	1.09109
0.45	1.08447	1.07066	1.06109	1.05378	1.04793	1.04308	1.03901	1.03553	1.03249	1.02983	1.11979
0.50	1.11091	1.09361	1.08150	1.07220	1.06470	1.05849	1.05319	1.04866	1.04471	1.04122	1.15470
0.55	1.14325	1.12165	1.10642	1.09468	1.08518	1.07729	1.07052	1.06470	1.05964	1.05516	1.19737
0.60	1.18294	1.15590	1.13680	1.12202	1.11002	1.10002	1.09147	1.08406	1.07761	1.07191	1.25000
0.65	1.23202	1.19795	1.17386	1.15517	1.13998	1.12730	1.11649	1.10709	1.09890	1.09167	1.31590
0.70	1.29353	1.24999	1.21918	1.19538	1.17587	1.15967	1.14589	1.13393	1.12351	1.11433	1.40028
0.75	1.37208	1.31498	1.27459	1.24333	1.21803	1.19699	1.17915	1.16377	1.15042	1.13870	1.51186
0.80	1.47485	1.39643	1.34106	1.29846	1.26427	1.23614	1.21254	1.19253	1.17532	1.16040	1.66667
0.85	1.61211	1.49428	1.41199	1.35089	1.30381	1.26652	1.23617	1.21106	1.19008	1.17217	1.89832
0.90	1.78461	1.58300	1.45970	1.37659	1.31665	1.27139	1.23595	1.20709	1.18386	1.16466	2.29416
0.95	1.89390	1.57788	1.42223	1.32820	1.26493	1.21858	1.18377	1.15807	1.13856	1.12306	3.20256
1.00	1.00000	1.00000	1.00000	1.00000	1.00000	1.00000	1.00000	1.00000	1.00000	1.00000	2.16780

Table 8.3 Leaky-mode correction factors for fibres with an index-exponent of 3

X: r/a	0.05	0.10	0.15	0.20	0.25	0.30	0.35	0.40	0.45	0.50	$Z = 0$
0.00	1.00000	1.00000	1.00000	1.00000	1.00000	1.00000	1.00000	1.00000	1.00000	1.00000	1.00000
0.05	1.00027	1.00012	1.00006	1.00003	1.00002	1.00002	1.00002	1.00002	1.00002	1.00002	1.00125
0.10	1.00199	1.00121	1.00080	1.00055	1.00040	1.00029	1.00021	1.00015	1.00011	1.00010	1.00504
0.15	1.00569	1.00394	1.00290	1.00222	1.00174	1.00138	1.00111	1.00091	1.00076	1.00063	1.01144
0.20	1.01160	1.00862	1.00674	1.00542	1.00445	1.00371	1.00312	1.00265	1.00227	1.00196	1.02062
0.25	1.02000	1.01547	1.01255	1.01045	1.00883	1.00757	1.00655	1.00570	1.00501	1.00442	1.03280
0.30	1.03119	1.02485	1.02065	1.01754	1.01516	1.01323	1.01164	1.01033	1.00921	1.00825	1.04828
0.35	1.04529	1.03694	1.03123	1.02702	1.02365	1.02096	1.01867	1.01675	1.01513	1.01372	1.06752
0.40	1.06276	1.05211	1.04478	1.03916	1.03467	1.03100	1.02792	1.02527	1.02301	1.02104	1.09109
0.45	1.08460	1.07110	1.06171	1.05434	1.04859	1.04374	1.03967	1.03618	1.03315	1.03049	1.11979
0.50	1.11161	1.09453	1.08248	1.07318	1.06579	1.05963	1.05433	1.04980	1.04588	1.04239	1.15470
0.55	1.14460	1.12316	1.10785	1.09636	1.08689	1.07914	1.07239	1.06657	1.06155	1.05709	1.19737
0.60	1.18499	1.15817	1.13893	1.12468	1.11268	1.10290	1.09442	1.08701	1.08062	1.07496	1.25000
0.65	1.23495	1.20126	1.17719	1.15926	1.14415	1.13174	1.12107	1.11170	1.10359	1.09642	1.31590
0.70	1.29781	1.25491	1.22451	1.20159	1.18244	1.16660	1.15305	1.14116	1.13085	1.12174	1.40028
0.75	1.37856	1.32276	1.28332	1.25346	1.22868	1.20821	1.19070	1.17543	1.16222	1.15057	1.51186
0.80	1.48498	1.40979	1.35639	1.31595	1.28281	1.25559	1.23231	1.21230	1.19510	1.17997	1.66667
0.85	1.62891	1.52011	1.44271	1.38356	1.33701	1.29924	1.26809	1.24185	1.21966	1.20065	1.89832
0.90	1.82909	1.63848	1.51478	1.42848	1.36607	1.31713	1.27867	1.24754	1.22146	1.19956	2.29416
0.95	1.99385	1.66529	1.49662	1.39197	1.32189	1.27086	1.23034	1.19844	1.17385	1.15447	3.20256
1.00	1.00000	1.00000	1.00000	1.00000	1.00000	1.00000	1.00000	1.00000	1.00000	1.00000	2.16780

coordinate system transverse to the axial direction z, the general result is (Morse and Feshbach, 1953; Matsuhara, 1973):

$$\beta^2 = \frac{\int_A [k^2 n^2(r)\psi^2 - (\nabla_t \psi)^2] \, dA}{\int_A \psi^2 \, dA} \tag{8.267}$$

where the subscript t indicates the transverse part of the operator, and the integrations are carried out over the entire area A of the waveguide; $n(r)$ is the refractive index distribution, assumed to be only a function of radial position r. Equation (8.267) may be shown to be a stationary expression for β^2 by a similar procedure to that used in Chapter 5 (Morse and Feshbach, 1953). It follows that the generalized form of equation (5.123) for the relationship between phase velocity V_p and group velocity V_g in a dispersive medium is given by:

$$\frac{c^2}{V_p V_g} = \frac{\int_A n(r) m(r) \psi^2 \, dA}{\int_A \psi^2 \, dA} \tag{8.268a}$$

$$= \frac{\int_0^\infty n(r) m(r) \psi^2(r) r \, dr}{\int_0^\infty \psi^2(r) r \, dr} \tag{8.268b}$$

where $m(r)$ is the group index, and the form (8.268b) applies specifically to fibres of circular symmetry.

For fibres with a parabolic index-profile in the core and a uniform cladding, Geshiro et al. (1978a) have used a trial function $\psi(r)$ based on the Laguerre–Gaussian form (8.100) found for infinitely-extended parabolic-index media:

$$\psi(r) = \left(\frac{2(N!)}{(l+N)!}\right)^{1/2} \frac{1}{\zeta} \left(\frac{r}{\zeta}\right)^l \exp\left(-\frac{r^2}{2\zeta^2}\right) L_N^l \left(\frac{r^2}{\zeta^2}\right) \tag{8.269}$$

where L_N^l is the Laguerre polynomial, and the factors independent of r ensure correct normalization of ψ. The quantity ζ appearing in (8.269) is defined as:

$$\zeta^2 = \frac{1}{k n_1 s^{1/2}} \tag{8.270}$$

where s is a parameter to be determined from the variational principle. Geshiro et al. (1978a) confine attention to index profiles of the form:

$$n^2(r) = \begin{cases} n_1^2 [1 - 2\rho\Delta(r/a)^2], & 0 \leq r \leq a \\ n_1^2 [1 - 2\Delta] \equiv n_2^2, & r > a. \end{cases} \tag{8.271}$$

These are cladded parabolic profiles with a discontinuity at the interface $r = a$, the sign of which depends on whether ρ is less than or greater than unity; the profiles are indicated schematically in Figure 8.29. For profiles given by (8.271), substituting (8.269) into (8.267) yields:

$$\beta^2 = \hat{\beta}^2 + k^2 n_1^2 \int_0^\infty h(r) \psi^2(r) r \, dr \tag{8.272}$$

where

$$h(r) = \begin{cases} (s - 2\rho\Delta/a^2) r^2, & 0 \leq r \leq a \\ (sr^2 - 2\Delta), & r > a \end{cases} \tag{8.273}$$

and

$$\hat{\beta}^2 = k^2 n_1^2 - \frac{2(2N + l + 1)}{\zeta^2}. \tag{8.274}$$

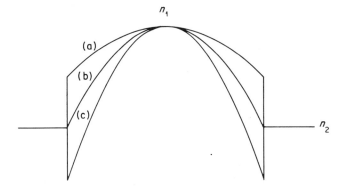

Figure 8.29 Refractive index profile for the cladded-parabolic profile as defined in equation (8.271): (a) $\rho < 1$; (b) $\rho = 1$; (c) $\rho > 1$. (From Adams (1978). Reproduced by permission of Chapman and Hall)

The value of the parameter s which satisfies the stationary property is obtained from the condition:

$$\frac{d\beta^2}{ds} = 0. \tag{8.275}$$

The solution of this equation gives the corresponding result for the eigenvalue β which may then be used to determine modal dispersion properties in the usual way.

As the parameter ρ varies in the index profile defined by (8.271), then the number of guided modes changes. This effect leads to a variation of the intermodal pulse dispersion with ρ which has been investigated in detail by Geshiro et al. (1978a). In general a very few modes close to cut-off for a cladded parabolic profile will have group velocities which are strongly influenced by the refractive index of the cladding since a large proportion of their modal power distributions lies in the cladding. Hence we may understand the effect of varying ρ, at least qualitatively, by noting that an optimum value of ρ might occur for which the velocity of these near-cut-off modes lies within the range of velocities of the other guided modes rather than outside this range. Geshiro et al. (1978a) have confirmed numerically that this is in fact the case and that the optimum value of ρ depends also on the normalized frequency v. In a practical multimode fibre, however, it is well known that the modes close to cut-off are strongly attenuated due to cladding absorption and also to bending losses. Hence if these losses are taken into account, the effect of the parameter ρ is somewhat reduced and is unlikely to play a major role in the optimization of intermodal pulse dispersion except for the case of fibres which support very few modes.

Another variational analysis of parabolic index-profile fibres has been performed by Yamada et al. (1977) who formulated the variational principle in such a way as to retain the full vector nature of the fields. The results for eigenvalues of the HE_{11} and HE_{12} modes were found to be in good agreement (errors less than a few parts in a thousand) with those from a zeroth-order perturbation approach (Yamada and Inabe, 1974) which is itself somewhat similar to the Kurtz–Streifer approach outlined in Section 8.1. Heyke and Kuhn (1973) have used the variational principle to examine the effect of higher-order terms in a power-series expansion of the index profile, using similar trial functions to those in (8.269). These authors have also considered the effects of lossy outer cladding layers in step-index fibres using the variational approach. In addition they have drawn attention to the advantages of using different trial field distributions for different frequency ranges of interest (cf. the remarks concerning the perturbation approach for cladded-parabolic profiles in Section 5.1.4 and Figure 5.2). Clearly a trial function which is accurate near the guide axis results in a good approximation far from cut-off, whereas small errors near cut-off require a good trial function in the cladding region. Hotate and Okoshi (1978) have used the Bessel function fields of the step-index fibre as trial functions in developing a variational analysis of approximate LP mode cut-offs in fibres of arbitrary index profile.

8.3.2 An alternative approach

For this method, due to Okoshi and Okamoto (1974), we commence with the Kurtz–Streifer equation (8.92) for the 'transverse field function' $\Phi_j(j = 1, 2)$. Writing $R(r)$ for $\Phi_1(r)$ and $\Phi_2(r)$ in the core-region of a cladded graded-index waveguide, with profile function $f(r)$ as in (8.1), then equation (8.92) may be written as:

$$\frac{d}{dr}\left(r\frac{dR(r)}{dr}\right) + \left[\frac{v^2}{a^2}(1-b-f(r)) - \frac{l^2}{r^2}\right]rR(r) = 0 \qquad (8.276)$$

where

$$l = \begin{cases} v+1 & \text{for } j = 1 \\ v-1 & \text{for } j = 2. \end{cases}$$

For a uniform cladding, the appropriate solution of the Kurtz–Streifer equation will be the modified Bessel function $K_l(vb^{1/2}r/a)$ as discussed in Chapter 7:

$$[\Phi_j(r)]_{\text{cladding}} = R(a)\frac{K_l(vb^{1/2}r/a)}{K_l(vb^{1/2})} \qquad (r \geqslant a, j = 1, 2). \qquad (8.277)$$

The boundary condition at $r = a$ in the weakly-guiding approximation is thus given by:

$$\left[\frac{1}{R(r)}\frac{dR(r)}{dr}\right]_{r=a} = \phi_\beta \qquad (8.278)$$

where

$$a\phi_\beta = \begin{cases} -l - vb^{1/2}\dfrac{K_{l-1}(vb^{1/2})}{K_l(vb^{1/2})}, & \text{for } j = 1 \ (l \neq 0) & (8.279a) \\[2mm] l - vb^{1/2}\dfrac{K_{l+1}(vb^{1/2})}{K_l(vb^{1/2})}, & \text{for } j = 2 \ (l \neq 0) & (8.279b) \\[2mm] -vb^{1/2}\dfrac{K_1(vb^{1/2})}{K_0(vb^{1/2})}, & \text{for } l = 0. & (8.279c) \end{cases}$$

The stationary expression which corresponds to the solution of (8.276) subject to the boundary condition (8.278) is (Okoshi and Okamoto, 1974):

$$J(R) = R^2(a)a\phi_\beta - \int_0^a \left[\frac{dR(r)}{dr}\right]^2 r\,dr + \int_0^a \left[\frac{v^2}{a^2}(1-b-f(r)) - \frac{l^2}{r^2}\right]R^2(r)r\,dr. \qquad (8.280)$$

This functional is the three-dimensional analogue of that given in (5.130) for two-dimensional graded-index waveguides. The proof that (8.280) is a stationary expression proceeds in the same way as that given for (5.130), i.e. we put $R(r) = R_0(r) + \varepsilon\eta(r)$ where $\eta(r)$ is an arbitrary continuous function of r, ε denotes a small quantity, and $R_0(r)$ is assumed to be the appropriate function which minimizes $J(R)$. Putting this function into (8.280) and considering that $J(R)$ must be minimized for $\varepsilon = 0$, we find:

$$\left.\frac{dJ}{d\varepsilon}\right|_{\varepsilon=0} = 2R_0(a)a\phi_\beta\eta(a) - 2\int_0^a r\frac{dR_0}{dr}\frac{d\eta}{dr}dr$$

$$+ 2\int_0^a \left[\frac{v^2}{a^2}(1-b-f(r)) - \frac{l^2}{r^2}\right]rR_0(r)\eta(r)\,dr = 0. \qquad (8.281)$$

Integrating the second term on the r.h.s. of (8.281) by parts, we obtain:

$$\eta(a)aR_0(a)\left\{\phi_\beta - \left[\frac{1}{R_0}\frac{dR_0}{dr}\right]_{r=a}\right\}$$

$$+ \int_0^a \left\{\frac{d}{dr}\left(r\frac{dR_0}{dr}\right) + \left[\frac{v^2}{a^2}(1-b-f(r)) - \frac{l^2}{r^2}\right]rR_0\right\}\eta(r)\,dr = 0. \qquad (8.282)$$

Since $\eta(r)$ is an arbitrary function of r, this equation shows that $R_0(r)$ satisfies both the wave equation (8.276) and the boundary condition (8.278).

The stationary expression (8.280) may be solved by the Rayleigh–Ritz method by expressing $R(r)$ in terms of a set of orthogonal functions $F_{li}(r)$:

$$R(r) = \sum_{i=0}^{L} C_i F_{li}(r) \qquad (8.283)$$

$$F_{li}(r) = \frac{2^{1/2}}{a}\frac{J_l(\lambda_i r/a)}{J_l(\lambda_i)} \qquad (8.284)$$

where

$$\lambda_i = \begin{cases} j_{1,i-1}, & l = 0 \\ j_{l-1,i}, & l \neq 0 \end{cases} \qquad (8.285)$$

and $j_{n,i}$ denotes the ith root of $J_n(z) = 0$. The functions $F_{li}(r)$ defined in (8.284) satisfy the orthonormalizing condition

$$\int_0^\infty F_{li}(r)F_{lm}(r)r\,dr = \delta_{im} \qquad (8.286)$$

where δ_{im} is the Kronecker delta. The coefficients C_i in (8.283) are to be chosen so as to minimize the functional (8.280). Inserting (8.283) into (8.280) and using various Bessel function relations (Abramowitz and Stegun, 1964) yields:

$$J(R) = \frac{2}{a}\phi_\beta \sum_{i=1}^{L}\sum_{m=1}^{L} C_i C_m + \frac{v^2(1-b)}{a^2}\sum_{i=1}^{L}\sum_{m=1}^{L} C_i C_m \delta_{im} - \sum_{i=1}^{L}\sum_{m=1}^{L} C_i C_m \frac{(\lambda_1^2 \delta_{im} - 2l)}{a^2}$$

$$- \frac{v^2}{a^2}\sum_{i=1}^{L}\sum_{m=1}^{L} C_i C_m A_{im} \qquad (8.287)$$

where

$$A_{im} = \int_0^a f(r)F_{li}(r)F_{lm}(r)r\,dr. \qquad (8.288)$$

To minimize the functional $J(R)$ in (8.287) with respect to the parameters C_i, then we demand that for all i the following condition is satisfied:

$$\frac{\partial J}{\partial C_i} = \sum_{m=1}^{L} C_m S_{im} = 0 \qquad (8.289)$$

where

$$S_{im} = 2(a\phi_\beta + l) + [v^2(1-b) - \lambda_i^2]\delta_{im} - v^2 A_{im}. \qquad (8.290)$$

In order that a nontrivial solution of (8.289) exists for all i, it follows that:

$$\det(S_{im}) = 0. \qquad (8.291)$$

This equation is the eigenvalue equation which may be used to determine the normalized propagation constant b as a function of the normalized frequency v for arbitrary profile function $f(r)$ in A_{im}.

The details of the numerical routine for evaluation of the determinantal condition (8.291) need not be considered here, since we have already given a detailed account for the corresponding two-dimensional waveguide in Section 5.3. For step-index (uniform core) fibres this method yields the exact result for the weakly-guiding eigenvalue equation (7.91) when the limit $L \to \infty$ is taken in the summation of (8.283). For graded-index fibres, Okoshi and Okamoto have used (8.291) to calculate b as a function of v for a number of profiles, including the truncated parabolic profile. For numerical calculations of this type it was found that good agreement with other results reported in the literature was obtained for expansions up to $L = 10$ in (8.283).

In order to examine the dispersion characteristics of fibres, the group velocity–phase velocity relationship (8.268) may be used in conjunction with the Rayleigh–Ritz method discussed above.

For the profiles of (8.1), equation (8.268b) becomes:

$$\frac{c^2}{V_p V_g} = n_1 m_1 \left\{ 1 - 2\Delta\left(1 + \frac{\varepsilon}{4}\right) \frac{\int_0^\infty f(r)\psi^2(r)r\,dr}{\int_0^\infty \psi^2(r)r\,dr} \right\} \qquad (8.292)$$

where ε is as defined in (8.150). Using the expansion on the r.h.s. of (8.283) for $\psi(r)$ in the core and (8.277) for $\psi(r)$ in the cladding, equation (8.292) yields for modal transit time τ_n of the nth mode (Okamoto and Okoshi, 1977):

$$\tau_n = \frac{Lm_1}{c} \frac{[1 - 2\Delta(1 + \varepsilon/4)\Theta]}{[1 - 2\Delta(1-b)]^{1/2}} \qquad (8.293)$$

where

$$\Theta = \frac{\sum_{i=1}^{L} \sum_{m=1}^{L} C_i C_m [A_{im} + 1/\kappa_l(vb^{1/2}) - 1]}{\sum_{i=1}^{L} \sum_{m=1}^{L} C_i C_m [\delta_{im} + 1/\kappa_l(vb^{1/2}) - 1]} \qquad (8.294)$$

and, as in (7.107),

$$\kappa_l(vb^{1/2}) = \frac{K_l^2(vb^{1/2})}{K_{l-1}(vb^{1/2})K_{l+1}(vb^{1/2})}.$$

In evaluating the integrals in (8.292), tabulated integrals of the modified Bessel functions have been used as in (7.114).

Okamoto and Okoshi (1977) have applied this analysis to profile functions $f(r)$ of the form:

$$f(r) = \sum_{\mu=1}^{N} d_\mu \left(\frac{r}{a}\right)^{2\mu} \tag{8.295}$$

with the object of determining the optimum set of d_μ to minimize intermodal pulse-broadening. As a measure of the intermodal dispersion at frequencies where I modes propagate, they define:

$$Q(I) = \frac{\int_{v_{cl}}^{v_{cl}+1} \sigma^2(v)\,dv}{\int_{v_{cl}}^{v_{cl}+1} dv} \tag{8.296}$$

where σ is the r.m.s. deviation:

$$\sigma^2 = \frac{1}{I} \sum_{n=1}^{I} \left[\tau_n^2 - \left(\frac{1}{I}\sum_{n=1}^{I} \tau_n\right)^2 \right] \tag{8.297}$$

and τ_n is given by (8.293). The optimum profile is that which minimizes $Q(I)$ for a fibre transmitting I modes, and this is therefore found from the solution of the set of equations:

$$\frac{\partial Q(I)}{\partial d_\mu} = 0, \quad \mu = 1, 2, \ldots, N. \tag{8.298}$$

In a numerical study of profile optimization Okamoto and Okoshi have considered a total of $I = 10$ LP modes (corresponding to 34 degenerate modes) on a fibre with $N = 5$ in the profile

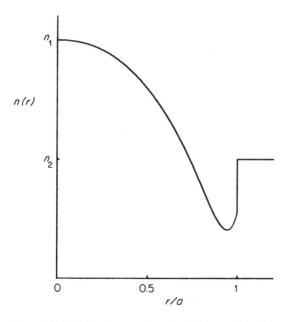

Figure 8.30 Optimum refractive index profile using a five-term polynomial for a fibre with 10 LP modes and profile dispersion $\varepsilon = 0.3$, as computed by Okamoto and Okoshi (1977). (Reproduced by permission of the IEEE)

expansion of (8.295). Using 3 sampling points between the v-limits on the integrals in (8.296), and a profile dispersion parameter $\varepsilon = 0.3$, they obtain for the optimum profile the coefficients (as quoted in Unger, 1977):

$$d_1 = 1.185 \quad d_4 = 13.418$$
$$d_2 = 3.396 \quad d_5 = -7.503.$$
$$d_3 = -9.050$$

This profile is illustrated in Figure 8.30 and shows a region close to the core–cladding interface where the index lies beneath the value in the cladding. This index depression has the effect, already noted, of ensuring that the group velocities of the modes close to cut-off lie within the range of the other guided modes' velocities, thus avoiding a large potential contribution of near-cut-off modes to the pulse-spreading. With the profile of Figure 8.30 the pulse-spreading is reduced to 30 ps/km for $\Delta = 0.01$. Recalling that in Section 8.2 an optimized α-profile was found to give an r.m.s. pulse-spreading of 15 ps/km for $\Delta = 0.01$ from equation (8.173), it would not appear that the present profile offers any advantages. However, it must be remembered that the WKB analysis leading to (8.173) does not include the effects of modes close to cut-off (since cladding effects are neglected), and that in fact if these modes were to be included the pulse-spreading would be greater than that given in (8.173). Once again we repeat the comment that modes close to cut-off will normally be heavily attenuated due to differential mode attenuation, and therefore calculations such as that above which take these modes into consideration may well overestimate their effect.

8.3.3 Results for alpha-profiles

An interesting and useful result for α-profiles based on the use of (8.292) and the wave equation (8.276) has been found by Krumbholz et al. (1980). With the usual definition of b, the l.h.s. of (8.292) may be re-written with the aid of (7.105) in the form:

$$\frac{c^2}{V_p V_g} = n_2 m_2 + \frac{1}{2}\left[b + \frac{d(bv)}{dv}\right](n_1 m_1 - n_2 m_2). \tag{8.299}$$

Assuming an α-profile as in (8.2) for the core ($0 \leqslant r \leqslant a$) and a uniform cladding of index n_2, then equation (8.292) becomes:

$$\frac{c^2}{V_p V_g} = n_1 m_1 \left\{1 - 2\Delta\left(1 + \frac{\varepsilon}{4}\right)\left[\frac{\int_0^a (r/a)^\alpha \psi^2(r)r\,dr}{\int_0^\infty \psi^2(r)r\,dr} + 1 - \Gamma\right]\right\} \tag{8.300}$$

where Γ is the ratio of power confined in the core to the total power. Furthermore, the definition of ε in (8.150) links $n_1 m_1$ and $n_2 m_2$ in the form:

$$n_1 m_1 - n_2 m_2 = n_1 m_1 2\Delta\left(1 + \frac{\varepsilon}{4}\right). \tag{8.301}$$

Combining equations (8.299), (8.300), and (8.301), we obtain (Krumbholz et al., 1980):

$$\Gamma = \frac{1}{2}\left[b + \frac{d(bv)}{dv}\right] + \frac{\int_0^a (r/a)^\alpha \psi^2(r)r\,dr}{\int_0^\infty \psi^2(r)r\,dr}. \tag{8.302}$$

To proceed further the integral in (8.302) must be eliminated. This can be achieved by assuming $\psi(r)$ is a solution of the wave equation (8.276) with $f = (r/a)^\alpha$; premultiplying (8.276) by $ra^2 d\psi/dr$ and integrating from 0 to a yields:

$$a^2 \int_0^a r \frac{d\psi}{dr} \frac{d}{dr}\left(r\frac{d\psi}{dr}\right) dr + v^2(1-b) \int_0^a r^2 \frac{d\psi}{dr} \psi \, dr - v^2 \int_0^a \left(\frac{r}{a}\right)^\alpha \psi \frac{d\psi}{dr} r^2 dr$$
$$- l^2 a^2 \int_0^a \psi \frac{d\psi}{dr} dr = 0.$$

Integrating by parts, we find:

$$a^4 \left(\frac{d\psi}{dr}\right)^2 \bigg|_{r=a} + v^2(1-b)\left[a^2\psi^2(a) - 2\int_0^a r\psi^2 dr\right] - v^2\left[a^2\psi^2(a) - (\alpha+2)\int_0^a \left(\frac{r}{a}\right)^\alpha \psi^2 r \, dr\right]$$
$$- l^2[\psi^2(a) - \psi^2(0)]a^2 = 0. \tag{8.303}$$

The final term $l^2\psi^2(0)$ vanishes since $\psi(0) = 0$ for $l > 0$. The derivative in the first term can be evaluated with the aid of the boundary condition (8.278), so that equation (8.303) may be rewritten in the form:

$$a^4 \phi_\beta^2 - v^2 ba^2 - l^2 a^2 - \frac{2v^2(1-b)}{\psi^2(a)} \int_0^a r\psi^2 dr + \frac{v^2(\alpha+2)}{\psi^2(a)} \int_0^a \left(\frac{r}{a}\right)^\alpha \psi^2 r \, dr = 0. \tag{8.304}$$

Noting, from (8.279) and (7.114), that:

$$a^2 \phi_\beta^2 = l^2 + \frac{v^2 b}{\kappa_l(vb^{1/2})} = l^2 + v^2 b\left[1 + \frac{2}{a^2\psi^2(a)} \int_a^\infty \psi^2 r \, dr\right],$$

then equation (8.304) becomes:

$$b \int_0^\infty \psi^2 r \, dr - \Gamma \int_0^\infty \psi^2 r \, dr + \frac{(\alpha+2)}{2} \int_0^a \left(\frac{r}{a}\right)^\alpha \psi^2 r \, dr = 0. \tag{8.305}$$

This last equation enables us to complete the analysis of (8.302). Combining (8.305) and (8.302) we obtain the result (Krumbholz et al., 1980):

$$\Gamma = \left(\frac{\alpha-2}{2\alpha}\right) b + \left(\frac{\alpha+2}{2\alpha}\right) \frac{d(vb)}{dv}. \tag{8.306}$$

Equation (8.306) is a general relation between Γ, b, and $d(bv)/dv$ for α-profiles. An equivalent result has been derived via the theorems of quantum mechanics by Pask (1979). Note that in the step-index limit $\alpha \to \infty$, (8.306) reduces to (7.116) which was derived from the LP solutions for step-index fibres. For the cladded-parabolic profile fibre ($\alpha = 2$), then (8.306) reduces to a particularly simple form:

$$\Gamma = \frac{d(vb)}{dv} \quad (\alpha = 2) \tag{8.307}$$

which says that the power confinement factor is equal to the normalized delay time for each mode.

Since equation (8.306) relates three important waveguide parameters, we may use it to investigate various propagation phenomena. For example, if it is combined with (8.299) and (8.301) we may find a simple expression relating modal delay time τ to the parameters b and Γ:

$$\tau = \left(\frac{Lm_1}{c}\right) \frac{\left\{1 - 2\Delta\left(1+\frac{\varepsilon}{4}\right)\left[\frac{2}{\alpha+2}(1-b) + \frac{\alpha}{\alpha+2}(1-\Gamma)\right]\right\}}{[1 - 2\Delta(1-b)]^{1/2}}. \tag{8.308}$$

Equation (8.308) reduces to the WKB result (8.151) when $\Gamma \to 1$; this is not surprising since the WKB approximation neglects cladding effects in calculating τ. In a sense equation (8.308) is more accurate than (8.151) since it includes the cladding effects in the final term on the r.h.s., although this term will usually be negligible for all but a few modes close to cut-off.

Okamoto et al. (1979) have applied the stationary expression (8.280) to α-profiles, including those with an index discontinuity at the core-cladding interface. By making the added assumption that the field in the core can be described approximately by a single Bessel function, i.e. only one of the $C_1, C_2, \ldots$ in an expansion like (8.283) is dominant and others can be neglected, these authors have derived a simple approximation for the eigenvalue equation. This derivation corrects an erroneous proof of the same relation given earlier by Okamoto and Okoshi (1976). The approximate relation is relatively accurate for high α-values, but the error increases appreciably for quadratic profiles ($\alpha = 2$). For α-profiles with no index discrepancy at $r = a$, the formula for cut-off frequency of the LP modes is (Okamoto et al., 1979):

$$v_c \simeq \left(\frac{\alpha+2}{\alpha}\right)^{1/2} j_{l-1,i} \qquad (8.309)$$

where $j_{l-1,i}$ denotes the ith root of $J_{l-1}(z) = 0$. For quadratic profile fibres the same authors have proposed a modified eigenvalue equation which gives somewhat better accuracy.

8.4 Perturbation Theory

In Section 5.1 we have discussed the application of first-order perturbation theory to finding eigenvalues and eigenfunctions of the scalar wave equation. In the present section we are concerned with extending the method to a few relevant problems in the analysis of graded-index fibres. If we restrict attention to circularly symmetric perturbations of the refractive index, and denote the general form of such perturbations by $\delta n^2(r)$, then the appropriate first-order results for use here are:

$$\beta'^2 = \beta_{Nl}^2 + k^2 \int_0^\infty \psi_{Nl}^2(r)\delta n^2(r) r \, dr \qquad (8.310)$$

$$\phi(r) = \psi_{Nl}(r) + \sum_{\substack{M \\ (M \neq N)}} \frac{\psi_{Ml}(r) \int_0^\infty \psi_{Ml}(r)\psi_{Nl}(r)\delta n^2(r) r \, dr}{(\beta_{Nl}^2 - \beta_{Ml}^2)} \qquad (8.311)$$

where β', $\phi(r)$ are the eigenvalue and eigenfunction of state (N, l) of the perturbed profile, and β_{Nl}, ψ_{Nl} are the corresponding quantities for the unperturbed profile $n(r)$. In writing equations (8.310) and (8.311), which are the cylindrical polar analogues of (5.39) and (5.40) in Cartesian coordinates, we have assumed the $\psi_{Nl}(r)$ are normalized to unity over the space $0 < r < \infty$. We will now apply these equations to three problems in graded-index fibre analysis:

(i) fourth-order polynomial profiles,
(ii) small departures from desired profiles, and
(iii) cladded α-profiles.

Perturbation theory has also been used to investigate the solutions of the vector wave equation viewed as a perturbation of the scalar equation (Steiner, 1973a; Matsuhara, 1973a; Yamada and Inabe, 1974a; Thyagarajan and Ghatak, 1974). Snyder and Young (1978) have given a treatment where the eigenvalues of a weakly-guiding fibre are expressed as perturbed forms of the solutions for the equivalent $n_1 = n_2$ waveguide. A recent approach (Ishikawa et al., 1978) takes account of the ∇n^2 term in the wave equation, the cladding, and material dispersion for an α-profile with an additional fourth-order term and an index discontinuity at the core–cladding interface. For a fibre with $v < 40$ it was found that the ultimate width of the impulse response is determined by the ∇n^2 term. The optimization of such a profile yields an intermodal pulse-broadening which is five times smaller than that of the conventional optimum α-profile.

8.4.1 Fourth-order polynomial profiles

Here we consider profiles which are in general of the form given in (8.1a), except that the range of r is unrestricted with the profile function $f(r)$ defined as:

$$f(r) = \left(\frac{r}{a}\right)^2 - K\left(\frac{r}{a}\right)^4 \quad \text{(all } r\text{)} \tag{8.312}$$

where K is a constant, independent of r.

Thus the unperturbed profile may be assumed as parabolic for all r, and the perturbation $\delta n^2(r)$ takes the form:

$$\delta n^2(r) = 2\Delta K \left(\frac{r}{a}\right)^4 n_1^2. \tag{8.313}$$

The eigenfunctions $\psi_{Nl}(r)$ of the unperturbed parabolic profile are given, with appropriate normalization, by the forms of (8.100):

$$\psi_{Nl}(r) = \left[\frac{2vN!}{(N+l)!}\right]^{1/2} \frac{x^{l/2}}{a} e^{-x/2} L_N^l(x) \tag{8.314}$$

where $x = v(r/a)^2$. The corresponding unperturbed eigenvalues are given by equation (8.101). Using (8.313) and (8.314), the perturbed eigenvalues are given from (8.310) as:

$$\beta'^2 - \beta_{Nl}^2 = n_1^2 k^2 \int_0^\infty \frac{N!}{(N+l)!} x^{l+2} e^{-x} \left[L_N^l(x)\right]^2 \frac{2\Delta K}{v^2} dx. \tag{8.315}$$

The integral on the r.h.s. of (8.315) may be evaluated with the aid of the recurrence relation and the orthogonality property of the generalized Laguerre polynomials (Abramowitz and Stegun, 1964):

$$xL_N^l(x) = (2N+l+1)L_N^l(x) - (N+1)L_{N+1}^l(x) - (N+l)L_{N-1}^l(x) \tag{8.316}$$

$$\int_0^\infty \frac{N!}{(N+l)!} L_N^l(x) L_M^l(x) x^l e^{-x} dx = \delta_{NM} \tag{8.317}$$

where δ_{NM} is the Kronecker delta. Hence the result for b, the normalized propagation constant of the perturbed system becomes (Kawakami and Nishizawa, 1968):

$$b = 1 - \frac{2(2N+l+1)}{v} + \frac{K}{v^2}\left[6N^2 + 6N(l+1) + (l+1)(l+2)\right]. \tag{8.318}$$

Higher-order perturbation corrections to this expression have been given by Jacobsen and Ramskov Hansen (1979). The normalized group dispersion follows from (8.318) in the form:

$$\frac{d(vb)}{dv} = 1 - \frac{K}{v^2}\left[6N^2 + 6N(l+1) + (l+1)(l+2)\right]. \tag{8.319}$$

The group delay τ of mode (N, l) in travelling a length L of fibre may be calculated from the generally applicable formulae (8.299) and (8.301) given in the previous section. Combining these equations we find:

$$\tau = \left(\frac{Lm_1}{c}\right) \frac{\left[1 - 2\Delta\left(1 + \frac{\varepsilon}{4}\right)\left(1 - \frac{b}{2} - \frac{1}{2}\frac{d(bv)}{dv}\right)\right]}{[1 - 2\Delta(1-b)]^{1/2}}. \qquad (8.320)$$

Substituting the results (8.318) and (8.319) into (8.320), and expanding in powers of Δ, yields, to order Δ^2:

$$\tau \simeq \left(\frac{Lm_1}{c}\right)\left\{1 - \frac{\varepsilon\Delta(2N+l+1)}{2v} + \frac{2\Delta^2}{v^2}\left[\left(1 - \frac{3K}{4\Delta} - \frac{\varepsilon}{2}\right)(2N+l+1)^2 + \frac{K}{4\Delta}(l^2 - 1)\right]\right\}. \qquad (8.321)$$

The delays given by equation (8.321) may be equalized in the case $\varepsilon = 0$ by choosing K such that (Kawakami and Nishizawa, 1968):

$$K = \frac{4\Delta(2N+l+1)^2}{[3(2N+l+1)^2 + 1 - l^2]}. \qquad (8.322)$$

For modes with $l = 0$ (corresponding to meridional rays) and $N \gg 1$ equation (8.322) reduces to $K = 4\Delta/3$. For modes with $N = 0$ and $l \gg 1$ (corresponding to helical rays) equation (8.322) reduces to $K = 2\Delta$. In general the total impulse response width may be calculated from (8.321) by allowing the mode numbers N, l to vary over their allowed ranges for guided modes, and taking the difference between the extreme values of τ. Arnaud (1975a) and Cook (1977) have shown that for the case $\varepsilon = 0$ the minimum total impulse response width occurs for $K = 4\Delta/3$ and is equal to $Lm_1\Delta^2/(6c)$. When unattenuated leaky modes are included and it is assumed that all modes are excited equally, Arnaud (1975) has shown that the minimum pulse width occurs for $K = 1.82\Delta$. The impulse response shapes have been calculated by Nemoto and Yip (1975, 1977) from equation (8.321) with $\varepsilon = 0$ in the cases (1) of equal excitation of all guided modes, and (2) of equal excitation of only the $l = 0$ modes, with no other modes being excited. For case (2) the results are identical to those obtained from a meridional ray analysis by Bouillie et al. (1974). The responses of a fibre with fourth-order polynomial profile to rectangular and Gaussian input pulses have also been calculated in cases (1) and (2) by Nemoto and Yip (1978). The response to a Gaussian input pulse for $l = 0$ modes, but including the effects of profile dispersion, material dispersion, and various input conditions, has been calculated by Gambling and Matsumura (1973).

The r.m.s. width σ of the impulse response may be calculated in the usual way from (8.321). Restricting attention to the intermodal contribution to pulse spreading we may use the first term on the r.h.s. of equation (8.167) to investigate optimal profiles:

$$\sigma^2_{\text{intermodal}} \simeq \langle\tau^2\rangle - \langle\tau\rangle^2. \qquad (8.323)$$

If the number of modes is sufficiently large that the sums required for the averages may be replaced by integrals, then we find (Ikeda, 1978) for equal excitation of all modes:

$$\langle\tau\rangle \simeq \left(\frac{Lm_1}{c}\right)\frac{16}{v^2}\int_0^{v/4} dl \int_0^{(v/2-l)/2} \tau \, dN \simeq \left(\frac{Lm_1}{c}\right)\Delta^2\left(\frac{1}{4} - \frac{K}{6\Delta}\right) \qquad (8.324)$$

$$\langle\tau^2\rangle \simeq \left(\frac{Lm_1}{c}\right)^2 \frac{16}{v^2}\int_0^{v/4} dl \int_0^{(v/2-l)/2} \tau^2 \, dN$$

$$\simeq \left(\frac{Lm_1}{c}\right)^2 \Delta^4\left(\frac{1}{12} - \frac{K}{9\Delta} + \frac{3K^2}{80\Delta^2}\right). \qquad (8.325)$$

Using the results (8.324) and (8.325) in (8.323) yields (Ikeda, 1978; Cook, 1977):

$$\sigma_{\text{intermodal}} \simeq \left(\frac{Lm_1}{c}\right)\frac{\Delta^2}{2}\left[\frac{1}{12} - \frac{K}{9\Delta} + \frac{7K^2}{180\Delta^2}\right]^{1/2}. \tag{8.326}$$

To minimize the intermodal pulse-broadening by optimal choice of K, we impose the condition:

$$\frac{d\sigma^2_{\text{intermodal}}}{dK} = 0$$

which yields the result $K = 10\Delta/7$. The corresponding value of $\sigma_{\text{intermodal}}$ is

$$(\sigma_{\text{intermodal}})_{\text{opt}} = \frac{Lm_1\Delta^2}{12c\sqrt{7}}. \tag{8.327}$$

The numerical coefficient of $Lm_1\Delta^2/c$ in this expression is 0.0315 as compared with 0.0288 for the corresponding result (8.173) in the case of an optimized α-profile. Ikeda (1978) has shown that the optimum value of K is not altered very much by including the effects of profile dispersion and material dispersion. If the fourth-order polynomial profile considered here and the α-profiles discussed in Section 8.2 are considered as independent modifications to the parabolic profile, then the question arises as to whether an appropriate combination of these profiles would provide an improved optimization of intermodal pulse-spreading. Cook (1977) has shown that this is indeed the case, with the new optimum profile given by ($\varepsilon = 0$):

$$n^2(r) = n_1^2\left[1 - 2\Delta\left(\frac{r}{a}\right)^{2(1-2\Delta/3)} + \frac{4\Delta^2}{3}\left(\frac{r}{a}\right)^4\right]. \tag{8.328}$$

For uniform excitation of all modes this profile yields an intermodal r.m.s. pulse width of $0.0216\,Lm_1\Delta^2/c$.

8.4.2 Small departures from desired profiles

In Section 8.2 we have already commented on the difficulties of fabricating desired index profiles. The fabrication process is frequently responsible for fine structure in the profile which may be viewed as a small perturbation from the distribution which was intended. For example, the CVD process involves deposition of layers of different refractive index on the inside of a substrate tube; as a consequence the index profile will show a stairlike distribution. Although this structure will be 'smoothed' to some extent by diffusion which may occur during the preform collapse and fibre drawing stages, many multimode fibre profiles show layer structure. In order to investigate the effects of this perturbation on graded-index fibre pulse dispersion, we commence by considering a suitable model for the layer structure. Let us assume that a constant volume of material is deposited in each layer, so that the fibre cross-section is composed of M equal area layers with radii $r_1, r_2, \ldots, r_M$. The area of the first layer (a circular disc) is πr_1^2, that of the second and all subsequent layers (annular rings) is given by $\pi(r_j^2 - r_{j-1}^2)$ for $j = 2, 3, \ldots, M$. Equating these expressions (for all j) to the known area of each ring, which must be $\pi a^2/M$, yields the result (French et al., 1976) $r_j/a = (j/M)^{1/2}$. Hence the desired graded-index profile will be modified by a 'staircase' distribution with the radius of the jth step proportional to $j^{1/2}$. After diffusion during subsequent stages of the fabrication process the staircase will appear smoothed, and Behm (1976) has suggested that a sinusoidal perturbation provides a useful model for the resultant distribution. Thus for a perturbed parabolic profile we may assume:

$$n^2(r) = n_1^2 \left[1 - 2\Delta \left(\frac{r}{a}\right)^2 + \frac{h\Delta}{M} \sin\left(\frac{2M\pi r^2}{a^2}\right) \right] \qquad (8.329)$$

where h is a constant related to the amount of diffusion. If $n^2(r)$ always decreases with increasing r then h must be less than $1/\pi$. The profile (8.329) is shown in Figure 8.31 for the case $M = 5$, $h = 1/\pi$; obviously, for large values of M the perturbation becomes less significant.

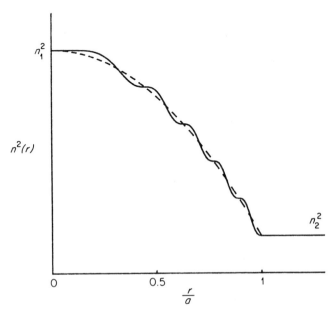

Figure 8.31 Parabolic profile perturbed by layer structure according to equation (8.329) with $M = 5$, $h = 1/\pi$. Broken line shows the unperturbed profile. (From Behm, (1976). Reproduced by permission of S. Hirzel Verlag, Stuttgart)

Use of the sinusoidal perturbation term from (8.329) and the Laguerre–Gaussian functions from (8.314) in the first-order perturbation expression (8.310) yields (Behm, 1976):

$$b = 1 - \frac{2(2N + l + 1)}{v} + b^{(1)} \qquad (8.330)$$

where

$$b^{(1)} = \frac{h}{2M} \frac{(2N+l)!}{N!(N+l)!} \frac{(-1)^N x_0^{l+1}}{(1+x_0^2)^{(2N+l+1)/2}} \sin\left[(2N+l+1)\tan^{-1}(1/x_0)\right]$$
$$\times F[-N, -N; -2N-l; 1+x_0^2]. \qquad (8.331)$$

In (8.331) F is the Gaussian hypergeometric function (Gradshteyn and Ryzhik, 1965) and $x_0 = v/(2M\pi)$. A somewhat complicated analytic expression for the group delay of each mode has been calculated by Behm (1976) by differentiation of (8.331). The group delay is found to oscillate with the number M of layers, tending asymptotically to the delay for an unperturbed parabolic profile as $M \to \infty$. The effect of the perturbation on the impulse response is to broaden the response corresponding to a parabolic index profile by an amount which decreases as M becomes larger. The total impulse response width for equal excitation of all modes in a fibre with $v = 65$ and $M = 40$ is about 1.5 times that in a parabolic index fibre of the same v-value. When the number of layers is sufficiently large that the perturbation oscillates rapidly by comparison with the field oscillations of the fibre then the excess pulse broadening effectively vanishes. In a somewhat similar analysis, Olshansky (1976b) has shown that this occurs for $M \simeq 0.3v$ in a parabolic-profile fibre with sinusoidal perturbation.

An additional refractive index depression at the boundary of each layer is encountered in addition to the diffused-staircase structure for fibres doped with germania. This additional structure is caused by the different deposition rates of GeO_2 and SiO_2, and by selective evaporation of GeO_2 in the hot zone. Thus for GeO_2-doped fibres Behm (1978) has extended the above theory to include profiles of the form:

$$n^2(r) = n_1^2 \left[1 - 2\Delta \left(\frac{r}{a}\right)^2 - \frac{h_s \Delta}{M_s} \sin \left(\frac{2M_s \pi r^2}{a^2} \right) \right.$$
$$\left. - h_L 2\Delta \exp\left(-\frac{d_L r^2}{a^2}\right) \cos\left(\frac{2M_L \pi r^2}{a^2}\right) \right] \quad (8.332)$$

where M_s is the number of profile steps and M_L is the number of layers (allowing for the fact that each step may contain more than one layer). By choosing the parameters h_s, h_L, and d_L to match measured profile perturbations, Behm (1978) has achieved good agreement between theory and experiment for the dependence of intermodal pulse-broadening on M_s. Note that the sign of the profile perturbation in (8.332) is opposite to that in (8.329). This is in order to correct the fact that (8.329) gives a profile where the cross-sectional area of the first and last layer is only half that of the other layers, whereas in practice all layers have the same area. The pulse broadening is hardly influenced by this change of sign.

Another index profile feature which has already been mentioned in an earlier section is the existence of an index depression on-axis as a consequence of dopant evaporation during preform collapse. For multimode fibres this feature may be modelled quite accurately by a Gaussian dip at the core centre, so that the perturbation $\delta n^2(r)$ may be written:

$$\delta n^2(r) = -n_1^2 2\delta_1 \exp(-r^2/d^2) \quad (8.333)$$

where δ_1 gives a measure of the depth of the dip. Figure 8.32 illustrates the effect of a Gaussian dip on a parabolic profile with $\delta_1 = \Delta/2$ and $d/a = 0.1$; this perturbation is usually a very good approximation to the actual form of observed dips (Khular et al., 1977). Using the perturbation from (8.333) together with the Laguerre–Gaussian functions from (8.314) in the perturbation expression (8.310) yields the normalized propagation constant b again as in (8.330), with the perturbation $b^{(1)}$ given this time by (Khular et al., 1977):

$$b^{(1)} = -\frac{\delta_1}{\Delta} \frac{(2N+l)!}{N!(N+l)!} \frac{y_0^{l+1}}{(1+y_0)^{2N+l+1}} F[-N, -N; -2N-l; 1-y_0^2] \quad (8.334)$$

where $y_0 = vd^2/a^2$. Differentiation of this expression produced a lengthy but analytic result for the group delay of mode (N, l) and hence permits pulse-broadening to be calculated. For a fibre with parabolic index profile and $\Delta = 0.012$, the maximum impulse response width is given from (8.152) as $Lm_1\Delta^2/(2c)$ or about 0.37 ns/km. For a 10 per cent index dip on-axis and a fibre v-value of 30, Khular et al. (1977) find that this width increases to 0.66 ns/km, 1.16 ns/km, and 1.73 ns/km for $d/a = 0.04$, 0.06, and 0.08, respectively.

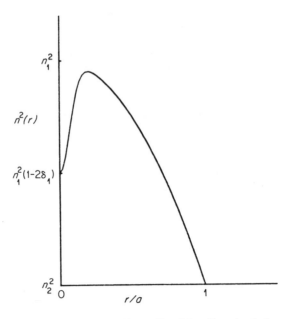

Figure 8.32 Parabolic profile with a Gaussian index dip at core centre, as described by equation (8.333) with $\delta_1 = \Delta/2$ and $d/a = 0.1$

In a similar study, Olshansky (1976b) has investigated the effect of a Gaussian dip or bump centred at various positions across the core radius, rather than just at the centre. His results indicate (i) that narrower perturbations have a more pronounced effect on r.m.s. pulse width than wider perturbations (except when positioned at core centre), and (ii) that perturbations located at larger radii have a much greater influence than those near the core centre. This latter effect is a consequence of the fact that a perturbation $\delta n^2(r)$ changes the mode volume by an amount proportional to $\int \delta n^2(r) 2\pi r dr$. Thus perturbations at radius r are weighted by the factor $2\pi r$ as a result of the cylindrical geometry. For an 8 per cent dip with r.m.s. width of 5 per cent of the core radius and a realistic power distribution, the pulse-broadening is about 0.3 ns/km when the dip occurs on-axis ($v = 60$, $\Delta = 0.01$). When the dip is centred at $r/a = 0.6$, the broadening becomes 2.2 ns/km, almost an order of magnitude larger. This underlines the necessity for accurate control of profile perturbations during fibre fabrication, especially at high radii.

The effect of the on-axis index depression on monomode fibre performance has also been investigated using perturbation theory (Khular et al., 1977; Sammut, 1979a; Pal et al., 1980). In this case the unperturbed guide is assumed as step-index with the appropriate Bessel function fields, and the perturbation is either assumed Gaussian as in (8.333) or described by a power law:

$$\delta n^2(r) = -n_1^2 2\delta_2 (1 - r/a)^\alpha \quad (0 < r < a). \tag{8.335}$$

The two different forms of index dip are illustrated in Figure 8.33. For the power-law of (8.335) Sammut (1979a) has performed the integral in (8.310) to obtain analytic results for propagation constants and cut-off frequencies. These results are in good agreement with those from a series solution of the wave equation (Gambling et al., 1977, 1978), even when the dip is quite large. Using the Gaussian dip (8.333), Khular et al. (1977) give analytic results for modal propagation constants, and Pal et al. (1980) have extended this treatment to investigate

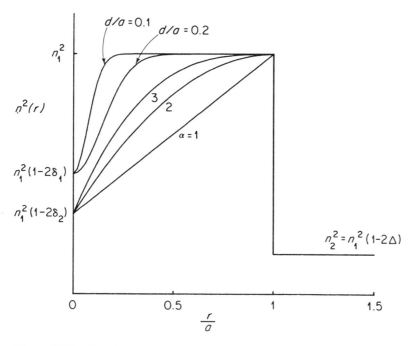

Figure 8.33 Gaussian and power-law on-axis index depressions in a step-index fibre, as described by equations (8.333) and (8.335)

the effects of the dip on the wavelength of zero total dispersion in monomode fibres. As a result of the dip the zero dispersion wavelength shifts to lower values by an amount which increases as the depth of the dip increases. For the fibre discussed in Section 7.2.4, with 13.5 m/0 GeO_2 in the core, pure SiO_2 cladding, and core radius 2 μm, the zero dispersion wavelength ($\simeq 1.55$ μm) shifts by an amount between 1 nm and 35 nm for values of d/a in the range 0.1–0.5 with δ_1/Δ between 0.1 and 0.5.

8.4.3 Cladded α-profiles

Perturbation theory may be used to study the influence of a uniform cladding (index n_2) on the α-profiles defined in (8.1) and (8.2). Following the suggestion of Someda and Zoboli (1975) for cladded parabolic profiles, Sammut and Ghatak (1978) have analysed cladded α-profiles as perturbations of the step-index profile defined by:

$$n^2(r) = \begin{cases} n_1^2(1-\sigma), & 0 \leq r < a \\ n_2^2, & r > a. \end{cases} \quad (8.336)$$

The graded-core fibre profile and the reference step-index profile are illustrated schematically in Figure 8.34. The parameter σ in (8.336) is to be determined from the condition that a mode must have the same cut-off frequency in both the graded- and step-index fibres. Thus the value of σ will depend on the specific mode under consideration. In this way the reference fibre is always chosen such that the field functions are reasonable first approximations to those required for the graded-core fibre. It follows from equation (8.336) that the perturbation formed by an α-profile is given by:

$$\delta n^2(r) = n_1^2\left[\sigma - 2\Delta\left(\frac{r}{a}\right)^\alpha\right] \quad (0 \leq r \leq a). \quad (8.337)$$

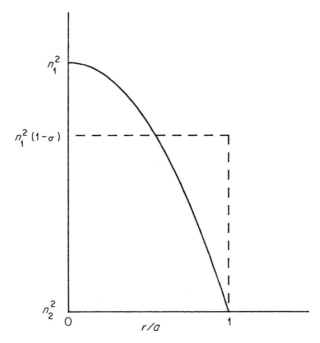

Figure 8.34 Graded-index fibre profile (solid line) and step-index reference profile as defined by equation (8.336)

For the reference step-index fibre the field functions (properly normalized) are given from Chapter 7 as:

$$\psi_{Nl}(r) = \begin{cases} \dfrac{U}{V}(2\kappa_l)^{1/2}\dfrac{J_l(Ur/a)}{J_l(U)}, & 0 \leqslant r \leqslant a \\ \dfrac{U}{V}(2\kappa_l)^{1/2}\dfrac{K_l(Wr/a)}{K_l(W)}, & a \leqslant r \end{cases} \quad (8.338)$$

where the capital letters U, V, W are the usual normalized variables for the reference step-index fibre (n_1^2 is replaced by $n_1^2(1-\sigma)$ in the usual definitions); we reserve the notation u, v, w for the graded-index fibre. In (8.338), κ_l is defined as in (7.107):

$$\kappa_l = \dfrac{K_l^2(W)}{K_{l-1}(W)K_{l+1}(W)}.$$

Using the functions (8.338) and the perturbation (8.337) in the formula (8.310) yields (Sammut and Ghatak, 1978):

$$\dfrac{\beta'^2 - \beta_{Nl}^2}{k^2 n_1^2} = \sigma\left(\dfrac{U}{V}\right)^2\left[\left(\dfrac{W}{U}\right)^2 + \kappa_l\right] - 4\Delta\kappa_l\left(\dfrac{U}{V}\right)^2 I_\alpha(U) \quad (8.339)$$

where

$$I_\alpha(U) = \dfrac{1}{U^{\alpha+2}}\int_0^U x^{\alpha+1}\dfrac{J_l^2(x)}{J_l^2(U)}\,dx \quad (8.340)$$

and β_{Nl} is the propagation constant of the reference step-index fibre. The parameter σ is evaluated from the condition that at cut-off $\beta' = \beta_{Nl} = kn_2$. Imposing this limit in (8.339) yields:

$$\sigma = 4\Delta I_\alpha(V_c) \tag{8.341}$$

where V_c is the cut-off frequency of the reference step-index fibre. Propagation constants calculated from (8.339) for the HE_{11} mode have been shown to be in good agreement with a series solution (Gambling and Matsumura, 1978) for small v-values; for example when $\alpha = 2$ the error is less than 2.4 per cent at $v = 2.5$. As α increases the error rapidly decreases (Sammut and Ghatak, 1978).

From the definition of the reference profile in (8.336) and the usual definition of v for the graded profile, it follows that:

$$v^2 - V^2 = a^2 k^2 n_1^2 \sigma = 2v^2 I_\alpha(V_c).$$

Hence the cut-off frequency v_c of the graded-core fibre is related to that of the reference fibre (V_c) by:

$$v_c^2 = \frac{V_c^2}{[1 - 2I_\alpha(V_c)]}. \tag{8.342}$$

Expressions for the evaluation of the integral I_α defined in (8.340) have been given by Sammut and Ghatak (1978). For the special case $\alpha = 2$ the result (8.342) assumes the simple form (Someda and Zoboli, 1975):

$$v_c = V_c^2 \left[\frac{1.5}{V_c^2 - l(l-1)} \right]^{1/2}. \tag{8.343}$$

Results for v_c of the LP_{11} mode as a function of profile exponent α, calculated from (8.342), have been given by Sammut and Ghatak (1978) and compared with those from a series solution by Gambling et al. (1977a). For $\alpha = 2$ the perturbation theory gives a value for v_c of 3.642, which is approximately 3.5 per cent higher than the series solution of 3.518; as α increases the agreement improves rapidly.

For modes far away from cut-off it is clear that the perturbation approach just discussed will no longer be sufficiently accurate since the Bessel function fields (8.338) are not a good approximation to the real field distributions. In this case a cladded parabolic profile is best viewed as a perturbation on the infinitely-extended square-law medium. This approach, using Laguerre–Gaussian modes for the unperturbed fields, yields analytic expressions for propagation constants and group delays of the azimuthally-symmetric HE_{1N} modes far from cut-off (Kumar and Ghatak, 1976). The method may be extended to include calculations of power confinement factor Γ (Rosenbaum, 1980) and pulse dispersion (Rosenbaum and Coren, 1980) for cladded-parabolic profiles with an index discontinuity at the core–cladding boundary.

To consider modes both near and far from cut-off in general α-profile fibres, a perturbation theory is required which guarantees good approximate field distributions over a wide range of frequencies. Such a theory has been developed by Meunier et al. (1980, 1980a) who take the unperturbed fibre profile as cladded-parabolic:

$$n^2(r) = \begin{cases} n_1^2(1-\sigma)\left[1 - 2\Delta_0\left(\frac{r}{a}\right)^2\right], & 0 \leq r \leq a \\ n_2^2 \equiv n_1^2(1-2\Delta), & r > a \end{cases} \tag{8.344}$$

where

$$2\Delta_0 = \frac{n_1^2(1-\sigma) - n_2^2}{n_1^2(1-\sigma)}.$$

The parameter σ is determined from the condition that a guided mode must have the same frequency in both the reference and graded-index fibres. For this profile the unperturbed fields are described by Kummer functions as mentioned in Section 8.1; substitution into (8.310) and numerical integration then yields propagation constants, cut-offs, and group velocities for fibres of arbitrary core profile. For α-profiles Meunier *et al.* (1980) have found good agreement with the series solution (Gambling *et al.*, 1977a), the maximum error occurring for $\alpha \to \infty$ when the method gives 2.487, an error of approximately 3.4 per cent.

8.5 Summary of Methods for Graded-index Fibre Analysis

In this final section we will attempt to summarize results which have been obtained by a variety of methods, most of which have been encountered already in earlier sections of the book. We will concentrate especially on methods which are appropriate for real index-profiles as measured on graded fibres, and which yield results likely to be influential in fibre design. Fibres of circular cross-section only will be considered here, although a number of methods have been developed for fibres of arbitrary cross-section (Yeh *et al.*, 1975, 1979; Mur and Fondse, 1979; Eyges *et al.*, 1979; Kuester and Pate, 1980). Whilst some of these techniques, in particular the finite-element approach (Yeh *et al.*, 1975, 1979) may be applicable to graded-core circular fibres, their high degree of complexity renders them outside the scope of the present work. The structure of the present discussion will be to consider first those methods best suited to multimode fibre analysis and secondly those more appropriate for monomode fibres.

8.5.1 Multimode fibres

The index profile of a CVD fibre, measured by the near-field scanning technique (Sladen *et al.*, 1976) is shown in Figure 8.35. This profile shows several features which have already been discussed, e.g. the central dip and layer structure characteristic of the CVD fabrication process. Here we are concerned with methods for calculating fibre properties corresponding

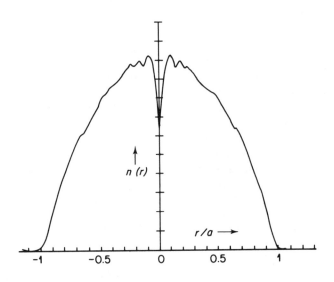

Figure 8.35 Measured refractive index profile for a CVD fibre

to such a profile; our principal concern is with intermodal dispersion and its prediction from profile measurements. Perhaps the most obvious approach to this problem is to fit an α-profile to the experimentally-measured index distribution and then use the standard results for α-profiles which were derived in Section 8.2. This approach is certainly attractive as a first approximation but there are many practical difficulties in assigning a value of α to measured profiles, especially those with the high degree of resolution seen in Figure 8.35. A straightforward r.m.s. fit to the index distribution to obtain α is usually meaningless in view of (a) the axial index depression, (b) structure associated with CVD layers, and (c) uncertainties in the position of the core–cladding interface. One common solution to this problem is to choose suitable inner and outer radii and fit the data only between these radii, thus eliminating the effects of (a) and (c) above. In more sophisticated versions (Presby et al., 1978) a search procedure is used to fit such parameters as the core radius and index difference as well as the α-value. Another alternative is to fit a local α-value to the profile (Hazan et al., 1978). Perhaps a more realistic approach is to use a weighted r.m.s. fit with a factor dependent on the optical intensity at each radius (White, 1979) in order to emphasize those radii where the majority of the optical power is carried. In addition the position of the data points chosen for the fit will implicitly affect the weighting; a choice of points at equal intervals along the profile curve gives the best results in this respect (White, 1979).

In general, however, the α-fitting procedure is less satisfactory than a numerical approach using ray-tracing or the WKB method. Integration of the ray equations, for example in the forms given by (8.30)–(8.32), was first performed for polynomial index profiles by Streifer and Paxton using the technique of algebraic computing to yield analytic rather than numerical results for the ray trajectories (Streifer and Paxton, 1971; Paxton and Streifer, 1971). The approach was designed for application in focusing rods rather than communications fibres, and hence no results for pulse dispersion were obtained. More recently numerical ray-tracing techniques have been combined with the modal resonance condition (8.40) to calculate propagation constants and group delays in graded-index fibres (Checcacci et al., 1975). When applied to profiles with central index-dips, the method (Checcacci et al., 1977) yields results similar to those of Olshansky (1976b) which were discussed in Section 8.4; the small effect of a central dip on overall pulse-spreading is interpreted in terms of the shift of caustics for the ray paths.

Calculations of impulse response from measured index profiles using WKB theory have been presented by Ramskov Hansen and Nicolaisen (1978). The results were compared with experimental measurements of pulse-spreading where the fibre was excited by a small spot placed at different positions on the end-face. Numerical predictions were also made using (i) perturbation theory and (ii) an α-fitting procedure as discussed above together with standard α-profile results. The superiority of WKB theory over both methods (i) and (ii) was demonstrated by the accuracy of agreement with the measured responses. Perturbation theory gave a qualitative agreement but failed in the case of large deviations from the parabolic profile, whilst the α-profile approximation gave incorrect results.

In a series of papers, Marcuse et al. (Marcuse, 1979b,c, 1980; Marcuse and Presby, 1979, 1979a; Presby et al., 1979) have developed an efficient computational procedure for calculating impulse responses within the framework of WKB theory. Basically the method involves the numerical evaluation of equation (8.136) for modal transit time τ corresponding to modes determined by the resonance equation (8.131). The numerical evaluation of (8.136) is complicated by the existence of pole contributions which must be evaluated analytically in the vicinity of the caustics at r_1 and r_2. Once the transit times τ are found for all the guided modes, then a weighting factor of 4 is applied to each mode with $l \neq 0$, and a factor of 2 is applied to each mode with $l = 0$. This factor accounts for the fourfold degeneracy (two polarizations and sine or cosine azimuthal dependence) of $l \neq 0$ modes and the twofold degeneracy (polarization) of $l = 0$ modes. The interval of pulse arrival times is subdivided into 15 time slots, the number of pulses arriving in each slot is counted, and a constant value proportional to the mode count is assigned to the impulse response over the region of each time slot. The method has been applied to profiles with central dips, sinusoidal ripples, departures from optimum near the core–cladding boundary, and bulge deformations at various radial positions. For many of these deformations the best achievable bandwidth is reduced by factors of 10–100 from that predicted for an optimized α-profile. Less severe degradation of bandwidth occurs for sinusoidal perturbations whose phases vary as a function of fibre length z (Marcuse, 1979c). Comparisons of bandwidths predicted from index profile measurements with experimental measurements have been presented (Presby et al., 1979; Marcuse, 1980) for fibres with bandwidths in the range 200–1000 MHz.km. The theoretical results agree with experiment for the lower bandwidths, but tend to be lower than experiment (by up to 50 per cent) for the higher bandwidths in this range.

The difficulty associated with the numerical evaluation of (8.136) which was referred to above, viz. the existence of poles at r_1, r_2, has stimulated the development of other methods for calculating transit times within the ray/WKB approximation. Barrell and Pask (1980a) have solved the problem by approximating the index profile by a series of parabolic sections, thus reducing the evaluation of (8.136) to a sum of standard integrals, and the determination of the caustics r_1, r_2 to the solution of a quadratic equation. Another approach, which has been developed by Ankiewicz (1978a) and Irving and Karbowiak (1979), is to use the Chebyshev approximation (Abramowitz and Stegun, 1964) to evaluate the integral in (8.136). This approximation yields the result:

$$\tau = \left(\frac{Lm_1}{c}\right)\left(\frac{kn_1}{\beta}\right)\left[1 - \Delta\left(2 + \frac{\varepsilon}{2}\right)\frac{\sum_{j=1}^{M} G(r_j)f(r_j)}{\sum_{j=1}^{M} G(r_j)}\right] \qquad (8.345)$$

where

$$G(r) = \frac{(r_2 - r)^{1/2}(r - r_1)^{1/2}}{q(r)} \qquad (8.346)$$

and

$$r_j = \left(\frac{r_1 + r_2}{2}\right) + \left(\frac{r_2 - r_1}{2}\right)\cos\left[\frac{(2j-1)\pi}{2M}\right]. \qquad (8.347)$$

The functions $q(r)$ and $f(r)$ are defined in (8.126) and (8.1), respectively. With the aid of this technique great accuracy in τ can be obtained for relatively small values of M. Karbowiak and Irving (1979) have pointed out the superiority of this method to that of fitting α-profiles or polynomial profiles to measured index data.

Hartog et al. (to be published) have applied the Chebyshev approximation, as expressed in (8.345)–(8.347) to the prediction of bandwidths as a function of wavelength from experimental measurements of the profile function $f(r)$ and the profile dispersion parameter $\varepsilon(\lambda)$. The fibre impulse response is found from the modal transit times in a similar manner to that described above for Marcuse's calculations. A value for M in (8.345) and (8.347) of about 40 is usually adequate, and equal increments in mode parameters b and $(l/v)^2$ were chosen, each of size 0.01. An example of an impulse response generated by this method is shown in Figure 8.36, in which 32 time slots were used. The transfer function and hence the bandwidth may then be found by Fourier transform. Figure 8.37 compares predicted with measured

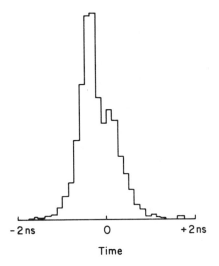

Figure 8.36 Impulse response calculated by the numerical technique discussed in the text; 32 time slots were used to build up the response from computed ray transit times

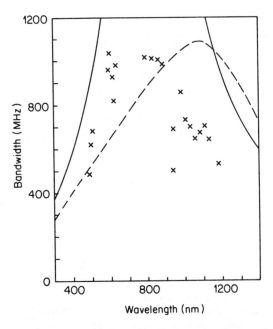

Figure 8.37 Predicted and measured variation of bandwidth with wavelength: solid line, theory for best-fit α-profile; broken line, calculated by numerical ray-tracing for the measured profile; crosses, measured bandwidths

bandwidths as a function of wavelength for a multimode graded-index fibre designed for operation at the GaAs emission wavelength. The crosses indicate the experimental values obtained by using a tunable dye-laser source and measuring the pulse-spreading experienced by short pulses after travelling through 1 km of fibre. The solid line gives the predicted variation of bandwidth using the theory given in Section 8.2 together with the best-fit α-profile for the measured fibre index distribution. The measured variation of profile dispersion ε (see Figure 8.13) was also used in calculating the solid line in Figure 8.37. The experimental data show the characteristic strong dependence on operating wavelength leading to maxima in the range 600–900 nm. However, the values lie considerably below those predicted by the α-profile theory, presumably as a consequence of the detailed structure of the real index profile. Also shown on Figure 8.37 is the result of numerically computing all the ray transit times to estimate the bandwidth at each wavelength from the measured $f(r)$ and ε (broken line). This result is much closer to the experimental data although it peaks at a rather different wavelength. This discrepancy is at present attributed to a small variation of refractive index profile along the length of the fibre.

As an alternative to WKB theory, a method developed by Langer (1949) and McKelvey (1959) may be applied to the problem of computing modal delay times. Streifer and Kurtz (1967) used the method to analyse α-profiles, Poschl–Teller, and Epstein-type profiles (Kurtz and Streifer, 1969a), and Olshansky (1979) applied the technique to double α-profiles. Whilst being completely analogous to the WKB theory, the McKelvey–Langer approach approximates the solutions to the wave equation with Bessel functions of appropriately modified argument, rather than by trigonometric and exponential functions as in the WKB case. Thus the method is designed specifically for problems of circular symmetry and eliminates the necessity for a turning-point at r_1. The McKelvey–Langer eigenvalue equation is given by (Streifer and Kurtz, 1967):

$$(2N + l + 1)\frac{\pi}{2} = \int_0^{r_0} (k^2 n^2(r) - \beta^2)^{1/2} \, dr \qquad (8.348)$$

where r_0 is the solution of $kn(r_0) = \beta$. Olshansky (1979a) has shown that for α-profiles the results for eigenvalues from (8.348) are equivalent to those from WKB theory (see, for example, equation (8.140)). The method has recently been applied to the calculation of multimode fibre impulse response by Chu (1980), who has obtained analytic results in the cases (a) of a profile composed of a power-law and a linear function, and (b) of a power-law profile with a linear dip at the centre.

The 'staircase' or stratification method (Clarricoats and Chan, 1970) which was discussed in some detail for two-dimensional waveguides in Section 5.4.2, has been applied to analysis of graded-index multimode fibres by a number of authors (Tanaka and Suematsu, 1976; Yeh and Lindgren, 1977; Bianciardi and Rizzoli, 1977; Arnold et al., 1977). The method consists of approximating the given index profile by a series of layers in each of which the index is constant and hence the solutions of the wave equation may be expressed as appropriate superpositions of Bessel functions. The continuity of the tangential fields at the boundaries of the layers then leads to an eigenvalue equation given by the vanishing of a characteristic determinant. The number of layers required to ensure accurate convergence of the method varies with the mode considered and the v-value of the fibre. Typical values range from 5 (Clarricoats and Chan, 1970; Yeh and Lindgren, 1977) to 20 (Bianciardi and Rizzoli, 1977), although calculations of group delay can show large anomalous variations for increasing radial wave number even when 20 layers are used.

Direct numerical integration of the wave equation is an attractive technique which is sometimes used for multimode fibres. The wave equation, e.g. in the form (8.276), may be written as a pair of first-order equations for $R(r)$ and the auxiliary function $S(r)$ (Arnaud and Mammel, 1976):

$$\frac{dR}{dr} = \frac{S}{r} \qquad (8.349a)$$

$$\frac{dS}{dr} = rR\left[\frac{l^2}{r^2} - \frac{v^2}{a^2}(1 - b - f(r))\right]. \qquad (8.349b)$$

To solve these equations, a value of b is chosen in the range 0–1, and the integration proceeds from the initial condition

$$\left.\frac{S}{R}\right|_{r=0} = l. \qquad (8.350)$$

This equation follows from the fact that, for small r, the r.h.s. of (8.349b) is approximately $l^2 R/r$, and therefore $R(r) = r^l$, $S(r) = lr^l$ is a solution of equation (8.349). The numerical integration of equation (8.349) may be carried out by standard methods, e.g. Runge–Kutta, predictor–corrector, etc. The value of b is varied until the condition $R \to 0$ as $r \to \infty$ is achieved; alternatively, the usual boundary conditions on R and S at the core–cladding interface may be imposed. With the aid of this direct integration technique, Arnaud and Mammel (1976) considered the problem of stairlike refractive index profiles and demonstrated that the anomalous group delay results referred to above for the stratification method could only be avoided by using a very large number of layers (certainly greater than 40). Subsequently Arnold et al. (1977) compared the computation time required by the two methods (i.e. direct integration and staircase) and showed that the direct integration approach required substantially less computer time. Arnold (1977) has also proved that the staircase method is formally equivalent to solution by the Euler method, but involves far greater computation since a great deal of redundant information is generated. It is therefore established that direct numerical integration of the wave equation is superior to the staircase method for multimode fibre analysis.

The variational analysis described in Section 8.3.2 has been modified recently for use with measured refractive index profiles of real fibres by Okamoto (1979). The new approach is based upon the functional $J(R)$ as defined in (8.280) but, instead of solution by the Rayleigh–Ritz method, Okamoto (1979) used a one-dimensional finite element technique in order to save computer time and use measured profiles as direct input data. The region between $r = 0$ and $r = a$ is divided into a suitable number of divisions (typically 25) and the functions $R(r)$ and $n^2(r)$ expressed as sets of linear approximations. The requirement that $J(R)$ be stationary with respect to the parameters $R(r_j)$ (where r_j are the discrete points of division of the range 0–a) results in a series of linear equations. This leads in the usual way to the vanishing of a determinant as a condition for the existence of a nontrivial solution, and hence to an algorithm for finding the propagation constant. The group delay may then be found by application of the result (8.268) connecting group and phase velocities. Okamoto (1979) has used this finite-element/variational approach to calculate impulse responses from measured index profiles of multimode fibres and has compared the results with measured responses. The relative errors in pulse widths lie in the range 2.6–10 per cent with fairly good agreement of pulse shapes; the index profiles were measured by interference microscopy and the impulse responses with the aid of a semiconductor laser source. Allowance for measured differential mode attenuaton was made in calculating the impulse responses. In general the fibres used had relatively low bandwidths and were non-optimal at the source wavelength of 840 nm.

Finally in this subsection we mention the so-called propagating beam method which has been applied to the analysis of multimode fibres (Feit and Fleck, 1978, 1979, 1980, 1980a) and geodesic lenses (Van der Donk and Lagasse, 1980). The method is developed from one used in quantum mechanical problems, and solves the scalar wave equation by repeated application of a marching algorithm along the length of the fibre. Thus the method is equivalent to replacing the fibre by a periodic array of thin lenses separated by a homogeneous medium. The field is then calculated as a two-dimensional Fourier series with a finite number of terms. The propagating beam method has been applied to parabolic profiles, profiles with central dips, and α-profiles under a range of excitation conditions and is able to demonstrate the effects of the fibre cladding on modes near cut-off in a similar way to that discussed earlier in this chapter. A somewhat related technique which deals specifically with the propagation of Gaussian beams in graded-index multimode fibres has been developed by Yeh et al. (1978).

8.5.2 Monomode fibres

Figure 8.38 shows the measured refractive index profile of a monomode-fibre preform (Sasaki et al., 1980); the measurement has been made in two ways, using a spatial-filtering

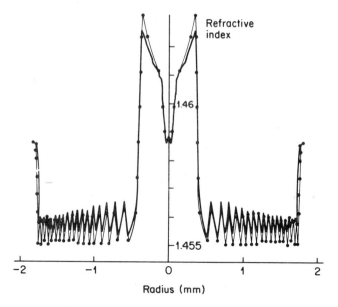

Figure 8.38 Refractive index profile measured on a single-mode fibre preform. Heavy line, measured by spatial-filtering technique; fine lines and dots, measured by thin-slice interferometry. (From Sasaki et al. (1980). Reproduced by permission of the IEE)

technique (heavy lines) and thin-slice interferometry (fine lines and dots). Whilst the profile in the fibre produced from this preform may be slightly different as a consequence of diffusion during the drawing process, it is believed that these changes will be small and that the main features shown in the figure are preserved in the fibre profile. The preform in question was fabricated by the CVD process and shows the characteristic central dip as well as layer structure in the cladding. In this final subsection we will address the problem of calculating the properties of fibres with index profiles such as that of Figure 8.38. In particular we are interested in the calculation of practical parameters such as single-mode cut-off or total chromatic dispersion. Other features such as bending loss, microbending loss, etc., although of great importance, must regrettably be deemed to fall outside the scope of the present work.

The numerical technique which has perhaps found the widest application in monomode fibre analysis is the power series method which was discussed for planar waveguides in section 5.4.1. The modifications required for the circularly-symmetric waveguide of arbitrary profile are straightforward and the method presents no serious problems for well-behaved profiles. However, it should be recognized at the outset that the series method may only be applied to profiles which may be represented by a polynomial in the radius r (Dil and Blok, 1973; Kirchhoff, 1973) and hence is not really suitable for treating real profiles of the type shown in Figure 8.38. Thus the method has been applied to calculating the properties of parabolic profiles (Midwinter, 1975), α-profiles (Gambling et al., 1977a; Gambling and Matsumura, 1978), and profiles with a power-law central dip as in equation (8.335) (Gambling et al., 1977, 1978); further discussion of these topics will be found in Section 8.4. The α-profile results have been extended from integral α-values to include arbitrary values of the exponent α by means of a simple transformation (Love, 1979). In addition the method has been used in calculations of the v-value for zero waveguide dispersion (Snyder and Sammut, 1979) and of the conditions for zero total chromatic dispersion (Gambling et al., 1979, 1979a) as discussed in Section 7.2. The effects of cladding and core losses have also been explored in graded-index fibres with the aid of the series solution (Gronthoud and Blok, 1978).

The variational/finite element approach discussed in the previous subsection with reference to multimode fibres has also been developed as a vectorial method suitable for use in monomode fibres. Okamoto and Okoshi (1978) have derived a stationary expression corresponding to the two coupled Kurtz–Streifer equations (8.73) and (8.74). The solution by a one-dimensional finite element technique follows the lines described in Section 8.5.1. For α-profiles with $\Delta = 0.01$, Okamoto and Okoshi (1978) have shown that the error in normalized propagation constant b caused by the scalar wave approximation is of the order of 0.1 per cent as compared with the vectorial analysis. For relative group delay the corresponding error is approximately 0.01 ns/km for $\alpha = 2$ and 0.8 ns/km for $\alpha = 10$.

A useful comparison of cut-off frequencies of the first higher-order mode for α-profile fibres calculated by some of the methods already discussed has been presented by Meunier et al. (1980a). Table 8.4 reproduces this comparison together with an extra column showing the results of the approximation (8.309) due to Okamoto et al. (1979). The table gives values of cut-off frequency v_c as a function of α as calculated by (i) the power-series method (Gambling et al., 1977a), (ii) the approximate variational method (Okamoto et al., 1979), (iii) the numerical variational method (Hotate and Okoshi, 1978), (iv) perturbation theory using a step-index profile as the unperturbed system (Sammut and Ghatak, 1978)—see equation (8.342)—and (v) perturbation theory using a cladded-parabolic profile as the unperturbed system (Meunier et al., 1980, 1980a). Column (i) is treated as the exact solution and it is immediately clear that the approximate result (8.309) given in column (ii) is the least accurate of the results quoted. Columns (iii) and (iv) give rather similar results, which is not surprising since both these methods use the step-index fibre eigenfunctions in calculating the cut-offs. The result of column (iv) for $\alpha \to \infty$ is exact since this condition is a starting-point for the calculation (see Section 8.4.3). Similarly the result of column (v) for the case $\alpha = 2$ is exact, and this method gives good results for low α with the error increasing as $\alpha \to \infty$.

As regards other methods of solving the full vector problem in monomode fibres we may consider again the staircase and direct integration approaches. The former has been applied successfully to calculations of total chromatic dispersion for measured profiles of real fibres, as mentioned in Section 8.2 (Sugimura et al., 1980). The results were in good agreement with dispersion measurements as a function of wavelength; for this application the scalar wave

Table 8.4 Comparison of cut-off frequencies calculated as discussed in the text. (From Meunier et al. (1980a). Reproduced by permission of the Institution of Electrical Engineers)

α	(i)	(ii)	(iii)	(iv)	(v)
1	4.381	4.166	4.572	4.611	4.390
2	3.518	3.401	3.613	3.642	3.518
3	3.181	3.105	3.238	3.262	3.184
4	3.000	2.946	3.036	3.057	3.008
5	2.886	2.846	2.910	2.929	2.900
10	2.649	2.635	2.650	2.665	2.686
20	2.527	2.522	2.520	2.531	2.583
∞	2.405	2.405	2.397	2.405	2.487

analysis was deemed sufficiently accurate. Vector wave analysis by direct numerical integration has been used in fibre analysis for many years (Vigants and Schlesinger, 1962; Dil and Blok, 1973; Vassell, 1974a; Yip and Ahmew, 1974; Armenise et al., 1976) and a useful review will be found in a paper by Parriaux and Gardiol (1976). However, for the most part the progress in this field has been towards handling larger values of the ratio (core radius/wavelength) with reasonable computer time and thus the ingenious techniques developed are not usually necessary for monomode fibre analysis. For the scalar wave equation the method outlined in the previous subsection is usually sufficiently accurate, although sometimes finite-difference methods may be used for the integration (Marcuse, 1978). More recent approaches have also invoked coordinate transformations for rapid convergence of the numerical procedures (North, 1979) as well as further work on including the gradient of the refractive index in the scalar wave equation (Kokubun and Iga, 1980).

The evanescent field theory discussed in Section 5.4.3 for two-dimensional waveguides has been applied also to graded-index fibre analysis (Choudhary and Felsen, 1977). Whilst the original formulation of the method gave results with rather less accuracy close to the core centre than for relatively large radii, a modification due to Jacobsen (1978) makes it possible to calculate very accurate field distributions provided that values of the propagation constants are supplied, e.g. by another method of analysis. The results for modal fields of near-parabolic fibres have been compared with those of perturbation theory by Ramskov Hansen and Jacobsen (1979); results for propagation constants and group delays have been compared with those of perturbation theory and the WKB approximation (Jacobsen and Ramskov Hansen, 1979). For lower-order modes a modified evanescent wave theory (Jacobsen and Ramskov Hansen, 1979a) yields more accurate results than WKB theory and third-order perturbation theory. Hence this method would appear suitable for application to monomode fibre analysis. However it seems that the reduced errors of the evanescent field theory are only achieved for profiles of the form:

$$n^2(r) = n_1^2 - a_0^2 r^2 (1 + a_1 r^2)^2$$

where a_0 is a real positive constant and a_1 is a real constant.

In view of the complications involved in some of the numerical methods applied to monomode fibre analysis, there has recently been a move towards a simplified approach where the details of the fibre profile are recognized as being of less importance than was previously thought. The philosophy behind this approach is based on the fact that the field distributions of the lowest-order mode of many fibres are quite similar (Gambling and Matsumura, 1977, 1978; Marcuse, 1977, 1978) and in many cases closely resemble a Gaussian distribution. This fact has prompted the idea that many of the interesting properties of monomode fibres (cut-offs, spot-size, dispersion, etc.) might be calculated direct from measurements of other fibre properties which are not so difficult to measure as the refractive index profile. Thus, for example, Brinkmeyer (1979) has developed a method for determining the spot-size (width of the fundamental mode) from a measurement of the first minimum in the diffraction pattern produced by transverse illumination of a fibre.

From a rather more fundamental point of view, Snyder and Sammut (1979a) have shown how to replace a given graded-index fibre by an equivalent step-index fibre; results for second-order mode cut-offs and fundamental mode fields are then obtained directly from the step-index results. The method uses a variational technique where the step-index fibre whose field is most like that of the graded-index fibre is the one which maximizes the value of the propagation constant β. For a parabolic-profile fibre, the error in cut-off frequency is 0.3 per cent by this method. Matsumura *et al.* (1980) have developed a somewhat similar method of determining the equivalent step-index fibre by matching the mode spot-size to that of the given fibre profile. An alternative version of the Snyder/Sammut theory replaces the graded index fibre by one with an extended parabolic profile and corresponding Gaussian fundamental mode (Sammut and Snyder, 1980). As an example of the accuracy of the procedure, the value of u for a step-index fibre given by the equivalent parabolic-profile fibre is in error by less than 1 per cent in the range $2.1 \leqslant v \leqslant 3$. The two parameters which characterize an equivalent fibre in these approaches are effective values of normalized frequency v and core radius a. Pask and Sammut (1980) have suggested that these parameters for equivalent step-index fibres might be obtained from measurements of the far-field radiation pattern, although errors of up to 10 per cent in the value of core radius are predicted by this method for parabolic-profile fibres. An alternative approach, again using two effective parameters to characterize a graded-index monomode fibre, has been developed by Stewart (1980).

References

Abramowitz, M., and I. A. Stegun (1964). *Handbook of Mathematical Functions*, Dover, New York.
Abrams, R. L., and W. B. Bridges (1973). 'Characteristics of sealed-off waveguide CO_2 lasers.' *IEEE J. Quantum Electron*, **QE-9**, 940–946.
Adams, M. J. (1969). 'Theoretical effects of exponential band tails on the properties of the injection laser.' *Solid State Electron.*, **12**, 661–669.
Adams, M. J. (1977). 'Loss calculations in weakly-guiding dielectric waveguides.' *Optics Communications*, **23**, 105–108.
Adams, M. J. (1978). 'The cladded parabolic-index profile waveguide: analysis and application to stripe-geometry lasers.' *Optical and Quantum Electron.*, **10**, 17–29.
Adams, M. J., and M. Cross (1971). 'Electromagnetic theory of heterostructure injection lasers.' *Solid State Electron.*, **14**, 865–883.
Adams, M. J., and M. Cross (1971a). 'On the polarisation of radiation from double-heterostructure injection lasers.' *Electron. Lett.*, **7**, 569–570.
Adams, M. J., and P. T. Landsberg (1969). 'The theory of the injection laser.' In C. H. Gooch (Ed.), *Gallium Arsenide Lasers*, Wiley, Chichester.
Adams, M. J., D. N. Payne, and C. M. Ragdale (1979). 'Birefringence in optical fibres with elliptical cross-section.' *Electron. Lett.*, **15**, 298–299.
Adams, M. J., D. N. Payne, and F. M. E. Sladen (1975). 'Leaky rays on optical fibres of arbitrary (circularly symmetric) index profiles.' *Electron. Lett.*, **11**, 238–240.
Adams, M. J., D. N. Payne, and F. M. E. Sladen (1975a). 'Mode transit times in near-parabolic-index optical fibres.' *Electron. Lett.*, **11**, 389–391.
Adams, M. J., D. N. Payne, and F. M. E. Sladen (1976). 'Splicing tolerances in graded-index fibers.' *Appl. Phys. Letts.*, **28**, 524–526.
Adams, M. J., D. N. Payne, and F. M. E. Sladen (1976a). 'Correction factors for the determination of optical-fibre refractive-index profiles by the near-field scanning technique.' *Electron. Lett.*, **12**, 281–283; erratum in *Electron. Lett.*, **12**, 348.
Adams, M. J., D. N. Payne, and F. M. E. Sladen (1976b). 'Length-dependent effects due to leaky modes on multimode graded-index optical fibres.' *Optics Communications*, **17**, 204–209.
Adams, M. J., D. N. Payne, F. M. E. Sladen, and A. H. Hartog (1978). 'Optimum operating wavelength for chromatic equalisation in multimode optical fibres.' *Electron. Lett.*, **14**, 64–66.
Adams, M. J., D. N. Payne, F. M. E. Sladen, and A. H. Hartog (1978a). 'Wavelength-dispersive properties of glasses for optical fibres: the germania enigma.' *Electron. Lett.*, **14**, 703–705.
Aiki, K., M. Nakamura, T. Kuroda, J. Umeda, R. Ito, N. Chinone, and M. Maeda (1978). 'Transverse mode stabilized $Al_x Ga_{1-x} As$ injection lasers with channeled-substrate-planar structure.' *IEEE J. Quantum Electron.*, **QE-14**, 89–94.
Aiki, K., M. Nakamura, and J. Umeda (1976). 'Frequency multiplexing light source with monolithically integrated distributed-feedback diode lasers.' *Appl. Phys. Letts.*, **29**, 506–508.
Aiki, K., M. Nakamura, J. Umeda, A. Yariv, A. Katzir, and H. W. Yen (1975). 'GaAs–GaAlAs distributed-feedback diode lasers with separate optical and carrier confinement.' *Appl. Phys. Letts.*, **27**, 145–146.
Albertin, F., P. di vita, and R. Vannucci (1974). 'Geometrical theory of energy launching and pulse distortion in dielectric optical waveguides.' *Opto-electronics*, **6**, 369–386.
Alferov, Zh. I., V. M. Andreev, D. Z. Garbuzov, Yu. V. Shilyaev, A. P. Morozov, E. L. Portnoi, and V. G. Triofim (1970a). 'Investigation of the influence of the AlAs–GaAs heterostructure parameters on the laser threshold current and the realisation of continuous emission at room temperature.' *Sov. Phys.—Semiconductors*, **4**, 1573–1575; and *Fiz. Tekh. Poluprov.*, **4**, 1826–1828.
Alferov, Zh. I., V. M. Andreev, V. I. Korol'Kov, E. L. Portnoi, and D. N. Tret'Yakov (1969). 'Coherent radiation of epitaxial heterojunction structures in the AlGaAs–GaAs systems.' *Sov. Phys.—Semiconductors*, **2**, 1288–1291; and *Fiz. Tekh. Poluprov.*, **2**, 1545–1547 (1968).

Alferov, Zh. I., V. M. Andreev, E. L. Portnoi, and M. K. Trukan (1970): AlAs–GaAs heterojunction injection lasers with a low room-temperature threshold.' *Sov. Phys.—Semiconductors*, **3**, 1107–1110; and *Fiz. Tekh. Poluprov.*, **3**, 1328–1332 (1969).

Alferov, Zh. I., and R. F. Kazarinov (1963). Author's certificate No. 181737, Claim 950840 of March 30 (as cited in Alferov *et al.*, 1970).

Anderson, D. B. (1965). 'Application of semiconductor technology to coherent optical transducers and spatial filters.' In J. Tippett (Ed.), *Optical and Electro-Optical Information Processing*, M.I.T. Press, Cambridge, Mass. pp. 221–234.

Anderson, W. W. (1965). 'Mode confinement and gain in junction lasers.' *IEEE J. Quantum Electron.*, **QE-1**, 228–236.

Ankiewicz, A. (1978). 'Comparison of wave and ray techniques for solution of graded index optical waveguide problems.' *Optica Acta*, **25**, 361–373.

Ankiewicz, A. (1978a). 'Geometric optics theory of graded index optical fibres.' *Ph.D. Thesis*, Australian National University, Canberra.

Ankiewicz, A. (1979). 'Ray theory of graded non-circular optical fibres.' *Optical and Quantum Electron.*, **11**, 197–203.

Ankiewicz, A. (1979a). 'Core fields in graded fibres with non-circular index contours.' *Optical and Quantum Electron.*, **11**, 525–539.

Ankiewicz, A., M. J. Adams, and N. J. Parsons (1979). 'Modes in graded non-circular multimode optical fibres.' *Optics Letts.*, **4**, 414–416.

Ankiewicz, A., and C. Pask (1977). 'Geometric optics approach to light acceptance and propagation in graded index fibres.' *Optical and Quantum Electron.*, **9**, 87–109.

Ankiewicz, A. and C. Pask (1978). 'Tunnelling rays in graded-index fibres.' *Optical and Quantum Electron.*, **10**, 83–93.

Armenise, M. N. R. de Leo, and M. de Sario (1976). 'Numerical analysis of propagating modes in inhomogeneous optical fibres.' *Alta Frequenza*, **45**, 230–234.

Arnaud, J. A. (1973). 'Hamiltonian theory of beam mode propagation.' In E. Wolf (Ed.), *Progress in Optics*, Vol. XI, North-Holland, Amsterdam. pp. 247–304.

Arnaud, J. A. (1974). 'Pulse spreading in multimode, planar optical fibres.' *Bell System Tech. J.*, **53**, 1599–1618.

Arnaud, J. A. (1974a). 'Transverse coupling in fiber optics. Part II: Coupling to mode sinks.' *Bell System Tech. J.*, **53**, 675–696.

Arnaud, J. A. (1975). 'Pulse broadening in multimode optical fibres.' *Bell System Tech. J.*, **54**, 1179–1205.

Arnaud, J. A. (1975a). 'Pulse broadening in near-square-law graded-index fibres.' *Electron. Lett.*, **11**, 447–448.

Arnaud, J. A. (1976). *Beam and Fiber Optics*, Academic Press, New York.

Arnaud, J. A. (1977). 'Optimum profiles for dispersive multimode fibres.' *Optical and Quantum Electron.*, **9**, 111–119.

Arnaud, J. A., and R. M. Derosier (1976). 'Novel technique for measuring the index profile of optical fibers.' *Bell System Tech. J.*, **55**, 1489–1508.

Arnaud, J. A., and J. W. Fleming (1976). 'Pulse broadening in multimode optical fibres with large $\Delta n/n$: numerical results.' *Electron. Lett.*, **12**, 167–169.

Arnaud, J. A., and W. Mammel (1975). 'Application of a property of the Airy function to fiber optics.' *IEEE Trans. Microwave Theory and Techniques*, **MTT-23**, 927–929.

Arnaud, J. A., and W. Mammel (1976). 'Dispersion in optical fibres with stairlike refractive-index profiles.' *Electron. Lett.*, **12**, 6–8.

Arnold, J. M. (1977). 'Stratification methods in the numerical analysis of optical waveguide transmission parameters.' *Electron. Lett.*, **13**, 660–661.

Arnold, J. M. (1977a). 'Attenuation of an optical fibre immersed in a high-index surrounding medium.' *Microwaves, Optics and Acoustics*, **1**, 93–102.

Arnold, J. M. (1977b). 'Asymptotic evaluation of the normalised cut-off frequencies of an optical waveguide with quadratic index variation.' *Microwaves, Optics and Acoustics*, **1**, 203–208.

Arnold, J. M., G. A. E. Crone, and P. J. B. Clarricoats (1977). 'Comparison of numerical computations of optical waveguide transmission parameters.' *Electron. Lett.*, **13**, 273–274.

Asbeck, P. M., D. A. Cammack, and J. J. Daniele (1978). 'Non-Gaussian fundamental mode patterns in narrow-stripe-geometry lasers.' *Appl. Phys. Letts.*, **33**, 504–506.

Asbeck, P. M., D. A. Cammack, J. J. Daniele, and V. Klebanoff (1979). 'Lateral mode behaviour in narrow stripe lasers.' *IEEE J. Quantum Electron.*, **QE-15**, 727–733.

Avrillier, S., and J. Verdonck (1977). 'Coupling losses in laser resonators containing a hollow rectangular dielectric waveguide.' *J. Appl. Phys.*, **48**, 4937–4941.
Ayant, Y., G. H. Chartier, and P. C. Jaussaud (1977). 'Etude a l'aide de l'optique intégrée des processus de diffusion et d'échange d'ions dans un verre alcalin.' *Journal de Physique*, **38**, 1089–1096.
Barrell, K. F., and C. Pask (1978). 'The effect of cladding loss in graded-index fibres.' *Optical and Quantum Electron.*, **10**, 223–231.
Barrell, K. F., and C. Pask (1979). 'Ray launching and observation in graded-index optical fibres.' *J. Opt. Soc. Am.*, **69**, 294–300.
Barrell, K. F., and C. Pask (1979a). 'Geometric optics analysis of non-circular graded-index fibres.' *Optical and Quantum Electron.*, **11**, 237–251.
Barrell, K. F., and C. Pask (1980). 'Leaky ray correction factors for elliptical multimode fibres.' *Electron. Lett.*, **16**, 532–534.
Barrell, K. F., and C. Pask (1980a). 'Pulse dispersion in optical fibres of arbitrary refractive-index profile.' *Applied Optics*, **19**, 1298–1305.
Batchman, T. E., and K. A. McMillan (1977). 'Measurement on positive-permittivity metal-clad waveguides.' *IEEE J. Quantum Electron.*, **QE-13**, 187–192.
Batchman, T. E., and S. C. Rashleigh (1972). 'Mode-selective properties of a metal-clad dielectric-slab waveguide for integrated optics.' *IEEE J. Quantum Electron.*, **QE-8**, 848–850.
Beattie, J. R. (1957). 'The anomalous skin effect and the infra-red properties of silver and aluminium.' *Physica*, **23**, 898–902.
Behm, K. (1976). 'Group delay in CVD-fabricated fibres with diffused stairlike index profiles.' *A.E.U.*, **30**, 329–331.
Behm, K. (1977). 'Dispersion in CVD-fabricated fibres with a refractive index dip on the fibre axis.' *A.E.U.*, **31**, 45–48.
Behm, K. (1978). 'Dispersion of CVD-fabricated fibres with a finite number of layers, theory and measurement.' *A.E.U.*, **32**, 403–408.
Belanov, A. S., E. M. Dianov, G. I. Ezhov and A. M. Prokhorov (1976). 'Propagation of normal modes in multilayer optical waveguides. I: Component fields and dispersion characteristics.' *Sov. J. Quantum Electron.*, **6**, 43–50.
Belanov, A. S., G. I. Ezhov, and W. W. Tscherny (1973). 'Wellenausbreitung in runden leitungen aus geschichteten dielektrika.' *A.E.U.*, **27**, 494–496.
Bianciardi, E., and V. Rizzoli (1977). 'Propagation in graded-core fibres: a unified numerical description.' *Optical and Quantum Electron.*, **9**, 121–133.
Biernson, G. and D. J. Kingsley (1965). 'Generalized plots of mode patterns in a cylindrical dielectric waveguide applied to retinal cores.' *IEEE Trans. Microwave Theory and Techniques*, **MTT-13**, 345–356.
Bird, T. S. (1977). 'Propagation and radiation characteristics of rib waveguide.' *Electron. Lett.*, **13**, 401–403.
Black, P. W., J. Irven, K. Byron, I. S. Few, and R. Worthington (1974). 'Measurements on waveguide properties of GeO_2–SiO_2-cored optical fibres.' *Electron. Lett.*, **10**, 239–240.
Blum, J. M., J. C. McGroddy, P. G. McMullin, K. K. Shih, A. W. Smith, and J. F. Ziegler (1975). 'Oxygen-implanted double-heterojunction GaAs/GaAlAs injection lasers.' *IEEE J. Quantum Electron.*, **QE-11**, 413–418.
Bogdankevich, O. V., V. S. Letokhov, and A. F. Suchkov (1969). 'Theory of the effects of excitation inhomogeneity on semiconductor lasers pumped with an electron beam.' *Sov. Phys.—Semiconductors*, **3**, 566–570; and *Fiz. Tekh. Poluprov.*, **3**, 665–670 (1969).
Bohm, D. (1951). *Quantum Theory*, Prentice-Hall, N. J.
Bond, W. L., B. G. Cohen, R. C. C. Leite, and A. Yariv (1963). 'Observation of the dielectric waveguide mode of light propagation in p–n junction.' *Appl. Phys. Letts.*, **2**, 57–59.
Born, M., and E. Wolf (1970). *Principles of Optics*, 4th edition, Pergamon Press, London.
Borner, M. (1966). 'Mehrstufiges Übertragungssystem für in Pulscodemodulation dargestellte Nachrichten.' DBP 1 254 513, Vol. 21, as quoted in Unger (1977).
Botez, D. (1978). 'Analytical approximation of the radiation confinement factor for the TE_0 mode of a double heterojunction laser.' *IEEE J. Quantum Electron.*, **QE-14**, 230–232.
Botez, D. (1978a). 'Near and far-field analytical approximations for the fundamental mode in symmetric waveguide DH lasers.' *R.C.A. Review*, **39**, 577–603.
Botez, D. (1979). 'Optimal cavity design for low-threshold-current-density operation of double-heterojunction diode lasers.' *Appl. Phys. Letts.*, **35**, 57–60.
Bouillie, R., A. Cozannet, K.-H. Steiner, and M. Treheux (1974). 'Ray delay in gradient waveguides

with arbitrary symmetric refractive profile.' *Applied Optics*, **13**, 1045–1049.
Brekhovskikh, L. M. (1960). *Waves in Layered Media*, Academic Press, New York.
Brinkmeyer, E. (1979). 'Spot size of graded-index single-mode fibers: profile-independent representation and new determination method.' *Applied Optics*, **18**, 932–937.
Brown, J. (1966). 'Electromagnetic momentum associated with waveguide modes.' *Proc. IEE.*, **113**, 27–34.
Butler, J. K., and J. B. Delaney (1978). 'A rigorous boundary value solution for the lateral modes of stripe geometry injection lasers.' *IEEE J. Quantum Electron.*, **QE-14**, 507–513.
Butler, J. K., and H. Kressel (1972). 'Transverse mode selection in injection lasers with widely spaced heterojunctions.' *J. Appl. Phys.*, **43**, 3403–3411.
Butler, J. K., and H. Kressel (1977). 'Design curves for double-heterojunction laser diodes.' *R.C.A. Review*, **38**, 542–558.
Butler, J. K., H. Kressel, and I. Ladany (1975). 'Internal optical losses in very thin c.w. heterojunction laser diodes.' *IEEE J. Quantum Electron.*, **QE-11**, 402–408.
Butler, J. K., and H. S. Sommers, Jr. (1978). 'Asymmetric modes in oxide stripe heterojunction lasers.' *IEEE J. Quantum Electron.*, **QE-14**, 413–417.
Butler, J. K., and C.-S. Wang (1976). 'Threshold behaviour of c.w. GaAs–AlGaAs injection lasers.' *IEEE J. Quantum Electron.*, **QE-12**, 165–168.
Butler, J. K., C. S. Wang, and J. C. Campbell (1976). 'Modal characteristics of optical stripline waveguides.' *J. Appl. Phys.*, **47**, 4033–4043.
Buus, J. (1977). 'The theory of the dielectric slab waveguide with complex refractive index applied to GaAs lasers.' *Proc. Seventh European Microwave Conference, Copenhagen*, Microwave Exhibitions and Publishers Ltd, Sevenoaks. pp. 29–33.
Buus, J. (1978). 'Detailed field model for DH stripe lasers.' *Optical and Quantum Electron.*, **10**, 459–474.
Buus, J. (1979). 'A theoretical investigation of semiconductor lasers.' *Ph.D. Thesis*, Technical University of Denmark.
Buus, J. (1979a). 'A model for the static properties of DH lasers.' *IEEE J. Quantum Electron.*, **QE-15**, 734–739.
Buus, J. (1980). 'Dispersion of TE modes in slab waveguides with reference to double heterostructure semiconductor lasers.' *Applied Optics*, **19**, 1987–1989.
Buus, J., and M. J. Adams (1979). 'Phase and group indices for double heterostructure lasers.' *Solid State and Electron Devices*, **3**, 189–195.
Bykov, V. P. and L. A. Vaĭnshteĭn (1965). 'Geometric optics of open resonators.' *Sov. Phys.—JETP*, **20**, 338–344.
Čada, M., J. Čtyroký, I. Gregora, and J. Schröfel (1979). 'WKB analysis of guided and semileaky modes in graded-index anisotropic optical waveguides.' *Optics Communications*, **28**, 59–63.
Carpenter, D. J., and C. Pask (1977). 'Geometric optics approach to optical fibre excitation by partially coherent sources.' *Optical and Quantum Electron.*, **9**, 373–382.
Carruthers, J. R., I. P. Kaminow, and L. W. Stulz (1974). 'Diffusion kinetics and optical waveguiding properties of outdiffused layers in lithium niobate and lithium tantalate.' *Applied Optics*, **13**, 2333–2342.
Case, K. M. (1972). 'On Wave propagation in inhomogeneous media.' *J. Math. Phys.*, **13**, 360.
Casey, H. C., Jr. (1978). 'Room-temperature threshold-current dependence of GaAs–Al_xGa_{1-x}As double-heterostructure lasers on x and active-layer thickness.' *J. Appl. Phys.*, **49**, 3684–3692.
Casey, H. C., Jr., and M. B. Panish (1975). 'Influence of Al_xGa_{1-x}As layer thickness on threshold current density and differential quantum efficiency for GaAs–Al_xGa_{1-x}As DH lasers.' *J. Appl. Phys.*, **46**, 1393–1395.
Casey, H. C., Jr., M. B. Panish, and J. L. Merz (1973). 'Beam divergence of the emission from double-heterostructure injection lasers.' *J. Appl. Phys.*, **44**, 5470–5475.
Casey, H. C., Jr., M. B. Panish, W. O. Schlosser, and T. L. Paoli (1974). 'GaAs–Al_xGa_{1-x}As heterostructure laser with separate optical and carrier confinement.' *J. Appl. Phys.*, **45**, 322–333.
Casey, H. C., Jr., S. Somekh, and M. Illegems (1975). 'Room-temperature operation of low-threshold separate-confinement heterostructure injection laser with distributed feedback.' *Appl. Phys. Letts.*, **27**, 142–144.
Casperson, L. W. (1976). 'Beam modes in complex lenslike media and resonators.' *J. Opt. Soc. Am.*, **66**, 1373–1379.
Casperson, L. W., and U. Ganiel (1977). 'The stability of modes in a laser resonator which contains an active medium.' *IEEE J. Quantum Electron.*, **QE-13**, 58–59.

Casperson, L. W., and T. S. Garfield (1979). 'Guided beams in concave metallic waveguides.' *IEEE J. Quantum Electron.*, **QE-15**, 491–496.
Caton, W. M. (1974). 'Propagation constants in diffused planar optical waveguides.' *Applied Optics*, **13**, 2755–2757.
Chang, C. T. (1979). 'Minimum dispersion in a single-mode step-index optical fibre.' *Applied Optics*, **18**, 2516–2522.
Chang, C. T. (1979a). 'Minimum dispersion at 1.55 μm for single-mode step-index fibres.' *Electron. Lett.*, **15**, 765–767.
Chang, W. S. G., and K. W. Loh (1972). 'Theoretical design of guided wave structure for electro-optical modulation at 10.6 μm.' *IEEE J. Quantum Electron.*, **QE-8**, 463–470.
Chaudhuri, B. B., and D. K. Paul (1978). 'Wave propagation through a hollow rectangular anisotropic dielectric guide.' *IEEE J. Quantum Electron.*, **QE-14**, 557–560.
Checcacci, P. F., R. Falciai, and A. M. Scheggi (1975). 'Ray-tracing technique for evaluating the dispersion characteristics of graded-index cylindrical fibres.' *Electron. Lett.*, **11**, 633–635.
Checcacci, P. F., R. Falciai, and A. M. Scheggi (1977). 'Influence of a central depression on the dispersion characteristics of a graded-index profile.' *Electron. Lett.*, **13**, 378–379.
Checcacci, P. F., R. Falciai, and A. M. Scheggi (1979). 'Whispering and bouncing modes in elliptical optical fibres.' *J. Opt. Soc. Am.*, **69**, 1255–1259.
Cherin, A. H., and E. J. Murphy (1975). 'An analysis of the effect of lossy coatings on the transmission energy in a multimode optical fiber.' *Bell System Tech. J.*, **54**, 1531–1548.
Cherny, V. V., G. A. Juravlev, A. I. Kirpa, I. L. Rylov, and V. P. Tjoy (1979). 'Self-filtering multilayer S-waveguides with absorption and radiation losses.' *IEEE J. Quantum Electron.*, **QE-15**, 1401–1404.
Chinone, N. (1977). 'Nonlinearity in power-output–current characteristics of stripe-geometry injection lasers.' *J. Appl. Phys.*, **48**, 3237–3243.
Cho, A. Y., A. Yariv, and P. Yeh (1977). 'Observation of confined propagation in Bragg waveguides.' *Appl. Phys. Letts.*, **30**, 471–472.
Choudhary, S., and L. B. Felsen (1977). 'Guided modes in graded index optical fibres.' *J. Opt. Soc. Am.*, **67**, 1192–1196.
Choudhary, S., and L. B. Felsen (1978). 'Asymptotic theory of ducted propagation.' *J. Acoust. Soc. Am.*, **63**, 661–666.
Chu, P. L. (1980). 'Calculation of impulse response of multimode optical fibre.' *Electron. Lett.*, **16**, 429–431.
Chua, S. J., and B. Thomas (1977). 'Optical waveguiding in semiconductor injection lasers and integrated optics.' *Optical and Quantum Electron.*, **9**, 15–32.
Citerne, J. (1980). 'Comment on "Higher-mode cutoff in elliptical dielectric waveguides."' *Electron. Lett.*, **16**, 13.
Clarricoats, P. J. B., and K. B. Chan (1970). 'Electromagnetic-wave propagation along radially inhomogeneous dielectric cylinders.' *Electron. Lett.*, **6**, 694–695.
Clarricoats, P. J. B., and K. B. Chan (1973), 'Propagation behaviour of cylindrical-dielectric-rod waveguides.' *Proc. IEE*, **120**, 1371–1378.
Cohen, L. G. (1976). 'Pulse transmission measurements for determining near optimal profile gradings in multimode borosilicate optical fibers.' *Applied Optics*, **15**, 1808–1814.
Cohen, L. G., I. P. Kaminow, H. W. Astle, and L. W. Stulz (1978). 'Profile dispersion effects on transmission bandwidths in graded index optical fibers.' *IEEE J. Quantum Electron.*, **QE-14**, 37–41.
Cohen, L. G., and Chinlon Lin (1977). 'Pulse delay measurements in the zero material dispersion wavelength region for optical fibers.' *Applied Optics*, **16**, 3136–3139.
Cohen, L. G., Chinlon Lin, and W. G. French (1979). 'Tailoring zero chromatic dispersion into the 1.5–1.6 μm low loss spectral region of single-mode fibers.' *Electron. Lett.*, **15**, 334–335.
Cohen, L. G., W. L. Mammel, and H. M. Presby (1980). 'Correlation between numerical predictions and measurements of single-mode fiber dispersion characteristics.' *Applied Optics*, **19**, 2007–2010.
Collin, R. E. (1960). *Field Theory of Guided Waves*, McGraw-Hill, New York.
Conwell, E. M. (1973). 'Modes in optical waveguides formed by diffusion.' *Appl. Phys. Letts.*, **23**, 328–329.
Conwell, E. M. (1974). 'Modes in anisotropic optical waveguides formed by diffusion.' *IEEE J. Quantum Electron.*, **QE-10**, 608–612.
Conwell, E. M. (1975). 'WKB approximation for optical guide modes in a medium with exponentially varying index.' *J. Appl. Phys.*, **46**, 1407.
Conwell, E. M. (1975a). ' "Buried modes" in planar media with nonmonotonically varying index of

refraction.' *IEEE J. Quantum Electron.*, **QE-11**, 217–218.
Cook, D. D., and F. R. Nash (1975). 'Gain-induced guiding and astigmatic output beam of GaAs lasers.' *J. Appl. Phys.*, **46**, 1660–1672.
Cook, J. S. (1977). 'Minimum impulse response in graded-index fibers.' *Bell System Tech. J.*, **56**, 719–728.
Cozens, J. R. and R. B. Dyott (1979). 'Higher-mode cutoff in elliptical dielectric waveguides.' *Electron. Lett.*, **15**, 558–559.
Crank, J. (1956). *The Mathematics of Diffusion*, Oxford University Press, Oxford.
Cross, M., and M. J. Adams (1972). 'Wave-guiding properties of stripe-geometry double heterostructure injection lasers.' *Solid State Electron.*, **15**, 919–921.
Cross, M., and M. J. Adams (1974). 'Effects of doping and free carriers on the refractive index of direct-gap semiconductors.' *Opto-electronics*, **6**, 199–216.
Daikoku, K., and A. Sugimura (1978). 'Direct measurement of wavelength dispersion in optical fibres—difference method.' *Electron. Lett.*, **14**, 149–151.
Dakss, M. L., L. Kuhn, P. F. Heidrich, and B. A. Scott (1970). 'Grating coupler for efficient excitation of optical guided waves in thin films.' *Appl. Phys. Letts.*, **16**, 523–525.
D'Asaro, L. A. (1973). 'Advances in GaAs junction lasers with stripe geometry.' *J. Luminescence*, **7**, 310–337.
Degnan, J. J. (1973). 'Waveguide laser mode patterns in the near and far field.' *Applied Optics*, **12**, 1026–1030.
Degnan, J. J. (1976). 'The waveguide laser: a review.' *Appl. Phys.*, **11**, 1–33.
Delaney, J. B., and J. K. Butler (1979). 'The effect of device geometry on lateral mode content of stripe geometry lasers.' *IEEE J. Quantum Electron.*, **QE-15**, 750–755.
Deschamps, G. A. (1971). 'Gaussian beam as a bundle of complex rays.' *Electron. Lett.*, **7**, 684–685.
Di Domenico, M., Jr. (1972). 'Material dispersion in optical fiber waveguides.' *Applied Optics*, **11**, 652–654.
Dil, J. G., and H. Blok (1973). 'Propagation of electromagnetic surface waves in a radially inhomogeneous optical waveguide.' *Opto-electronics*, **5**, 415–428.
Dumke, W. P. (1975). 'The angular beam divergence in double-heterojunction lasers with very thin active regions.' *IEEE J. Quantum Electron.*, **QE-11**, 400–402.
Dupuis, R. D., and P. D. Dapkus (1978). 'Room-temperature operation of distributed-Bragg-confinement $Ga_{1-x}Al_xAs$–GaAs lasers grown by metalorganic chemical vapor deposition.' *Appl. Phys. Letts.*, **33**, 68–69.
Dyment, J. C. (1967). 'Hermite–Gaussian mode patterns in GaAs junction lasers.' *Appl. Phys. Letts.*, **10**, 84–86.
Dyment, J. C. and L. A. D'Asaro (1967). 'Continuous operation of GaAs junction lasers on diamond heat sinks at 200 K.' *Appl. Phys. Letts.*, **11**, 292–294.
Dyment, J. C., L. A. D'Asaro, J. C. North, B. I. Miller, and J. E. Ripper (1972). 'Proton-bombardment formation of stripe-geometry heterostructure lasers for 300 K c.w. operation.' *Proc. IEEE*, **60**, 726–728.
Dyment, J. C., F. R. Nash, C. J. Hwang, G. A. Rozgonyi, R. L. Hartman, H. M. Marcos, and S. E. Haszko (1974). 'Threshold reduction by the addition of phosphorus to the ternary layers of double-heterostructure GaAs lasers.' *Appl. Phys. Letts.*, **24**, 481–484.
Dyott, R. B., and J. R. Stern (1971). 'Group delay in glass-fibre waveguide.' *Electron. Lett.*, **7**, 82–84.
Dyott, R. B., J. R. Cozens, and D. G. Morris (1979). 'Preservation of polarisation in optical fibre waveguides with elliptical cores.' *Electron. Lett.*, **15**, 380–382.
Eaglesfield, C. C. (1962). 'Optical pipeline: a tentative assessment.' *Proc. IEE*, **B109**, 26–32.
Epstein, P. S. (1930). 'Reflection of waves in an inhomogeneous absorbing medium.' *Proc. Nat. Acad. Sci. U.S.A.*, **16**, 627–637.
Eve, M. (1976). 'Rays and time dispersion in multimode graded core fibres.' *Optical and Quantum Electron.*, **8**, 285–293.
Eve, M. (1977). 'Statistical model for the prediction of the bandwidth of an optical route.' *Electron. Lett.*, **13**, 315–316.
Eve, M. (1978). 'Multipath time dispersion theory of an optical network.' *Optical and Quantum Electron.*, **10**, 41–51.
Eve, M., A. Hartog, R. Kashyap, and D. N. Payne (1978). 'Wavelength dependence of light propagation in long fibre links.' *Proc. Fourth European Conference on Optical Communication, Genoa*. pp. 58–63, Instituto Internazionale delle Communicazioni, Genova, Italy.
Eyges, L. (1978). 'Fiber optic guides of noncircular cross section.' *Applied Optics*, **17**, 1673–1674.

Eyges, L., P. Gianino, and P. Wintersteiner (1979). 'Modes of dielectric waveguides of arbitrary cross sectional shape.' *J. Opt. Soc. Am.*, **69**, 1226–1235.

Feit, M. D. (1979). 'Minimal dispersion refractive index profiles.' *Applied Optics*, **18**, 2927–2929.

Feit, M. D., and J. A. Fleck, Jr. (1978). 'Light propagation in graded-index optical fibers.' *Applied Optics*, **17**, 3990–3998.

Fiet, M. D., and J. A. Fleck, Jr. (1979). 'Calculation of dispersion in graded index multi-mode fibers by a propagating beam method.' *Applied Optics*, **18**, 2843–2851.

Feit, M. D., and J. A. Fleck, Jr. (1980). 'The computation of mode properties in optical fiber waveguides by a propagating beam method.' *Applied Optics*, **19**, 1154–1164.

Feit, M. D., and J. A. Fleck, Jr. (1980a). 'The computation of mode eigenfunctions in graded index optical fibers by the propagating beam method.' *Applied Optics*, **19**, 2240–2246.

Felsen, L. B. (1976). 'Evanescent waves.' *J. Opt. Soc. Am.*, **66**, 751–760.

Findakly, T., and C.-L. Chen (1978). 'Diffused optical waveguides with exponential profile: effects of metal-clad and dielectric overlay.' *Applied Optics*, **17**, 469–474.

Fink, H. J. (1976). 'Propagation of waves in optical waveguides with various dielectric and metallic claddings.' *IEEE J. Quantum Electron.*, **QE-12**, 365–367.

Fleming, J. W. (1976). 'Material and mode dispersion in Ge O_2.B_2O_3.SiO_2 glasses.' *J. Am Ceram. Soc.*, **59**, 503–507.

Fleming, J. W. (1978). 'Material dispersion in lightguide glasses.' *Electron Lett.*, **14**, 326–328.

Fletcher, A., T. Murphy, and A. Young (1954). 'Solutions of two optical problems.' *Proc. Roy. Soc., A.*, **223**, 216–225.

Fox, A. J. (1974). 'The grating guide—a component for integrated optics.' *Proc. IEEE*, **62**, 644–645.

French, W. G., J. P. MacChesney, P. B. O'Connor, and G. W. Tasker (1974). 'Optical waveguides with very low losses.' *Bell System Tech. J.*, **53**, 951–954.

French, W. G., G. W. Tasker, and J. R. Simpson (1976). 'Graded index fiber waveguides with borosilicate composition: fabrication techniques.' *Applied Optics*, **15**, 1803–1807.

Fukuma, M., J. Noda, and H. Iwasaki (1978). 'Optical properties of titanium-diffused $LiNbo_3$ strip waveguides.' *J. Appl. Phys.*, **49**, 3693–3698.

Furuta, H., H. Noda, and A. Ihaya (1974). 'Novel optical waveguide for integrated optics.' *Applied Optics*, **13**, 322–326.

Furuya, K., Y. Suematsu, J. Nayyer, S. Ishikawa, and F. Tagami (1975). 'External higher-index mode filters for band widening of multimode optical fibers.' *Appl. Phys. Letts.*, **27**, 456–458.

Gallagher, J. G. (1979). 'Mode dispersion of trapezoidal cross-section dielectric optical waveguides by the effective-index method.' *Electron. Lett.*, **15**, 734–735.

Gallagher, J. G., and R. M. de la Rue (1977). 'TE and TM mode analysis of planar ion-exchanged waveguides.' *Microwaves, Optics and Acoustics*, **1**, 215–219.

Gambling, W. A., and P. J. R. Laybourn (1970). 'Limiting bandwidth of a glass-fibre transmission line.' *Electron. Lett.*, **6**, 661–662.

Gambling, W. A., and H. Matsumura (1973). 'Pulse dispersion in a lens-like medium.' *Opto-electronics*, **5**, 429–437.

Gambling, W. A., and H. Matsumura (1977). 'Simple characterisation factor for practical single-mode fibres.' *Electron. Lett.*, **13**, 691–693.

Gambling, W. A., and H. Matsumura (1978). 'Propagation in radially-inhomogeneous single-mode fibre.' *Optical and Quantum Electron.*, **10**, 31–40.

Gambling, W. A., H. Matsumura, and C. M. Ragdale (1978). 'Wave propagation in a single-mode fibre with dip in the refractive index.' *Optical and Quantum Electron.*, **10**, 301–309.

Gambling, W. A., H. Matsumura, and C. M. Ragdale (1979). 'Zero total dispersion in graded-index single-mode fibres.' *Electron. Lett.*, **15**, 474–476.

Gambling, W. A., H. Matsumura, and C. M. Ragdale (1979a). 'Mode dispersion, material dispersion and profile dispersion in graded-index single-mode fibres.' *Microwaves, Optics and Acoustics*, **3**, 239–246.

Gambling, W. A., D. N. Payne, and H. Matsumura (1977). 'Effect of dip in the refractive index on the cut-off frequency of a single-mode fibre.' *Electron. Lett.*, **13**, 174–175.

Gambling, W. A., D. N. Payne, and H. Matsumura (1977a). 'Cut-off frequency in radially inhomogeneous single-mode fibre.' *Electron. Lett.*, **13**, 139–140.

Gandrud, W. B. (1971). 'Reduced modulator drive-power requirements for 10.6 μm guided waves.' *IEEE J. Quantum Electron.*, **QE-7**, 580–581.

Ganiel, U., and Y. Silberberg (1975). 'Stability of optical resonators with an active medium.' *Applied Optics*, **14**, 306–309.

Garmire, E. M. (1976). 'Propagation of i.r. light in flexible hollow waveguides: further discussion.' *Applied Optics*, **15**, 3037–3039.
Garmire, E. M., T. McMahon, and M. Bass (1976). 'Propagation of infrared light in flexible hollow waveguides.' *Applied Optics*, **15**, 145–150.
Garmire, E. M., T. McMahon, and M. Bass (1976a). 'Flexible infrared-transmission metal waveguides.' *Appl. Phys. Letts.*, **29**, 254–256.
Garmire, E. M., T. McMahon, and M. Bass (1977). 'Low-loss optical transmission through bent hollow metal waveguides.' *Appl. Phys. Letts.*, **31**, 92–94.
Garmire, E., T. McMahon, and M. Bass (1980). 'Flexible infrared waveguides for high-power transmission.' *IEEE J. Quantum Electron.*, **QE-16**, 23–32.
Garmire, E. M., and H. Stoll (1972). 'Propagation losses in metal-film-substrate optical waveguides.' *IEEE J. Quantum Electron.*, **QE-8**, 763–766.
Garside, B. K., T. K. Lim, and J. P. Marton (1980). 'Propagation characteristics of parabolic-index fiber modes: linearly polarized approximation.' *J. Opt. Soc. Am.*, **70**, 395–400.
Geckeler, S. (1975). 'Dispersion in optical waveguides with graded refractive index.' *Electron. Lett.*, **11**, 139–140.
Geckeler, S. (1978). 'Dispersion in optical fibers: new aspects.' *Applied Optics*, **17**, 1023–1029.
Geckeler, S. (1979). 'Compensation of profile dispersion in graded-index optical fibres.' *Electron. Lett.*, **15**, 682–683.
Gedeon, A. (1974). 'Comparison between rigorous theory and WKB-analysis of modes in graded-index waveguides.' *Optics Communications*, **12**, 329–332.
Geshiro, M., M. Matsuhara, and N. Kumagai (1978a). 'Truncated parabolic-index fiber with minimum mode dispersion.' *IEEE Trans. Microwave Theory and Techniques*, **MTT-26**, 115–119.
Geshiro, M., M. Ohtaka, M. Matsuhara, and N. Kumagai (1978). 'Modal analysis of strip-loaded diffused optical waveguides by a variational method.' *IEEE J. Quantum Electron.*, **QE-14**, 259–263.
Ghatak, A. K., and L. A. Kraus (1974). 'Propagation of waves in a medium varying transverse to the direction of propagation.' *IEEE J. Quantum Electron.*, **QE-10**, 465–467.
Gloge, D. (1971). 'Weakly guiding fibers.' *Applied Optics*, **10**, 2252–2258.
Gloge, D. (1971a). 'Dispersion in weakly-guiding fibers.' *Applied Optics*, **10**, 2442–2445.
Gloge, D. (1975). 'Propagation effects in optical fibers.' *IEEE Trans. Microwave Theory and Techniques*, **MTT-23**, 106–120.
Gloge, D. (1975a). 'Optical-fiber packaging and its influence on fiber straightness and loss.' *Bell System Tech. J.*, **54**, 245–262.
Gloge, D. (1979). 'Effect of chromatic dispersion on pulses of arbitrary coherence.' *Electron. Lett.*, **15**, 686–687; errata in *Electron Lett.*, **16**, 240 (1980).
Gloge, D., I. P. Kaminow, and H. M. Presby (1975). 'Profile dispersion in multimode fibers: measurement and analysis.' *Electron Lett.*, **11**, 469–471.
Gloge, D., and E. A. J. Marcatili (1973). 'Multimode theory of graded-core fibers.' *Bell System Tech. J.*, **52**, 1563–1578.
Gloge, D., and E. A. J. Marcatili (1973a). 'Impulse response of fibers with ring-shaped parabolic index distribution.' *Bell System Tech. J.*, **52**, 1161–1168.
Gloge, D., E. A. J. Marcatili, D. Marcuse, and S. D. Personick (1979). 'Dispersion properties of fibers.' In S. E. Miller and A. G. Chynoweth (Eds), *Optical Fiber Telecommunications*, Academic Press, New York.
Goell, J. E. (1969). 'A circular-harmonic computer analysis of rectangular dielectric waveguides.' *Bell System Tech. J.*, **48**, 2133–2160.
Goell, J. E. (1973). 'Rib waveguide for integrated optical circuits.' *Applied Optics*, **12**, 2797–2798.
Goodfellow, R. C., A. C. Carter, I. Griffith, and R. R. Bradley (1979). 'GaInAsP/InP fast, high-radiance, 1.05–1.3 μm wavelength LED's with efficient lens coupling to small numerical aperture silica optical fibers.' *IEEE Trans. Electron. Devices*, **ED-26**, 1215–1220.
Gordon, J. P. (1966). 'Optics of general guiding media.' *Bell System Tech. J.*, **45**, 321–332.
Gradshteyn, I. S., and I. W. Ryzhik (1965). *Table of Integrals, Series, and Products*, Academic Press, New York.
Gray, D. E. (Ed.) (1963). *American Institute of Physics Handbook*, McGraw-Hill, New York.
Gronthoud, A. G., and H. Blok (1978). 'The influence of bulk losses and bulk dispersion on the propagation properties of surface waves in a radially inhomogeneous optical waveguide.' *Optical and Quantum Electron.*, **10**, 95–106.
Gubanov, A. I., and B. N. Sharapov (1970). 'Threshold current of an injection laser with a p–n heterojunction.' *Sov. Phys.—Semiconductors*, **4**, 367–371; and *Fiz. Tekh. Poluprov.*, **4**, 433–438 (1970).

Hakki, B. W. (1973). 'Carrier and gain spatial profiles in GaAs stripe geometry lasers.' *J. Appl. Phys.*, **44**, 5021–5028.

Hakki, B. W. (1975). 'GaAs double heterostructure lasing behaviour along the junction plane.' *J. Appl. Phys.*, **46**, 292–302.

Hakki, B. W., and C. J. Hwang (1974). 'Mode control in GaAs large-cavity double-heterostructure lasers.' *J. Appl. Phys.*, **45**, 2168–2173.

Hakki, B. W., and T. L. Paoli (1975). 'Gain spectra in GaAs double-heterostructure injection lasers.' *J. Appl. Phys.*, **46**, 1299–1306.

Hall, D., A. Yariv, and E. Garmire (1970). 'Optical guiding and electro-optic modulation in GaAs epitaxial layers.' *Optics Communications*, **1**, 403–405.

Hall, D. R., E. K. Gorton, and R. M. Jenkins (1977). '10 μm propagation losses in hollow dielectric waveguides.' *J. Appl. Phys.*, **48**, 1212–1216.

Hall, R. N., G. E. Fenner, J. D. Kingsley, T. J. Soltys, and R. O. Carlson (1962). 'Coherent light emission from GaAs junctions.' *Phys. Rev. Letts.*, **9**, 366–368.

Hall, R. N., and D. J. Olechna (1963). 'Wave propagation in an active dielectric slab.' *J. Appl. Phys.*, **34**, 2565–2566.

Hamilton, M. C., D. A. Wille, and W. J. Miceli (1977). 'An integrated optical RF spectrum analyzer.' *Optical Engineering*, **16**, 475–478.

Hammer, J. M. (1976). 'Metal diffused stripe waveguides: approximate closed form solution for lower order modes.' *Applied Optics*, **15**, 319–320.

Hartog, A. H. (1979). 'Influence of waveguide effects on pulse-delay measurements of material dispersion in optical fibres.' *Electron. Lett.*, **15**, 632–634.

Hartog, A. H., and M. J. Adams (1977). 'On the accuracy of the WKB approximation in optical dielectric waveguides.' *Optical and Quantum Electron.*, **9**, 223–232.

Hartog, A. H., M. J. Adams, F. M. E. Sladen, A. Ankiewicz, and D. N. Payne (to be published). 'Comparison of measured and predicted bandwidth of graded-index multimode fibres.'

Hashimoto, M. (1975). 'The effect of an outer layer on propagation in a parabolic-index optical waveguide.' *Int. J. Electronics*, **39**, 579–582.

Hashimoto, M. (1976). 'Propagation of cladded inhomogeneous dielectric waveguides.' *IEEE Trans. Microwave Theory and Techniques*, **MTT-24**, 404–409.

Hashimoto, M. (1976a). 'A perturbation method for the analysis of wave propagation in inhomogeneous dielectric waveguides with perturbed media.' *IEEE Trans. Microwave Theory and Techniques*, **MTT-24**, 559–566.

Hashimoto, M., S. Nemoto, and T. Makimoto (1977). 'Analysis of guided waves along the cladded optical fiber: parabolic-index core and homogeneous cladding.' *IEEE Trans. Microwave Theory and Techniques*, **MTT-25**, 11–17.

Hass, G. (1965). In R. Kingslake (Ed.), *Applied Optics and Optical Engineering*, Vol. 3, Academic Press, New York. pp. 316–317.

Hatz, J., and E. Mohn (1967). 'Calculation of intrinsic threshold for TE and TM modes in GaAs laser diodes.' *IEEE J. Quantum. Electron.*, **QE-3**, 656–662.

Haus, H. A., and H. Kogelnik (1976). 'Electromagnetic momentum and momentum flow in dielectric waveguides.' *J. Opt. Soc. Am.*, **66**, 320–327.

Haus, H. A., and R. V. Schmidt (1976). 'Approximate analysis of optical waveguide grating coupling coefficients.' *Applied Optics.*, **15**, 774–781.

Hayashi, I., M. B. Panish, and P. W. Foy (1969). 'A low-threshold room-temperature injection laser.' *IEEE J. Quantum Electron.*, **QE-5**, 211–212.

Hayashi, I., M. B. Panish, P. W. Foy, and S. Sumski (1970). 'Junction lasers which operate continuously at room temperature.' *Appl. Phys. Lett.*, **17**, 109–111.

Hayashi, I., M. B. Panish, and F. K. Reinhart (1971). 'GaAs–Al$_x$Ga$_{1-x}$As double heterostructure injection lasers.' *J. Appl. Phys.*, **42**, 1929–1940.

Hazan, J. P., J. J. Bernard, and D. Kuppers (1977). 'Medium-numerical-aperture low-pulse-dispersion fibre.' *Electron Lett.*, **13**, 540–542.

Hazan, J. P., J. P. Cabanie, and J. J. Bernard (1978). 'Method of assessing index profile data.' *Electron. Lett.*, **14**, 416–418.

Heading, J. (1962). *An Introduction to Phase-Integral Methods*, Methuen, London.

Heyke, H. J, and M. H. Kuhn (1973). 'Dispersion characteristics of general gradient fibres.' *A.E.U.*, **27**, 235–238.

Hildebrand, F. B. (1956). *Introduction to Numerical Analysis*, McGraw-Hill, New York.

Hocker, G. B. (1976). 'Strip-loaded diffused optical waveguides.' *IEEE J. Quantum Electron.*, **QE-12**, 232–236.

Hocker, G. B., and W. K. Burns (1975). 'Modes in diffused optical waveguides of arbitrary index profile.' *IEEE J. Quantum Electron.*, **QE-11**, 270–276.
Hocker, G. B., and W. K. Burns (1977). 'Mode dispersion in diffused channel waveguides by the effective index method.' *Applied Optics*, **16**, 113–118.
Hondros, D., and P. Debye (1910). 'Elektromagnetishe Wellen on dielektrischen Drähten.' *Ann. Physik*, **32**, 465–476.
Horiguchi, M., Y. Ohmori, and T. Miya (1979). 'Evaluation of material dispersion using a nanosecond optical pulse radiator.' *Applied Optics*, **18**, 2223–2228.
Horiguchi, M., Y. Ohmori, and H. Takata (1980). 'Profile dispersion characteristics in high-bandwidth graded-index optical fibres.' *Applied Optics*, **19**, 3159–3167.
Horiguchi, M., and H. Osanai (1976). 'Spectral losses of low-OH-content optical fibres.' *Electron. Lett.*, **12**, 310–312.
Hotate, K., and T. Okoshi (1978). 'Formula giving single mode limit of optical fibre having arbitrary refractive-index profile.' *Electron. Lett.*, **14**, 246–248.
Hsieh, J. J., J. A. Rossi, and J. P. Donelly (1976). 'Room-temperature c.w. operation of GaInAsP/InP double-heterostructure diode lasers emitting at 1.1 μm.' *Appl. Phys. Letts.*, **28**, 709–711.
Huxley, L. G. H. (1947). *A Survey of the Principles and Practice of Waveguides*, Cambridge University Press, Cambridge.
Ikeda, M. (1974). 'Propagation characteristics of multimode fibres with graded core index.' *IEEE J. Quantum Electron.*, **QE-10**, 362–371.
Ikeda, M. (1978). 'Minimum total dispersion in multimode graded-index fibres.' *Optical and Quantum Electron.*, **10**, 1–8.
Ikuno, H. (1978). 'Analysis of wave propagation in inhomogeneous dielectric slab waveguides.' *IEEE Trans. Microwave Theory and Techniques*, **MTT-26**, 261–266.
Ikuno, H. (1979). 'Propagation constants of guided modes in graded-index fibre with polynomial-profile core.' *Electron. Lett.*, **15**, 762–763.
Irving, D. H., and A. E. Karbowiak (1979). 'Method of calculating the modal delay for optical fibres with arbitrary refractive-index profiles.' *Electron. Lett.*, **15**, 160–162.
Ishikawa, S., K. Furuya, and Y. Suematsu (1978). 'Vector wave analysis of broadband multimode optical fibres with optimum refractive-index distribution.' *J. Opt. Soc. Am.*, **68**, 577–583.
Jacobs, I. (1978). 'Atlanta fiber system experiment: overview.' *Bell Syst. Tech. J.*, **57**, 1717–1721.
Jacobsen, G. (1978). 'Exact field expressions for guided modes of a general class of graded-index fibres.' *Electron. Lett.*, **14**, 464–465.
Jacobsen, G., and J. J. Ramskov Hansen (1979). 'Propagation constants and group delays of guided modes in graded-index fibers: a comparison of three theories.' *Applied Optics*, **18**, 2837–2842.
Jacobsen, G., and J. J. Ramskov Hansen (1979a). 'Modified evanescent wave theory for evaluation of propagation constants and group delays of graded index fibers.' *Applied Optics*, **18**, 3719–3720.
Jacobsen, G., and J. J. Ramskov Hansen (1980). 'Detailed error estimates for first order WKB calculation method.' *Electron. Lett.*, **16**, 540–541.
Jacomme, L. (1975). 'A model for ray propagation in a multimode graded-index fibre.' *Optics Communications*, **14**, 134–138.
Jacomme, L. (1977). 'Approximations in the evaluation of the impulse response of nearly parabolic fibres.' *Optical and Quantum Electron.*, **9**, 197–202.
Jacomme, L. (1978). 'Wavelength dependent propagation in optical multimode fibers.' *Optics Communications*, **24**, 213–216.
Jacomme, L. and D. Rossier (1977). 'Parallel-beam impulse response of multimode fibres: numerical results.' *Optical and Quantum Electron.*, **9**, 203–208.
Janta, J., and J. Čtyroký (1978). 'On the accuracy of WKB analysis of TE and TM modes in planar graded-index waveguides.' *Optics Communications*, **25**, 49–52.
Jeunhomme, L. (1979). 'Dispersion minimisation in single-mode fibres between 1.3 μm and 1.7 μm.' *Electron. Lett.*, **15**, 478–479.
Johnson, M. (1977). 'Diffused waveguides: mode perturbation by thin-film cladding.' *Electron. Lett.*, **13**, 67–68.
Johnston, W. D., Jr. (1971). 'Characteristics of optically pumped platelet lasers of ZnO, CdS, CdSe, and $CdS_{0.6}Se_{0.4}$ between 300° and 80° K.' *J. Appl. Phys.*, **42**, 2731–2740.
Jürgensen, K. (1974). 'Comment on "Dispension minimisation in dielectric waveguides".' *Applied Optics*, **13**, 1289–1290.
Jürgensen, K. (1975). 'Dispersion optimised optical single-mode glass-fiber waveguides.' *Applied Optics*, **14**, 163–168.
Jürgensen, K. (1979). 'Dispersion minimum of monomode fibres.' *Applied Optics*, **18**, 1259–1261.

Kaiser, P. and H. W. Astle (1974). 'Low-loss single-material fibers made from pure fused silica.' *Bell System Tech. J.*, **53**, 1021–1039.

Kaiser, P., E. A. J. Marcatili, and S. E. Miller (1973). 'A new optical fiber.' *Bell System Tech. J.*, **52**, 265–269.

Kaminow, I. P., W. L. Mammel, and H. P. Weber (1974). 'Metal-clad optical waveguides: analytical and experimental study.' *Applied Optics*, **13**, 396–405.

Kaminow, I. P., and H. M. Presby (1977). 'Profile synthesis in multicomponent glass optical fibers.' *Applied Optics*, **16**, 108–112.

Kaminow, I. P., H. M. Presby, J. B. MacChesney, and P. B. O'Connor (1977). 'Ternary fiber glass composition for minimum modal dispersion over a range of wavelengths.' *Technical Digest (Post-Deadline Papers) of the Topical Meeting on Optical Fibre Transmission II, Williamsburg, Virginia, Paper PD5-1*, Optical Society of America, Washington, D.C.

Kaminow, I. P., J. R. Simpson, H. M. Presby, and J. B. MacChesney (1979). 'Strain birefringence in single-polarisation germanosilicate optical fibres.' *Electron. Lett.*, **15**, 677–679.

Kao, K. C., and G. A. Hockham (1966). 'Dielectric-fibre surface waveguides for optical frequencies.' *Proc. IEE*, **113**, 1151–1158.

Kapany, N. S., and J. J. Burke (1961). 'Fiber optics. IX. Waveguide effects.' *J. Opt. Soc. Am.*, **51**, 1067–1078.

Kapany, N. S., and J. J. Burke (1972). *Optical Waveguides*, Academic Press, New York.

Kapron, F. P. (1977). 'Maximum information capacity of fibre-optic waveguides.' *Electron. Lett.*, **13**, 96–97.

Kapron, F. P., and D. B. Keck (1971). 'Pulse transmission through a dielectric optical waveguide.' *Applied Optics*, **10**, 1519–1523.

Kapron, F. P., D. B. Keck, and R. D. Maurer (1970). 'Radiation losses in glass optical waveguides.' *Appl. Phys. Letts.*, **17**, 423–425.

Karbowiak, A. E., and D. H. Irving (1979). 'Modelling optical fibres with arbitrary refractive-index profiles.' *Optical and Quantum Electron.*, **11**, 507–516.

Kashima, N., and N. Uchida (1977). 'Excess loss caused by a lossy outer layer in multimode optical fibers.' *Applied Optics*, **16**, 1038–1040.

Kashima, N., and N. Uchida (1977a). 'Excess loss caused by an outer layer in multimode step-index fibres: experiment.' *Applied Optics*, **16**, 1320–1322.

Kashima, N., and N. Uchida (1978). 'Transmission characteristics of graded-index optical fibers with a lossy outer layer.' *Applied Optics*, **17**, 1199–1207.

Kashima, N., N. Uchida, and Y. Ishida (1977). 'Excess loss caused by the outer layer in a multimode step-index optical fiber: theory.' *Applied Optics*, **16**, 2732–2737.

Kashima, N., N. Uchida, N. Susa, and S. Seikai (1978). 'The influence of a fibre structure on optical loss in a graded-index fibre.' *Electron. Lett.*, **14**, 78–80.

Kawakami, S. (1975). 'Relation between dispersion and power-flow distribution in a dielectric waveguide.' *J. Opt. Soc. Am.*, **65**, 41–45.

Kawakami, S., and S. Nishida (1974). 'Characteristics of cutoff modes in a symmetrical five-layer slab optical waveguide with low-index intermediate layers.' *Electronics and Communications in Japan*, **57-C**, 121–128.

Kawakami, S., and S. Nishida (1974a). 'Anomalous dispersion of new doubly clad optical fibre.' *Electron. Lett.*, **10**, 38–40.

Kawakami, S., and S. Nishida (1974b). 'Characteristics of a doubly clad optical fiber with a low-index inner cladding.' *IEEE J. Quantum Electron.*, **QE-10**, 879–887.

Kawakami, S., and S. Nishida (1975). 'Perturbation theory of a doubly clad optical fiber with a low-index inner cladding.' *IEEE. J. Quantum Electron.*, **QE-11**, 130–138.

Kawakami, S., S. Nishida, and M. Sumi (1976). 'Transmission characteristics of W-type optical fibres.' *Proc. IEE*, **123**, 586–590.

Kawakami, S., and J.-I. Nishizawa (1968). 'An optical waveguide with the optimum distribution of the refractive index with reference to waveform distortion.' *IEEE Trans. Microwave Theory and Techniques*, **MTT-16**, 814–818.

Kawakami, S., and K. Ogusu (1978). 'Variational expression of the propagation constant for leaky modes.' *Electron. Lett.*, **14**, 73–75.

Keck, D. B., and R. Bouillie (1978). 'Measurements on high-bandwidth optical waveguides.' *Optics Communications*, **25**, 43–48.

Keck, D. B., R. D. Maurer, and P. C. Schultz (1973). 'On the ultimate lower limit of attenuation in glass optical waveguides.' *Appl. Phys. Letts.*, **22**, 307–309.

Keller, J. B. (1956). 'A geometrical theory of diffraction.' *Proc. Symp. Appl. Math.*, **8**, 27–52.
Keller, J. B., and S. I. Rubinow (1960). 'Asymptotic solution of eigenvalue problems.' *Ann. Phys.*, **9**, 24–75.
Keller, J. B., and W. Streifer (1971). 'Complex rays with an application to Gaussian beams.' *J. Opt. Soc. Am.*, **61**, 40–43.
Kennedy, D. P., and P. C. Murley (1966). 'Calculations of impurity atom diffusion through a narrow diffusion mask opening.' *I.B.M. Journal*, **10**, 6–12.
Kharadly, M. M. Z., and J. E. Lewis (1969). 'Properties of dielectric-tube waveguides.' *Proc. IEE*, **116**, 214–224.
Khular, E., A. Kumar, and A. K. Ghatak (1977). 'Effect of the refractive index dip on the propagation characteristics of step-index and graded index fibers.' *Optics Communications*, **23**, 263–267.
Kirchhoff, H. (1970). 'Optical wave propagation in self-focussing fibres.' *Proc. Conference on Trunk Telecommunications by Guided Waves, London*, IEE Conference Publication No. 71. pp. 69–72.
Kirchhoff, H. (1972). 'The solution of Maxwell's equation for inhomogeneous dielectric slabs.' *A.E.U.*, **26**, 537–541.
Kirchhoff, H. (1973). 'Wave propagation along radially inhomogeneous glass fibres.' *A.E.U.*, **27**, 13–18.
Kirkby, P. A., A. R. Goodwin, G. H. B. Thompson, and P. R. Selway (1977). 'Observations of self-focussing in stripe geometry semiconductor lasers and the development of a comprehensive model of their operation.' *IEEE J. Quantum Electron.*, **QE-13**, 705–719.
Kirkby, P. A., and G. H. B. Thompson (1972). 'The effect of double heterojunction waveguide parameters on the far field emission patterns of lasers.' *Opto-electronics*, **4**, 323–334.
Kirkby, P. A., and G. H. B. Thompson (1976). 'Channeled substrate buried heterostructure GaAs–(GaAl)As injection lasers.' *J. Appl. Phys.*, **47**, 4578–4589.
Kitayama, K.-I., S. Seikai, Y. Kato, N. Uchida, O. Fukuda, and K. Inada (1979). 'Transmission characteristics of long spliced graded-index optical fibers at 1.27 μm.' *IEEE J. Quantum Electron.*, **QE-15**, 638–642.
Knox, R. M., and P. P. Toulios (1970). 'Integrated circuits for the millimeter through optical frequency range.' In J. Fox (Ed.), *Proceedings of the MRI Symposium on Submillimeter Waves*, Polytechnic Press, Brooklyn. pp. 497–516.
Kobayashi, S., N. Shibata, S. Shibata and T. Izawa (1978). 'Characteristics of optical fibres in infrared wavelength region.' *Review of ECL*, **26**, 453–467.
Kogelnik, H. (1965). 'On the propagation of Gaussian beams of light through lenslike media including those with a loss or gain variation.' *Applied Optics*, **4**, 1552–1569.
Kogelnik, H. (1975). 'Theory of dielectric waveguides.' In T. Tamir (Ed.), *Integrated Optics*, Springer-Verlag, New York, pp. 13–81.
Kogelnik, H., and V. Ramaswamy (1974). 'Scaling rules for thin-film optical waveguides.' *Applied Optics*, **13**, 1857–1862.
Kogelnik, H., and C. V. Shank (1971). 'Stimulated emission in a periodic structure.' *Appl. Phys. Letts.*, **18**, 152–154.
Kogelnik, H., and C. V. Shank (1972). 'Coupled-wave theory of distributed feedback lasers.' *J. Appl. Phys.*, **43**, 2327–2335.
Kogelnik, H., T. P. Sosnowski, and H. P. Weber (1973). 'A ray-optical analysis of thin-film polarization converters.' *IEEE J. Quantum Electron.*, **QE-9**, 795–800.
Kogelnik, H., and H. P. Weber (1974). 'Rays, stored energy, and power flow in dielectric waveguides.' *J. Opt. Soc. Am.*, **64**, 174–185.
Kokubun, Y., and K. Iga (1980). 'Mode analysis of graded index optical fibers using a scalar wave equation including gradient-index terms and direct numerical integration.' *J. Opt. Soc. Am.*, **70**, 388–394.
Kornhauser, E. T., and S. D. Yaghjian (1967). 'Model solution of a point source in a strongly focussing medium.' *Radio Science*, **2**, 299–310.
Krammer, H. (1976). 'Field configurations and propagation constants of modes in hollow rectangular dielectric waveguides.' *IEEE J. Quantum Electron.*, **QE-12**, 505–507.
Krammer, H. (1977). 'Propagation of modes in curved hollow metallic waveguides for the infrared.' *Applied Optics*, **16**, 2163–2165.
Kressel, H., and J. K. Butler (1977). *Semiconductor lasers and Heterojunction LED's*, Academic Press, New York.
Kressel, H., J. K. Butler, F. Z. Hawrylo, H. F. Lockwood and M. Ettenberg (1971). 'Mode guiding in

symmetrical (AlGa)As–GaAs heterojunction lasers with very narrow active regions.' *R.C.A. Review*, **32**, 393–401.

Kressel, H., and M. Ettenberg (1976). 'Low-threshold double heterojunction AlGaAs/GaAs laser diodes: theory and experiment.' *J. Appl. Phys.*, **47**, 3533–3537.

Kressel, H., H. F. Lockwood, and F. Z. Hawrylo (1971a). 'Low threshold LOC GaAs injection lasers.' *Appl. Phys. Letts.*, **18**, 43–45.

Kressel, H., and H. Nelson (1969). 'Close confinement gallium arsenide p–n junction lasers with reduced optical loss at room temperature.' *R.C.A. Review*, **30**, 106–113.

Kroemer, H. (1963). 'A proposed class of heterojunction injection lasers.' *Proc. IEEE*, **51**, 1782–1783.

Krumbholz, D., E. Brinkmeyer, and E.-G. Neumann (1980). 'Core/cladding power distribution, propagation constant, and group delay: Simple relation for power-low graded-index fibers.' *J. Opt. Soc. Am.*, **70**, 179–183.

Krupka, D. C. (1975). 'Selection of modes perpendicular to the junction plane in GaAs large-cavity double-heterostructure lasers.' *IEEE J. Quantum Electron.*, **QE-11**, 390–400.

Kuester, E. F., and R. C. Pate (1980). 'Fundamental mode propagation on dielectric fibres of arbitrary cross-section.' *IEE Proc.*, **127**, Part H, 41–51.

Kuhn, L., P. F. Heidrich, and E. G. Lean (1971). 'Optical guided wave mode conversion by an acoustic surface wave.' *Appl. Phys. Letts.*, **19**, 428–430.

Kuhn, M. H. (1974). 'The influence of the refractive index step due to the finite cladding of homogeneous fibres on the hybrid properties of modes.' *A.E.U.*, **28**, 393–401.

Kuhn, M. H. (1975). 'Optimum attenuation of cladding modes in homogeneous single mode fibres.' *A.E.U.*, **29**, 201–204.

Kuhn, M. H. (1975a). 'Lossy jacket design for multimode cladded core fibres.' *A.E.U.*, **29**, 353–355.

Kumar, A., and A. K. Ghatak (1976). 'Perturbation theory to study the guided propagation through cladded parabolic index fibres.' *Optica Acta*, **23**, 413–419.

Kumar, A., and E. Khular (1978). 'A perturbation analysis for modes in diffused waveguides with a Gaussian profile.' *Optics Communications*, **27**, 349–352.

Kumar, A., K. Thyagarajan, and A. K. Ghatak (1974). 'Modes in inhomogeneous slab waveguides.' *IEEE J. Quantum Electron.*, **QE-10**, 902–904.

Kuroda, T., M. Nakamura, K. Aiki, and J. Umeda (1978). 'Channeled-substrate-planar structure $Al_xGa_{1-x}As$ lasers: an analytical waveguide study.' *Applied Optics*, **17**, 3264–3267.

Kurtz, C. N. (1975). 'Scalar and vector mode relations in gradient-index light guides.' *J. Opt. Soc. Am.*, **65**, 1235–1240.

Kurtz, C. N., and W. Streifer (1969). 'Guided waves in inhomogeneous focussing media. Part I: Formulation, solution for quadratic inhomogeneity.' *IEEE Trans. Microwave Theory and Techniques*, **MTT-17**, 11–15.

Kurtz, C. N., and W. Streifer (1969a). 'Guided waves in inhomogeneous focussing media. Part II: Asymptotic solution for general weak inhomogeneity.' *IEEE Trans. Microwave Theory and Techniques*, **MTT-17**, 250–253.

Laakmann, K. D., and W. H. Steier (1976). 'Waveguides: characteristic modes of hollow rectangular dielectric waveguides.' *Applied Optics*, **15**, 1334–1340.

Lang, R. (1979). 'Lateral transverse mode instability and its stabilization in stripe geometry injection lasers.' *IEEE J. Quantum Electron.*, **QE-15**, 718–726.

Langer, R. E. (1949). 'The asymptotic solutions of ordinary linear differential equations of the second order, with special reference to a turning point.' *Trans. Amer. Math. Soc.*, **67**, 461–490.

Lasher, G. J. (1963). 'Threshold relations and diffraction loss for injection lasers.' *IBM Journal*, **7**, 58–61.

Lasher, G. J., and F. Stern (1964). 'Spontaneous and stimulated recombination radiation in semiconductors.' *Phys. Rev.*, **133**, A553–A563.

Laybourn, P. J. R. (1968). 'Group velocity of dielectric waveguide modes.' *Electron. Lett.*, **4**, 507–509.

Laybourn, P. J. R., and W. A. Gambling (1973). 'Bandwidths of single-mode and multimode optical fibre.' *Optics Communications*, **8**, 195–200.

Lee, C. P., S. Margalit, and A. Yariv (1978). 'Waveguiding in an exponentially decaying gain medium.' *Optics Communications*, **25**, 1–4.

Lee, T. P., C. A. Burrus, B. I. Miller, and R. A. Logan (1975). '$Al_xGa_{1-x}As$ double-heterostructure rib waveguide injection laser.' *IEEE J. Quantum Electron.*, **QE-11**, 432–435.

Lenham, A. P., and D. M. Treherne (1966). 'Optical constants of transition metals in the infrared.' *J. Opt. Soc. Am.*, **56**, 1137–1138.

Lewin, L., D. C. Chang, and E. F. Kuester (1977). *Electromagnetic waves and curved structures*, Peter Peregrinus (on behalf of IEE), Stevenage.

Lewis, J. E., and G. Deshpande (1979). 'Modes on elliptical cross-section dielectric tube waveguides.' *Microwaves, Optics and Acoustics*, **3**, 147–155.

Li, T. (1978). 'Optical fiber communication—the state of the art.' *IEEE Trans. Communications*, **COM-26**, 946–955.

Lim, T. K., B. K. Garside, and J. P. Marton (1979). 'Guided modes in fibres with parabolic-index core and homogeneous cladding.' *Optical and Quantum Electron.*, **11**, 329–344.

Lim, T. K., and J. W. Y. Lit (1978). 'A general approach for the analysis of optical waveguides.' *Optics Communications*, **26**, 36–40.

Lin, Chinlon, L. G. Cohen, W. G. French, and V. A. Foertmeyer (1978). 'Pulse delay measurements in the zero-material-dispersion region for germanium- and phosphorus-doped silica fibres.' *Electron. Lett.*, **14**, 170–172.

Lockwood, H. F., H. Kressel, H. S. Sommers, Jr., and F. Z. Hawrylo (1970). 'An efficient large optical cavity injection laser.' *Appl. Phys. Letts*, **17**, 499–502.

Logan, R. A., and F. K. Reinhart (1975). 'Integrated GaAs–$Al_xGa_{1-x}As$ double-heterostructure laser with independently controlled optical output divergence.' *IEEE J. Quantum Electron.*, **QE-11**, 461–464.

Lotsch, H. K. V. (1968). 'Reflection and refraction of a beam of light at a plane interface.' *J. Opt. Soc. Am.*, **58**, 551–561.

Lotspeich, J. F. (1975). 'Explicit general eigenvalue solutions for dielectric slab waveguide.' *Applied optics,*, **14**, 327–335.

Lotspeich, J. F. (1976). 'A perturbation analysis of modes in diffused optical waveguides with Gaussian index profile.' *Optics Communications*, **18**, 567–572.

Love, J. D. (1979). 'Power series solutions of the scalar wave equation for cladded, power-law profiles of arbitrary exponent.' *Optical and Quantum Electron.*, **11**, 464–466.

Love, J. D., and A. K. Ghatak (1979). 'Exact solutions for TM modes in graded index slab waveguides.' *IEEE J. Quantum Electron.*, **QE-15**, 14–16.

Love, J. D., and C. Pask (1976). 'Universal curves for power attenuation in ideal multimode fibres.' *Electron. Lett.*, **12**, 254–255.

Love, J. D., C. Pask, and C. Winkler (1979). 'Rays and modes on step-index multimode elliptical waveguides.' *Microwaves, Optics and Acoustics*, **3**, 231–238.

Love, J. D., R. A. Sammut, and A. W. Snyder (1979a). 'Birefringence in elliptically deformed optical fibres.' *Electron. Lett.*, **15**, 615–616.

Love, J. D., and A. W. Snyder (1976). 'Optical fiber eigenvalue equation; plane wave derivation.' *Applied Optics*, **15**, 2121–2125.

Love, J. D., and C. Winkler (1977). 'Attenuation and tunnelling coefficients for leaky rays in multilayered optical waveguides.' *J. Opt. Soc. Am.*, **67**, 1627–1633.

Love, J. D., and C. Winkler (1978). 'The step index limit of power law refractive index profiles for optical waveguides.' *J. Opt. Soc. Am.*, **68**, 1188–1191.

Luke, Y. L. (1962). *Integrals of Bessel Functions*, McGraw-Hill, New York.

Lukowski, T. I., and F. P. Kapron (1977). 'Parabolic fiber cutoffs: a comparison of theories.' *J. Opt. Soc. Am.*, **67**, 1185–1187.

Luther-Davies, B., D. N. Payne, and W. A. Gambling (1975). 'Evaluation of material dispersion in low-loss phosphosilicate-core optical fibres.' *Optics Communications*, **13**, 84–88.

Lyubimov, L. A., G. I. Veselov, and N. A. Bei (1961). 'Dielectric waveguide with elliptical cross section.' *Radio Eng. and Electron., USSR*, **6**, 1668–1677.

Maeda, M., and S. Yamada (1977). 'Leaky modes on W-fibers: mode structure and attenuation.' *Applied Optics*, **16**, 2198–2203.

Mallitson, I. H. (1965). 'Interspecimen comparison of the refractive index of fused silica.' *J. Opt. Soc. Am.*, **55**, 1205–1209.

Marcatili, E. A. J. (1964). 'Modes in a sequence of thick astigmatic lens-like focusers.' *Bell Syst. Tech. J.*, **43**, 2887–2904.

Marcatili, E. A. J. (1969). 'Dielectric rectangular waveguide and directional coupler for integrated optics.' *Bell Syst. Tech. J.*, **48**, 2071–2102.

Marcatili, E. A. J. (1974). 'Slab-coupled waveguides.' *Bell Syst. Tech. J.*, **53**, 645–674.

Marcatili, E. A. J. (1977). 'Modal dispersion in optical fibers with arbitrary numerical aperture and profile dispersion.' *Bell Syst. Tech. J.*, **56**, 49–63.

Marcatili, E. A. J., and R. A. Schmeltzer (1964). 'Hollow metallic and dielectric waveguides for long

distance optical transmission and lasers.' *Bell Syst. Tech. J.*, **43**, 1783–1809.
Marcatili, E. A. J., and S. E. Miller (1969). 'Improved relations describing directional control in electromagnetic wave guidance.' *Bell Syst. Tech. J.*, **48**, 2161–2187.
Marcuse, D. (1970). 'Modes and pseudomodes in dielectric waveguides.' *IEEE Trans. Microwave Theory and Techniques*, **MTT-18**, 62–63.
Marcuse, D. (1971). 'The coupling of degnerate modes in two parallel dielectric waveguides.' *Bell Syst. Tech. J.*, **50**, 1791–1816.
Marcuse, D. (1972). *Light Transmission Optics*, Van Nostrand Reinhold, New York.
Marcuse, D. (1972a). 'Hollow dielectric waveguide for distributed feedback lasers.' *IEEE J. Quantum Electron.*, **QE-8**, 661–669.
Marcuse, D. (1973). 'The impulse response of an optical fiber with parabolic index profile.' *Bell Syst. Tech. J.*, **52**, 1169–1174.
Marcuse, D. (1973a). 'The effect of the ∇n^2 term on the modes of an optical square-law medium.' *IEEE J. Quantum Electron.*, **QE-9**, 958–960.
Marcuse, D. (1973b). 'TE modes of graded-index slab waveguides.' *IEEE J. Quantum Electron.*, **QE-9**, 1000–1006.
Marcuse, D. (1974). *Theory of Dielectric Optical Waveguides*, Academic Press, New York.
Marcuse, D. (1974a). 'Theory of single-material fiber.' *Bell Syst. Tech. J.*, **53**, 1619–1641.
Marcuse, D. (1976). 'Elementary derivation of the phase shift at a caustic.' *Applied Optics*, **15**, 2949–2950.
Marcuse, D. (1976a). 'Steady-state losses of optical fibers and fiber resonators.' *Bell Syst. Tech. J.*, **55**, 1445–1462.
Marcuse, D. (1976b). 'Scattering and absorption losses of multimode optical fibers and fiber lasers.' *Bell Syst. Tech. J.*, **55**, 1463–1488.
Marcuse, D. (1977). 'Loss analysis of single-mode fiber splices.' *Bell Syst. Tech. J.*, **56**, 703–718.
Marcuse, D. (1978). 'Gaussian approximation of the fundamental modes of graded-index fibers.' *J. Opt. Soc. Am.*, **68**, 103–109.
Marcuse, D. (1979). 'Interdependence of waveguide and material dispersion.' *Applied Optics*, **18**, 2930–2932.
Marcuse, D. (1979a). 'Multimode fiber with z-dependent a-value.' *Applied optics*, **18**, 2229–2231.
Marcuse, D. (1979b). 'Calculation of bandwidth from index profiles of optical fibers. 1: Theory.' *Applied Optics*, **18**, 2073–2080.
Marcuse, D. (1979c). 'Multimode delay compensation in fibers with profile distortions.' *Applied Optics*, **18**, 4003–4005.
Marcuse, D. (1980). 'Calculation of bandwidth from index profiles of optical fibers: correction.' *Applied Optics*, **19**, 188–189.
Marcuse, D. (1980a). 'Pulse distortion in single-mode fibers.' *Applied Optics*, **19**, 1653–1660.
Marcuse, D. and W. L. Mammel (1973). 'Tube waveguide for optical transmission.' *Bell Syst. Tech. J.*, **52**, 423–435.
Marcuse, D., and H. M. Presby (1979). 'Fiber bandwidth-spectrum studies.' *Applied Optics*, **18**, 3242–3248.
Marcuse, D., and H. M. Presby (1979a). 'Effects of profile deformations on fiber bandwidth.' *Applied Optics*, **18**, 3758–3763.
Marhic, M. E. (1979). 'Polarization and losses of whispering-gallery waves along twisted trajectories.' *J. Opt. Soc. Am.*, **69**, 1218–1226.
Marhic, M. E., L. I. Kwan, and M. Epstein (1978). 'Optical surface waves along a toroidal metallic guide.' *Appl. Phys. Letts.*, **33**, 609–611.
Marhic, M. E., L. I. Kwan, and M. Epstein (1978a). 'Invariant properties of helical-circular metallic waveguides.' *Appl. Phys. Letts.*, **33**, 874–876.
Masaki, Y., M. Matsuhara, and N. Kumagai (1978). 'Effect of lossy cladding on modal dispersion characteristics of parabolic-index fiber.' *IEEE Trans. Microwave Theory and Techniques*, **MTT-26**, 852–855.
Masuda, M., and J. Koyama (1977). 'Effects of a buffer layer on TM modes in a metal-clad optical waveguide using Ti-diffused LiNbo$_3$ C-plate.' *Applied Optics*, **16**, 2994–3000.
Masuda, M., A. Tanji, Y. Ando, and J. Koyama (1977). 'Propagation losses of guided modes in an optical graded-index slab waveguide with metal cladding.' *IEEE Trans. Microwave Theory and Techniques*, **MTT-25**, 773–776.
Matsuhara, M. (1973). 'Analysis of TEM modes in dielectric waveguides by a variational method.' *J. Opt. Soc. Am.*, **63**, 1514–1517.

Matsuhara, M. (1973a). 'Analysis of electromagnetic-wave modes in lens-like media.' *J. Opt. Soc. Am.*, **63**, 135–138.

Matsumoto, T., and K. Nakagawa (1979). 'Wavelength dependence of spliced graded-index multimode fibers.' *Applied Optics*, **18**, 1449–1454.

Matsumura, H. (1975). 'The light acceptance angle of a graded index fibre.' *Optical and Quantum Electron.*, **7**, 81–86.

Matsumura, H., T. Suganuma, and T. Katsuyama (1980). 'Simple normalization of single mode fibres with arbitrary index profile.' *Proc. Sixth European Conference on Optical Communications, York, England.* pp. 103–106, I.E.E. Conference Publication Number 190.

McKelvey, R. (1959). 'Solution about a singular point of a linear differential equation involving a large parameter.' *Trans. Amer. Math. Soc.*, **91**, 410–424.

McLachlan, N. W. (1947). *Theory and Applications of Mathieu Functions*, Oxford, Clarendon Press.

McLevige, W. V., T. Itoh, and R. Mittra (1975). 'New waveguide structures for millimeter-wave and optical integrated circuits.' *IEEE Trans. Microwave Theory and Techniques*, **MTT-23**, 788–794.

McWhorter, A. L., H. J. Zeiger, and B. Lax (1963). 'Theory of semiconductor maser of GaAs.' *J. Appl. Phys.*, **34**, 235–236.

Meunier, J. P., J. Pigeon, and J. N. Massot (1980). 'Analyse perturbative des caractéristiques de propagation des fibres optiques à gradient d'indice quasi-parabolique.' *Optical and Quantum Electron.*, **12**, 41–49.

Meunier, J. P., J. Pigeon, and J. N. Massot (1980a). 'Perturbation theory for the evaluation of the normalised cut-of frequencies in radially inhomogeneous fibres.' *Electron. Lett.*, **16**, 27–29.

Midwinter, J. E. (1975). 'Propagating modes of a lens-like medium.' *Optical and Quantum Electron.*, **7**, 289–296.

Midwinter, J. E. (1979). *Optical Fibres for Transmission*, Wiley, New York.

Midwinter, J. E. (1979a). 'Progress in fibre optic transmission systems.' *Paper 3.2.1, Part II, Third World Telecommunications Forum*, International Telecommunications Union, Geneva.

Midwinter, J. E., and M. H. Reeve (1974). 'A technique for the study of mode cut-offs in multimode optical fibres.' *Opto-electronics*, **6**, 411–416.

Mikoshiba, K., and H. Kajioka (1978). 'Transmission characteristics of multimode W-type optical fiber: experimental study of the effect of the intermediate layer.' *Applied Optics*, **17**, 2836–2841.

Miles, R. O., and R. W. Grow (1978). 'Characteristics of a hollow-core distributed feedback CO_2 laser.' *IEEE J. Quantum Electron.*, **QE-14**, 275–283.

Miles, R. O., and R. W. Grow (1979). 'Propagation losses at 10.6 μm in hollow-core rectangular waveguides for distributed feedback applications.' *IEEE J. Quantum Electron.*, **QE-15**, 1396–1401.

Miller, S. E. (1965). 'Light propagation in generalized lenslike media.' *Bell Syst. Tech. J.*, **44**, 2017–2064.

Miller, S. E. (1969). 'Integrated optics: an introduction.' *Bell Syst. Tech. J.*, **48**, 2059–2069.

Miya, T., Y. Terunuma, T. Hosaka, and T. Miyashita (1979). 'Ultimate low-loss single-mode fibre at 1.55 μm.' *Electron. Lett.*, **15**, 106–108.

Miyagi, M., and S. Nishida (1977). 'Peculiar properties of cutoff frequencies in a dielectric optical tube waveguide.' *Proc. IEEE*, **65**, 1411–1412.

Miyagi, M., and S. Nishida (1979). 'An approximate formula for describing dispersion properties of optical dielectric slab and fiber waveguides.' *J. Opt. Soc. Am.*, **69**, 291–293.

Miyagi, M., and S. Nishida (1979a). 'Further discussions of approximate formulas describing dispersion properties of dielectric optical waveguides.' *J. Opt. Soc. Am.*, **69**, 1373–1376.

Miyagi, M., and S. Nishida (1979b). 'Pulse spreading in a single-mode fiber due to third-order dispersion.' *Applied Optics*, **18**, 678–682.

Miyagi, M., and S. Nishida (1979c). 'Pulse spreading in a single-mode optical fiber due to third-order dispersion: effect of optical source bandwidth.' *Applied Optics*, **18**, 2237–2240.

Miyamoto, T. (1980). 'Numerical analysis of a rib optical waveguide with trapezoidal cross section.' *Optics Communications*, **34**, 35–38.

Miyashita, T., M. Horiguchi, and A. Kawana (1977). 'Wavelength dispersion in a single-mode fibre.' *Electron. Lett.*, **13**, 227–228.

Mogensen, G. (1980). 'Wide-band optical fibre local distribution systems.' *Optical and Quantum Electron.*, **12**, 353–381.

Morishita, K., Y. Kondoh, and N. Kumagai (1980). 'On the accuracy of scalar approximation technique in optical fiber analysis.' *IEEE Trans. Microwave Theory and Techniques*, **MTT-28**, 33–36.

Morse, P. M. and H. Feshbach (1953). *Methods of Theoretical Physics*, McGraw-Hill, New York.

Moshkun, I., T.. J. M. Boyd, and R. H. C. Newton (1978). 'Guided modes of a four-layer optical waveguide with a metal buffer.' *Electron. Lett.*, **14**, 587–588.

Mur, G., and P. J. Fondse (1979). 'Computation of electromagnetic fields guided by a cylindrical inhomogeneity in a homogeneous medium of infinite extent.' *Microwaves, Optics, and Acoustics*, **3**, 224–230.

Muska, W. M., T. Li, T. P. Lee, and A. G. Dentai (1977). 'Material-dispersion-limited operation of high bit-rate optical-fibre data links using L.E.D.'s.' *Electron. Lett.*, **13**, 605–607.

Nakahara, M., S. Sudo, N. Inagaki, K. Yoshida, S. Shibaya, K. Kokura, and T. Kuroha (1980). 'Ultra wide bandwidth V. A. D. fibre.' *Electron. Lett.*, **16**, 391–392.

Nakamura, M., K. Aiki, J. Umeda, and A. Yariv (1975). 'C. W. operation of distributed-feedback GaAs–GaAlAs diode lasers at temperatures up to 300 K.' *Appl. Phys. Letts.*, **27**, 403–405.

Nakamura, M., H. W. Yen, A. Yariv, E. Garmire, and S. Somekh (1973). 'Laser oscillation in epitaxial GaAs waveguide with corrugation feedback.' *Appl. Phys. Letts.*, **23**, 224–225.

Nash, F. R. (1973). 'Mode guidance parallel to the junction plane of double-heterostructure GaAs lasers.' *J. Appl. Phys.*, **44**, 4696–4707.

Nash, F. R., W. R. Wagner, and R. L. Brown (1976). 'Threshold current variations and optical scattering losses in (Al,Ga)As double-heterostructure lasers.' *J. Appl. Phys.*, **47**, 3992–4005.

Nathan, M. I., W. P. Dumke, G. Burns, F. H. Dill, and G. J. Lasher (1962). 'Stimulated emission of radiation from GaAs p–n junctions.' *Appl. Phys. Letts.*, **1**, 62–64.

Nelson, D. F., and J. McKenna (1967). 'Electromagnetic modes of anisotropic dielectric waveguides at p–n junctions.' *J. Appl. Phys.*, **38**, 4057–4074.

Nelson, D. F., and F. K. Reinhart (1964). 'Light modulation by the electro-optic effect in reverse-biassed GaP p–n junctions.' *Appl. Phys. Letts.*, **5**, 148–150.

Nemoto, S., and T. Makimoto (1976). 'A relationship between phase and group indices of guided modes in dielectric waveguides.' *Int. J. Electronics*, **40**, 187–190.

Nemoto, S., and T. Makimoto (1976a). 'On the upper bound to a group index in slab dielectric waveguides.' *J. Opt. Soc. Am.*, **66**, 500–501.

Nemoto, S., and T. Makimoto (1977). 'Relation between phase and group indices in anisotropic inhomogeneous guiding media.' *J. Opt. Soc. Am.*, **67**, 124–126.

Nemoto, S., and G. L. Yip (1975). 'Impulse response of a self-focusing optical fibre.' *Electron. Lett.*, **11**, 191–192.

Nemoto, S., and G. L. Yip (1977). 'Impulse response of a self-focusing optical fiber.' *Applied Optics*, **16**, 705–710.

Nemoto, S., and G. L. Yip (1978). 'Propagation of rectangular and Gaussian pulses in a self-focusing optical fiber.' *Applied Optics*, **17**, 57–62.

Nishihara, H., T. Inoue, and J. Koyama (1974). 'Low-loss parallel-plate waveguide at 10.6 μm.' *Appl. Phys. Letts.*, **25**, 391–393.

Noda, J., S. Zembutsu, S. Fukunishi, and N. Uchida (1978). 'Strip-loaded waveguide formed in a graded-index $LiNbO_3$ planar waveguide.' *Applied Optics*, **17**, 1953–1958.

Norman, S. R., D. N. Payne, M. J. Adams, and A. M. Smith (1979). 'Fabrication of single-mode fibres exhibiting extremely low polarisation birefringence.' *Electron. Lett.*, **15**, 309–311.

North, D. O. (1979). 'Dielectric-waveguide-modal properties: a new analysis of the one-dimensional wave equation.' *IEEE J. Quantum Electron.*, **QE-15**, 17–26.

Nosu, K., and J. Hamasaki (1976). 'The influence of the longitudinal plasma wave on the propagation characteristics of a metal-clad dielectric-slab waveguide.' *IEEE J. Quantum Electron.*, **QE-12**, 745–748.

Nunes, F. D., N. B. Patel, J. G. Mendoza Alvarez, and J. E. Ripper (1979). 'Refractive-index profile and resonant modes in GaAs lasers.' *J. Appl. Phys.*, **50**, 3852–3857.

Ogusu, K. (1977). 'Numerical analysis of the rectangular dielectric waveguide and its modifications.' *IEEE Trans. Microwave Theory and Techniques*, **MTT-25**, 874–885.

Ogusu, K., S. Kawakami, and S. Nishida (1979). 'Optical strip waveguide: an analysis.' *Applied Optics*, **18**, 908–914; correction in *Applied Optics*, **18**, 3725.

Ohmori, Y., K. Chida, M. Horiguchi, and I. Hatakeyama (1978). 'Optimum profile parameter on graded-index optical fibre at 1.27 μm wavelength.' *Electron. Lett.*, **24**, 764–765.

Ohtaka, Y., S. Kawakami, and S. Nishida (1974). 'Transmission characteristics of a multi-layer dielectric slab optical waveguide with strongly evanescent wave layers.' *Trans. Inst. Electron. Commun. Eng., Japan*, **57-C**, 187–194.

Ohtaka, M., M. Matsuhara, and N. Kumagai (1976). 'Analysis of the guided modes in slab-coupled waveguides using a variational method.' *IEEE J. Quantum Electron.*, **QE-12**, 378–382.

Okamoto, K. (1979). 'Comparison of calculated and measured impulse responses of optical fibers.' *Applied Optics*, **18**, 2199–2206.

Okamoto, K., and T. Okoshi (1976). 'Analysis of wave propagation in optical fibers having core with α-power refractive-index distribution and uniform cladding.' *IEEE Trans. Microwave Theory and Techniques*, **MTT-24**, 416–421.

Okamoto, K., and T. Okoshi (1977). 'Computer-aided synthesis of the optimum refractive-index profile for a multimode fiber.' *IEEE Trans. Microwave Theory and Techniques*, **MTT-25**, 213–221.

Okamoto, K., and T. Okoshi (1978). Vectorial wave analysis of inhomogeneous optical fibers using finite element method.' *IEEE Trans. Microwave Theory and Techniques*, **MTT-26**, 109–114.

Okamoto, K., T. Okoshi, and K. Hotate (1979). 'A closed-form approximate dispersion formula for α-power graded-core fibers.' *Fiber and Integrated Optics*, **2**, 127–143.

Okoshi, T., and K. Okamoto (1974). 'Analysis of wave propagation in inhomogeneous optical fibers using a variational method.' *IEEE Trans. Microwave Theory and Techniques*, **MTT-22**, 938–945.

Oliner, A. A. (1976). 'Acoustic surface waveguides and comparisons with optical waveguides.' *IEEE Trans. Microwave Theory and Techniques*, **MTT-24**, 914–920.

Oliner, A. A., and S. T. Peng (1978). 'Effects of metal overlays on 3-D optical waveguides.' *Applied Optics*, **17**, 2866–2867.

Olshansky, R. (1976). 'Microbending loss of single mode fibres.' *Proc. Second European Conference on Optical Fibre Communication, Paris, paper IV.3*. pp. 101–103, S.E.E., Paris.

Olshansky, R. (1976a). 'Leaky modes in graded index optical fibers.' *Applied Optics*, **15**, 2773–2777.

Olshansky, R. (1976b). 'Pulse broadening caused by deviations from the optimal index profile.' *Applied Optics*, **15**, 782–788.

Olshansky, R. (1977). 'Effect of the cladding on pulse broadening in graded-index optical waveguides.' *Applied Optics*, **16**, 2171–2174.

Olshansky, R. (1978). 'Optical waveguides with low pulse dispersion over an extended spectral range.' *Electron Lett.*, **14**, 330–331.

Olshansky, R. (1979). 'Multiple-α index profiles.' *Applied Optics*, **18**, 683–689.

Olshansky, R. (1979a). 'Propagation in glass optical waveguides.' *Rev. Mod. Phys.*, **51**, 341–367.

Olshansky, R., and D. B. Keck (1976). 'Pulse broadening in graded-index optical fibers.' *Applied Optics*, **15**, 483–491.

Olshansky, R., and D. A. Nolan (1976). 'Mode-dependent attentuation of optical fibers: excess loss.' *Applied Optics*, **15**, 1045–1047.

Olshansky, R., and D. A. Nolan (1977). 'Mode dependent attentuation in parabolic optical fibers.' *Applied Optics*, **16**, 1639–1641.

Onoda, S., T. P. Tanaka, and M. Sumi (1976). 'W fiber design considerations.' *Applied Optics*, **15**, 1930–1935.

Osinski, M. (1976). 'Dielectric-slab and Epstein-layer electromagnetic models of injection lasers.' *A.E.U.*, **30**, 223–224.

Osinski, M. (1977). 'Epstein-layer and dielectric-slab electromagnetic models of semiconductor injection lasers.' *Optical and Quantum Electron.*, **9**, 361–377.

Osinski, M., and P. G. Eliseev (1979). 'Three-dimensional analysis of the mode properties of stripe-geometry d.h. lasers.' *Solid State and Electron Devices*, **3**, 215–219.

Osterberg, H., and L. W. Smith (1964). 'Transmission of optical energy along surface: Parts I and II.' *J. Opt. Soc. Am.*, **54**, 1073–1084.

Otto, A., and W. Sohler (1971). 'Modification of the total reflection modes in a dielectric film by one metal boundary.' *Optics Communications*, **3**, 254–258.

Pal, B. P., A. Kumar, and A. K. Ghatak (1980). 'Effect of axial refractive-index dip on zero total dispersion wavelength in single-mode fibres.' *Electron. Lett.*, **16**, 505–507.

Panish, M. B., I. Hayashi, and S. Sumski (1969). 'A technique for the preparation of low-threshold room-temperature GaAs laser diode structures.' *IEEE J. Quantum Electron.*, **QE-5**, 210–211.

Paoli, T. (1977). 'Waveguiding in a stripe-geometry junction laser.' *IEEE J. Quantum Electron.*, **QE-13**, 662–668.

Paoli, T., B. W. Hakki, and B. I. Miller (1973). 'Zero-order transverse mode operation of GaAs double-heterostructure lasers with thick waveguides.' *J. Appl. Phys.*, **44**, 1276–1280.

Parriaux, O., and F. E. Gardiol (1976). 'Propagation of guided electromagnetic waves in symmetrical circularly cylindrical structures.' *Wave Electron.*, **1**, 363–380.

Pask, C. (1978). 'On the derivation and interpretation of the Marcatili profile condition for optical fibres.' *Electron Lett.*, **14**, 13–15.

Pask, C. (1978a). 'Equal excitation of all modes on an optical fiber.' *J. Opt. Soc. Am.*, **68**, 572–576.

Pask, C. (1978b). 'Generalized parameters for tunneling ray attentuation in optical fibres.' *J. Opt. Soc. Am.*, **68**, 110–116.

Pask, C. (1979). 'Exact expressions for scalar modal eigenvalues and group delays in power-law optical fibers.' *J. Opt. Soc. Am.*, **69**, 1599–1603.

Pask, C. (1980). 'Pulse propagation in optical fibres with index profiles slowly varying along their length.' *Optical and Quantum Electron.*, **12**, 281–290.

Pask, C., and R. A. Sammut (1980). 'Experimental characterisation of graded-index single-mode fibres.' *Electron. Lett.*, **16**, 310–311.

Pask, C., A. W. Snyder, and D. J. Mitchell (1975). 'Number of modes on optical waveguides.' *J. Opt. Soc. Am.*, **65**, 356–357.

Paxton, K. B., and W. Streifer (1971). 'Analytic solution of ray equations in cylindrically inhomogeneous guiding media—Part 2: Skew rays.' *Applied Optics*, **10**, 1164–1171.

Payne, D. N. and W. A. Gambling (1974). 'New silica-based low-loss optical fibre.' *Electron. Lett.*, **10**, 289–290.

Payne, D. N., and W. A. Gambling (1974a). 'Preparation of water-free silica-based optical-fibre waveguide.' *Electron. Lett.*, **10**, 335–337.

Payne, D. N., and W. A. Gambling (1975). 'Zero material dispersion in optical fibres.' *Electron. Lett.*, **11**, 176–178.

Payne, D. N., and A. H. Hartog (1977). 'Determination of the wavelength of zero material dispersion in optical fibres by pulse-delay measurements.' *Electron. Lett.*, **13**, 627–629.

Pelosi, M., P. Vandenbulcke, C. D. W. Wilkinson, and R. M. de la Rue (1978). 'Propagation characteristics of trapezoidal cross-section ridge optical waveguides: an experimental and theoretical investigation.' *Applied Optics*, **17**, 1187–1193.

Peng, S. T., and A. A. Oliner (1977). 'Leakage and resonance effects on strip waveguides for integrated optics.' *Proc. International Conference on Integrated Optics and Optical Fibre Communications, Tokyo, Paper A2-4.* pp. 29–32, Institute of Electronics and Communication Engineers of Japan, Tokyo.

Personick, S. D. (1973). 'Receiver design for digital fiber-optic communication systems. Parts I and II.' *Bell Syst. Tech. J.*, **52**, 843–886.

Personick, S. D. (1973a). 'Baseband linearity and equalization in fiber optic digital communication systems.' *Bell Syst. Tech. J.*, **52**, 1175–1194.

Petermann, K. (1975). 'The design of W-fibres with graded index core.' *A.E.U.*, **29**, 485–487.

Petermann, K. (1975a). 'The mode attenuation in general graded core multimode fibres.' *A.E.U.*, **29**, 345–348.

Petermann, K. (1976). 'Theory of single-mode single-material fibres.' *A.E.U.*, **30**, 147–153.

Petermann, K. (1976a). 'Theory of microbending loss in monomode fibres with arbitrary refractive index profile.' *A.E.U.*, **30**, 337–342.

Petermann, K. (1977). 'Leaky mode behaviour of optical fibres with non-circularly symmetric refractive index profile.' *A.E.U.*, **31**, 201–204.

Petermann, K. (1977a). 'Uncertainties of the leaky mode correction for near-square-law optical fibres.' *Electron. Lett.*, **13**, 513–514.

Petermann, K. (1978). 'Modes in active waveguides with inhomogeneous gain profiles as applied to injection lasers.' *A.E.U.*, **32**, 313–320.

Piefke, G. (1964). 'Grundlagen sur Berechnung der Übertragungseigenschaften elliptischer Wellenleiter.' *A.E.U.*, **18**, 4–8.

Polky, J. N., and G. L. Mitchell (1974). 'Metal-clad planar dielectric waveguide for integrated optics.' *J. Opt. Soc. Am.*, **64**, 274–279.

Pöschl, G. and E. Teller (1933). 'Bemerkungen zur Quantenmechanik des Anharmonischen Oszillators.' *Zeits. Phys.*, **83**, 143–151.

Pratesi, R., and L. Ronchi (1978). 'Wave propagation in a nonparabolic graded-index medium.' *IEEE Trans. Microwave Theory and Techniques*, **MTT-26**, 856–858.

Pregla, R. (1974). 'A method for the analysis of coupled rectangular dielectric waveguides.' *A.E.U.*, **28**, 349–357.

Presby, H. M. (1977). 'Axial refractive index depression in preforms and fibers.' *Fiber and Integrated Optics*, **2**, 111–126.

Presby, H. M., and I. P. Kaminow (1976). 'Binary silica optical fibers: refractive index and profile dispersion measurements.' *Applied Optics*, **15**, 3029–3036; corrigenda in *Applied Optics*, **17**, 3530–3531 (1978).

Presby, H. M., D. Marcuse, and H. W. Astle (1978). 'Automatic refractive-index profiling of optical fibers.' *Applied Optics*, **17**, 2209–2214.
Presby, H. M., D. Marcuse, and L. G. Cohen (1979). 'Calculation of bandwidth from index profiles of optical fibers. 2: Experiment.' *Applied Optics*, **18**, 3249–3255.
Quist, T. M., R. H. Rediker, R. J. Keyes, W. E. Krag, B. Lax, A. L. McWhorter, and H. J. Zeiger (1962). 'Semiconductor maser of GaAs.' *Appl. Phys. Letts.*, **1**, 91–92.
Ramaswamy, V. (1974). 'Ray model of energy and power flow in anisotropic film waveguides.' *J. Opt. Soc. Am.*, **64**, 1313–1320.
Ramaswamy, V. (1974a). 'Strip-loaded film waveguide.' *Bell Syst. Tech. J.*, **53**, 697–704.
Ramaswamy, V., W. G. French, and R. D. Standley (1978a). 'Polarisation characteristics of noncircular-core single-mode fibres.' *Applied Optics*, **17**, 3014–3017.
Ramaswamy, V., I. P. Kaminow, P. Kaiser, and W. G. French (1978b). 'Single polarization optical fibres: exposed cladding technique.' *Appl. Phys. Letts.*, **33**, 814–816.
Ramaswamy, V., R. D. Standley, P. Sze, and W. G. French (1978). 'Polarisation effects in short length, single-mode fibres.' *Bell Syst. Tech. J.*, **57**, 635–651.
Ramskov Hansen, J. J. (1978). 'Pulse broadening in near-optimum graded index fibres.' *Optical and Quantum Electron.*, **10**, 521–526.
Ramskov Hansen, J. J., M. J. Adams, A. Ankiewicz, and F. M. E. Sladen (1980a). 'Near-field correction factors for elliptical fibres.' *Electron. Lett.*, **16**, 580–581.
Ramskov Hansen, J. J., A. Ankiewkz, and M. J. Adams (1980). 'Attenuation of leaky modes in graded noncircular multimode fibres.' *Electron. Lett.*, **16**, 94–96.
Ramskov Hansen, J. J., and G. Jacobsen (1979). 'Modal fields in graded-index fibers: comparison between perturbation theory and evanescent field theory.' *Applied Optics*, **18**, 3–5.
Ramskov Hansen, J. J., and E. Nicolaisen (1978). 'Propagation in graded-index fibers: comparison between experiment and three theories.' *Applied Opitics*, **17**, 2831–2835.
Rashleigh, S. C. (1976). 'Positive-permittivity-metal cladding; its effect on the modes of dielectric optical waveguides.' *Applied Optics*, **15**, 2804–2811.
Rashleigh, S. C. (1976a). 'Four-layer metal-clad thin film optical waveguide.' *Optical and Quantum Electron.*, **8**, 49–60.
Rashleigh, S. C. (1976b). 'Equal-phase-velocity propagation for the TE and TM modes of planar metal-clad optical waveguides.' *Optical and Quantum Electron.*, **8**, 241–253.
Rashleigh, S. C. (1976c). 'A partially reflecting metal cladding: its effect on the modes of dielectric optical waveguides.' *Optical and Quantum Electron.*, **8**, 433–452.
Rashleigh, S. C., and R. Ulrich (1978). 'Polarization mode dispersion in single-mode fibers.' *Optics Letters*, **3**, 60–62.
Rawson, E. G., D. R. Herriott, and J. McKenna (1970). 'Analysis of refractive index distributions in cylindrical, graded index glass rods (GRIN rods) used as image relays.' *Applied Optics*, **9**, 753–759.
Reeve, M. H., and J. E. Midwinter (1975). 'Studies of tunnelling from the guided modes of a multimode fibre.' *Proc. First European Conference on Optical Fibre Communications, London.* pp. 16–18, I.E.E. Conference Publication Number 132.
Reinhart, F. K., and R. A. Logan (1974). 'Monolithically integrated AlGaAs double heterostructure optical components.' *Appl. Phys. Letts.*, **25**, 622–624.
Reinhart, F. K., and R. A. Logan (1975). 'GaAs–AlGaAs double heterostructure lasers with taper-coupled passive waveguides.' *Appl. Phys. Letts.*, **26**, 516–518.
Reinhart, F. K., and R. A. Logan (1975a). 'Integrated electro-optic intracavity frequency modulation of double-heterostructure injection laser.' *Appl. Phys. Letts.*, **27**, 532–534.
Reinhart, F. K., R. A. Logan, and T. P. Lee (1974). 'Transmission properties of rib waveguides formed by anodization of epitaxial GaAs on $Al_x Ga_{1-x}As$ layers.' *Appl. Phys. Letts.*, **24**, 270–272.
Reinhart, F. K., and B. I. Miller (1972). 'Efficient $GaAs–Al_xGa_{1-x}As$ double heterostructure light modulators.' *Appl. Phys. Letts.*, **20**, 36–38.
Reinhart, F. K., J. C. Shelton, R. A. Logan, and B. W. Lee (1980). 'MOS rib waveguide polarizers.' *Appl. Phys. Letts.*, **36**, 237–240.
Reisinger, A. (1973). 'Characteristics of optical guided modes in lossy waveguides.' *Applied Optics*, **12**, 1015–1025.
Reisinger, A. (1973a). 'Attenuation properties of optical waveguides with a metal boundary.' *Appl. Phys. Letts.*, **23**, 237–239.
Rengarajan, S. R., and J. E. Lewis (1980). 'First higher-mode cutoff in two-layer elliptical fibre waveguides.' *Electron. Lett.*, **16**, 263–264.

Rengarajan, S. R., and J. E. Lewis (1980a). 'Propagation characteristics of elliptical dielectric-tube waveguides.' *IEE Proc.*, **127**, Part H, 121–126.

Rigrod, W. W., J. H. McFee, M. A. Pollack, and R. A. Logan (1975). 'Index-profile determination of heterostructure GaAs planar waveguides from mode-angle measurements at 10.6 μm wavelength.' *J. Opt. Soc. Am.*, **65**, 46–55.

Ripper, J. E., J. C. Dyment, L. A. D'Asaro, and T. L. Paoli (1971). 'Stripe-geometry double heterostructure junction lasers: mode structure and c.w. operation above room temperature.' *Appl. Phys. Letts.*, **18**, 155–157.

Roberts, R. (1970). 'Propagation characteristics of multimode dielectric waveguides at optical frequencies.' *Proc. Conference on Trunk Telecommunications by Guided Waves*, IEE Conference Publication Number 71 (London). pp. 39–44.

Roberts, R. (1975). 'Attenuation characteristics of multimode optical fibre with lossy cladding and lossy jacket.' *Electron. Lett.*, **11**, 529–530.

Rollke, K. H., and W. Sohler (1977). 'Metal-clad waveguide as cutoff polarizer for integrated optics.' *IEEE J. Quantum Electron.*, **QE-13**, 141–145.

Ronchi, L., R. Pratesi, and G. Pieraccini (1980). 'Wave propagation in a slab of transversally inhomogeneous medium, with gain or loss variations.' *J. Opt. Soc. Am.*, **70**, 191–197.

Rosenbaum, J. (1980). 'Perturbation analysis of clad parabolic-index waveguides.' *Optical and Quantum Electron.*, **12**, 51–56.

Rosenbaum, J., and R. Coren (1980). 'Pulse dispersion in a clad lens-like medium.' *Optical and Quantum Electron.*, **12**, 109–118.

Rousseau, M., and J. Arnaud (1977). 'Microbending loss of multimode square-law fibres: a ray theory.' *Electron. Lett.*, **13**, 265–267.

Rousseau, M., and J. Arnaud (1978). 'Ray theory of microbending.' *Optics Communications*, **25**, 333–336.

Rozzi, T., T. Itoh, and L. Grun (1977). 'Two-dimensional analysis of the GaAs double hetero stripe-geometry laser.' *Radio Science*, **12**, 543–549.

Rozzi, T. E., J. H. C. Van Heuven, and G. H. In'tveld (1978). 'A new D. H. Laser configuration with passive transverse field confinement.' *Electron. Lett.*, **14**, 87–88.

Rudolph, H. D., and E. G. Neumann (1976). 'Approximations for the eigenvalues of the fundamental mode of a step-index glass fibre waveguide.' *Nachrichtentech. Z.*, **29**, 328–329.

Rütze, U. (1977). 'Rigorous analysis of the optical rib-guide using rectangular waveguide modes.' *A.E.U.*, **31**, 88–90.

Safaai-Jazi, A., and G. L. Yip (1978). 'Cut-off conditions in three-layer cylinder dielectric waveguides.' *IEEE Trans. Microwave Theory and Techniques*, **MTT-26**, 898–903.

Sammut, R. A. (1977). 'Pulse dispersion in partially-excited graded-index fibres.' *Optical and Quantum Electron.*, **9**, 61–74.

Sammut, R. A. (1978). 'Range of monomode operation of W-fibres.' *Optical and Quantum Electron.*, **10**, 509–514.

Sammut, R. A. (1979). 'Analysis of approximations for the mode dispersion in monomode fibres.' *Electron. Lett.*, **15**, 590–591.

Sammut, R. A. (1979a). 'Perturbation theory of low-v optical fibres with dip in the refractive index.' *Optical and Quantum Electron.*, **11**, 147–151.

Sammut, R. A. (1980). 'Birefringence in graded-index monomode fibres with elliptical cross-section.' *Electron. Lett.*, **16**, 156–157.

Sammut, R. A., and A. K. Ghatak (1978). 'Perturbation theory of optical fibres with power-law core profile.' *Optical and Quantum Electron.*, **10**, 475–482.

Sammut, R. A., and A. W. Snyder (1980). 'Graded monomode fibres and planar waveguides.' *Electron. Lett.*, **16**, 32–34.

Sasaki, I., D. N. Payne, and M. J. Adams (1980). 'Measurement of refractive-index profiles in optical-fibre preforms by spatial-filtering technique.' *Electron. Lett.*, **16**, 219–221.

Savatinova, I., and E. Nadjakov (1975). 'Modes in diffused optical waveguides (parabolic and Gaussian models).' *Appl. Phys.*, **8**, 245–250.

Schechter, R. S. (1967). *The Variational Method in Engineering*, McGraw-Hill, N.Y.

Scheggi, A. M., P. F. Checcacci, and R. Falciai (1975). 'Dispersion characteristics in quasi-step and graded-index slab waveguides by ray-tracing technique.' *J. Opt. Soc. Am.*, **65**, 1022–1026.

Schiff, L. I. (1955). *Quantum Mechanics*, McGraw-Hill, N.Y.

Schlosser, W. O. (1964). 'Der rechtige dielektrische Draht.' *A.E.U.*, **18**, 403–410.

Schlosser, W. O. (1965). 'Die Störung der Eigenwerte des runden dielektrischen Drahtes bei schmacher elliptischer Deformation der Randkontur.' *A.E.U.*, **19**, 1–8.

Schlosser, W. O. (1972). 'Delay distortion in weakly guiding optical fibres due to elliptic deformation of the boundary.' *Bell Syst. Tech. J.*, **51**, 487–492.

Schlosser, W. O. (1973). 'Gain-induced modes in planar structures.' *Bell Syst. Tech. J.*, **52**, 887–905.

Schlosser, W. O., and H.-G. Unger (1966). 'Partially filled waveguides and surface waveguides of rectangular cross-section.' In *Advances in Microwaves*, Vol. 1 (Ed. L. Young), Academic Press, N. Y. pp. 319–387.

Schneider, H., H. Harms, A. Papp, and H. Aulich (1978). 'Low birefringence single-mode optical fibres: preparation and polarization characteristics.' *Applied Optics*, **17**, 3035–3037.

Scifres, D. R., W. Streifer, and R. D. Burnham (1976). 'Leaky wave room-temperature double heterostructure GaAs:GaAlAs diode laser.' *Appl. Phys. Letts.*, **29**, 23–25.

Selway, P. R. (1976). 'Semiconductor lasers for optical communications.' *Proc. IEE*, **123**, 609–618.

Selway, P. R., and A. R. Goodwin (1972). 'The properties of double heterostructure lasers with very narrow active regions.' *J. Phys D.*, **5**, 904–915.

Senmoto, S., and K. Okura (1976). 'Optical fiber cable transmission systems applied to telecommunication networks in Japan.' *Paper XII.5, presented at the Second European Conference on Optical Fibre Communications, Paris*, S.E.E., Paris.

Sharapov, B. N. (1970). 'Threshold currents of injection lasers with one and two heterojunctions.' *Sov. Phys.—Semiconductors*, **4**, 948–953; and *Fiz. Tekh. Poluprov.*, **4**, 1121–1129.

Sharma, A. B., J. Saijonmaa, and S. J. Halme (1980). 'Comparison of pulse responses of step-index fibres.' *Electron. Lett.*, **16**, 557–559.

Shellan, J. B., W. Ng, P. Yeh, A. Yariv, and A. Cho (1978). 'Transverse Bragg-reflector injection lasers.' *Optics Letters*, **2**, 136–138.

Shelton, J. C., F. K. Reinhart, and R. A. Logan (1979). 'Characteristics of rib waveguides in AlGaAs.' *J. Appl. Phys.*, **50**, 6675–6687.

Shen, C. C., J. J. Hsieh, and T. A. Lind (1977). '1500-h continuous c.w. operation of double-heterostructure GaInAsP/InP lasers.' *Appl. Phys. Letts.*, **30**, 353–354.

Shevchenko, V. V. (1974). 'Waves in a focussing optical dielectric waveguide.' *Sov. Radio Engineering and Electronic Physics*, **19**, 1–7.

Shore, K. A., and M. J. Adams (1976). 'Theory of the double heterostructure laser: II. Waveguide model incorporating carrier-concentration-dependent refractive index.' *Optical and Quantum Electron.*, **8**, 373–381.

Silberberg, Y., and U. Levy (1979). 'Modal treatment of an optical fiber with a modified hyperbolic secant index distribution.' *J. Opt. Soc. Am.*, **69**, 960–963.

Sladen, F. M. E. (1978). 'Fibres for Optical Communications.' *Ph.D. Thesis*, University of Southampton.

Sladen, F. M. E., D. N. Payne, and M. J. Adams (1976). 'Determination of optical fiber refractive index profiles by a near-field scanning technique.' *Appl. Phys. Letts.*, **28**, 255–258.

Sladen, F. M. E., D. N. Payne, and M. J. Adams (1977). 'Measurement of profile dispersion in optical fibres: a direct technique.' *Electron. Lett.*, **13**, 212–213.

Sladen, F. M. E., D. N. Payne, and M. J. Adams (1978). 'Profile dispersion measurements for optical fibres over the wavelength range 350 nm to 1900 nm.' *Proc. Fourth European Conference on Optical Communications, Genoa.* pp. 48–57, Instituto Internazionale Delle Communicazioni, Genova, Italy.

Sladen, F. M. E., D. N. Payne, and M. J. Adams (1979). 'Definitive profile-dispersion data for germania-doped silica fibres over an extended wavelength range.' *Electron. Lett.*, **15**, 469–470.

Smith, A. G. (1978). 'Polarisation and magneto-optic properties of single-mode optical fibre.' *Applied Optics*, **17**, 52–56.

Smith, G. E. (1968). 'Phase matching in four-layer optical waveguides.' *IEEE J. Quantum Electron.*, **QE-4**, 288–289.

Smith, L., and E. Snitzer (1973). 'Dispersion minimisation in dielectric waveguides.' *Applied Optics*, **12**, 1592–1599.

Smith, L., and E. Snitzer (1974). 'Author's reply to comments on "Dispersion minimisation in dielectric waveguides".' *Applied Optics*, **13**, 1290.

Smith, P. W. (1971). 'A waveguide gas laser.' *Appl. Phys. Letts.*, **19**, 132–134.

Snitzer, E. (1961). 'Cylindrical dielectric waveguide modes.' *J. Opt. Soc. Am.*, **51**, 491–498.

Snitzer, E., and H. Osterberg (1961). 'Observed dielectric waveguide modes in the visible spectrum.' *J. Opt. Soc. Am.*, **51**, 499–504.

Snyder, A. W. (1969). 'Asymptotic expressions for eigenfunctions and eigenvalues of a dielectric or optical waveguide.' *IEEE Trans. Microwave Theory and Techniques*, **MTT-17**, 1130–1138.

Snyder, A. W. (1972). 'Power loss on optical fibres.' *Proc. IEEE*, **60**, 757–758.

Snyder, A. W. (1974). 'Leaky-ray theory of optical waveguides of circular cross-section.' *Appl. Phys.*, **4**, 273–298.

Snyder, A. W., and J. D. Love (1975). 'Reflection at a curved dielectric interface—electromagnetic tunneling.' *IEEE Trans. Microwave Theory and Techniques*, **MTT-23**, 134–141.

Snyder, A. W., and J. D. Love (1976). 'Attenuation coefficient for tunnelling leaky rays in graded fibres.' *Electron. Lett.*, **12**, 324–326.

Snyder, A. W. and J. D. Love (1976a). 'Attenuation coefficient for rays in graded fibres with absorbing cladding.' *Electron. Lett.*, **12**, 255–257.

Snyder, A. W., and D. J. Mitchell (1974). 'Leaky rays on circular optical fibers.' *J. Opt. Soc. Am.*, **64**, 599–607.

Snyder, A. W., D. J. Mitchell, and C. Pask (1974). 'Failure of geometric optics for analysis of circular optical fibers.' *J. Opt. Soc. Am.*, **64**, 608–614.

Snyder, A. W., and R. A. Sammut (1979). 'Dispersion in graded single-mode fibres.' *Electron. Lett.*, **15**, 269–270.

Snyder, A. W., and R. A. Sammut (1979a). 'Fundamental (HE_{11}) modes of graded optical fibers.' *J. Opt. Soc. Am.*, **69**, 1663–1671.

Snyder, A. W., and W. R. Young (1978). 'Modes of optical waveguides.' *J. Opt. Soc. Am.*, **68**, 297–309.

Sohler, W. (1973). 'Light-wave coupling to optical waveguides by a tapered cladding medium.' *J. Appl. Phys.*, **44**, 2343–2345.

Someda, C. G., and M. Zoboli (1975). 'Cut off wavelengths of guided modes in optical fibres with square-law core profile.' *Electron. Lett.*, **11**, 602–603.

South, C. R. (1979). 'Total dispersion in step-index monomode fibres.' *Electron. Lett.*, **15**, 394–395.

Southwell, W. H. (1977). 'Inhomogeneous optical waveguide lens analysis.' *J. Opt. Soc. Am.*, **67**, 1004–1009.

Spain, B., and M. G. Smith (1970). *Functions of Mathematical Physics*, Van Nostrand Reinhold, London.

Standley, R. D., and V. Ramaswamy (1974). 'Nb-diffused $LiTaO_3$ optical waveguides: Planar and embedded strip guides.' *Appl. Phys. Letts.*, **25**, 711–713.

Steinberg, R. A., and T. G. Giallorenzi (1976). 'Performance limitations imposed on optical waveguide switches and modulators by polarization.' *Applied Optics*, **15**, 2440–2453.

Steinberg, R. A., and T. G. Giallorenzi (1977). 'Modal fields of anisotropic channel waveguides.' *J. Opt. Soc. Am.*, **67**, 523–533.

Steiner, K.-H. (1973). 'Multimode waveguides.' *Nachrichtentechn. Z.*, **26**, 468–473.

Steiner, K.-H. (1973a). 'A perturbation method for treating dielectric gradient waveguides.' *A.E.U.*, **27**, 87–90.

Steiner, K.-H. (1974). 'A delay formula for arbitrary ray paths in graded-index media.' *N.T.Z.*, **27**, 250–252.

Steiner, K.-H. (1974a). 'Ray delay in gradient waveguides with asymmetric transverse refractive index profiles.' *Opto-electronics*, **6**, 401–409.

Stern, F. (1973). 'Gain-current relation for GaAs lasers with n-type and undoped active layers.' *IEEE J. Quantum Electron.*, **QE-9**, 291–294.

Stewart, G., C. A. Millar, P. J. R. Laybourn, C. D. W. Wilkinson, and R. M. de la Rue (1977). 'Planar optical waveguides formed by silver-ion migration in glass.' *IEEE J. Quantum Electron.*, **QE-13**, 192–200.

Stewart, W. J. (1975). 'End launching of, and emission from, leaky modes in graded fibres.' *Electron. Lett.*, **11**, 516–518.

Stewart, W. J. (1975a). 'Leaky modes in graded fibres.' *Electron. Lett.*, **11**, 321–322.

Stewart, W. J. (1975b). 'A new technique for determining the v values and refractive index profiles of optical fibres.' *Paper Tu D8-1, Technical Digest of the OSA/IEEE Topical Meeting on Optical Fiber Transmission, Williamsburg, Virginia*, Optical Society of America, Washington, D.C.

Stewart, W. J. (1975c). 'Measured mode spectra in multimode optical fibres and comparison with theory.' *Paper PD6-1, Technical Digest of the OSA/IEEE Topical Meeting on Optical Fiber Transmission, Williamsburg, Virginia*, Optical Society of America, Washington, D.C.

Stewart, W. J. (1980). 'Simplified parameter-based analysis of single-mode optical guides.' *Electron. Lett.*, **16**, 380–382.

Stolen, R. H. (1975). 'Modes in fiber optical waveguides with ring index profiles.' *Applied Optics*, **14**, 1533–1537.

Stolyarov, S. N. (1970). 'The effect of absorption on waveguide properties of layers with variable parameters.' *Izv. VUZ Radiofiz.*, **13**, 749–756.

Stolyarov, S. N. (1972). 'Influence of waveguide properties of heterojunction layers on the principal characteristics of injection lasers.' *Sov. Quantum Electron.*, **2**, 144–149; and *Kvontanaya Elektronika*, **2**, 69–76.
Stratton, J. A. (1941). *Electromagnetic Theory*, McGraw-Hill, N.Y.
Streifer, W., R. D. Burnham, and D. R. Scifres (1976). 'Substrate radiation losses in GaAs heterostructure lasers.' *IEEE J. Quantum Electron.*, **QE-12**, 177–182.
Streifer, W., R. D. Burnham, and D. R. Scifres (1976a). 'Analysis of grating-coupled radiation in GaAs:GaAlAs lasers and waveguides—II. Blazing effects.' *IEEE J. Quantum Electron.*, **QE-12**, 494–499.
Streifer, W., R. D. Burnham, and D. R. Scifres (1979). 'Symmetrical and asymmetrical waveguiding in very narrow conducting stripe lasers.' *IEEE J. Quantum Electron.*, **QE-15**, 136–141.
Streifer, W., and E. Kapon (1979). 'Application of the equivalent-index method to DH diode lasers.' *Applied Optics*, **18**, 3724–3725.
Streifer, W., and C. N. Kurtz (1967). 'Scalar analysis of radially inhomogeneous guiding media.' *J. Opt. Soc. Am.*, **57**, 779–786.
Streifer, W., and K. B. Paxton (1971). 'Analytic solution of ray equations in cylindrically inhomogeneous guiding media—1: Meridional rays.' *Applied Optics*, **10**, 769–775.
Streifer, W., D. R. Scifres, and R. D. Burnham (1978). 'Analysis of gain-induced waveguiding in stripe geometry diode lasers.' *IEEE J. Quantum Electron.*, **QE-14**, 418–427.
Stutius, W., and W. Streifer (1977). 'Silicon nitride films on silicon for optical waveguides.' *Applied Optics*, **16**, 3218–3222.
Suematsu, Y., and K. Furuya (1972). 'Propagation mode and scattering loss of a two-dimensional dielectric waveguide with gradual distribution of refractive index.' *IEEE Trans. Microwave Theory and Techniques*, **MTT-20**, 524–531.
Suematsu, Y., and K. Furuya (1975). 'Quasi-guided modes and related radiation losses in optical dielectric waveguides with external higher index surroundings.' *IEEE Trans. Microwave Theory and Techniques*, **MTT-23**, 170–175.
Suematsu, Y., M. Hakuta, K. Furuya, K. Chiba, and R. Hasumi (1972). 'Fundamental transverse electric field (TE_0) mode selection for thin-film asymmetric light guides.' *Appl. Phys. Letts.*, **21**, 291–293.
Suematsu, Y., and M. Yamada (1973). 'Transverse mode control in semiconductor lasers.' *IEEE J. Quantum Electron.*, **QE-9**, 305–311.
Suematsu, Y., and M. Yamada (1974). 'Oscillation-modes and mode-control in semiconductor lasers with stripe-geometry.' *Trans. IECE*, **57-C**, 434–440.
Sugimura, A., K. Daikoku, N. Imoto, and T. Miya (1980). 'Wavelength dispersion characteristics of single-mode fibers in low-loss region.' *IEEE J. Quantum Electron.*, **QE-16**, 215–225.
Suhara, T., Y. Handa, H. Nishihara, and J. Koyama (1979). 'Analysis of optical channel waveguides and directional couplers with graded-index profile.' *J. Opt. Soc. Am.*, **69**, 807–815.
Sumner, G. T. (1977). 'A new technique for refractive index profile measurement in multimode optical fibres.' *Optical and Quantum Electron.*, **9**, 79–82.
Sun, M. J., and M. W. Muller (1977). 'Measurements on four-layer isotropic waveguides.' *Applied Optics*, **16**, 814–815.
Sutton, L. E., and O. N. Stavroudis (1961). 'Fitting refractive index data by least squares.' *J. Opt. Soc. Am.*, **51**, 901–905.
Szczepanek, P. S., and J. W. Berthold III (1978). 'Side launch excitation of selected modes in graded-index optical fibers.' *Applied Optics*, **17**, 3245–3247.
Takano, T., and J. Hamasaki (1972). 'Propagating modes of a metal-clad-dielectric slab waveguide for integrated optics.' *IEEE J. Quantum Electron.*, **QE-8**, 206–212.
Tanaka, T., and Y. Suematsu (1976). 'An exact analysis of cylindrical fiber with index distribution by matrix method and its application to focusing fiber.' *Trans. IECE, Japan*, **E59**, 1–8.
Tanaka, T. P., S. Onoda, and M. Sumi (1976). 'W-type optical fiber: relation between refractive index difference and transmission bandwidth.' *Applied Optics*, **15**, 1121–1122.
Tasker, G. W., W. G. French, J. R. Simpson, P. Kaiser, and H. M. Presby (1978). 'Low-loss single-mode fibers with different B_2O_3–SiO_2 compositions.' *Applied Optics*, **17**, 1836–1842.
Taylor, H. F. (1976). 'Dispersion characteristics of diffused channel waveguides.' *IEEE J. Quantum Electron.*, **QE-12**, 748–752.
Thompson, G. H. B. (1972). 'A theory of filamention in semiconductor lasers including the dependence of dielectric constant on injected carrier density.' *Opto-electronics*, **4**, 257–310.
Thompson, G. H. B. (1980). *Physics of Semiconductor Laser Devices*, Wiley, Chichester.

Thompson, G. H. B., G. D. Henshall, G. E. A. Whiteaway, and P. A. Kirkby (1976). 'Narrow-beam five layer (GaAl)As/GaAs heterostructure lasers with low threshold and high peak power.' *J. Appl. Phys.*, **47**, 1501–1514.

Thompson, G. H. B., and P. A. Kirkby (1973). '(GaAl)As lasers with a heterostructure for optical confinement and additional heterojunctions for extreme carrier confinement.' *IEEE J. Quantum Electron.*, **QE-9**, 311–318.

Thompson, G. H. B., and P. A. Kirkby (1973a). 'Low threshold-current density in 5-layer-heterostructure (GaAl)As/GaAs localized-gain-region injection lasers.' *Electron. Lett.*, **9**, 295–296.

Thyagarajan, K., and A. K. Ghatak (1974). 'Perturbation theory for studying the effect of the $\nabla \varepsilon$ term in lens-like media.' *Optics Communications*, **11**, 417–421.

Thyagarajan, K., and A. K. Ghatak (1977). 'Pulse broadening in optical fibers.' *Applied Optics*, **16**, 2583–2585.

Tien, P. K. (1971). 'Light waves in thin films and integrated optics.' *Applied Optics*, **10**, 2395–2413.

Tien, P. K. (1977). 'Integrated optics and new wave phenomena.' *Rev. Mod. Phys.*, **49**, 361–420.

Tien, P. K., J. P. Gordon, and J. R. Whinnery (1965). 'Focusing of a light beam of Gaussian field distribution in a continuous and periodic lens-like media.' *Proc. IEEE*, **53**, 129–136.

Tien, P. K., R. J. Martin, and S. Riva-Sanseverino (1975). 'Novel metal-clad optical components and method of insulating high-index substrates for forming integrated optical circuits.' *Appl. Phys. Letts.*, **27**, 251–253.

Tien, P. K., R. J. Martin, and G. Smolinsky (1973). 'Formation of light guiding interconnections in an integrated optical circuit by composite taper film coupling.' *Applied Optics*, **12**, 1909–1916.

Tien, P. K., R. J. Martin, R. Wolfe, R. C. Lecraw, and S. L. Blank (1972). 'Switching and modulation of light in magneto-optic waveguides of garnet films.' *Appl. Phys. Letts.*, **21**, 394–396.

Tien, P. K., S. Riva-Sanseverino, R. J. Martin, A. A. Ballman, and H. Brown (1974). 'Optical waveguide modes in single-crystalline $LiNbO_3$–$LiTaO_3$ solid solution films.' *Appl. Phys. Letts.*, **24**, 503–506.

Tien, P. K., and R. Ulrich (1970). 'Theory of prism-film coupler and thin-film light guides.' *J. Opt. Soc. Am.*, **60**, 1325–1337.

Tien, P. K., R. Ulrich, and R. J. Martin (1969). 'Modes of propagating light waves in thin deposited semiconductor films.' *Appl. Phys. Letts.*, **14**, 291–294.

Timmermann, C. C. (1974). 'Material dispersion in optical glass fibres.' *A.E.U.*, **28**, 144–145.

Timmermann, C. C. (1974a). 'Mode distribution and impulse response of general graded multimode fibres.' *A.E.U.*, **28**, 186–188.

Timmermann, C. C. (1974b). 'The influence of deviations from the square-law refractive index profile of gradient core fibres on mode dispersion.' *A.E.U.*, **28**, 344–346.

Timmermann, C. C. (1975). 'Airy function approximation of graded index fibre modes.' *A.E.U.*, **29**, 186–188.

Tjaden, D. L. A. (1978). 'First-order correction to "weak-guidance" approximation in fibre optics theory.' *Philips J. Res.*, **33**, 103–112.

Tjaden, D. L. A. (1978a). 'Birefringence in single-mode optical fibres due to core ellipticity.' *Philips J. Res.*, **33**, 254–263.

Tomlinson, W. J. (1977). 'Wavelength multiplexing in multimode optical fibers.' *Applied Optics*, **16**, 2180–2194.

Tsang, W. T. (1978). 'The effects of lateral current spreading, carrier out-diffusion and optical mode losses on the threshold current density of $GaAs$–$Al_xGa_{1-x}As$ stripe-geometry lasers.' *J. Appl. Phys.*, **49**, 1031–1044.

Tsang, W. T., and R. A. Logan (1979). '$GaAs$–$Al_xGa_{1-x}As$ strip buried heterostructure lasers.' *IEE J. Quantum Electron.*, **QE-15**, 451–469.

Tsang, W. T., R. A. Logan, and M. Ilegems (1978). 'High-power fundamental-transverse-mode strip buried heterostructure lasers with linear light–current characteristics.' *Appl. Phys. Letts.*, **32**, 311–314.

Tsuchiya, H., and N. Imoto (1979). 'Dispersion-free single-mode fibre in 1.5 μm wavelength region.' *Electron Lett.*, **15**, 476–478.

Tsuchiya, H., and J.-I. Sakai (1974). 'Characteristics of single-mode optical fiber with elliptical core' (in Japanese). *I.E.C.E.J. Trans.*, OQE **86**, 21–30.

Tsukada, T. (1974). '$GaAs$–$Ga_{1-x}Al_xAs$ buried-heterostructure injection lasers.' *J. Appl. Phys.*, **45**, 4899–4906.

Tsukada, T., H. Nakashima, J. Umeda, S. Nakamura, N. Chinone, R. Ito, and O. Nakada (1972).

'Very-low-current operation of mesa-stripe-geometry double heterostructure injection lasers.' *Appl. Phys. Letts.*, **20**, 344–345.

Uchida, N. (1976). 'Optical waveguide loaded with high refractive-index strip film.' *Applied Optics*, **15**, 179–182.

Uchida, N., O. Mikami, S. Uehara, and J. Noda (1976). 'Optical field distribution in a waveguide loaded with high refractive-index film: modulation efficiency improvement in a planar-type modulator.' *Applied Optics*, **15**, 455–458.

Ulrich, R., and W. Prettl (1973). 'Planar leaky light-guides and couplers.' *Appl. Phys.*, **1**, 55–68.

Unger, H.-G. (1971). 'The minimum threshold current of double-heterojunction lasers.' *A.E.U.*, **25**, 539–540.

Unger, H.-G. (1977). *Planar Optical Waveguides and Fibres*, Oxford University Press, Oxford.

Unger, H.-G. (1977a). 'Optical pulse distortion in glass fibres at the wavelength of minimum dispersion.' *A.E.U.*, **31**, 518–519.

Unger, K. (1967). 'Moden in einem Halbleiterlaser.' *Ann. Physik.*, **19**, 64–75.

Vali, V., and R. W. Shorthill (1977). 'Ring interferometer 950 m long.' *Applied Optics*, **16**, 290–291.

Van der Donk, J., and P. E. Lagasse (1980). 'Analysis of geodesic lenses by beam propagation method.' *Electron. Lett.*, **16**, 292–294.

Vassallo, C. (1979). 'Radiating normal modes of lossy planar waveguides.' *J. Opt. Soc. Am.*, **69**, 311–316.

Vassell, M. O. (1974). 'Structure of optical guided modes in planar multilayers of optically anisotropic materials.' *J. Opt. Soc. Am.*, **64**, 166–173.

Vassell, M. O. (1974a). 'Calculation of propagating modes in a graded-index optical fibre.' *Optoelectronics*, **6**, 271–286.

Vigants, A., and S. P. Schlesinger (1962). 'Surface waves on radially inhomogeneous cylinders.' *IRE Trans. Microwave Theory and Techniques*, **MTT-10**, 375–382.

Vincent, D., and J. W. Y. Lit (1976). 'Effects of a thin overlying film on optical waveguides and couplers.' *J. Opt. Soc. Am.*, **66**, 226–232.

Vincent, D., and J. W. Y. Lit (1977). 'Thin film beam splitter for integrated optics.' *J. Opt. Soc. Am.*, **67**, 533–538.

Voges, E. (1974). 'Losses and transverse mode selection in stripe-geometry double-heterojunction lasers.' *A.E.U.*, **28**, 183–186.

Voges, E. (1976). 'Variational calculation of dielectric bulge guides and couplers for integrated optics.' *A.E.U.*, **30**, 176–179.

Wang, S. (1973). 'Proposal of periodic layered waveguide structures for distributed lasers.' *J. Appl. Phys.*, **44**, 767–780.

Weierholt, A. (1979). 'Modal dispersion of optical fibres with a composite α-profile graded-index core.' *Electron. Lett.*, **15**, 733–734.

Weller, J. F., and T. G. Giallorenzi (1975). 'Indiffused waveguides: effects of thin film overlays.' *Applied Optics*, **14**, 2329–2330.

Werts, A. (1966). 'Propagation de la lumière cohérente dans les fibres optiques.' *L'Onde Electrique*, **45**, 967–980.

White, I. A., and C. Pask (1977). 'Effect of Goos–Hänchen shifts on pulse widths in optical waveguides.' *Applied Optics*, **16**, 2353–2355.

White, K. I. (1979). 'Practical application of the refracted near-field technique for the measurement of optical fibre refractive index profiles.' *Optical and Quantum Electron.*, **11**, 185–196.

White, K. I., and B. P. Nelson (1979). 'Zero total dispersion in step-index monomode fibres at 1.30 and 1.55 μm.' *Electron. Lett.*, **15**, 396–397.

Wilmot, D. W. and E. R. Schineller (1966). 'Optical waveguide modes in a bisected dielectric slab.' *J. Opt. Soc. Am.*, **56**, 839–840.

Yamada, R., and Y. Inabe (1974). 'Guided waves along graded index dielectric rod.' *IEEE Trans. Microwave Theory and Techniques*, **MTT-22**, 813–814.

Yamada, R., and Y. Inabe (1974a). 'Guided waves in an optical square-law medium.' *J. Opt. Soc. Am.*, **64**, 964–968.

Yamada, R., T. Meiri, and N. Okamoto (1977). 'Guided waves along an optical fiber with parabolic index profile.' *J. Opt. Soc. Am.*, **67**, 96–103.

Yamamoto, Y., T. Kamiya, and H. Yanai (1975). 'Propagation characteristics of a partially metal-clad optical guide: metal-clad optical strip line.' *Applied Optics*, **14**, 322–326.

Yamamoto, Y., T. Kamiya, and H. Yanai (1975a). 'Characteristics of optical guided modes in multilayer metal-clad planar optical guide with low-index dielectric buffer layer.' *IEEE J. Quantum Electron.*, **QE-11,** 729–736.

Yamanouchi, K., T. Kamiya, and K. Shibayama (1978). 'New leaky surface waves in anisotropic metal-diffused optical wave guides.' *IEEE Trans. Microwave Theory and Techniques*, **MTT-26,** 298–305.

Yariv, A., and R. C. C. Leite (1963). 'Dielectric waveguide mode of light propagation in p–n junction.' *Appl. Phys. Letts.*, **2,** 55–57.

Yasuura, K., K. Shimohara, and T. Miyamoto (1980). 'Numerical analysis of a thin-film waveguide by mode-matching method.' *J. Opt. Soc. Am.*, **70,** 183–191.

Yeh, C. (1962). 'Elliptical dielectric waveguides.' *J. Appl. Phys.*, **33,** 3235–3243.

Yeh, C. (1976). 'Modes in weakly guiding elliptical optical fibres.' *Optical and Quantum Electron.*, **8,** 43–47.

Yeh, C., L. Casperson, and B. Szejn (1978). 'Propagation of truncated Gaussian beams in multimode fiber guides.' *J. Opt. Soc. Am.*, **68,** 989–993.

Yeh, C., S. B. Dong, and W. Oliver (1975). 'Arbitrarily shaped inhomogeneous optical fiber or integrated optical waveguides.' *J. Appl. Phys.*, **46,** 2125–2129.

Yeh, C., K. Ha, S. B. Dong, and W. P. Brown (1979). 'Single-mode optical waveguides.' *Applied Optics*, **18,** 1490–1504.

Yeh, C., and G. Lindgren (1977). 'Computing the propagation characteristics of radially stratified fibers: an efficient method.' *Applied Optics*, **16,** 483–493.

Yeh, P., and A. Yariv (1976). 'Bragg reflection waveguides.' *Optics Communications*, **19,** 427–430.

Yeh, P., A. Yariv, and A. Y. Cho (1978). 'Optical surface waves in periodic layered media.' *Appl. Phys. Letts.*, **32,** 104–105.

Yeh, P., A. Yariv, and C.-S. Hong (1977). 'Electromagnetic propagation in periodic stratified media. I. General theory.' *J. Opt. Soc. Am.*, **67,** 423–438.

Yeh, P., A. Yariv, and E. Marom (1978a). 'Theory of Bragg fiber.' *J. Opt. Soc. Am.*, **68,** 1196–1201.

Yip, G. L., and Y. H. Ahmew (1974). 'Propagation characteristics of radially inhomogeneous optical fibre.' *Electron. Lett.*, **10,** 37–38.

Yip, G. L., and S. Nemoto (1975). 'The relations between scalar modes in a lenslike medium and vector modes in a self-focusing optical fiber.' *IEEE Trans. Microwave Theory and Techniques*, **MTT-23,** 260–263.

Yonezu, H., I. Sakuma, K. Kobayashi, T. Kamejima, M. Ueno, and Y. Nannichi (1973). 'A GaAs–$Al_xGa_{1-x}As$ double heterostructure planar stripe laser.' *Japan. J. Appl. Phys.*, **12,** 1585–1592.

Yust, M., N. Bar-Chaim, S. H. Izadpanah, S. Margalit, I. Ury, D. Wilt, and A. Yariv (1979). 'A monolithically integrated optical repeater.' *Appl. Phys. Letts.*, **35,** 795–797.

Zachos, T. H., and J. E. Ripper (1969). 'Resonant modes of GaAs junction lasers.' *IEEE J. Quantum Electron.*, **QE-5,** 29–37.

Zehe, A., and G. Röpke (1973). 'Light propagation in an inhomogeneously absorbing medium: near field of GaAs-light emitting diodes.' *Ann. Physik.*, **29,** 351–357.

Zemon, S., and D. Fellows (1976). 'Tunneling leaky modes in a parabolic index fiber.' *Applied Optics*, **15,** 1936–1941.

Index

Absorption 25, 240–241, 275
Acoustic surface waveguides 178
AlGaAs, lasers 43–44, 77–78, 81, 84
   refractive index 48
Alpha (power-law) profiles 278
   monomode 367
   multiple-alpha 314
   theory of 302–315, 348–350
Angle of acceptance, *see* Numerical aperture
Anisotropic dielectric media 18, 82, 158, 172, 187, 201
Anti-guidance 51–52, 58
Attenuation coefficient 25
   asymmetric metal-clad guides 64, 66, 68
   differential mode attenuation 242, 348
   fibres 241
   four-layer metal-clad guides 79–81
   hollow guides with metallic walls 27, 262–264
   leaky modes 292, 330–339
   symmetric metal-clad guides 70, 71
   W-fibres 273
   W-guides 88–89, 155
   weakly-guiding planar guides 28, 41, 159

Bandwidth xiv, 40, 223, 249, 352–353, 361–364
   alpha-profiles 303–320
   leaky mode effects 328–330, 354–355
   W-fibres 276
   see also Dispersion, Pulse-spreading
Beam-splitter (thin-film) 211
Beam waist (spot-size) 99–100, 102, 105, 109, 210
Bending loss 71, 241, 265
Birefringence 256, 258–260
Boundary conditions
   cladded parabolic-index guides 295
   dielectric interface 11–12
   dielectric slab guides 30, 31
   exponential profiles 116, 118, 120
   linear profiles 124–126
   mirrors/conducting walls 3, 7, 180–181, 219
   multilayer guides 76, 268–269
   rectangular dielectric guides 184
   step-index fibres 225
   and variational method 161, 344
   WKB approximation 148–150, 153–154
Bragg waveguides 75, 277
Brewster angle 14, 18, 19

Caustic 96, 115, 134, 146–150, 172, 214, 215, 251–252, 284–286, 298–300, 320, 331–332, 338–339
Channel waveguides 178, 202–206
Chemical vapour deposition (CVD) 239–240, 250, 275, 353, 355, 360–361, 366
   dopants 240, 275, 306–307
   see also Refractive index
Confinement factor (filling-factor) 41
   applied to heterostructure lasers 46–48, 209–210
   approximate result 48
   cladded-parabolic profile 167–169
   Epstein-layer profile 133
   exponential profile 121
   five-layer lasers 84–85
   gain-guidance 57
   linear profile 127–128
   and normalized variables 348–349
   parabolic-index media 103
   and phase/group velocity relation 159
   step-index fibres 237–239, 241, 245–247
   W-guides 86–87, 276
Connection formulae (WKB) 146–148, 150, 152–153, 299, 330–331
Conversion, mode 242, 314, 337
Critical angle 15, 18, 19, 276, 320
Cut-offs
   cladded-parabolic profiles 167–168, 295, 343, 357, 359
   conducting-wall guides 6
   dielectric slabs 21, 31, 32, 36, 52, 53, 195
   elliptical fibres 255, 256
   exponential profiles 115–121
   graded-index fibres 289, 290
   graded-index rectangular guides 203–204
   linear profiles 122, 124–128
   multilayer fibres 269–272, 274
   parabolic-index media 91, 100, 110
   single-mode fibres 366–369
   step-index fibres 226–228, 232, 236–237
   variational results 348, 350
   W-guides 86, 87

Detectors xii
Dielectric permittivity, complex 49, 104–105, 208–209
Diffusion 178, 201–207
   equation 203
Dip, central 305, 355–357, 360–361, 366–367

Dispersion
  chromatic xiii, 239, 245–249, 275, 308–315, 356–357
  intermodal 237, 242–243, 255, 259, 278, 285, 295, 303–315, 328–330, 343, 347–348, 352–355, 360–364
  intramodal 308–320
  material xiii, 38, 93, 223, 234, 239, 243–249, 275, 308–315, 352
  polarization 223, 256–260
  profile 38, 239, 247–249, 303–320, 352, 363–364
  waveguide 38, 39, 223, 234–237, 239, 245–249, 275, 351, 367
Distributed feedback (DFB) lasers xii, 74, 86

Effective index 37, 56, 183, 188–189, 191–193, 195–197, 201–206, 212, 234
Eigenvalue equation (dispersion relation)
  alpha-profiles 302
  asymmetric metal-clad guides 61
  cladded-parabolic profiles 143, 144, 166–167, 177
  conducting-wall guides 9, 180–182, 214–217, 219–220
  dielectric slabs 22, 28, 30, 31, 36, 164, 193
  elliptical fibres 255
  Epstein-layer profiles 131–132
  exponential profiles 115–119
  five-layer guides 83–86
  four-layer guides 76–77, 192, 194
  gain-guidance 53
  graded-index media 285, 294, 296
  hollow guides 72, 199–200
  leaky modes 331–332
  linear profiles 122, 124–127
  McKelvey–Langer 364
  metal-clad circular dielectric guides 261
  multilayer guides 269, 273
  parabolic-index media 96, 97, 100, 102, 110, 294, 296
  perturbation theory 350, 351, 354, 355
  polynomial profiles 142
  rectangular dielectric guides 186–187, 189
  'staircase' method 172
  step-index fibres 225–227, 231, 233
  symmetric metal-clad guides 69
  variational method 162, 342
  WKB approximation 148, 149, 151, 299
Eikonal equation 280
Eikonal function 145, 297
Electro-optic effect xii, 68, 78
Elliptical dielectric waveguides 223, 250–260, 338–339
Epitaxial growth xii, 43
Epstein-layer profile 113, 128–135, 364
  applied to lasers 134
Equivalent step-index fibre 369

Evanescent field theory 172–174, 301, 368
Extinction coefficient 25, 50, 264

Fabry–Perot cavity 44
Faraday-effect current transducer 258
Fibres, optical
  graded-index xiii, 240, 243, 276, 278–366
  single-material 195–197
  step-index 223–239
Field distributions
  asymmetric metal-clad guides 66–67
  conducting-wall guides 7, 8, 219–222
  elliptical fibres 251–254
  exponential profiles 117
  graded-index guides 292–295
  Hermite–Gauss 100, 101, 296
  hollow waveguides 266–267
  leaky modes 330–331, 333
  monomode fibres 368–369
  multilayer guides 268
  parabolic-index media 100, 101, 296
  perturbation theory 350, 351, 358, 359
  step-index fibre 224–225, 228–230
  WKB approximation 298–301
Filling factor, see Confinement factor
Finite element techniques 166, 187, 193, 196–197, 206, 360, 365, 367
Flexible waveguides (for CO_2 radiation) 59, 70, 201, 213, 260, 265
Focal length (parabolic-index media) 95
Frequency, normalized 9, 31, 193, 196, 202, 203, 205
  complex normalized 49, 53, 71
Fresnel's law 11–13

Gain, of a laser 44–46, 56–57
  effects of loss and gain 49–58, 104–112, 208–212
Gain-guiding 49
  cut-offs 52
  in p–n junctions 42
  in stripe-geometry lasers 56–58, 109, 210
  theory 53–55, 112
Gallium arsenide (GaAs) xii, 19–20
  laser threshold currents 42
  lattice constant 43
  refractive index 19, 44
Geometrical optics, see Rays
Goos–Haenchen shift 16–18, 22–23, 38–40, 74, 236–237, 264
Grating coupler xii
Group delay (transit time)
  alpha profiles 302–303, 307
  dielectric slabs 38–40
  Epstein-layer profiles 134–135
  fibres 238, 259–260, 285, 289, 315–317
  graded-index media 93
  leaky modes 328–330

Group delay (transit time) (*continued*)
   parabolic-index media 95, 97, 103
   perturbation theory 352, 354–355
   polynomial profile 137–138
   variational method 346, 349
   WKB approximation 300–301, 361, 362
Group index 24, 38, 93, 159, 234, 243, 285
   effective 234, 235
   related to phase index 235, 238, 239, 302
   of SiO_2 246
Group velocity 1, 2, 24–25, 38, 41, 182
   related to phase velocity 157–159, 223, 235, 238, 342, 346, 348
Gyroscope, fibre 259

Helical rays 284–285
Hermite–Gaussian modes 100, 105, 175, 210, 212, 278, 296, 297
Hollow waveguides
   circular (hollow pipes) 213–223
   dielectric circular 264
   dielectric planar 52, 59, 61, 71–74
   dielectric planar with metallic walls 26–27, 61
   rectangular 198–201

Impedance, characteristic 3
Impulse response 307, 328–330, 352, 355, 362–363
InGaAsP xiii, 244
Integrated optics xii, 66, 81, 198, 259
Intensity distribution 1, 220, 222, 228, 332, 337–339, 361
Invariant, ray 281–284

Joints in fibres xiii

Laguerre–Gaussian modes 278, 294–297, 300–301, 342, 351, 354, 355, 359
Lasers, heterostructure xii, 42–48
   buried heterostructure 56
   channelled substrate planar (CSP) 194, 211
   four-layer 81
   large optical cavity (LOC) 77–78
   leaky-wave 81
   localized-gain-region (LGR) 84
   rib 195
   separate confinement heterostructure (SCH) 84
   strip buried heterostructure (SBH) 195
   stripe geometry (analysis) 208–212
Lasers, semiconductor xii
   first reports 42
   new structures (analysis) 198
   optimum active-layer width 48
   stripe-geometry 55–58, 109, 187
   threshold currents 41, 42, 47, 57, 85
Lasers, waveguide 59, 74, 201, 260, 264

Leaky modes/rays
   fibres 242, 289, 297, 320–339, 352
   four-layer guides 81
   hollow guides 52, 61, 71–72
   metal-clad guides 61
   multilayer fibres 270, 272, 276
   variational method 157
   W-guides 88–89
   WKB approximation 152–155
Light-emitting diodes (LED's) xiii, 238, 245, 310
Linearly-polarized (LP) modes 213, 223, 228–239, 267, 295
Localized plane waves 5, 91, 92, 213–216, 218, 283–284, 321–322
Loss, dielectric planar guides 25–28
   effects of loss and gain 49–58
   end-loss of a laser 45, 46, 57
   fibres xii, 213, 239–242
   hollow metal guides 264
   hollow waveguides 73–74, 201
   leaky modes 337
   multilayer fibres 272–276

Maxwell's equations 1, 28, 29, 98, 180, 181, 184, 217–218, 223, 290
McKelvey–Langer theory 364
Meridional rays 35, 250, 252, 275, 276, 284, 322
Metal-clad waveguides, circular 260–267
   four-layer 78–82
   graded-index 119, 126
   planar 59–71
Microbending loss 242
Mode, waveguide 1
   'buried' 149–152
   degeneracy 226, 231–233, 249–250, 256, 295, 361–362
   graded-index media 295
   hybrid 224, 226–233, 254
   metal-clad dielectric guide 261–262, 264–265
   number 6, 31–32, 36, 220, 225–227, 231, 232, 249, 253–254, 289, 295, 296, 301, 302, 325–328, 333
Mode analyzers (filters) 68, 88–89, 276
Multilayer waveguides, circular 267–277
   five-layer 82–89
   four-layer 75–82
   planar 75–89
   W-guides 86–89, 260

Normalized variables 1, 7, 35, 84, 115, 181, 187, 189, 218, 223, 228, 233–235, 237–238, 267, 270, 291, 315, 335–338, 348–349, 351, 354
   complex 50, 71
Numerical aperture 196, 276

Numerical aperture (*continued*)
   leaky modes 322–323
   local 287–289
   wavelength dependence 305–307

Orthonormalization of eigenfunctions 140, 345
Overmoded guides 26

P–n junction xii, 42
Parabolic-index media 89–111, 278–297
Paraxial approximation 93–95
Perturbation theory 27, 136–145, 187, 209
   corrections to variational result 157
   graded-index fibres 350–361, 367–368
   hollow circular waveguides 262
   multilayer fibres 273–275
Phase change, at caustic 96, 216, 285
   on reflection at conducting wall 5, 62, 214–216
   on reflection at dielectric interface 15–16, 19–21, 77, 80, 114–115, 117–118, 233
   transverse 6, 97, 285
Phase fronts 50, 280
   radius of curvature 105, 108–109, 210
Phase index, effective 235
Phase velocity 1, 2, 5, 41, 80, 182
   related to group velocity 157–159, 223, 235, 238, 342, 346, 348
Polarization 1, 184, 197
   see also Birefringence
Polarizer 80, 82
Pöschl–Teller potential 113, 130, 364
Power distribution (modal) 305, 308, 309, 313, 323–328, 352
Power-flow, time averaged 2, 12–13, 32–35, 158
Poynting vector 2, 32, 33, 220–221, 237, 275
Prism coupler xii, 66, 332
Profile, refractive index
   cladded-parabolic 91, 143, 165–169, 174–177, 342–343, 357, 359
   complementary error function 207
   complex parabolic 104–105, 208
   complex slab 25
   CVD (measured) 360, 366
   Epstein-layer 128, 132, 174, 211, 212
   exponential 114, 119, 202, 207
   five-layer slab 82
   four-layer slab 75
   Gaussian 142, 207
   linear 122, 126
   multiple-alpha 317–320
   optimum (variational) 347
   parabolic 90, 174, 279–297
   planar step-index 20
   polynomial 137, 141, 174, 351, 353
   power-law 211, 278
   squared-tangent 211
   with central dip 355–357
   with steps 354–355
Profile parameter (alpha) 278, 279, 290, 303–305, 315–320, 356, 357, 361–363
Propagation constant, complex longitudinal 54, 262, 273, 320, 331–332
   complex normalized 50, 54, 72, 110, 112
   and effective index method 189
   longitudinal 5, 6, 21, 29–30, 281, 302, 358
   normalized 8–9, 36–37, 100, 131, 188, 203, 267, 354–355
   see also Eigenvalue equation
Pulse-spreading 40, 197, 213, 237, 242, 243, 259, 285, 303–320, 328–330, 347–348, 352–355, 361–364

Quaternary semiconductors xiii, 244

Radiative recombination 44, 46
Rayleigh–Ritz method 156, 157, 161–163, 206, 345, 346
Rayleigh scattering 240
Ray path, *see* Trajectory of a ray
Rays, circular hollow waveguides 213–217
   complex rays and Gaussian beams 173
   elliptical fibres 251–253
   Epstein-layer profiles 134–135
   exponential profiles 114–121
   graded-index media 279–290
   hollow dielectric slabs 73–74
   leaky rays/modes 321–339
   linear profiles 122–128
   parabolic-index media 91–97
   perturbation solution 136–139
   planar conducting-wall guides 1, 3–5
   planar dielectric guides 16–18, 20–28
   step-index fibres 233, 236
   tracing 361–364
   see also Helical rays, Meridional rays, and Skew rays
Reflection coefficient 12, 13, 19, 26–28, 64, 73, 263, 265
Refracted modes (radiation modes) 71, 270
Refractive index, AlGaAs 19, 44
   CVD materials 243–245, 306–307, 315–320
   GaAs 19, 44
   metals 60
   SiO_2 197, 246
Repeaters, optical xii
Resonance condition, transverse 5, 21, 97, 114, 122
   WKB 148, 149, 299, 315, 332, 361, 364
Rib waveguides 178, 190, 195–197
   polarizers 80
Room-temperature c.w. operation of semiconductor lasers xii, 245

Scattering loss 190, 240
Sellmeier equation 243, 306
Series solution 166–169, 360, 367

SiO$_2$, buffer layer 81
    in fibres 240–241
        refractive index 197, 244, 246
Single-mode fibres xiii, 223, 227, 241, 366–369
    central dip 356–357
    chromatic dispersion 245–249
    elliptical 256–260
    multilayer 272–275
Skew rays 214, 287, 320
Skin depth 1
Slab-coupled guides 183, 190–198
Snell's law 4, 11
Spectral width xiii, 238, 307–315
Spectrum analyser, R.F. xii
Spot-size 369
    see also Beam waist
Stability of modes 49–52, 106–108
'Staircase' approximation 169–172, 353–355, 364–365, 367
Stationary expression (for eigenvalues) 156–157, 160–161, 344–345
Stratification method, *see* 'Staircase' approximation
Strip-loaded guides 178, 190–195, 206–207
    metal-clad optical strip-line 178, 197
Surface plasma wave 64–67, 69, 70, 80, 82, 207

Taper coupler xii, 788
Threshold condition of a laser 44–45, 57
Trajectory of a ray, elliptical fibres 251–253
    graded-index media 280–287
    hollow circular waveguides 214, 215
    parabolic-index media 93–95
    ray-tracing 361–364
Transfer function 313–314
Transit time, *see* Group delay
Transmission coefficient 12–14, 25, 155, 332
Transmission line 1–3
Transverse electric (TE) modes
    cladded-parabolic profiles 143–144
    complex dielectric slabs 51
    conducting-wall guides 4, 6–7, 180, 217–223
    dielectric slabs 29–30
    elliptical 256
    Epstein-layer profiles 130–133
    exponential profiles 116–121
    five-layer slabs 83
    four-layer metal-clad slabs 79
    graded-index guides 291, 295
    hollow slabs 71
    linear profiles 123–128
    metal-clad dielectric guides 261, 264–265
    multilayer slabs 76
    parabolic-index media 99–102
    perturbation theory 139–144
    polynomial profiles 141–143
    step-index fibres 224–228
    variational method 157–159

Transverse electromagnetic (TEM) wave 1–3, 6, 8, 178, 232
Transverse magnetic (TM) modes
    conducting-wall guides 4, 7, 181, 217–223
    dielectric fibres 30–31
    elliptical fibres 256
    five-layer slabs 83
    four-layer metal-clad slabs 79–80
    graded-index guides 291, 295
    metal-clad dielectric guides 261, 264–265
    multilayer slabs 76
    parabolic-index media 102
    perturbation theory 139–141
    step-index fibres 224–228
    TM_0 and TM_1 in metal-clad slabs 65–66, 70
    variational method 157–159
Tunnelling coefficient 152–155, 332
Turning-point, *see* Caustic

Variational methods 155–166, 187, 193, 196, 201, 206, 207, 339–350, 365, 367–369
Vector field theory 290–293, 343–344, 367–368

W-waveguides, slabs 86–89, 155
    fibres 260, 267–277
Wave equation, elliptical fibres 251, 252
    evanescent field theory 173–174
    graded-index planar guides 97–99, 209
    hollow conducting-wall guides 217
    perturbation theory 139–141
    planar guides 29
    rectangular coordinates 178–179
    scalar, graded-index media 279–280, 294, 296, 297
    step-index fibres 223–224
    variational method 156–157, 344
    vector, graded-index media 292
    WKB approximation 145–146, 298–299
Wave number 4
Wave vector 213–216, 218, 283, 322
Waveguide lens 78
Weakly-guiding condition, alpha-profiles 303
    complex dielectric slab 51
    dielectric slab 27, 35–42
    elliptical fibres 254–256
    fibres 213, 223, 226, 228–239
    five-layer slab 84, 86
    graded-index guides 291, 294–295
    multilayer guides 267
    parabolic-index media 90, 102, 103
Wentzel–Kramers–Brillouin (WKB) approximation 145–155, 174, 175, 205, 206, 218, 276, 277, 285, 361–364, 368
    analysis of graded-index fibres 297–339
Width, effective 22–24, 26–28, 32–35, 37
    normalized effective 37